Modern Birkhäuser Classics

Many of the original research and survey monographs in pure and applied mathematics published by Birkhäuser in recent decades have been groundbreaking and have come to be regarded as foundational to the subject. Through the MBC Series, a select number of these modern classics, entirely uncorrected, are being re-released in paperback (and as eBooks) to ensure that these treasures remain accessible to new generations of students, scholars, and researchers.

Classic Papers in Combinatorics

Ira Gessel
Gian-Carlo Rota

Reprint of the 1987 Edition

Birkhäuser
Boston • Basel • Berlin

Editors
Ira Gessel
Brandeis University
Department of Mathematics
Waltham, MA 02454
USA

Gian-Carlo Rota (Deceased)
Massachusetts Institute of Technology (MIT)
Department of Mathematics
Cambridge, MA 02139
USA

ISBN: 978-0-8176-4841-1 e-ISBN: 978-0-8176-4842-8
DOI: 10.1007/978-0-8176-4842-8

Library of Congress Control Number: 2008939075

Mathematics Subject Classification (2000): 68Rxx, 68R05, 05-xx

Printed on acid-free paper.

9 8 7 6 5 4 3 2 1

www.birkhauser.com

Classic Papers in Combinatorics

Edited by

Ira Gessel Gian-Carlo Rota

Birkhäuser
Boston · Basel · Stuttgart

Ira Gessel
Department of Mathematics
Brandeis University
Waltham, MA 02454
U.S.A.

Gian-Carlo Rota
Department of Mathematics
Massachusetts Institute of Technology
Cambridge, MA 02139
U.S.A.

Library of Congress Cataloging in Publication Data
Classic papers in combinatorics.
1. Combinatorial analysis. I. Gessel, Ira.
II. Rota, Gian-Carlo, 1932– .
QA164.C56 1987 511'.6 87-5138

CIP-Kurztitelaufnahme der Deutschen Bibliothek
Classic papers in combinatorics / ed. by Ira
Gessel and Gian-Carlo Rota.—Boston ; Basel ;
Stuttgart : Birkhäuser, 1987
ISBN 3-7643-3364-2 (Basel . . .)
ISBN 0-8176-3364-2 (Boston)
NE: Gessel, Ira [Hrsg.]

ISBN 0-8176-3364-2
ISBN 3-7643-3364-2

Printed and bound by Quinn-Woodbine Inc., Woodbine, New Jersey.
Printed in the U.S.A.

9 8 7 6 5 4 3 2 1

We wish to thank the following persons who have made valuable suggestions of papers to be included in this volume: George Andrews, Richard Brualdi, Andrew Gleason, Phil Hanlon, David Jackson, Jeff Kahn, Adalbert Kerber, Joseph Kung, and Richard Stanley.

Contents

Introduction

This volume surveys the development of combinatorics since 1930 by presenting in chronological order the fundamental results of the subject proved in the orginal papers.

We begin with the celebrated theorem of Ramsey [1930], originally developed to settle a special case of the decision problem for the predicate calculus with equality. It remains to this day the fundamental generalization of the classical pigeonhole principle. The paper by Erdös and Szekeres [1935a] was one of the first applications of Ramsey's theorem, and it is still one of the most elegant. Through the partition calculus of Erdös and Rado [1956a], Ramsey's theorem made inroads into set theory, where nowadays it holds the limelight. The next major advance along the lines initiated by Ramsey came with the work of Hales and Jewett [1963a], a result which has served as a foundation for much further work in the area. The categorical underpinning of Ramsey theory was worked out by Graham, Leeb, and Rothschild [1972a]. Here the original ideas of Ramsey are cleverly blended with the contribution of Hales and Jewett.

Whitney's paper [1932] marks the beginning of what is now the theory of matroids. Three years later the theory makes its appearance fully clad in another paper of Whitney [1935c] which remains the basic reference on the subject. The theory of matroids was also in the backround of Tutte's paper [1947]. Tutte's paper is couched in the language of graphs and was later generalized to arbitrary matroids by Brylawski. The motivation behind much of the work of the two outstanding graph theorists of the day, Whitney and Tutte, was the coloring problem for graphs. Two short and elegant results on coloring problems are Brooks's theorem [1941] relating the chromatic number of a graph to its maximal degree and Lovász's theorem [1972b] characterizing perfect graphs.

Philip Hall's paper [1935b] was the first in what is now called matching theory. A very short proof of Hall's marriage theorem was given by Halmos and Vaughan [1950b]. In the same year Dilworth [1950a] proved his famous decomposition theorem for partially ordered sets, which generalizes Hall's theorem. Several other minimax combinatorial theorems can be viewed as variants or generalizations of the marriage theorem. Such are Tutte's definitive work on factors in graphs [1952], Ford and Fulkerson's theory of flows in networks [1956b], and Edmonds's [1965] efficient algorithm for matching in graphs. Gale [1957a] used network flow theory to prove a result on matrices of 0's and 1's with given row and column sums also proved directly by Ryser [1957b].

De Bruijn and van Aardenne-Ehrenfest [1951], taking their lead from the early work of Kirchhoff, obtained a definitive result, now called the BEST theorem (de Bruijn-Ehrenfest-Stone-Tutte) concerning the enumeration of spanning trees and Eulerian circuits of a graph by determinants. Several years later, Kasteleyn [1961a] succeeded in solving a packing problem for dimers on a lattice by reducing the problem to the evaluation of Pfaffians.

Pólya's paper on picture-writing [1956c] foreshadows the notion of the incidence algebra, a term introduced years later by Rota [1964] in his theory of Möbius functions. Rota's work was substantially extended by Crapo [1968]. The mystery of the characteristic polynomial, defined in terms of the Möbius function of a partially ordered set, motivates Stanley's beautiful result [1973a] on acyclic orientations of graphs. Geissinger's three papers [1973b–d] are the definitive presentation of the theory of Möbius functions.

The subject that is now called extremal set theory is represented by Katona's paper [1966a]. The main result, independently proved by Kruskal, harks back to a theorem of Macaulay, and was generalized by Clements and Lindström [1969]. Lubell's blitz proof of Sperner's theorem [1966b] has been extensively generalized and applied to many problems. Kleitman's solution [1970b] of a long-standing problem of Erdoš related to the Littlewood-Offord problem shows the power of a simple, but far from obvious, induction argument.

Brooks, Smith, Stone, and Tutte's paper [1940] on the decomposition of rectangles into squares was the first to use Kirchhoff's laws for the solution of a problem in combinatorics, a technique that has since become standard.

Kaplansky's solution [1943] of the *problème des ménages* using the inclusion-exclusion principle has developed into what is now the theory of permutations with restricted position. Lovász's contribution [1972c] to the Ulam reconstruction problem is another ingenious use of the inclusion-exclusion principle.

Erdös's paper on graph theory and probability [1959] is the first paper to show how probabilistic methods can lead to combinatorial existence theorems. A substantial number of theorems in combinatorics, for which no explicit construction is known, can be given existence proofs by this method.

Schensted's bijection [1961b] between permutations and pairs of standard Young diagrams has proved central in the seemingly unrelated topics of plane partitions and representations of the symmetric group.

Schützenberger's paper [1962] lays the foundation of what is now the theory of rational and algebraic power series in noncommutative variables.

Nash-Williams's striking proof [1963b] that finite trees form a well-quasi-ordered set has blossomed both in logic and in graph theory.

I. J. Good's short proof [1970a] of a conjecture of Dyson has since been widely generalized but his approach is still the good one.

Ira M. Gessel
Gian-Carlo Rota

Classic Papers in Combinatorics

ON A PROBLEM OF FORMAL LOGIC

By F. P. RAMSEY.

[Received 28 November, 1928.—Read 13 December, 1928.]

This paper is primarily concerned with a special case of one of the leading problems of mathematical logic, the problem of finding a regular procedure to determine the truth or falsity of any given logical formula*. But in the course of this investigation it is necessary to use certain theorems on combinations which have an independent interest and are most conveniently set out by themselves beforehand.

I.

The theorems which we actually require concern finite classes only, but we shall begin with a similar theorem about infinite classes which is easier to prove and gives a simple example of the method of argument.

THEOREM A. *Let Γ be an infinite class, and μ and r positive integers; and let all those sub-classes of Γ which have exactly r members, or, as we may say, let all r-combinations of the members of Γ be divided in any manner into μ mutually exclusive classes C_i ($i = 1, 2, \ldots, \mu$), so that every r-combination is a member of one and only one C_i; then, assuming the axiom of selections, Γ must contain an infinite sub-class Δ such that all the r-combinations of the members of Δ belong to the same C_i.*

Consider first the case $\mu = 2$. (If $\mu = 1$ there is nothing to prove.) The theorem is trivial when r is 1, and we prove it for all values of r by induction. Let us assume it, therefore, when $r = \rho - 1$ and deduce it for $r = \rho$, there being, since $\mu = 2$, only two classes C_i, namely C_1 and C_2.

* Called in German the *Entscheidungsproblem*; see Hilbert und Ackermann, *Grundzüge der Theoretischen Logik*, 72–81.

It may happen that Γ contains a member x_1 and an infinite sub-class Γ_1, not including x_1, such that the ρ-combinations consisting of x_1 together with any $\rho-1$ members of Γ_1, all belong to C_1. If so, Γ_1 may similarly contain a member x_2 and an infinite sub-class Γ_2, not including x_2, such that all the ρ-combinations consisting of x_2 together with $\rho-1$ members of Γ_2, belong to C_1. And, again, Γ_2 may contain an x_3 and a Γ_3 with similar properties, and so on indefinitely. We thus have two possibilities: either we can select in this way two infinite sequences of members of Γ $(x_1, x_2, \ldots, x_n, \ldots)$, and of infinite sub-classes of Γ $(\Gamma_1, \Gamma_2, \ldots, \Gamma_n, \ldots)$, in which x_n is always a member of Γ_{n-1}, and Γ_n a sub-class of Γ_{n-1} not including x_n, such that all the ρ-combinations consisting of x_n together with $\rho-1$ members of Γ_n, belong to C_1; or else the process of selection will fail at a certain stage, say the n-th, because Γ_{n-1} (or if $n=1$, Γ itself) will contain no member x_n and infinite sub-class Γ_n not including x_n such that all the ρ-combinations consisting of x_n together with $\rho-1$ members of Γ_n belong to C_1. Let us take these possibilities in turn.

If the process goes on for ever let Δ be the class $(x_1, x_2, \ldots, x_n, \ldots)$. Then all these x's are distinct, since if $r>s$, x_r is a member of Γ_{r-1} and so of $\Gamma_{r-2}, \Gamma_{r-3}, \ldots$, and ultimately of Γ_s which does not contain x_s. Hence Δ is infinite. Also all ρ-combinations of members of Δ belong to C_1; for if x_s is the term of such a combination with least suffix s, the other $\rho-1$ terms of the combination belong to Γ_s, and so form with x_s a ρ-combination belonging to C_1. Γ therefore contains an infinite sub-class Δ of the required kind.

Suppose, on the other hand, that the process of selecting the x's and Γ's fails at the n-th stage, and let y_1 be any member of Γ_{n-1}. Then the $(\rho-1)$-combinations of members of $\Gamma_{n-1}-(y_1)$ can be divided into two mutually exclusive classes C_1' and C_2' according as the ρ-combinations formed by adding to them y_1 belong to C_1 or C_2, and by our theorem (A), which we are assuming true when $r=\rho-1$ (and $\mu=2$), $\Gamma_{n-1}-(y_1)$ must contain an infinite sub-class Δ_1 such that all $(\rho-1)$-combinations of the members of Δ_1 belong to the same C_i'; *i.e.* such that the ρ-combinations formed by joining y_1 to $\rho-1$ members of Δ_1 all belong to the same C_i. Moreover, this C_i cannot be C_1, or y_1 and Δ_1 could be taken to be x_n and Γ_n and our previous process of selection would not have failed at the n-th stage. Consequently the ρ-combinations formed by joining y_1 to $\rho-1$ members of Δ_1 all belong to C_2. Consider now Δ_1 and let y_2 be any of its members. By repeating the preceding argument $\Delta_1-(y_2)$ must contain an infinite sub-class Δ_2 such that all the ρ-combinations got by joining y_2 to $\rho-1$ members of Δ_2 belong to the same C_i.

And, again, this C_i cannot be C_1, or, since y_2 is a member and Δ_2 a sub-class of Δ_1 and so of Γ_{n-1} which includes Δ_1, y_2 and Δ_2 could have been chosen as x_n and Γ_n and the process of selecting these would not have failed at the n-th stage. Now let y_3 be any member of Δ_2; then $\Delta_2-(y_3)$ must contain an infinite sub-class Δ_3 such that all ρ-combinations consisting of y_3 together with $\rho-1$ members of Δ_3, belong to the same C_i, which, as before, cannot be C_1 and must be C_2. And by continuing in this way we shall evidently find two infinite sequences $y_1, y_2, \ldots, y_n, \ldots$ and $\Delta_1, \Delta_2, \ldots, \Delta_n, \ldots$ consisting respectively of members and sub-classes of Γ, and such that y_n is always a member of Δ_{n-1}, Δ_n a sub-class of Δ_{n-1} not including y_n, and all the ρ-combinations formed by joining y_n to $\rho-1$ members of Δ_n belong to C_2; and if we denote by Δ the class $(y_1, y_2, \ldots, y_n, \ldots)$ we have, by a previous argument, that all ρ-combinations of members of Δ belong to C_2.

Hence, in either case, Γ contains an infinite sub-class Δ of the required kind, and Theorem A is proved for all values of r, provided that $\mu=2$. For higher values of μ we prove it by induction; supposing it already established for $\mu=2$ and $\mu=\nu-1$, we deduce it for $\mu=\nu$.

The r-combinations of members of Γ are then divided into ν classes C_i $(i=1, 2, \ldots, \nu)$. We define new classes C'_i for $i=1, 2, \ldots, \nu-1$ by

$$C'_i = C_i \quad (i = 1, 2, \ldots, \nu-2),$$

$$C'_{\nu-1} = C_{\nu-1}+C_\nu.$$

Then by the theorem for $\mu=\nu-1$, Γ must contain an infinite sub-class Δ such that all r-combinations of the members of Δ belong to the same C'_i. If, in this C'_i, $i \leqslant \nu-2$, they all belong to the same C_i, which is the result to be proved; otherwise they all belong to $C'_{\nu-1}$, *i.e.* either to $C_{\nu-1}$ or to C_ν. In this case, by the theorem for $\mu=2$, Δ must contain an infinite sub-class Δ' such that the r-combinations of members of Δ' either all belong to $C_{\nu-1}$ or all belong to C_ν; and our theorem is thus established.

Coming now to finite classes it will save trouble to make some conventions as to notation. Small letters other than x and y, whether Italic or Greek (*e.g.* n, r, μ, m) will always denote finite cardinals, positive unless otherwise stated. Large Greek letters (*e.g.* Γ, Δ) will denote classes, and their suffixes will indicate the number of their members (*e.g.* Γ_m is a class with m members). The letters x and y will represent members of the classes Γ, Δ, etc., and their suffixes will be used merely to distinguish them. Lastly, the letter C will stand, as before, for classes of combinations, and its suffixes will not refer to the number of members, but serve merely to distinguish the different classes of combinations considered.

Corresponding to Theorem A we then have

THEOREM B. *Given any r, n, and μ we can find an m_0 such that, if $m \geqslant m_0$ and the r-combinations of any Γ_m are divided in any manner into μ mutually exclusive classes C_i $(i = 1, 2, \ldots, \mu)$, then Γ_m must contain a sub-class Δ_n such that all the r-combinations of members of Δ_n belong to the same C_i.*

This is the theorem which we require in our logical investigations, and we should at the same time like to have information as to how large m_0 must be taken for any given r, n, and μ. This problem I do not know how to solve, and I have little doubt that the values for m_0 obtained below are far larger than is necessary.

To prove the theorem we begin, as in Theorem A, by supposing that $\mu = 2$. We then take, not Theorem B itself, but the equivalent

THEOREM C. *Given any r, n, and k such that $n+k \geqslant r$, there is an m_0 such that, if $m \geqslant m_0$ and the r-combinations of any Γ_m are divided into two mutually exclusive classes C_1 and C_2, then Γ_m must contain two mutually exclusive sub-classes Δ_n and Λ_k such that all the combinations formed by r members of $\Delta_n + \Lambda_k$ which include at least one member from Δ_n belong to the same C_i.*

That this is equivalent to Theorem B with $\mu = 2$ is evident from the fact that, for any given r, Theorem C, for n and k, asserts more than Theorem B for n, but less than Theorem B for $n+k$.

The proof of Theorem C must be performed by mathematical induction, and can conveniently be set out as a demonstration that it is possible to define by recursion a function $f(r, n, k)$ which will serve as m_0 in the theorem.

If $r = 1$, the theorem is evidently true with m_0 equal to the greater of $2n-1$ and $n+k$, so that we may define

$$f(1, n, k) = \max(2n-1, n+k) \quad (n \geqslant 1,\ k \geqslant 0).$$

For other values of r we define $f(r, n, k)$ by recursion formulae involving an auxiliary function $g(r, n, k)$. Suppose that $f(r-1, n, k)$ has been defined for a certain $r-1$, and all n, k such that $n+k \geqslant r-1$, then we define it for r by putting

$$f(r, 1, k) = f(r-1, k-r+2, r-2)+1 \quad (k+1 \geqslant r),$$
$$g(r, 0, k) = \max(r-1, k),$$
$$g(r, n, k) = f\{r, 1, g(r, n-1, k)\} \quad (n \geqslant 1),$$
$$f(r, n, k) = f\{r, n-1, g(r, n, k)\} \quad (n > 1).$$

These formulae can be easily seen to define $f(r, n, k)$ for all positive values of r, n and k satisfying $n+k \geqslant r$, and $g(r, n, k)$ for all values of r greater than 1, and all positive values of n and k; and we shall prove that Theorem C is true when we take m_0 to be this $f(r, n, k)$. We know that this is so when $r=1$, and we shall therefore assume it for all values up to $r-1$ and deduce it for r.

When $n=1$, and $m \geqslant m_0 = f(r-1, k-r+2, r-2)+1$, we may take any member x of Γ_m to be sole member of Δ_1 and there remain at least $f(r-1, k-r+2, r-2)$ members of $\Gamma_m-(x)$; the $(r-1)$-combinations of these members of $\Gamma_m-(x)$ can be divided into classes C_1' and C_2' according as they belong to C_1 or C_2 when x is added to them, and, by our theorem for $r-1$, $\Gamma_m-(x)$ must contain two mutually exclusive classes Δ_{k-r+2}, Λ_{r-} such that every combination of $r-1$ terms from $\Delta_{k-r+2}+\Lambda_{r-2}$ (since one of its terms must come from Δ_{k-r+2}, Λ_{r-2} having only $r-2$ members) belongs to the same C_i'. Taking Λ_k to be this $\Delta_{k-r+2}+\Lambda_{r-2}$ all combinations consisting of x, together with $r-1$ members of Λ_k, belong to the same C_i. The theorem is therefore true for r when $n=1$.

For other values of n we prove it by induction, assuming it for $n-1$ and deducing it for n. Taking

$$m \geqslant m_0 = f(r, n, k) = f\{r, n-1, g(r, n, k)\},$$

Γ_m must, by the theorem for $n-1$, contain a Δ_{n-1} and a $\Lambda_{g(r,n,k)}$ such that every combination of r members of $\Delta_{n-1}+\Lambda_{g(r,n,k)}$, at least one term of which comes from Δ_{n-1}, belongs to the same C_i, say to C_1. If, now, $\Lambda_{g(r,n,k)}$ contains a member x and a sub-class Λ_k not including x, such that every combination of x and $r-1$ members of Λ_k belongs to C_1, then, taking Δ_n to be $\Delta_{n-1}+(x)$ and Λ_k to be this Λ_k, our theorem is true. If not, there can be no member of $\Lambda_{g(r,n,k)}$ which has a sub-class of k members of $\Lambda_{g(r,n,k)}$ connected with it in this way. But since

$$g(r, n, k) = f\{r, 1, g(r, n-1, k)\},$$

$\Lambda_{g(r,n,k)}$ must contain a member x_1 and a sub-class $\Lambda_{g(r,n-1,k)}$, not including x_1, such that x_1 combined with any $r-1$ members of $\Lambda_{g(r,n-1,k)}$ gives a combination belonging to the same C_i, which cannot be C_1, or x_1 and any k members of $\Lambda_{g(r,n-1,k)}$ could have been taken as the x and Λ_k above. Hence the combinations formed by x_1 together with any $r-1$ members of $\Lambda_{g(r,n-1,k)}$ all belong to C_2. But now

$$g(r, n-1, k) = f\{r, 1, g(r, n-2, k)\},$$

and $\Lambda_{g(r,n-1,k)}$ must contain an x_2 and a $\Lambda_{g(r,n-2,k)}$, not including x_2, such that the combinations formed by x_2 and $r-1$ members of $\Lambda_{g(r,n-2,k)}$ all belong to the same C_i, which must, as before, be C_2, since x_2 and $\Lambda_{g(r, n-2, k}$ are both contained in $\Lambda_{g(r, n, k)}$ and $g(r, n-2, k) \geqslant k$. Continuing in this way we can find n distinct terms $x_1, x_2, \ldots, x_n$ and a $\Lambda_{g(r, 0, k)}$ such that every combination of r terms from $(x_1, x_2, \ldots, x_n)+\Lambda_{g(r, 0, k)}$ belongs to C_2, provided that at least one term of the combination comes from $(x_1, x_2, \ldots, x_n)$. Since $g(r, 0, k) \geqslant k$ this proves our theorem, taking Δ_n to be $(x_1, x_2, \ldots, x_n)$ and Λ_k to be any k terms of $\Lambda_{g(r, 0, k)}$.

Theorem C is therefore established for all values of r, n, and k, with m_0 equal to $f(r, n, k)$. It follows that, if $\mu = 2$, Theorem B is true for all values of r and n with m_0 equal to $f(r, n-r+1, r-1)$, which we shall also call $h(r, n, 2)$.

For other values of μ we prove Theorem B by induction, taking m_0 to be $h(r, n, \mu)$, where

$$h(r, n, 2) = f(r, n-r+1, r-1)$$

$$h(r, n, \mu) = h\{r, h(r, n, \mu-1), 2\} \quad (\mu > 2).$$

For, assuming the theorem for $\mu-1$, we prove it for μ by defining new classes of combinations

$$C_1' = C_1,$$

$$C_2' = \sum_{i=2}^{\mu} C_i.$$

If then $m \geqslant h(r, n, \mu) = h\{r, h(r, n, \mu-1), 2\}$, by the theorem for $\mu = 2$, Γ_m must contain a $\Gamma_{h(r, n, \mu-1)}$ the r-combinations of whose members belong either all to C_1' or all to C_2'. In the first case there is no more to prove; in the second we have only to apply the theorem for $\mu-1$ to $\Gamma_{h(r, n, \mu-1)}$.

In the simplest case in which $r = \mu = 2$ the above reasoning gives m_0 equal to $h(2, n, 2)$, which is easily shown to be $2^{n(n-1)/2}$. But for this case there is a simple argument which gives the much lower value $m_0 = n\,!$, and shows that our value $h(r, n, \mu)$ is altogether excessive.

For, taking Theorem C first, we can prove by induction with regard to n that, for $r = 2$, we may take m_0 to be $k\,.\,(n+1)\,!$. (k is here supposed greater than or equal to 1.) For this is true when $n = 1$, since, if $m \geqslant 2k$, of the $m-1$ pairs obtained by combining any given member of Γ_m with the others, at least k must belong to the same C_i. Assuming it, then, for $n-1$, let us prove it for n.

If $m \geqslant k\,.\,(n+1)\,! = k(n+1)\,.\,n\,!$, Γ_m must, by the theorem for $n-1$, contain two mutually exclusive sub-classes Δ_{n-1} and $\Lambda_{k(n+1)}$ such that all pairs from $\Delta_{n-1}+\Lambda_{k(n+1)}$, at least one term of which comes from Δ_{n-1}, belong to the same C_i, say C_1. Now consider the members of $\Lambda_{k(n+1)}$; in the first place, there may be one of these, x say, which is such that there are k other members of $\Lambda_{k(n+1)}$ which combined with x give pairs belonging to C_1. If so, the theorem is true, taking Δ_n to be $\Delta_{n-1}+(x)$; if not, let x_1 be any member of $\Lambda_{k(n+1)}$. Then there are at most $k-1$ other members of $\Lambda_{k(n+1)}$ which combined with x_1 give pairs belonging to C_1, and $\Lambda_{k(n+1)}-(x_1)$ must contain a Λ_{kn} any member of which gives when combined with x_1 a pair belonging to C_2. Let x_2 be any member of Λ_{kn}, then, since x_2 and Λ_{kn} are both contained in $\Lambda_{k(n+1)}$, there are at most $k-1$ other members of Λ_{kn} which when combined with x_2 give pairs belonging to C_1. Hence $\Lambda_{kn}-(x_2)$ contains a $\Lambda_{k(n-1)}$ any member of which combined with x_2 gives a pair belonging to C_2. Continuing in this way we obtain $x_1, x_2, \ldots, x_n$ and Λ_k, such that every pair x_i, x_j and every pair consisting of an x_i and a member of Λ_k belongs to C_2. Theorem C is therefore proved.

Theorem B for n then follows, with the m_0 of Theorem C for $n-1$ and 1, *i.e.* with m_0 equal to $n!$*; and it is an easy extension to show that, if in Theorem B $r=2$ but $\mu \neq 2$, we can take m_0 to be $n!!!, \ldots$, where the process of taking the factorial is performed $\mu-1$ times.

II.

We shall be concerned with logical formulae containing variable propositional functions, *i.e.* predicates or relations, which we shall denote by Greek letters ϕ, χ, ψ, etc. These functions have as arguments individuals denoted by x, y, z, etc., and we shall deal with functions with any finite number of arguments, *i.e.* of any of the forms

$$\phi(x), \quad \chi(x, y), \quad \psi(x, y, z), \quad \ldots.$$

In addition to these variable functions we shall have the one constant function of identity

$$x = y \quad \text{or} \quad = (x, y).$$

By operating on the values of ϕ, χ, ψ, ..., and $=$ with the logical operations

$\sim$	meaning	*not*,
$\vee$	,,	*or*,
.	,,	*and*,
(x)	,,	*for all x*.
(Ex)	,,	*there is an x for which*,

* But this value is, I think, still much too high. It can easily be lowered slightly even when following the line of argument above, by using the fact that if k is even it is impossible for every member of an odd class to have exactly $k-1$ others with which it forms a pair of C_1, for then twice the number of these pairs would be odd; we can thus start when k is even with a $\Lambda_{k(n+1)-1}$ instead of a $\Lambda_{k(n+1)}$

we can construct expressions such as

$$[(x, y)\{\phi(x, y) \vee x = y\}] \vee \{(Ez)\chi(z)\}$$

in which all the individual variables are made "apparent" by prefixes (x) or (Ex), and the only real variables left are the functions ϕ, χ, ... Such an expression we shall call a *first order formula.*

If such a formula is true for all interpretations* of the functional variables ϕ, χ, ψ, etc., we shall call it *valid*, and if it is true for no interpretations of these variables we shall call it *inconsistent*. If it is true for some interpretations (whether or not for all) we shall call it *consistent*†.

The *Entscheidungsproblem* is to find a procedure for determining whether any given formula is valid, or, alternatively, whether any given formula is consistent; for these two problems are equivalent, since the necessary and sufficient condition for a formula to be consistent is that its contradictory should not be valid. We shall find it more convenient to take the problem in this second form as an investigation of *consistency*. The consistency of a formula may, of course, depend on the number of individuals in the universe considered, and we shall have to distinguish between formulae which are consistent in every universe and those which are only consistent in universes with some particular numbers of members. Whenever the universe is infinite we shall have to assume the axiom of selections.

The problem has been solved by Behmann‡ for formulae involving only functions of one variable, and by Bernays and Schönfinkel§ for formulae involving only two individual apparent variables. It is solved below for the further case in which, when the formula is written in "normal form", there are any number of prefixes of generality (x) but none of existence (Ex)‖. By "normal form"¶ is here meant that all the prefixes stand at the beginning, with no negatives between or in front of them, and have scopes extending to the end of the formula.

---

* To avoid confusion we call a constant function substituted for a variable ϕ, not a value but an *interpretation* of ϕ; the *values* of $\phi(x, y, z)$ are got by substituting constant individuals for x, y, and z.

† German *erfüllbar*.

‡ H. Behmann, "Beiträge zur Algebra der Logik und zum Entscheidungsproblem", *Math. Annalen*, 86 (1922), 163–229.

§ P. Bernays und M. Schönfinkel, "Zum Entscheidungsproblem der mathematischen Logik", *Math. Annalen*, 99 (1928), 342–372. These authors do not, however, include identity in the formulae they consider.

‖ Later we extend our solution to the case in which there are also prefixes of existence provided that these all precede all the prefixes of generality.

¶ Hilbert und Ackermann, *op. cit.*, 63–4.

The formulae to be considered are thus of the form

$$(x_1, x_2, \ldots, x_n) F(\phi, \chi, \psi, \ldots, =, x_1, x_2, \ldots, x_n),$$

where the matrix F is a truth-function of values of the functions ϕ, χ, ψ, etc., and $=$ for arguments drawn from $x_1, x_2, \ldots, x_n$.

This type of formula is interesting as being the general type of an axiom system consisting entirely of "general laws"*. The axioms for order, betweenness, and cyclic order are all of this nature, and we are thus attempting a general theory of the consistency of axiom systems of a common, if very simple, type.

If identity does not occur in F the problem is trivial, since in this case whether the formula is consistent or not can be shown to be independent of the number of individuals in the universe, and we have only the easy task of testing it for a universe with one member only†.

But when we introduce identity the question becomes much more difficult, for although it is still obvious that if the formula is consistent in a universe U it must be consistent in any universe with fewer members than U, yet it may easily be consistent in the smaller universe but not in the larger. For instance,

$$(x_1, x_2)[x_1 = x_2 \vee \{\phi(x_1) . \sim \phi(x_2)\}]$$

is consistent in a universe with only one member but not in any other.

We begin our investigation by expressing F in a special form. F is a truth-function of the values of $\phi, \chi, \psi, \ldots$, and $=$ for arguments drawn from $x_1, x_2, \ldots, x_n$. If ϕ is a function of r variables there will be n^r‡ values of ϕ which can occur in F, and F will be a truth-function of Σn^r values of $\phi, \chi, \psi, \ldots$, and $=$, which we shall call *atomic propositions*. With regard to these Σn^r atomic propositions there are $2^{\Sigma n^r}$ possibilities of truth and falsity which we shall call *alternatives*, each alternative being a conjunction of Σn^r propositions which are either atomic propositions or their contradictions. In constructing the alternatives all the Σn^r atomic propositions are to be used whether or not they occur in F. F can then be expressed as a disjunction of some of these alternatives, namely those with which it is compatible. It is well known that such an

* C. H. Langford, "Analytic completeness of postulate sets", *Proc. London Math. Soc.* (2), 25 (1926), 115–6.

† Bernays und Schönfinkel, *op. cit.*, 359. We disregard altogether universes with no members.

‡ Here and elsewhere numbers are given not because they are relevant to the argument, but to enable the reader to check that he has in mind the same class of entities as the author.

expression is possible; indeed, it is the dual of what Hilbert and Ackermann call the "ausgezeichnete konjunktive Normalform"*, and is fundamental also in Wittgenstein's logic. The only exception is when F is a self-contradictory truth-function, in which case our formula is certainly not consistent.

F having been thus expressed as a disjunction of alternatives (in our special sense of the word), our next task is to show that some of these alternatives may be able to be removed without affecting the consistency or inconsistency of the formula. If all the alternatives can be removed in this way the formula will be inconsistent; otherwise we shall have still to consider the alternatives that remain.

In the first place an alternative may violate the laws of identity by containing parts of any of the following forms :—

$$x_i \neq x_i\dagger,$$

$$x_i = x_j \,.\, x_j \neq x_i \qquad (i \neq j),$$

$$x_i = x_j \,.\, x_j = x_k \,.\, x_i \neq x_k \quad (i \neq j,\ j \neq k,\ k \neq i),$$

or by containing $x_i = x_j$ $(i \neq j)$ and values of a function ϕ and its contradictory $\sim \phi$ for sets of arguments which become the same when x_i is substituted for x_j [*e.g.* $x_1 = x_2 \,.\, \phi(x_1, x_2, x_3) \,.\, \sim \phi(x_2, x_1, x_3)$].

Any alternative which violates these laws must always be false and can evidently be discarded without affecting the consistency of the formula. The remaining alternatives can then be classified according to the number of x's they make to be different, which may be anything from 1 up to n.

Suppose that for a given alternative this number is ν, then we can derive from it what we will call the corresponding y alternative by the following process :—

For x_1, wherever it occurs in the given alternative, write y_1; next, if in the alternative $x_2 = x_1$, for x_2 write y_1 again, if not for x_2 write y_2. In general, if x_i is in the given alternative identical with any x_j with j less than i, write for x_i the y previously written for x_j; otherwise write for x_i, y_{k+1}, where k is the number of y's already introduced. The expression which results contains ν y's all different instead of n x's, some of which are identical, and we shall call it the y alternative corresponding to the given x alternative.

---

* *Op. cit.*, 16.

† We write $x \neq y$ for $\sim (x = y)$.

Thus to the alternative

$$\phi(x_1) . \sim \phi(x_2) . \phi(x_3) . \sim \phi(x_4) . x_1 = x_3 . x_2 = x_4 . x_1 \neq x_2 \text{*}$$

corresponds the y alternative

$$\phi(y_1) . \sim \phi(y_2) . y_1 \neq y_2 .$$

We call two y alternatives *similar* if they contain the same number of y's and can be derived from one another by permuting those y's, and we call two x alternatives *equivalent* if they correspond to similar (or identical) y alternatives.

Thus

$$\phi(x_1) . \sim \phi(x_2) . \phi(x_3) . \sim \phi(x_4) . x_1 = x_3 . x_2 = x_4 . x_1 \neq x_2, \qquad (\alpha)$$

is equivalent to

$$\sim \phi(x_1) . \phi(x_2) . \phi(x_3) . \phi(x_4) . x_1 \neq x_2 . x_2 = x_3 = x_4, \qquad (\beta)$$

since they correspond to the similar y alternatives

$$\phi(y_1) . \sim \phi(y_2) . y_1 \neq y_2$$

$$\sim \phi(y_1) . \phi(y_2) . y_1 \neq y_2$$

derivable from one another by interchanging y_1 and y_2, although (α) and (β) are not so derivable by permuting the x's.

We now see that we can discard any alternative contained in F unless F also contains all the alternatives equivalent to it; *e.g.* if F contains (α) but not (β), (α) may be discarded from it. For omitting alternatives clearly cannot make the formula consistent if it was not so before; and we can easily prove that, if it was consistent before, omitting *these* alternatives cannot make it inconsistent.

For suppose that the formula is consistent, *i.e.* that for some particular interpretation of $\phi, \chi, \psi, \ldots, F$ is true for every set of x's, and let p be an alternative contained in F, q an alternative equivalent to p but not contained in F. Then for every set of x's one and only one alternative in F will (on this interpretation of $\phi, \chi, \psi, \ldots$) be the true one, and this alternative can never be p. For if it were p, the corresponding y alternative would be true for some set of y's, and the similar y alternative corresponding to q would be true for a set of y's got by permuting this last set. Giving the x's suitable values in terms of the y's, q would then

* We take one function of one variable only for simplicity; also to save space we omit expressions which may be taken for granted, such as $x_1 = x_1$, $x_1 \neq x_4$.

be true for a certain set of x's and F would be false for these x's contrary to hypothesis. Hence p is never the true alternative and may be omitted without affecting the consistency of the formula.

When we have discarded all these alternatives from F, the remainder will fall into sets each of which is the complete set of all alternatives equivalent to a given alternative. To such a set of x alternatives will correspond a complete set of similar y alternatives, and the disjunction of such a complete set of similar y alternatives (*i.e.* of all permutations of a given y alternative) we shall call a *form**. A form containing ν y's we shall denote by an Italic capital with suffix ν, *e.g.* A_ν, B_ν.

The force of our formula can now be represented by the following conjunction, which we shall call P.

$$\left.\begin{array}{ll}
\text{For every } y_1, & A_1 \text{ or } B_1 \text{ or } \ldots \\
\text{For every distinct } y_1, y_2, & A_2 \text{ or } B_2 \text{ or } \ldots \\
\ldots \quad \ldots \quad \ldots \quad \ldots & \ldots \quad \ldots \quad \ldots \\
\text{For every distinct } y_1, y_2, \ldots, y_\nu, & A_\nu \text{ or } B_\nu \text{ or } \ldots \\
\ldots \quad \ldots \quad \ldots \quad \ldots & \ldots \quad \ldots \quad \ldots \\
\text{For every distinct } y_1, y_2, \ldots, y_n, & A_n \text{ or } B_n \text{ or } \ldots
\end{array}\right\} \quad (P)$$

where A_ν, B_ν†, etc., are the forms corresponding to the x alternatives still remaining in F. If for any ν there are no such forms, *i.e.* if no alternatives with ν different x's remain in F, our formula implies that there are no such things as ν distinct individuals, and so cannot be consistent in a world of ν or more members.

We have now to define what is meant by saying that one form is *involved* in another. Consider a form A_ν and take one of the y alternatives contained in it. This y alternative is a conjunction of the values of $\phi, \chi, \psi, \ldots$, and their negatives for arguments drawn from $y_1, y_2, \ldots, y_\nu$. (We may leave out the values of identity and difference, since it is taken for granted that y's are always different.) If $\mu < \nu$ we can select μ of these y's in any way and leave out from the alternative all the terms in it which contain any of the $\nu - \mu$ y's not selected. We have left an alternative in μ y's which we can renumber $y_1, y_2, \ldots, y_\mu$, and the form E_μ to which this new alternative belongs we shall describe as being *involved* in the A_ν with which we started. Starting with one particular y alternative in A_ν we shall get a large number of different E_μ's by

* Cf. Langford, *op. cit.*, 116–120.

† The notation is partially misleading, since A_ν has no closer relation to A_μ than to B_μ.

T 2

choosing differently the μ y's which we select to preserve; and from whichever y alternative in A_ν we start, the E_μ's which we find to be involved in A_ν will be the same.

For example,

$$\{\phi(y_1, y_1) \cdot \phi(y_1, y_2) \cdot \phi(y_2, y_1) \cdot \sim \phi(y_2 \cdot y_2)\}$$
$$\vee \{\sim \phi(y_1, y_1) \cdot \phi(y_1, y_2) \cdot \phi(y_2, y_1) \cdot \phi(y_2, y_2)\}$$

is a form A_2 which involves the two E_1's

$$\phi(y_1, y_1)$$
$$\sim \phi(y_1, y_1).$$

It is clear that if for some distinct set of ν y's a form A_ν is true, then every form E_μ involved in A_ν will be true for some distinct set of μ y's contained in the ν.

We are now in a position to settle the consistency or inconsistency of our formula when N, the number of individuals in the universe, is less than or equal to n, the number of x's in our formula. In fact, if $N \leqslant n$, it is necessary and sufficient for the consistency of the formula that P should contain a form A_N together with all the forms E_μ involved in it for every μ less than N.

This condition is evidently necessary, since the N individuals in the universe must, taken as $y_1, y_2, \ldots, y_N$, have some form A_N in regard to any $\phi, \chi, \psi, \ldots$; and all forms involved in this A_N must be true for different selections of y's, and so contained in P if P is to be true for this $\phi, \chi, \psi, \ldots$.

Conversely, suppose that P contains a form A_N together with all forms involved in A_N; then, calling the N individuals in the universe $y_1, y_2, \ldots, y_N$, we can define functions $\phi, \chi, \psi, \ldots$ to make any assigned y alternative in A_N true; for any permutation of these N y's another alternative in A_N will be the true one, and for any subset of y's some y alternative in a form involved in A_N. Since all these y alternatives are by hypothesis contained in P, P will be true for these $\phi, \chi, \psi, \ldots$, and our formula consistent.

When, however, $N > n$ the problem is not so simple, although it clearly depends on the A_n's in P such that all forms involved in them are also contained in P. These A_n's we may call *completely contained* in P, and if there are no such A_n's a similar argument to that used when $N \leqslant n$ will show that the formula is inconsistent. But the converse argument, that if there is an A_n completely contained in P the formula must be consistent, no longer holds good; and to proceed further we have to introduce a new conception, the conception of a form being *serial*.

But before proceeding to explain this idea it is best to simplify matters by the introduction of new functions. Let ϕ be one of the variable functions in our formula, with, say, r arguments. Then, if $r < n$, ϕ will occur in P with all its arguments different [*e.g.* $\phi(y_1, y_2, \ldots, y_r)$] and also with some of them the same [*e.g.* $\phi(y_1, y_2, \ldots, y_{r-1}, y_1)$]; but we can conveniently eliminate values of the second kind by introducing new functions of fewer arguments than r, which, when all their arguments are different, take values equivalent to those of ϕ with some of its arguments identical.

E.g. we may put

$$\phi_1(y_1, y_2, \ldots, y_{r-1}) = \phi(y_1, y_2, \ldots, y_{r-1}, y_1).$$

In this way ϕ gives rise to a large number of functions with fewer arguments; each of these functions we define only for the case in which all its arguments are different, as is secured by these arguments being y's with different suffixes. If $r > n$, there is no difference except that ϕ can never occur with all its arguments different, and so is entirely replaced by the new functions.

If we do this for all the functions $\phi, \chi, \psi, \ldots$, and replace them by new functions wherever they occur in P with some of their arguments the same, P will contain a new set of variable functions (including all the old ones which have no more than n arguments), and these will never occur in P with the same argument repeated.

It is easy to see that this transformation does not affect the consistency of the formula, for, if it were consistent before, it must be consistent afterwards, since the new functions have simply to be replaced by their definitions. And if it is consistent afterwards it must have been so before, since any function of the old set has only to be given for any set of arguments the value of the appropriate function of the new set*.

---

* For instance, if $\phi(y_1, y_2, y_3)$ is a function of the old set, we have five new functions

$$\begin{aligned} \phi_0(y_1, y_2, y_3) &= \phi(y_1, y_2, y_3), \\ \chi_0(y_1, y_2) &= \phi(y_1, y_1, y_2), \\ \psi_0(y_1, y_2) &= \phi(y_1, y_2, y_1), \\ \pi_0(y_1, y_2) &= \phi(y_1, y_1, y_2), \\ \rho_0(y_1) &= \phi(y_1, y_1, y_1), \end{aligned}$$

and any value of ϕ is equivalent to a value of one and only one of the new functions. It must be remembered that the new functions are used only with all their arguments different; for otherwise they would not be independent, since we should have, for instance, $\chi_0(y_1, y_1)$ equivalent to $\rho_0(y_1)$. But $\chi_0(y_1, y_1)$ never occurs, and $\phi(y_1, y_1, y_1)$ is equivalent not to any value of χ_0 but only to $\rho_0(y_1)$.

In view of this fact we shall find it more convenient to take P in its new form, and denote the new set of functions by $\phi_0, \chi_0, \psi_0, \ldots$.

Suppose, then, that ϕ_0 is a function of r variables; there are

$$n(n-1)\ldots(n-r+1)$$

values of ϕ_0 with r different arguments drawn from $y_1, y_2, \ldots, y_n$ and every y alternative must contain each of these values or its contradictory. $r!$ of these values will have as arguments permutations of $y_1, y_2, \ldots, y_r$. Any other set of r y's can be arranged in the order of their suffixes as $y_{s_1}, y_{s_2}, \ldots, y_{s_r}$, $s_1 < s_2 < s_3 \ldots < s_r$, and it may happen that a given alternative contains the values of ϕ_0 for those and only those permutations of $y_{s_1}, y_{s_2}, \ldots, y_{s_r}$ which correspond (in the obvious way) to the permutations of $y_1, y_2, \ldots, y_r$ for which it (the alternative) contains the values of ϕ_0; *e.g.* if the alternative contains $\phi_0(y_1, y_2, \ldots, y_r)$ and $\phi_0(y_r, y_{r-1}, \ldots, y_1)$, but for every other permutation of $y_1, y_2, \ldots, y_r$ contains the corresponding value of $\sim\phi_0$, then it may happen that the alternative contains $\phi_0(y_{s_1}, y_{s_2}, \ldots, y_{s_r})$ and $\phi_0(y_{s_r}, y_{s_{r-1}}, \ldots, y_{s_1})$, but for every other permutation of $y_{s_1}, y_{s_2}, \ldots, y_{s_r}$ contains the corresponding value of $\sim\phi_0$.

If this happens, no matter how the set of r y's, $y_{s_1}, y_{s_2}, \ldots, y_{s_r}$ is chosen from $y_1, y_2, \ldots, y_n$, then we say that the *alternative* is *serial in* ϕ_0*, and if an alternative is serial in every function of the new set we shall call it *serial* simply.

Consider, for example, the following alternative, in which we may imagine ϕ_0 and ψ_0 to be derived from one "old" function ϕ by the definitions

$$\phi_0(y_i, y_k) = \phi(y_i, y_k),$$

$$\psi_0(y_i) = \phi(y_i, y_i),$$

$$\phi_0(y_1, y_2)\,.\sim\phi_0(y_2, y_1)\,.\,\phi_0(y_1, y_3)\,.\sim\phi_0(y_3, y_1)\,.\,\phi_0(y_2, y_3)\,.\sim\phi_0(y_3, y_2),$$

$$\psi_0(y_1)\,.\sim\psi_0(y_2)\,.\,\psi_0(y_3).$$

This is serial in ϕ_0, since we always have $\phi_0(y_{s_1}, y_{s_2})\,.\sim\phi_0(y_{s_2}, y_{s_1})$; but not in ψ_0, since we sometimes have $\psi_0(y_{s_1})$, but sometimes $\sim\psi_0(y_{s_1})$. Hence it is not a serial alternative.

We call a *form serial* when it contains at least one serial alternative, and can now state our chief result as follows.

---

* Thus, if ϕ_0 is a function of n variables, *all* alternatives are serial in ϕ_0.

THEOREM.—*There is a finite number m, depending on n, the number of functions ϕ, χ, ψ, ..., and the numbers of their arguments, such that the necessary and sufficient condition for our formula to be consistent in a universe with m or more members is that there should be a serial form A_n completely contained in P. For consistency in a universe of fewer than m members this condition is sufficient but not necessary.*

We shall first prove that, whatever be the number N of individuals in the universe, the condition is sufficient for the consistency of the formula. If $N \leqslant n$, this is a consequence of a previous result, since, if A_n is completely contained in P, so is any A_N involved in A_n.

If $N > n$, we suppose the universe ordered in a series by a relation R. (If N is infinite this requires the Axiom of Selections.) Let q be any serial alternative contained in A_n. If ϕ_0 is a function of r arguments, q will contain the values of either ϕ_0 or $\sim \phi_0$ (but not both) for every permutation of $y_1, y_2, \ldots, y_r$. Any such permutation can be written $y_{\rho_1}, y_{\rho_2}, \ldots, y_{\rho_r}$ where $\rho_1, \rho_2, \ldots, \rho_r$ are 1, 2, ..., r rearranged. We make a list of all those permutations $(\rho_1, \rho_2, \ldots, \rho_r)$ for which q contains the values of ϕ_0, and call this list Σ. We now give ϕ_0 the constant interpretation that $\phi_0(z_1, z_2, \ldots, z_r)$ is to be true if and only if the order of the terms $z_1, z_2, \ldots, z_r$ in the series R is given by one of the permutations $(\rho_1, \rho_2, \ldots, \rho_r)$ contained in Σ, in the sense that, for each i, z_i is the ρ_i-th of $z_1, z_2, \ldots, z_r$ as they are ordered by R.

Let us suppose now that $y_1, y_2, \ldots, y_n$ are numbered in the order in which they occur in R, *i.e.* that in the R series y_1 is the first of them, y_2 the second, and so on. Then we shall see that, if ϕ_0 is given the constant interpretation defined above, all the values of ϕ_0 and $\sim \phi_0$ in q will be true. Indeed, for values whose arguments are obtained by permuting $y_1, y_2, \ldots, y_r$ this follows at once from the way in which ϕ_0 has been defined. For $\phi_0(y_{\sigma_1}, y_{\sigma_2}, \ldots, y_{\sigma_r})$ is true if and only if the order of $y_{\sigma_1}, y_{\sigma_2}, \ldots, y_{\sigma_r}$ in the R series is given by a permutation $(\rho_1, \rho_2, \ldots, \rho_r)$ contained in Σ. But the order in the series of $y_{\sigma_1}, y_{\sigma_2}, \ldots, y_{\sigma_r}$ is in fact given (on our present hypothesis that the order of the y's is $y_1, y_2, \ldots, y_r$) by $(\sigma_1, \sigma_2, \ldots, \sigma_r)$, which is contained in Σ if and only if $\phi_0(y_{\sigma_1}, y_{\sigma_2}, \ldots, y_{\sigma_r})$ is contained in q. Hence values of ϕ_0 for arguments consisting of the first r y's are true when they are contained in q and false otherwise, *i.e.* when the corresponding values of $\sim \phi_0$ are contained in q.

For sets of arguments not confined to the first r y's our result follows from the fact that q is serial, *i.e.* that if $s_1 < s_2 < \ldots < s_r$, so that $y_{s_1}, y_{s_2}, \ldots, y_{s_r}$ are in the order given by the R series, q contains the values of ϕ_0 for just those permutations of $y_{s_1}, y_{s_2}, \ldots, y_{s_r}$ which correspond to the permutations of $y_1, y_2, \ldots, y_r$ for which it contains the values of ϕ_0, *i.e.* by the definition of ϕ_0 and the preceding argument, for just those permutations of $y_{s_1}, y_{s_2}, \ldots, y_{s_r}$ which make ϕ_0 true.

Hence all the values of ϕ_0 and $\sim \phi_0$ in q are true when $y_1, y_2, \ldots, y_n$ are in the order given by the R series.

If, then, we define analogous constant interpretations for χ_0, ψ_0, etc., and combine these with our interpretation of ϕ_0, the whole of q will be true provided that $y_1, y_2, \ldots, y_n$ are in the order given by the R series, and if $y_1, y_2, \ldots, y_n$ are in any other order the true alternative will be obtained from q by suitably permuting the y's, *i.e.* will be an alternative similar to q and contained in the same form A_n. Hence A_n is true for any set of distinct $y_1, y_2, \ldots, y_n$. Moreover, for any set of distinct $y_1, y_2, \ldots, y_\nu$ $(\nu < n)$ the true form will be one involved in A_n, and since A_n and all forms involved in it are contained in P, P will be true for these interpretations of $\phi_0, \chi_0, \psi_0, \ldots$, and our formula must be consistent.

Having thus proved our condition for consistency sufficient in any universe, we have now to prove it necessary in any infinite or sufficiently large finite universe, and for this we have to use the Theorem B proved in the first part of the paper.

Our line of argument is as follows: we have to show that, whatever $\phi_0, \chi_0, \psi_0, \ldots$ we take P will be false unless it completely contains a serial A_n. For this it is enough to show that, given any $\phi_0, \chi_0, \psi_0, \ldots$, there must be a set of n y's for which the true form is serial*, or, since a serial form is one which contains a serial alternative, that there must be a set of values of $y_1, y_2, \ldots, y_n$ for which the true alternative is serial.

Let us suppose that among our functions $\phi_0, \chi_0, \psi_0, \ldots$ there are a_1 functions of one variable, a_2 of two variables, ..., and a_n of n variables, and let us order the universe by a serial relation R.

The N individuals in the universe are divided by the a_1 functions of one variable into 2^{a_1} classes according to which of these functions they make true or false, and if $N \geqslant 2^{a_1} k_1$ we can find k_1 individuals which all belong to the same class, *i.e.* agree as to which of the a_1 functions they make true and which false, where k_1 is a positive integer to be assigned later. Let us call this set of k_1 individuals Γ_{k_1}.

Now consider any two distinct members of Γ_{k_1}, z_1 and z_2 say, and let z_1 precede z_2 in the R series. Then in regard to any of the a_2 functions of two variables, ϕ_0 say, there are four possibilities. We may either

* For then P can only be true for $\phi_0, \chi_0, \psi_0, \ldots$ by *completely* containing this true serial form.

have

$$(1)\quad \phi_0(z_1, z_2) . \phi_0(z_2, z_1),$$

or

$$(2)\quad \phi_0(z_1, z_2) . \sim \phi_0(z_2, z_1),$$

or

$$(3)\quad \sim \phi_0(z_1, z_2) . \phi_0(z_2, z_1),$$

or

$$(4)\quad \sim \phi_0(z_1, z_2) . \sim \phi_0(z_2, z_1).$$

ϕ_0 thus divides the combinations two at a time of the members of Γ_{k_1} into four distinct classes according to which of these four possibilities is realised when the combination is taken as z_1, z_2 in the order in which its terms occur in the R series; and the whole set of a_2 functions of two variables divide the combinations two at a time of the members of Γ_{k_1} into 4^{a_2} classes, the combinations in each class agreeing in the possibility they realise with respect to each of the a_2 functions. Hence, by Theorem B, if $k_1 = h(2, k_2, 4^{a_2})$, Γ_{k_1} must contain a sub-class Γ_{k_2} of k_2 members such that all the pairs out of Γ_{k_2} agree in the possibilities they realise with respect to each of the a_2 functions of two variables.

We continue to reason in the same way according to the following general form :—

Consider any r distinct members of $\Gamma_{k_{r-1}}$; suppose that in the R series they have the order $z_1, z_2, \ldots, z_r$. Then with respect to any function of r variables there are $2^{r!}$ possibilities in regard to $z_1, z_2, \ldots, z_r$, and the a_r functions of r variables divide the combinations r at a time of the members of $\Gamma_{k_{r-1}}$ into $2^{r!a_r}$ classes. By Theorem B, if $k_{r-1} = h(r, k_r, 2^{r!a_r})$*, $\Gamma_{k_{r-1}}$ must contain a sub-class Γ_{k_r} of k_r members such that all the combinations r at a time of the members of Γ_{k_r} agree in the possibilities they realise with respect to each of the a_r functions of r variables.

We proceed in this way until we reach $\Gamma_{k_{n-1}}$, all combinations $n-1$ at a time of whose members agree in the possibilities they realise with respect to each of the a_{n-1} functions of $n-1$ variables. We then determine that k_{n-1} shall equal n, which fixes k_{n-2} as $h(n-1, n, 2^{(n-1)!a_{n-1}})$ and so on back to k_1, every k_{r-1} being determined from k_r.

If, then, $N \geqslant 2^{a_1} k_1$, the universe must contain a class $\Gamma_{k_{n-1}}$ or Γ_n (since $k_{n-1} = n$) of n members which is contained in Γ_{k_r} for every r, $r = 1, 2, \ldots, n-1$. Let its n members be, in the order given them by R, $y_1, y_2, \ldots, y_n$. Then for every r less than n, $y_1, y_2, \ldots, y_n$ are contained in Γ_{k_r} and all r combinations of them agree in the possibilities they realise

* If $a_r = 0$ we interpret $h(r, k_r, 1)$ as k_r and identify $\Gamma_{k_{r-1}}$ and Γ_{k_r}.

with respect to each function of r variables. Let $y_{s_1}, y_{s_2}, \ldots, y_{s_r}$ $(s_1 < s_2 < \ldots < s_r)$ be such a combination, and χ_0 a function of r variables. Then $y_{s_1}, y_{s_2}, \ldots, y_{s_r}$ are in the order given them by R, and so are $y_1, y_2, \ldots, y_r$; consequently the fact that these two combinations agree in the possibilities which they realise with respect to χ_0 means that χ_0 is true for the same permutations of $y_{s_1}, y_{s_2}, \ldots, y_{s_r}$ as it is of $y_1, y_2, \ldots, y_r$. The true alternative for $y_1, y_2, \ldots, y_n$ is therefore serial in χ_0, and similarly it is serial in every other function of any number r of variables*; it is therefore a serial alternative.

Our condition is, therefore, shown to be necessary in any universe of at least $2^{a_1} k_1$ members where k_1 is given by

$$k_{n-1} = n,$$

$$\left.\begin{aligned} k_{r-1} &= h(r, k_r, 2^{r!a_r}) \quad \text{if} \quad a_r \neq 0 \\ &= k_r \qquad\qquad\qquad \text{if} \quad a_r = 0 \end{aligned}\right\} \quad (r = n-1, n-2, \ldots, 2).$$

For universes lying between n and $2^{a_1} k_1$ we have not found a necessary and sufficient condition for the consistency of the formula, but it is evidently possible to determine by trial whether any given formula is consistent in any such universe.

III.

We will now consider what our result becomes when our formula

$$(x_1, x_2, \ldots, x_n)\, F(\phi, \chi, \psi, \ldots, =, x_1, x_2, \ldots, x_n)$$

contains in addition to identity only one function ϕ of two variables.

In this case we have two functions ϕ_0, ψ_0 given by

$$\phi_0(y_i, y_k) = \phi(y_i, y_k) \quad (i \neq k),$$

$$\chi_0(y_i) \quad = \phi(y_i, y_i),$$

so that $a_1 = 1$, $a_2 = 1$, $a_r = 0$ when $r > 2$. Consequently

$$k_2 = k_3 = \ldots = k_{n-1} = n \quad \text{and} \quad k_1 = h(2, n, 4);$$

but the argument at the end of I shows that we may take instead $k_1 = n!!!$, and our necessary and sufficient condition for consistency applies to any universe with at least $2 . n!!!$ individuals.

In this simple case we can present our condition in a more striking form as follows.

* We have shown this when $r < n$; we may also have $r = n$, but then there is nothing to prove since in a function of n variables every alternative is serial.

It is necessary and sufficient for the consistency of the formula that it should be true when ϕ is replaced by at least one of the following types of function :—

(1) The universal function $x = x \,.\, y = y$.

(2) The null function $x \neq x \,.\, y \neq y$.

(3) Identity $x = y$.

(4) Difference $x \neq y$.

(5) A serial function ordering the whole universe in a series, *i.e.* satisfying

(*a*) $(x) \sim \phi(x, x)$,

(*b*) $(x, y)[x = y \vee \{\phi(x, y) \,.\sim \phi(y, x)\} \vee \{\phi(y, x) \,.\sim \phi(x, y)\}]$,

(*c*) $(x, y, z)\{\sim\phi(x, y) \vee \sim \phi(y, z) \vee \phi(x, z)\}$.

(6) A function ordering the whole universe in a series, but also holding between every term and itself, *i.e.* satisfying

(*a'*) $(x)\,\phi(x, x)$

and (*b*) and (*c*) as in (5).

Types (1)–(4) include only one function each; in regard to types (5) and (6) it is immaterial what function of the type we take, since if one satisfies the formula so, we shall see, do all the others*.

We have to prove this new form of our condition by showing that P will completely contain a serial A_n if and only if it is satisfied by functions of at least one of our six types. Now an alternative in n y's is serial in χ_0 if it contains

either (i) $\chi_0(y_1) \,.\, \chi_0(y_2) \,.\, \ldots \,.\, \chi_0(y_n)$ or, for short, $\prod_r \chi_0(y_r)$,

or (ii) $\sim \chi_0(y_1) \,.\, \ldots \,.\sim \chi_0(y_n)$ „ „ $\prod_r \sim \chi_0(y_r)$,

but not otherwise, and it will be serial in ϕ_0 if it contains

either (*a*) $\prod_{r<s} \phi_0(y_r, y_s) \,.\, \phi_0(y_s, y_r)$,

or (*b*) $\prod_{r<s} \phi_0(y_r, y_s) \,.\sim \phi_0(y_s, y_r)$,

or (*c*) $\prod_{r<s} \sim \phi_0(y_r, y_s) \,.\, \phi_0(y_s, y_r)$,

or (*d*) $\prod_{r<s} \sim \phi_0(y_r, y_s) \,.\sim \phi_0(y_s, y_r)$.

* A result previously obtained for type (5) by Langford, *op. cit.*

There are thus altogether eight alternatives serial in both ϕ_0 and χ_0 got by combining either of (i), (ii) with any of (*a*), (*b*), (*c*), (*d*); but these eight serial alternatives only give rise to six serial forms, since the alternatives (i) (*b*) and (i) (*c*) can be obtained from one another by reversing the order of the y's and so belong to the same form, and so do the alternatives (ii) (*b*) and (ii) (*c*).

It is also easy to see that any formula completely containing one of these six serial forms will be satisfied by all functions of one of the six types according to the scheme

Form	(i) (*a*)	(i) (*b* and *c*)	(i) (*d*)	(ii) (*a*)	(ii) (*b* and *c*)	(ii) (*d*)
Type of function	1	6	3	4	5	2

and that conversely a formula satisfied by a function of one of the six types must completely contain the corresponding form. For instance, a function of type 6 will satisfy the alternative (i) (*b*) when $y_1, y_2, \ldots, y_n$ are in their order in the series determined by the function, and when $y_1, y_2, \ldots, y_n$ are in any other order the function will satisfy an alternative of the same form.

In the language of the theory of postulate systems we can interpret our universe as a class K, and conclude that a postulate system on a base (K, R) consisting only of general laws involving at most n elements will be compatible with K having as many as $2 . n!!!$ members if and only if it can be satisfied by an R of one of our six types.

IV.

Let us, in conclusion, briefly indicate how to extend our method in order to determine the consistency or inconsistency of formulae of the more general type

$$(Ez_1, z_2, \ldots, z_m)(x_1, x_2, \ldots, x_n) F(\phi, \chi, \psi, \ldots, =, z_1, z_2, \ldots, z_m, x_1, x_2, \ldots, x_n)$$

which have in normal form both kinds of prefix, but satisfy the condition that all the prefixes of existence precede all those of generality.

As before, we can suppose F represented as a disjunction of alternatives and discard those which violate the laws of identity. Those left we can group according to the values of identity and difference for arguments drawn entirely from the z's. Such a set of values of identity and difference we can denote by $H_i\ (=, z_1, z_2, \ldots, z_m)$, and F can be put in the form

$$(H_1 . F_1) \vee (H_2 . F_2) \vee (H_3 . F_3) \vee \ldots,$$

and the whole formula is equivalent to a disjunction of formulae.

$$(Ez_1, z_2, \ldots, z_m)\{H_1(=, z_1, z_2, \ldots, z_m)$$
$$.\,(x_1, x_2, \ldots, x_n)\, F_1(\phi, \ldots, =, z_1, \ldots, z_m, x_1, \ldots, x_n)\}$$
$$\vee\,(Ez_1, z_2, \ldots, z_m)\{H_2(=, z_1, \ldots, z_m)$$
$$.\,(x_1, x_2, \ldots, x_n)\, F_2(\phi, \ldots, =, z_1, \ldots, z_m, x_1, \ldots, x_n)\}$$

$\vee$ etc.

Since if any one of these formulae is consistent so is their disjunction, and if their disjunction is consistent one at least of its terms must be consistent, it is enough for us to show how to determine the consistency of any one of them, say the first. In this $H_1(=, z_1, z_2, \ldots, z_m)$ is a consistent set of values of identity and difference for every pair of z's. We renumber the z's $z_1, z_2, \ldots, z_\mu$ using the same suffix for every set of z's that are identical in H_1, and our formula becomes

$$(Ez_1, z_2, \ldots, z_\mu)(x_1, x_2, \ldots, x_n)\, F_1(\phi, \chi, \ldots, =, z_1, z_2, \ldots, z_\mu, x_1, x_2, \ldots, x_n), \quad \text{(i)}$$

in which it is understood that two z's with different suffixes are always different.

Now supposing the universe to have at least $\mu+n$ members, we consider the different possibilities in regard to the x's being identical with the z's, and rewrite our formula

$$(Ez_1, z_2, \ldots, z_\mu)(x_1, x_2, \ldots, x_n)$$
$$\left\{\prod_{\substack{i=1,\ldots,n\\ j=1,\ldots,\mu}} x_i \neq z_j \rightarrow G(\phi, \chi, \ldots, =, z_1, \ldots, z_\mu, x_1, \ldots, x_n)\right\},$$

in which $\rightarrow$ means "if, then" and

$$G(\phi, \ldots, x_n) = \Pi\, F_1(\phi, \chi, \ldots, =, z_1, \ldots, z_\mu, \theta_1, \theta_2, \ldots, \theta_n),$$

the product being taken for

$$\theta_1 = x_1, z_1, z_2, \ldots, z_\mu,$$
$$\theta_2 = x_2, z_1, z_2, \ldots, z_\mu,$$
$$\ldots \quad \ldots \quad \ldots$$
$$\theta_n = x_n, z_1, z_2, \ldots, z_\mu,$$

and in G any term $x_i = z_j$ is replaced by a falsehood (*e.g.* $x_i \neq x_i$) not involving any z.

Next we modify G by introducing new functions. In G occur values of, *e.g.* ϕ, with arguments some of which are z's and some x's; from these we define functions of the x's only by simply regarding the z's as constants, and call these new functions $\phi_0, \chi_0, \ldots$. Values of $\phi, \chi, \psi, \ldots,$ which include no x's among their arguments, we replace by constant propositions $p, q, \ldots$. The only values of identity in G are of the form $x_i = x_j$ and these we leave alone. Suppose that by this process G turns into

$$L(\phi, \chi, \psi, \ldots, \phi_0, \chi_0, \ldots, p, q, \ldots, =, x_1, x_2, \ldots, x_n).$$

Then the consistency of formula (i) in a universe of N individuals is evidently equivalent to the consistency in a universe of $N-\mu$ individuals of the formula

$$(x_1, x_2, \ldots, x_n) L(\phi, \chi, \psi, \ldots, \phi_0, \chi_0, \ldots, p, q, \ldots, =, x_1, x_2, \ldots, x_n).$$

But this is a formula of the type previously dealt with, except for the variable propositions $p, q, \ldots$, which are easily eliminated by considering the different cases of their truth and falsity, the formula being consistent if it is consistent in one such case.

Reprinted from
Proc. London Math. Soc. **30** (1930), 264–286

NON-SEPARABLE AND PLANAR GRAPHS*

BY

HASSLER WHITNEY

Introduction. In this paper the structure of graphs is studied by purely combinatorial methods. The concepts of rank and nullity are fundamental. The first part is devoted to a general study of non-separable graphs. Conditions that a graph be non-separable are given; the decomposition of a separable graph into its non-separable parts is studied; by means of theorems on circuits of graphs, a method for the construction of non-separable graphs is found, which is useful in proving theorems on such graphs by mathematical induction. In the second part, a dual of a graph is defined by combinatorial means, and the paper ends with the theorem that a necessary and sufficient condition that a graph be planar is that it have a dual.

The results of this paper are fundamental in papers by the author on *Congruent graphs and the connectivity of graphs*† and on *The coloring of graphs*.‡

I. Non-separable graphs

1. **Definitions.**§ A *graph G* consists of two sets of symbols, finite in number: *vertices*, $a, b, c, \cdots, f$, and *arcs*, $\alpha(ab), \beta(ac), \cdots, \delta(cf)$. If an arc $\alpha(ab)$ is present in a graph, its *end vertices* a, b are also present. We may write an arc $\alpha(ab)$ or $\alpha(ba)$ at will; we may write it also ab or ba if no confusion arises,—if there is but a single arc joining a and b in G. We say the vertices a and b are *on* the arc $\alpha(ab)$, and the arc $\alpha(ab)$ is *on* the vertices a and b. The *null graph* is the graph containing no arcs or vertices.

The obvious geometrical interpretation of such a graph, or *abstract graph*, is a *topological graph*, let us say. Corresponding to each vertex of the abstract graph, we select a point in 3-space, a vertex of the topological graph. Corresponding to each arc $\alpha(ab)$ of the abstract graph, we select an arc joining the corresponding vertices of the topological graph. An arc is here a set of points in (1, 1) correspondence with the unit interval, its end vertices corresponding

* Presented to the Society, October 25, 1930; received by the editors February 2, 1931. An outline of this paper will be ound in the Proceedings of the National Academy of Sciences, vol. 17 (1931), pp. 125–127.

† American Journal of Mathematics, vol. 54 (1932), pp. 150–168.

‡ An outline will be found in the Proceedings of the National Academy of Sciences, vol. 17 (1931), pp. 122–125.

§ Compare Ste. Laguë, *Les Réseaux*, Mémorial des Sciences Mathématiques, fascicule 18, Paris, 1926.

with the ends of the interval. Moreover, we let no arc pass through other vertices or intersect other arcs. We shall consider topological graphs no further till we come to the section on planar graphs.

An *isolated vertex* is a vertex which is not on any arc. A *chain* is a set of one or more distinct arcs which can be ordered thus: ab, bc, cd, $\cdots$, ef, where vertices in different positions are distinct, i.e. the chain may not intersect itself. A *suspended chain* is a chain containing two or more arcs such that no vertex of the chain other than the first and last is on other arcs, and these two vertices are each on at least two other arcs. A *circuit* is a set of one or more distinct arcs which can be put in cyclic order, $ab, bc, \cdots, ef, fa$, vertices being distinct as in the case of the chain. A *k-circuit* is a circuit containing k arcs. Thus, the arc $\alpha(aa)$ is a 1-circuit; the two arcs $\alpha(ab)$, $\beta(ab)$ form a 2-circuit.

A graph is *connected* if any two of its vertices are joined by a chain. Obviously, if a and b are joined by a chain, and b and c are joined by a chain, then a and c are joined by a chain. Any graph consists of a certain number of *connected pieces* (one, if the graph is connected). In particular, an isolated vertex is one of the connected pieces of a graph. A graph is called *cyclicly connected* if any two of its vertices are contained in a circuit. If $G_1, G_2, \cdots, G_m$ are a set of graphs, no two of which have a common vertex (or arc, therefore), we say the graph G, formed of the arcs and vertices of all these graphs, is the *sum* of these graphs. Thus, a graph is the sum of its connected pieces. A *forest* is a graph containing no circuit. A *tree* is a connected forest. A *subgraph* H of G is a graph containing a subset (in particular, all or none), of the arcs of G, and those vertices of G which are on these arcs.

2. **Rank and nullity.*** Given a graph G which contains V vertices, E arcs, and P connected pieces, we define its *rank* R, and its *nullity* (or *cyclomatic number* or first Betti number) N, by the equations

$$R = V - P,$$
$$N = E - R = E - V + P.$$

If G contains the single arc ab, it is of rank 1, nullity 0, while if it contains the single arc aa, it is of rank 0, nullity 1.

The first two theorems follow immediately from the definitions of rank and nullity:

THEOREM 1. *If isolated vertices be added to or subtracted from a graph, the rank and nullity remain unchanged.*

* These are just the rank and nullity of the matrix H_1 of Poincaré. See Veblen's Colloquium Lectures, *Analysis Situs*.

THEOREM 2. *Let the graph G' be formed from the graph G by adding the arc ab. Then*

(1) *if a and b are in the same connected piece in G, then*

$$R' = R, \quad N' = N + 1;$$

(2) *if a and b are in different connected pieces in G, then*

$$R' = R + 1, \quad N' = N.$$

THEOREM 3. *In any graph G,*

$$R \geqq 0, \quad N \geqq 0.$$

For let G_1 be the graph containing the vertices of G but no arcs. Then if R_1 and N_1 are its rank and nullity,

$$R_1 = N_1 = 0.$$

We build up G from G_1 by adding the arcs one at a time. The theorem now follows from Theorem 2.

THEOREM 4. *A forest G is a graph of nullity 0, and conversely.*

Suppose first G contained a circuit P. We shall show that the nullity of G is >0. We build up G arc by arc, adding first the arcs of the circuit P. In adding the last arc of the circuit, the nullity is increased by 1, as this arc joins two vertices already connected. (This argument holds even if the circuit is a 1-circuit.) But in adding the rest of the arcs, the nullity is never decreased, by Theorem 2. Thus the nullity of G is >0.

Now suppose G is a forest, and therefore contains no circuit. Build up G arc by arc. Each arc we add joins two vertices formerly not connected. For otherwise, this arc, together with the arcs of a chain connecting the two vertices, would form a circuit. Therefore, by Theorem 2, the nullity remains always the same, and is thus 0.

3. Theorems on non-separable graphs. We introduce the following

Definitions. Let H_1, which contains the vertex a_1, and H_2, which contains the vertex a_2, be two graphs without common vertices. Let us rename a_1 a, and rename the arcs of H_1 on a_1 accordingly; that is, if a_1b is an arc on a_1, we rename it ab. Rename also a_2 a, and rename the arcs of H_2 accordingly. H_1 and H_2 have now the vertex a in common; they form the graph G, say. We say G is formed by letting the vertex a_1 of H_1 *coalesce* with the vertex a_2 of H_2, or, by joining H_1 and H_2 at a vertex. Geometrically, we pull the vertices a_1 and a_2 together to form the single vertex a.

Let G be a connected graph such that there exist no two graphs H_1 and

H_2, each containing at least one arc, which form G if they are joined at a vertex. Then G is called *non-separable.* Geometrically, a connected graph is non-separable if we cannot break it at a single vertex into two graphs, each containing an arc. For example, the graph consisting of the two arcs ab, bc is separable, as is the graph consisting of the two arcs $\alpha(aa)$, $\beta(aa)$. A graph containing but a single arc is non-separable, as is the graph containing only the arcs $\alpha(ab)$, $\beta(ab)$.

If G is not non-separable, we say G is *separable.* Thus, a graph that is not connected is separable. Suppose some connected piece G_1 of G is separable. If H_1 and H_2 joined at the vertex a form G_1, we say a is a *cut vertex* of G. We have consequently

THEOREM 5. *A necessary and sufficient condition that a connected graph be non-separable is that it have no cut vertex.*

THEOREM 6. *Let G be a connected graph containing no* 1-*circuit. A necessary and sufficient condition that the vertex a be a cut vertex of G is that there exist two vertices b, c in G, each distinct from a, such that every chain from b to c passes through a.*

First suppose a is a cut vertex of G. Then, by definition, H_1 and H_2, each containing at least one arc which is not a 1-circuit, form G if they are joined at a. Let b be a vertex of H_1 and c a vertex of H_2, each distinct from a. As a is the only vertex in both H_1 and H_2, every chain from b to c in G passes through a.

Suppose now every chain from b to c in G passes through a. Remove the vertex a and all the arcs on a. The resulting graph G' is not connected, b and c being in different connected pieces. Let H_1' be that connected piece of G' containing b, and let H_2' be the rest of G'. Replace a by the two vertices a_1 and a_2. Now put back the arcs we removed, letting them touch a_1 if their other end vertices are in H_1', and letting them touch a_2 otherwise. Let H_1 and H_2 be the resulting graphs. Then H_1 and H_2 each contain at least one arc, and they form G if the two vertices a_1, a_2 are made to coalesce. Hence, by definition, a is a cut vertex of G.

THEOREM 7. *Let G be a graph containing no* 1-*circuit and containing at least two arcs. A necessary and sufficient condition that G be non-separable is that it be cyclicly connected.**

If G is not connected, the theorem is obvious. Assume therefore G is connected.

* A similar theorem has been proved for more general continuous curves by G. T. Whyburn, Bulletin of the American Mathematical Society, vol. 37 (1931), pp. 429–433.

Suppose first G is separable. Then, by Theorem 5, G has a cut vertex a, and by Theorem 6, there are two vertices b, c in G such that every chain from b to c passes through a. Hence there is no circuit in G containing b and c.

Suppose now there exist two vertices b, c in G which are contained in no circuit. Let $bd, de, \cdots, gc$ be some chain from b to c.

Case 1. There exists a circuit containing b and d. In this case, let a be the last vertex of the chain which is contained in a circuit passing also through b. Let f be the next vertex of the chain. Then every chain from f to b passes through a. For suppose the contrary. Let C be a chain from f to b not passing through a. Let P be a circuit containing b and a. Follow C from f till we first reach a vertex of P. Follow the circuit P now as far as b if b was not the vertex we reached, and continue along P till we reach a. Passing from a to f along the arc af completes a circuit containing both b and f, contrary to hypothesis. Hence, by Theorem 6, a is a cut vertex of G, and therefore G is separable.

Case 2. There exists no circuit containing b and d. Then there is but a single arc joining b and d, and they are joined by no other chain. As G is connected and contains at least two arcs, there is either another arc on b or another arc on d, say the first. The other case is exactly similar. If we add a vertex b' and replace the arc bd by the arc $b'd$, b and d are no longer joined by a chain, and hence the resulting graph G' is not connected. Let H_1 be that part of G' containing the arc $b'd$, and let H_2 be the rest of G'. As there is still an arc on b, H_2 contains at least one arc. Letting the vertices b and b' coalesce forms G, and hence G is separable. The proof is now complete.

THEOREM 8. *A non-separable graph G containing at least two arcs contains no* 1*-circuit and is of nullity* >0. *Each vertex is on at least two arcs.*

Suppose G contained a 1-circuit. Call it H_1. Let H_2 be the rest of the graph. Then H_1 and H_2 have but a single vertex in common, and thus G is separable.

Next, by Theorem 7, G is cyclicly connected. As G contains no 1-circuit, G contains at least two vertices. Containing these there is a circuit. Therefore, by Theorem 4, the nullity of G is >0.

Finally, if there were a vertex on no arcs, G would not be connected. If there were a vertex a on the single arc ab, b would be a cut vertex of G.

THEOREM 9. *Let G be a graph of nullity* 1 *containing no isolated vertices, such that the removal of any arc reduces the nullity to* 0. *Then G is a circuit.*

By Theorem 4, G contains a circuit. Suppose G contained other arcs besides. Removing one o these, the nullity remains 1, as the circuit is still present, contrary to hypothesis. There are no other vertices in G, as G contains no isolated vertices. Hence G is just this circuit.

THEOREM 10. *A non-separable graph G of nullity 1 is a circuit.*

If G contains but a single arc, it is a 1-circuit, being of nullity 1. Suppose G contains at least two arcs. By Theorem 8, it contains no 1-circuit. By Theorem 7, it is cyclicly connected. Remove any arc ab from G; a and b are still connected, and therefore, by Theorem 2, the nullity of G is reduced to 0. Hence, by Theorem 9, G is a circuit.

The converses of the last two theorems are obviously true.

4. **Decomposition of separable graphs.** If the graph G contains a connected piece which is separable, we may separate that piece into two graphs, these graphs having formerly but a single vertex in common. We may continue in this manner until every resulting piece of G is non-separable. We say G is separated into its *components*.

LEMMA. *Let the connected separable graph G be decomposed into the two pieces H_1 and H_2 which had only the vertex a in common in G. Then every non-separable subgraph of G is contained wholly in either H_1 or H_2.*

Suppose the contrary. Then some non-separable subgraph I of G is not contained wholly in either H_1 or H_2. Let I_1 be that part of I in H_1, and I_2 that part in H_2; I_1 and I_2 have at most the vertex a in common. I_1 and I_2 each contain at least one arc. For otherwise, if I_1, say, contained no arc, as it contains a vertex distinct from a, it would not be connected. Thus I is separable into the pieces I_1 and I_2, a contradiction again.

THEOREM 11. *Every non-separable subgraph of G is contained wholly in one of the components of G.*

This follows upon repeated application of the above lemma.

THEOREM 12. *A graph G may be decomposed into its components in a unique manner.*

Suppose we could decompose G into the components $H_1, H_2, \cdots, H_m$, and also into the components $H_1', H_2', \cdots, H_n'$. We shall show that these sets are identical. Take any H_i. It is a non-separable subgraph of G, and thus is contained in some component H_j', by Theorem 11. Similarly, H_j' is contained in some component H_k. Thus H_i is contained in H_k, and they are therefore identical. Hence H_i and H_j' are identical. In this manner we show that each H_k is identical with some H_l', and each H_l' is identical with some H_k, proving the theorem.

THEOREM 13. *Let $H_1, H_2, \cdots, H_m$ be the components of G. Let $R_1, R_2, \cdots, R_m$, and $N_1, N_2, \cdots, N_m$ be their ranks and nullities. Then*

$$R = R_1 + R_2 + \cdots + R_m,$$
$$N = N_1 + N_2 + \cdots + N_m.$$

Let G' be G separated into its components, and let R' be the rank of G'. G is formed from G' by letting vertices of different components coalesce. Each time we join two pieces, the number of vertices and the number of connected pieces are each reduced by 1, so that the rank remains the same. Thus

$$R = R'.$$

Now

$$V' = V_1 + V_2 + \cdots + V_m,$$
$$P' = P_1 + P_2 + \cdots + P_m$$

(where each $P_i = 1$). Subtracting,

$$R = R' = R_1 + R_2 + \cdots + R_m.$$

As also

$$E = E_1 + E_2 + \cdots + E_m,$$

it follows that

$$N = N_1 + N_2 + \cdots + N_m.$$

For a converse of this theorem, see Theorem 17.

THEOREM 14. *Divide the arcs of the non-separable graph G into two groups, each containing at least one arc, forming the subgraphs H_1 and H_2, of ranks R_1 and R_2. Then*

$$R_1 + R_2 > R.$$

Let the connected pieces of H_1 be $H_{11}, \cdots, H_{1m}$ (there may be but one piece, H_{11}), and let those of H_2 be $H_{21}, \cdots, H_{2n}$. Then obviously

$$R_1 = R_{11} + \cdots + R_{1m},$$
$$R_2 = R_{21} + \cdots + R_{2n},$$

whence

$$R_1 + R_2 = R_{11} + \cdots + R_{1m} + R_{21} + \cdots + R_{2n}.$$

Let G' be the sum of the graphs $H_{11}, \cdots, H_{2n}$. Then G' is of rank $R_{11} + \cdots + R_{2n}$. We form G from G' by letting vertices of the graphs $H_{11}, \cdots, H_{2n}$ coalesce. Each time we let vertices of different connected pieces coalesce, the rank is unaltered. Each time we let vertices in the same connected piece coalesce, the rank is reduced by 1. This latter operation happens at least once. For otherwise, let a_1 and a_2 be the last two vertices we let coalesce. Then a_1 and a_2 were formerly in two different pieces, I_1 and I_2. Thus

I_1 and I_2 joined at a vertex form G, and G is separable, contrary to hypothesis. Thus the rank of G is less than the rank of G', that is,

$$R < R_{11} + \cdots + R_{2n}.$$

Hence

$$R_1 + R_2 > R.^*$$

Theorems 13 and 14 give

THEOREM 15. *A necessary and sufficient condition that a graph be non-separable is that there exist no division of its arcs into two groups H_1 and H_2, each containing at least one arc, so that*

$$R = R_1 + R_2.$$

5. **Circuits of graphs.** We shall say two non-separable graphs, each containing at least one arc, form a *circuit of graphs*, if they have at least two common vertices. (They may also have common arcs.) Thus the two graphs G_1: $\alpha(ab)$ and G_2: $\alpha(ab)$ (which are the same graph) form a circuit of graphs. However, the two graphs G_1: $\alpha(aa)$ and G_2: $\beta(aa)$, having but one common vertex, do not form a circuit of graphs. We shall say three or more non-separable graphs form a *circuit of graphs* if we can name them $G_1, G_2, \cdots, G_m$ in such a way that G_1 and G_2 have just the vertex a_1 in common, G_2 and G_3 have just the vertex a_2 in common, $\cdots$, G_m and G_1 have just the vertex a_m in common, these vertices are all distinct, and no other two of these graphs have a common vertex. Thus the three graphs G_1:ab, G_2:bc, G_3:ca form a circuit of graphs.

We note that there can be no 1-circuit in a circuit of graphs; also, no subset of the graphs in a circuit of graphs form a circuit of graphs. We may think of a circuit of graphs as forming a single graph.

THEOREM 16. *A circuit of graphs G is a non-separable graph.*

First suppose there are but two graphs, G_1 and G_2, present. Suppose G were separable. Then it is separable into at least two components $H_1, H_2, \cdots, H_k$. By Theorem 11, G_1 and G_2 are each contained wholly in one of these components. As G_1 and G_2 together form G, there are just two components, and they are G_1 and G_2. These, when joined at a vertex, form G. But this is contrary to the hypothesis that G_1 and G_2 have at least two vertices in common.

Next suppose there are more than two graphs present. Let C_1 be a chain in G_1 joining a_m and a_1, let C_2 be a chain in G_2 joining a_1 and a_2, $\cdots$, let C_m be a chain in G_m joining a_{m-1} and a_m. These chains taken together form a cir-

* This theorem may also be proved easily from Theorem 17.

cuit P passing through all the graphs. Now separate G into its components. By Theorem 11 (see the converse of Theorem 10), P is contained in one of these components. The same is true of each of the graphs $G_1, G_2, \cdots, G_m$, and hence these graphs are all contained in the same component. Thus G is itself this component, that is, G is non-separable.

THEOREM 17. *Let $G_1, \cdots, G_m$ be a set of non-separable graphs, each containing at least one arc, and let G be formed by letting vertices and arcs of different graphs coalesce. Then the following four statements are all equivalent:*

(1) *$G_1, \cdots, G_m$ are the components of G.*

(2) *No two of the graphs $G_1, \cdots, G_m$ have an arc in common, and there is no circuit in G containing arcs of more than one of these graphs.*

(3) *No subset of these graphs form a circuit of graphs.*

(4) *If $R, R_1, \cdots, R_m$ are the ranks of $G, G_1, \cdots, G_m$ respectively, then*

$$R = R_1 + \cdots + R_m.$$

We note that we cannot replace the word rank by the word nullity in (4). For let G be the graph containing the arcs $\alpha(ab), \beta(ab), \gamma(ab)$. Let G_1 contain α and β, and G_2, β and γ. Then the nullity of G is the sum of the nullities of G_1 and G_2, but G_1 and G_2 are not the components of G. We shall prove

(a) if (1) holds, (2) holds,

(b) if (2) holds, (3) holds,

(c) if (3) holds, (1) holds, establishing the equivalence of (1), (2) and (3);

(d) if (1) holds, (4) holds, and finally

(e) if (4) holds, (3) holds, establishing the equivalence of (4) and the other statements.

(a) If (1) holds, (2) holds. For first, in forming G from its components $G_1, \cdots, G_m$, we let vertices alone coalesce, and thus no two of the graphs have an arc in common. Also, there is no circuit in G containing arcs of more than one of the graphs; for each circuit, being a non-separable graph, is contained entirely in one of the components of G, by Theorem 11.

(b) If (2) holds, (3) holds. For suppose the contrary. If, first, some two graphs, say G_1 and G_2, form a circuit of graphs, they have at least two vertices in common, say a and b. Join a and b by a chain C in G_1 and by a chain D in G_2. By hypothesis, G_1 and G_2 have no arcs in common, and thus the arcs of C and D are distinct. From a follow along C till we first reach a vertex d of D. From d follow along D till we get back to a. We have formed thus a circuit containing arcs of both G_1 and G_2, contrary to hypothesis.

Now suppose the graphs $G_1, \cdots, G_k$, $k>2$, formed a circuit of graphs. In the proof of Theorem 16 we found a circuit passing through all the graphs of such a circuit of graphs, again contrary to hypothesis.

(c) If (3) holds, (1) holds. Assuming that no subset of the graphs $G_1, \cdots, G_m$ forms a circuit of graphs, we will show first that some one of these graphs has at most a single vertex in common with other of the graphs. For suppose each graph had at least two vertices in common with other graphs. Then G_1 has a vertex a_1 in common with some graph, say G_2. As G_2 has at least two vertices in common with other graphs, it has a vertex a_2, distinct from a_1, in common with another graph, say G_3. If we continue in this manner, we must at some point get back to a graph we have already considered.

Now starting with G_1, consider the graphs in order, and let G_i be the first one which has a vertex in common with one of the preceding graphs other than the vertex a_{i-1}, which we know already it has in common with G_{i-1}. Now of the graphs $G_{i-1}, G_{i-2}, \cdots, G_1$, let G_j be the first with which G_i has a common vertex, other than the vertex a_{i-1}. First suppose G_j is G_{i-1}. Then G_i and G_{i-1} have at least two vertices in common, and they form therefore a circuit of graphs, contrary to hypothesis. Next suppose G_j is not G_{i-1}. Then on account of the choice of G_i and G_j, G_j and G_{j+1} have just one common vertex a_j, G_{j+1} and G_{j+2} have just one common vertex $a_{j+1}, \cdots$, G_i and G_j have just one common vertex a_i (for otherwise G_i and G_j would form a circuit of graphs), and no other two of these graphs have a vertex in common. These vertices $a_j, a_{j+1}, \cdots, a_i$ are all distinct. For, on account of the construction of the chain of graphs, two succeeding vertices a_k and a_{k+1} are distinct. a_i and a_j are distinct, for otherwise G_i and G_{j+1} would have a common vertex, etc. These graphs $G_j, G_{j+1}, \cdots, G_i$ form therefore a circuit of graphs, contrary to hypothesis.

Some graph therefore, say G_1, has at most a single vertex in common with the other graphs. Thus either it is separated from them, or we can separate it at a single vertex. Now among the graphs $G_2, \cdots, G_m$, there is also no circuit of graphs, so again we can separate one of them, say G_2. Continuing, we have finally separated G into its components $G_1, G_2, \cdots, G_m$.

(d) If (1) holds, (4) holds. This is just Theorem 13.

(e) If (4) holds, (3) holds. Let G' be the sum of the graphs $G_1, \cdots, G_m$. We form G from G' by letting vertices and arcs of different graphs coalesce. Each time we let two vertices coalesce, either (α) the two vertices were formerly in different connected pieces, in which case the rank is unchanged, or (β) the two vertices were in the same connected piece, in which case the rank is reduced by 1. Letting arcs alone coalesce (their end vertices having already coalesced) does not alter the rank. Thus in any case, the rank is never increased. To begin with, the rank of G' is $G_1+ \cdots +G_m$, and by hypothesis, the rank of G is $G_1+ \cdots +G_m$. Thus the rank is never altered, and (β) never occurs. Hence, obviously, no circuit of graphs is formed in forming G from G'. This completes the proof of the theorem.

6. **Construction of non-separable graphs.** We prove the following theorem:

THEOREM 18. *If G is a non-separable graph of nullity $N>1$, we can remove an arc or suspended chain from G, leaving a non-separable graph G' of nullity $N-1$.*

Assume the theorem is true for all graphs of nullity $2, 3, \cdots, N-1$. We shall prove it for any graph of nullity N (including the case where $N=2$). This will establish the theorem in general.

Take any non-separable graph G of nullity $N>1$. It contains at least two arcs, and therefore, by Theorem 8, it contains no 1-circuit. Remove from G any arc ab, forming the graph G_1. If G_1 is non-separable, we are through. Suppose therefore G_1 is separable, and let its components be $H_1, H_2, \cdots, H_{m-1}$. G_1 is connected, for between any two vertices c, d there exists a circuit in G by Theorem 7, and therefore there is a chain joining them in G_1.

Let H_m consist of the arc ab. By Theorem 17, no subset of the graphs $H_1, \cdots, H_{m-1}$ form a circuit of graphs, while some subset of the graphs $H_1, \cdots, H_m$ form a circuit of graphs. We shall show that the whole set of graphs $H_1, \cdots, H_m$ form a circuit of graphs. Otherwise, some proper subset, which includes H_m, form a circuit of graphs.

Let H be the graph formed from this circuit of graphs by dropping out H_m. By Theorem 16, the circuit of graphs is a non-separable graph; hence H is connected. All the arcs in G_1 not in the circuit of graphs, form a graph I. Let I_1 be a connected piece of I. Then I_1 has at most a single vertex in common with the rest of G. For suppose I_1 had the two vertices c and d in common with H. From c follow along some chain towards d in H till we first reach a vertex e in I_1. From e follow back along some chain in I_1 to c. We have formed thus a circuit containing arcs of both H and I_1. But as H consists of a certain subset of the components of G_1, this circuit contains arcs of at least two components of G_1, contrary to Theorem 17. Thus I_1 has at most a single vertex in common with the rest of G, and hence G is separable, contrary to hypothesis. Thus $H_1, \cdots, H_m$ form a circuit of graphs, that is, G is formed of a circuit of graphs.

As we assumed G_1 was separable, $m \geqq 3$. Therefore we can order the graphs so that H_1 and H_2 have just the vertex a_1 in common, $\cdots$, H_{m-1} and H_m have just the vertex $a_{m-1}=b$ in common, and H_m and H_1 have just the vertex $a_m=a$ in common. Moreover, these vertices are all distinct, and no other two of the graphs $H_1, \cdots, H_m$ have a common vertex.

As the nullity of G was >1, the nullity of G_1 is >0. By Theorem 13, this is the sum of the nullities of $H_1, \cdots, H_{m-1}$. Therefore the nullity of some one of these graphs, say H_i, is >0.

Suppose first the nullity of H_i is 1. Then, by Theorem 10, H_i is a circuit, consisting of two chains joining a_{i-1} and a_i. Remove one of these chains from G. This leaves a graph G', which again is a circuit of graphs. For the graph H_i we replace by an ordered set of non-separable graphs, each consisting of one of the arcs of the chain we have left in H_i.

Suppose next the nullity of H_i is >1. It is less than N, as H_i is contained in G_1, whose nullity is $N-1$. Therefore, by induction, we can remove an arc or a suspended chain, leaving a non-separable graph H_i' of nullity one less. If neither a_{i-1} nor a_i has thus been removed, we again have a circuit of graphs. Suppose a_i but not a_{i-1} was removed. Replace that part of the chain we removed joining a_i and a vertex of H_i distinct from a_{i-1}. Here again we have a circuit of graphs, H_i being replaced by H_i' and a set of arcs. The case is the same if a_{i-1} but not a_i was removed. If finally, both a_i and a_{i-1} were in the chain we removed, we put back all of the chain but that part between these two vertices. Here again, the resulting graph G' is a circuit of graphs.

Thus in all cases we can drop out from G an arc or suspended chain, leaving a circuit of graphs. By Theorem 16, the resulting graph G' is non-separable. As also the nullity of G' is one less than the nullity of G, the theorem is now proved.

As a consequence of this theorem, Theorem 8, and Theorem 10, we have

THEOREM 19. *We can build up any non-separable graph containing at at least two arcs by taking first a circuit, then adding successively arcs or suspended chains, so that at any stage of the construction we have a non-separable graph.*

It is easily seen that, conversely, any graph built up in this manner is non-separable. For each time we add an arc or suspended chain, these arcs, each considered as a graph, together with the non-separable graph already present, form a circuit of graphs.

II. DUALS, PLANAR GRAPHS

7. Congruent graphs. We introduce the following

Definitions. Given two graphs G and G', if we can rename the vertices and arcs of one, giving distinct vertices and distinct arcs different names, so that it becomes identical with the other, we say the two graphs are *congruent.** (We used formerly the word "homeomorphic.")

* See the author's American Journal paper, cited in the introduction.

The geometrical interpretation is that we can bring the two graphs into complete coincidence by a (1, 1) continuous transformation.

Two graphs are called *equivalent* if, upon being decomposed into their components, they become congruent, except possibly for isolated vertices.

8. **Duals.** Given a graph G, if H_1 is a subgraph of G, and H_2 is that subgraph of G containing those arcs not in H_1, we say H_2 is the *complement* of H_1 in G.

Throughout this section, R, R', r, r', etc., will stand for the ranks of G, G', H, H', etc., respectively, with similar definitions for V, E, P, N.

Definition. Suppose there is a (1, 1) correspondence between the arcs of the graphs G and G', such that if H is any subgraph of G and H' is the complement of the corresponding subgraph of G', then

$$r' = R' - n.$$

We say then that G' is a *dual* of G.*

Thus, if the nullity of H is n, then H' (including all the vertices of G') is in n more connected pieces than G'.

THEOREM 20. *Let G' be a dual of G. Then*

$$R' = N,$$

$$N' = R.$$

For let H be that subgraph of G consisting of G itself. Then

$$n = N.$$

If H' is the complement of the corresponding subgraph of G', H' contains no arcs, and is the null graph. Thus

$$r' = 0.$$

But as G' is a dual of G,

$$r' = R' - n.$$

These equations give

$$R' = N.$$

The other equation follows when we note that $E' = E$.

THEOREM 21. *If G' is a dual of G, then G is a dual of G'.*

Let H' be any subgraph of G', and let H be the complement of the corresponding subgraph of G. Then, as G' is a dual of G,

* While this definition agrees with the ordinary one for graphs lying on a plane or sphere, a graph on a surface of higher connectivity, such as the torus, has in general no dual. (See Theorems 29 and 30.)

$$r' = R' - n.$$

By Theorem 20,

$$R' = N.$$

We note also,

$$e + e' = E.$$

These equations give

$$r = e - n = e - (R' - r') = e - N + (e' - n')$$
$$= E - N - n' = R - n'.$$

Thus G is a dual of G'.

Whenever we have shown that one graph is a dual of another graph, we may now call the graphs "dual graphs."

LEMMA. *If a graph G is decomposed into its components, the rank and nullity of any subgraph H is left unchanged.*

For each time we separate G at a vertex, H is either unchanged or is separated at a vertex. Hence neither its rank nor its nullity is altered. (See the proof of Theorem 13.)

THEOREM 22. *If G' and G'' are equivalent and G' is a dual of G, then G'' is a dual of G.*

Let H be any subgraph of G, and let H' be the complement of the corresponding subgraph of G'. Let G_1' and G_1'' be G' and G'' decomposed into their components. Then G_1' and G_1'' are congruent. H' turns into a subgraph H_1' of G'. Let H_1'' be the corresponding subgraph of G_1'', and H'' the same subgraph in G''. Then

$$r_1' = r_1''.$$

But by the above lemma,

$$r' = r_1', \qquad r'' = r_1''.$$

Hence

$$r' = r''.$$

As a special case of this equation, letting H' be the whole of G', we have

$$R' = R''.$$

As G' is a dual of G,

$$r' = R' - n.$$

Therefore

$$r'' = R'' - n,$$

and G'' is a dual of G.

The converse of this theorem is not true. For define the three graphs

G: $\alpha(ab)$, $\beta(ab)$, $\gamma(ac)$, $\delta(cb)$, $\epsilon(ad)$, $\zeta(db)$;

$G': \alpha'(a'b'), \beta'(c'd'), \gamma'(a'd'), \delta'(a'd'), \epsilon'(b'c'), \zeta'(b'c');$
$G'': \alpha''(a''b''), \beta''(b''c''), \gamma''(a''d''), \delta''(a''d''), \epsilon''(c''d''), \zeta''(c''d'').$
G' and G'' are both duals of G, but they are not congruent.*

THEOREM 23. *Let $G_1, \cdots, G_m$ and $G_1', \cdots, G_m'$ be the components of G and G' respectively, and let G_i' be a dual of G_i, $i=1, \cdots, m$. Then G' is a dual of G.*

Let H be any subgraph of G, and let the parts of H in $G_1, \cdots, G_m$ be $H_1, \cdots, H_m$. Let H_i' be the complement of the subgraph corresponding to H_i in G_i', $i=1, \cdots, m$, and let H' be the union of $H_1', \cdots, H_m'$ in G'. Then H' is the complement of the subgraph in G' corresponding to H in G. Using the proof of Theorem 13, we find that

$$r' = r_1' + \cdots + r_m',$$

and

$$n = n_1 + \cdots + n_m.$$

As also

$$R' = R_1' + \cdots + R_m'$$

and

$$r_i' = R_i' - n_i \qquad (i = 1, \cdots, m),$$

adding these last equations gives

$$r' = R' - n,$$

and hence G' is a dual of G.

THEOREM 24. *Let $G_1, \cdots, G_m$ and $G_1', \cdots, G_m'$ be the components of the dual graphs G and G', and let the correspondence between these two graphs be such that arcs in G_i correspond to arcs in G_i', $i=1, \cdots, m$. Then G_i and G_i' are duals, $i=1, \cdots, m$.*

Let H_1 be any subgraph of G_1, let H' be the complement of the corresponding subgraph in G', and let H_1' be the complement in G'. Then $H_1', G_2', \cdots, G_m'$ form H'. By Theorem 13, we find

$$R' = R_1' + R_2' + \cdots + R_m'$$

and

$$r' = r_1' + R_2' + \cdots + R_m'.$$

Now

$$r' = R' - n_1,$$

hence

$$r_1' = R_1' - n_1,$$

and G_1' is a dual of G_1. Similarly for $G_2', \cdots, G_m'$.

* See the author's American Journal paper, however.

THEOREM 25. *Let G and G' be dual graphs, and let $H_1, \cdots, H_m$ be the components of G. Let $H_1', \cdots, H_m'$ be the corresponding subgraphs of G'. Then $H_1', \cdots, H_m'$ are the components of G', and H_i' is a dual of H_i, $i=1, \cdots, m$.*

H_1 is the subgraph of G corresponding to H_1' in G'. Its complement is I_1, the graph formed of the arcs of $H_2, \cdots, H_m$. Obviously $H_2, \cdots, H_m$ are the components of I_1. Hence, by Theorem 13, the nullity of I_1 is $n_2+n_3+\cdots+n_m$. Thus, as G' is a dual of G,

$$r_1' = R' - (n_2 + n_3 + \cdots + n_m).*$$

Similarly,

$$r_2' = R' - (n_1 + n_3 + \cdots + n_m),$$

$$\cdots\cdots\cdots\cdots$$

$$r_m' = R' - (n_1 + n_2 + \cdots + n_{m-1}).$$

Adding these equations gives

$$r_1' + r_2' + \cdots + r_m' = mR' - (m-1)(n_1 + n_2 + \cdots + n_m).$$

As $H_1, H_2, \cdots, H_m$ are the components of G,

$$N = n_1 + n_2 + \cdots + n_m.$$

Also, as G and G' are duals, by Theorem 20,

$$R' = N.$$

Hence

$$\begin{aligned} r_1' + r_2' + \cdots + r_m' &= mR' - (m-1)R' \\ &= R'. \end{aligned}$$

Let now $H_{11}', \cdots, H_{1k_1}'$ be the components of H_1' (there may be but one) and similarly for $H_2', \cdots, H_m'$. Then, by Theorem 13,

$$r_1' = r_{11}' + \cdots + r_{1k_1}',$$

$$\cdots\cdots\cdots\cdots$$

$$r_m' = r_{m1}' + \cdots + r_{mk_m}'.$$

Adding these equations gives

$$\sum_{i,j} r_{ij}' = r_1' + \cdots + r_m' = R'.$$

As the graphs $H_{11}', \cdots, H_{mk_m}'$ are non-separable, Theorem 17 tells us that they are the components of G'. Hence G' has at least as many components as

* Which equals n_1.

G. Similarly, G has at least as many components as G'. They have therefore the same number, m, of components.

There are therefore m graphs in the set $H_{11}', \cdots, H_{mk_m}'$. But there is at least one such graph in each graph $H_1', \cdots, H_m'$, and there is therefore exactly one in each. Hence each graph H_{i1}' fills out the graph H_i', and the two sets of graphs $H_{11}', \cdots, H_{mk_m}'$ and $H_1', \cdots, H_m'$ are identical, that is, $H_1', \cdots, H_m'$ are the components of G'.

The rest of the theorem follows from Theorem 24.

As a special case of this theorem, we have

THEOREM 26. *A dual of a non-separable graph is non-separable.*

9. **Planar graphs.** Up till now, we have been considering abstract graphs alone. However, the definition of a planar graph is topological in character. This section may be considered as an application of the theory of abstract graphs to the theory of topological graphs.

Definitions. A topological graph is called *planar* if it can be mapped in a (1, 1) continuous manner on a sphere (or a plane). For the present, we shall say that an abstract graph is *planar* if the corresponding topological graph is planar. Having proved Theorem 29, we shall be justified in using the following purely combinatorial definition: *A graph is planar if it has a dual.*

We shall henceforth talk about "graphs" simply, the terms applying equally well to either abstract or topological graphs.

LEMMA. *If a graph can be mapped on a sphere, it can be mapped on a plane, and conversely.*

Suppose we have a graph mapped on a sphere. We let the sphere lie on the plane, and rotate it so that the new north pole is not a point of the graph. By stereographic projection from this pole, the graph is mapped on the plane. The inverse of this projection maps any graph on the plane onto the sphere.

By the *regions* of a graph lying on a sphere or in a plane is meant the regions into which the sphere or plane is thereby divided. A given region of the graph is characterized by those arcs of the graph which form its boundary. If the graph is in a plane, the outside region is the unbounded region.

LEMMA. *A planar graph may be mapped on a plane so that any desired region is the outside region.*

We map the graph on a sphere, and rotate it so that the north pole lies inside the given region. By stereographic projection, the graph is mapped onto the plane so that the given region is the outside region.

We return now to the work in hand.

THEOREM 27. *If the components of a graph G are planar, G is planar.*

Suppose the graphs G_1 and G_2 are planar, and G' is formed by letting the vertices a_1 and a_2 of G_1 and G_2 coalesce. We shall show that G' is planar. Map G_1 on a sphere, and map G_2 on a plane so that one of the regions adjacent to the vertex a_2 is the outside region. Shrink the portion of the plane containing G_2 so it will fit into one of the regions of G_1 adjacent to a_1. Drawing a_1 and a_2 together, we have mapped G' on the sphere.* The theorem follows as a repeated application of this process.

THEOREM 28. *Let G and G' be dual graphs, and let $\alpha(ab)$, $\alpha'(a'b')$ be two corresponding arcs. Form G_1 from G by dropping out the arc $\alpha(ab)$, and form G_1' from G' by dropping out the arc $\alpha'(a'b')$, and letting the vertices a' and b' coalesce if they are not already the same vertex. Then G_1 and G_1' are duals, preserving the correspondence between their arcs.*

Let H_1 be any subgraph of G_1 and let H_1' be the complement of the corresponding subgraph of G_1'.

Case 1. Suppose the vertices a' and b' were distinct in G'. Let H be the subgraph of G identical with H_1. Then

$$n = n_1.$$

Let H' be the complement in G' of the subgraph corresponding to H. Then

$$r' = R' - n.$$

Now H' is the subgraph in G' corresponding to H_1' in G_1', except that H' contains the arc $\alpha'(a'b')$, which is not in H_1'. Thus if we drop out $\alpha'(a'b')$ from H' and let a' and b' coalesce, we form H_1'. In this operation, the number of connected pieces is unchanged, while the number of vertices is decreased by 1. Hence

$$r_1' = r' - 1.$$

As a special case of this equation, if H' contains all the arcs of G', we find

$$R_1' = R' - 1.$$

These equations give

$$r_1' = R_1' - n_1.$$

Thus G_1' is a dual of G_1.

Case 2. Suppose a' and b' are the same vertex in G'. In this case, defining H and H' as before, we form H_1' from H' by dropping out the arc $\alpha'(a'a')$. This leaves the number of vertices and the number of connected pieces un-

* Here and in a few other places we are using point-set theorems which, however, are geometrically evident.

changed. Thus two of the equations in Case 1 are replaced by the equations

$$r_1' = r_1, \quad R_1' = R_1.$$

The other equations are as before, so we find again that G_1' is a dual of G_1. The theorem is now proved.

THEOREM 29. *A necessary and sufficient condition that a graph be planar is that it have a dual.*

We shall prove first the necessity of the condition. Given any planar graph G, we map it onto the surface of a sphere. If the nullity of G is N, it divides the sphere into $N+1$ regions. For let us construct G arc by arc Each time we add an arc joining two separate pieces, the nullity and the number of regions remain the same. Each time we add an arc joining two vertices in the same connected pieces, the nullity and the number of regions are each increased by 1. To begin with, the nullity was 0 and the number of regions was 1. Therefore, at the end, the number of regions is $N+1$.

We construct G' as follows: In each region of the graph G we place a point, a vertex of G'. Therefore G' contains $V'=N+1$ vertices. Crossing each arc of G we place an arc, joining the vertices of G' lying in the two regions the arc of G separates (which may in particular be the same region, in which case this arc of G' is a 1-circuit). The arcs of G and G' are now in (1, 1) correspondence.

G' is the dual of G in the ordinary sense of the word. We must show it is the dual as we have defined the term.

Let us build up G arc by arc, removing the corresponding arc of G' each time we add an arc to G. To begin with, G contains no arcs and G' contains all its arcs, and at the end of the process, G contains all its arcs and G' contains no arcs. We shall show

(1) each time the nullity of G is increased by 1 upon adding an arc, the number of connected pieces in G' is reduced by 1 in removing the corresponding arc, and

(2) each time the nullity of G remains the same, the number of connected pieces in G' remains the same.

To prove (1) we note that the nullity of G is increased by 1 only when the arc we add joins two vertices in the same connected piece. Let ab be such an arc. As a and b were already connected by a chain, this chain together with ab forms a circuit P. Let $a'b'$ be the arc of G' corresponding to ab. Before we removed it, a' and b' were connected. Removing it, however, disconnects them. For suppose there were still a chain C' joining them. As a' and b' are on opposite sides of the circuit P, C' must cross P, by the Jordan Theorem, that is, an arc of C' must cross an arc of P. But we removed this arc of C' when we put in the arc of P it crosses. (1) is now proved.

The total increase in the nullity of G during the process is of course just N. Therefore the increase in the number of connected pieces in G' must be at least N. But G' was originally in at least one connected piece, and is at the end of the process in $V=N+1$ connected pieces. Thus the increase in the number of connected pieces in G' is just N (hence, in particular, G' itself is connected) and therefore this number increases only when the nullity of G increases, which proves (2).

Let now H be any subgraph of G, let H' be the complement of the corresponding subgraph of G', and let H' include all the vertices of G'. We build up H arc by arc, at the same time removing the corresponding arcs of G'. Thus when H is formed, H' also is formed. By (1) and (2), the increase in the number of connected pieces in forming H' from G' equals the nullity of H, that is,

$$p' - P' = n.$$

But

$$r' = V' - p', \quad R' = V' - P',$$

as G' and H' contain the same vertices. Therefore

$$r' = R' - n,$$

that is, G' is a dual of G.

To prove the sufficiency of the condition, we must show that if a graph has a dual, it is planar. It is enough to show this for non-separable graphs. For if the separable graph G has a dual, its components have duals, by Theorem 25, hence its components are planar, and hence G is planar, by Theorem 27. This part of the theorem is therefore a consequence of the following theorem:

THEOREM 30. *Let the non-separable graph G have a dual G'. Then we can map G and G' together on the surface of a sphere so that*

(1) *corresponding arcs in G and G' cross each other, and no other pair of arcs cross each other, and*

(2) *inside each region of one graph there is just one vertex of the other graph.*

The theorem is obviously true if G contains a single arc. (The dual of an arc ab is an arc $a'a'$, and the dual of an arc aa is an arc $a'b'$.) We shall assume it to be true if G contains fewer than E arcs, and shall prove it for any graph G containing E arcs. By Theorem 8, each vertex of G is on at least two arcs.

Case 1. G contains a vertex b on but two arcs, ab and bc. As G is non-separable, there is a circuit containing these arcs. Thus dropping out one of them will not alter the rank, while dropping out both reduces the rank by 1. As G' is a dual of G, the arcs corresponding to these two arcs are each of nullity 0, while the two arcs taken together are of nullity 1. They are thus of the form $\alpha'(a'b')$, $\beta'(a'b')$, the first corresponding to ab, and the second, to bc.

Form G_1 from G by dropping out the arc bc and letting the vertices b and c coalesce, and form G_1' from G' by dropping out the arc $\beta'(a'b')$. By Theorem 28, G_1 and G_1' are duals, preserving the correspondence between the arcs. As these graphs contain fewer than E arcs,* we can, by hypothesis, map them together on a sphere so that (1) and (2) hold; in particular, $\alpha'(a'b')$ crosses ac. Mark a point on the arc ac of G_1 lying between the vertex c and the point where the arc $\alpha'(a'b')$ of G' crosses it. Let this be the vertex b, dividing the arc ac into the two arcs ab and bc. Draw the arc $\beta'(a'b')$ crossing the arc bc. We have now reconstructed G and G', and they are mapped on a sphere so that (1) and (2) hold.

Case 2. Each vertex of G is on at least three arcs. As then G contains no suspended chain, and G is not a circuit and therefore is of nullity $N>1$, we can, by Theorem 18, drop out an arc ab so that the resulting graph G_1 is non-separable. G' is non-separable, by Theorem 26, and hence the arc $a'b'$ corresponding to ab in G is not a 1-circuit. Drop it out and let the vertices a', b' coalesce into the vertex a_1', forming the graph G_1'. By Theorem 28, G_1 and G_1' are duals, and thus G_1' also is non-separable.

Consider the arcs of G' on a'. If we drop them out, the resulting graph G'' has a rank one less than that of G'. For if its rank were still less, G'' would be in at least three connected pieces, one of them being the vertex a'. Let c and d be vertices in two other connected pieces of G''. They are joined by no chain in G'', and hence every chain joining them in G' must pass through a', which contradicts Theorem 6. If we put back any arc, the rank is brought back to its original value, as a' is then joined to the rest of the graph. Hence, G' being a dual of G, the arcs of G corresponding to these arcs are together of nullity 1, while dropping out one of them reduces the nullity to 0. Therefore, by Theorem 9, these arcs form a circuit P. One of these arcs is the arc ab. The remaining arcs form a chain C. Similarly, the arcs of G corresponding to the arcs of G' on b' form a circuit Q, and this circuit minus the arc ab forms a chain D. C and D have the vertices a and b as end vertices. Also, the arcs of G_1 corresponding to the arcs of G_1' on a_1' form a circuit R. These arcs of G_1' are the arcs of G' on either a' or b', except for the arc $a'b'$ we dropped out. Thus the arcs of G_1 forming the circuit R are the arcs of the chains C and D.

As G_1 and G_1' contain fewer than E arcs, we can map them together on a

* Obviously G_1 is non-separable.

sphere so that properties (1) and (2) hold. a_1' lies on one side of the circuit R, which we call the inside. Each arc of R is crossed by an arc on a_1', and thus there are no other arcs of G_1' crossing R. There is no part of G_1' lying inside R other than a_1', for it could have only this vertex in common with the rest of G_1', and G_1' would be separable. Also, there is no part of G_1 lying inside R, for any arc would have to be crossed by an arc of G_1', and any vertex would have to be joined to the rest of G_1 by an arc, as G_1 is non-separable.

Let us now replace a_1' by the two vertices a' and b', and let those arcs abutting on a_1' that were formerly on a' be now on a', and those formerly on b', now on b'. As the first set of arcs all cross the chain C, and the second set all cross the chain D, we can do this in such a way that no two of the arcs cross each other. We may now join a and b by the arc ab, crossing none of these arcs. This divides the inside of R into two parts, in one of which a' lies, and in the other of which b' lies. We may therefore join a' and b' by the arc $a'b'$, crossing the arc ab. G and G' are now reconstructed, and are mapped on the sphere as required. This completes the proof of the theorem, and therefore of Theorem 29.

THEOREM 31. *A necessary and sufficient condition that a graph be planar is that it contain neither of the two following graphs as subgraphs:*

G_1. *This graph is formed by taking five vertices a, b, c, d, e, and joining each pair by an arc or suspended chain.*

G_2. *This graph is formed by taking two sets of three vertices, a, b, c, and d, e, f, and joining each vertex in one set to each vertex in the other set by an arc or suspended chain.*

This theorem has been proved by Kuratowski.* It would be of interest to show the equivalence of the conditions of the theorem and Theorem 29 directly, by combinatorial methods. We shall do part of this here, in the following theorem:†

THEOREM 32. *Neither of the graphs G_1 and G_2 has a dual.*

Suppose the graph G_1 had a dual. By Theorem 28, if G_1 contains a suspended chain, we can drop out one of its arcs and let the two end vertices coalesce, and the resulting graph will have a dual. Continuing, we see that the graph G_3, in which each pair of vertices of the set a, b, c, d, e are joined by an arc, must have a dual. Similarly, if G_2 has a dual, then the graph G_4, in which each vertex of the set a, b, c is joined to each vertex of the set d, e, f by an arc, must have a dual. Both of these are impossible.

* Fundamenta Mathematicae, vol. 15 (1930), pp. 271–283.

† The other half has recently been proved by the author. See Bulletin of the American Mathematical Society, abstract (38-1-39). (Note added in proof.)

(a) *The graph* G_3. To avoid subscripts, let us call it G. Suppose it had a dual, G'. Then

$$R = N' = 4,$$
$$N = R' = 6,$$
$$E = E' = 10.$$

If G' has isolated vertices, we drop them out, which does not alter its relation to G.

(1) There are no 1-circuits, 2-circuits or triangles in G'. For if there were, dropping out the corresponding arcs of G would have to reduce the rank of G. But we cannot reduce its rank without dropping out at least four arcs.

(2) G' contains at least five quadrilaterals. For if we drop out the four arcs on any vertex of G, the rank is reduced by 1, and if we put back any of these arcs, the rank is brought back to its original value; Theorem 9 now applies.

(3) At least two of these quadrilaterals have an arc in common, as there are but ten arcs in G'.

There are just two ways of forming two quadrilaterals out of fewer than eight arcs without forming any 2-circuits or triangles. One of these graphs, I_1', contains the arcs $a'b'$, $b'e'$, $a'c'$, $c'e'$, $a'd'$, $d'e'$. The other, I_2', contains the arcs $a'e'$, $e'f'$, $f'b'$, $b'a'$, $e'c'$, $c'd'$, $d'f'$. But there is no subgraph of the type I_1' in G', for this subgraph is of rank 4 and nullity 2, and there would have to be a subgraph of G of rank 2 and nullity 2, and such a graph contains a 1- or a 2-circuit, of which there are none in G. Hence G' contains a subgraph I_2'.

(4) Each vertex of G' is on at least three arcs, as there are no 1- or 2-circuits in G.

Each of the vertices a', b', c', d' of I_2' is on but two arcs. Hence there must be another arc on each of these vertices. As I_2' contains seven arcs, and G' contains but ten, one of the three arcs left must join two of these vertices. But if we add an arc $a'b'$ or $c'd'$, we would form a 2-circuit; if we add an arc $a'c'$ or $b'd'$, we would form a triangle; if we add an arc $a'd'$ or $b'c'$, we would form a graph of the type I_1'. As G' contains none of these graphs, we have a contradiction.

(b) *The graph* G_4. Let us call it G. If it has a dual G', then

$$R = N' = 5,$$
$$N = R' = 4,$$
$$E = E' = 9.$$

We proceed exactly as for the graph G_3. In outline:

(1) G' contains no 1- or 2-circuits.

(2) There is no subgraph of G' containing four vertices, each pair being joined by an arc. For this graph is of rank 3 and nullity 3, and G would have to contain a subgraph of rank 2 and nullity 1, that is, a 2-circuit.

(3) There are at least nine subgraphs of G' of rank 3 and nullity 2, and hence of the form $a'b'$, $a'c'$, $b'c'$, $b'd'$, $c'd'$, as there are nine quadrilaterals in G.

(4) As G' contains but nine arcs, two of these subgraphs have an arc in common. There is therefore a subgraph of one of the forms I_1' : $a'e'$, $a'b'$, $b'e'$, $a'c'$, $c'e'$, $a'd'$, $d'e'$, or I_2' : $a'e'$, $a'b'$, $b'e'$, $b'c'$, $c'e'$, $c'd'$, $d'e'$.

(5) Each vertex of G' is on at least four arcs.

Now each of the graphs I_1', I_2' contains seven arcs. We have but two arcs left which we must place so that each vertex of I_1' or I_2' is on at least four arcs. This cannot be done. The theorem is now proved.

Theorem 31 together with this theorem gives an alternative proof of the second part of Theorem 29. For suppose a graph G had a dual. Then it contains neither the graph G_1 nor G_2. For if it did, dropping out all the arcs of G but those forming one of these graphs, Theorem 28 tells us that this graph has a dual. But we have just seen that this is not so. Hence, by Theorem 31, G is planar.

Euler's formula. Map any connected planar graph G on a sphere, and construct its connected dual G' as described in the proof of Theorem 29. Then in each region of G there is a vertex of G'. Let F be the number of regions (or faces) in G. Then

$$R' = N,$$
$$R = V - 1,$$
$$R' = V' - 1,$$
$$V' = F,$$

and hence

$$V - E + F = R + 1 - E + N + 1 = 2,$$

which is Euler's formula.

HARVARD UNIVERSITY,
CAMBRIDGE, MASS.

Reprinted from
Trans. Amer. Math. Soc. **34** (1932), 339–362

A Combinatorial Problem in Geometry

by

P. Erdös and G. Szekeres
Manchester

Introduction.

Our present problem has been suggested by Miss Esther Klein in connection with the following proposition.

From 5 points of the plane of which no three lie on the same straight line it is always possible to select 4 points determining a convex quadrilateral.

We present E. Klein's proof here because later on we are going to make use of it. If the least convex polygon which encloses the points is a quadrilateral or a pentagon the theorem is trivial. Let therefore the enclosing polygon be a triangle ABC. Then the two remaining points D and E are inside ABC. Two of the given points (say A and C) must lie on the same side of the connecting straight line $\overline{DE}$. Then it is clear that $AEDC$ is a convex quadrilateral.

Miss Klein suggested the following more general problem. *Can we find for a given n a number $N(n)$ such that from any set containing at least N points it is possible to select n points forming a convex polygon?*

There are two particular questions: (1) does the number N corresponding to n exist? (2) If so, how is the least $N(n)$ determined as a function of n? (We denote the least N by $N_0(n)$.)

We give two proofs that the first question is to be answered in the affirmative. Both of them will give definite values for $N(n)$ and the first one can be generalised to any number of dimensions. Thus we obtain a certain preliminary answer to the second question. But the answer is not final for we generally get in this way a number N which is too large. Mr. E. Makai proved that $N_0(5) = 9$, and from our second demonstration, we obtain $N(5) = 21$ (from the first a number of the order 2^{10000})

Thus it is to be seen, that our estimate lies pretty far from the true limit $N_0(n)$. It is notable that $N(3) = 3 = 2 + 1$, $N_0(4) = 5 = 2^2+1$, $N_0(5) = 9 = 2^3 + 1$.

We might conjecture therefore that $N_0(n) = 2^{n-2} + 1$, but the limits given by our proofs are much larger.

It is desirable to extend the usual definition of convex polygon to include the cases where three or more consecutive points lie on a straight line.

First proof.

The basis of the first proof is a combinatorial theorem of Ramsey [1]). In the introduction it was proved that from 5 points it is always possible to select 4 forming a convex quadrangle. Now it can be easily proved by induction that n points determine a convex polygon if and only if any 4 points of them form a convex quadrilateral.

Denote the given points by the numbers 1, 2, 3, . . ., N, then any k-gon of the set of points is represented by a set of k of these numbers, or as we shall say, by a k-combination. Let us now suppose each n-gon to be concave, then from what we observed above we can divide the 4-combinations into two classes (i. e. into „convex" and „concave" quadrilaterals) such that every 5-combination shall contain at least one „convex" combination and each n-combination at least one concave one. (We regard one combination as contained in another, if each element of the first is also an element of the second.)

From Ramsey's theorem, it follows that this is impossible for a sufficiently large N.

Ramsey's theorem can be stated as follows:

Let k, l, i be given positive integers, $k \geq i$; $l \geq i$. Suppose that there exist two classes, α and β, of i-combinations of m elements such that each k-combination shall contain at least one combination from class α und each l-combination shall contain at least one combination from class β. Then for sufficiently great $m < m_i(k, l)$ this is not possible. Ramsey enunciated his theorem in a slightly different form.

In other words: if the members of α had been determined as above at our discretion and $m \geq m_i(k, l)$, then there must be at least one l-combination with every combination of order i belonging to class α.

---

1) F. P. Ramsey, Collected papers. On a problem of formal logic, 82—111. Recently Skolem also proved Ramsey's theorem [Fundamenta Math. **20** (1933), 254—261].

We give here a new proof of Ramsey's theorem, which differs entirely from the previous ones and gives for $m_i(k, l)$ slightly smaller limits.

a) If $i = 1$, the theorem holds for every k and l. For if we select out of m some determined elements (combinations of order 1) as the class α, so that every k-gon (this shorter denomination will be given to the combination of order k) must contain at least one of the α elements, there are at most $(k-1)$ elements which do not belong to the class α. Then there must be at least $(m-k+1)$ elements of α. If $(m-k+1) \geqq l$, then there must be an l-gon of the α elements and thus

$$m \leqq k + l - 2$$

which is evidently false for sufficiently great m.

Suppose then that $i > 1$.

b) The theorem is trivial, if k or l equals i. If, for example, $k = i$, then it is sufficient to choose $m = l$.

For $k = 1$ means that all i-gons are α combinations and thus in virtue of $m = l$ there is one polygon (i. e. the l-gon formed of all the elements), whose i-gons are all α-combinations.

The argument for $l = i$ runs similarly.

c) Suppose finally that $k > i$; and suppose that the theorem holds for $(i-1)$ and every k and l, further for $i, k, l-1$ and $i, k-1, l$. We shall prove that it will hold for i, k, l also and in virtue of (a) and (b) we may say that the theorem is proved for all i, k, l.

Suppose then that we are able to carry out the division of the i-polygons mentioned above. Further let k' be so great that if in every l-gon of k' elements there is at least one β combination, then there is one $(k-1)$-gon all of whose i-gons are β combinations. This choice of k' is always possible in virtue of the induction-hypothesis, we have only to choose $k' = m_i(k-1, l)$.

Similarly we choose l' so great that if each k-gon of l' elements contains at least one α combination, then there is one $(l-1)$-gon all of whose i-gons are α combinations.

We then take m larger than k' and l'; and let

$$(a_1, a_2, \ldots, a_{k'}) = A$$

be an arbitrary k'-gon of the first $(n-1)$ elements. By hypothesis each l-gon contains at least one β combination, hence owing to the choice of k', A contains one $(k-1)$-gon $(a_{m_1}, a_{m_2}, \ldots, a_{m_{k-1}})$ whose i-gons all belong to the class β. Since in $(a_{m_1}, \ldots, a_{m_{k-1}}, n)$

30

there is at least one α combination, it is clear that this must be one of the i-gons

$$(a_{p_1}, a_{p_2}, \ldots, a_{p_{i-1}}, n) \equiv B.$$

In just the same way we may prove by replacing the roles of k and l by k' and l' and of α by β, that if

$$(b_1, b_2, \ldots, b_{l'}) \equiv A'$$

is an arbitrary l'-gon of the first $(n-1)$ elements, then among the i-gons

$$(b_{r_1}, b_{r_2}, \ldots, b_{r_{i-1}}, n) \equiv B'$$

there must be a β combination.

Thus we can divide the $(i-1)$-gons of the first $(n-1)$ elements into classes α' and β' so that each k'-gon A shall contain at least one α' combination B and each l'-gon A' at least one β' combination B'. But, by the induction-hypotheses this is impossible for $m \geqq m_{i-1}(k'l') - 1$.

By following the induction, it is easy to obtain for $m_i(k, l)$ the following functional equation:

$$m_i(k, l) = m_{i-1}[m_i(k-1, l), \; m_i(k, l-1)] + 1. \qquad (1)$$

By this recurrence-formula and the initial values

$$\left.\begin{array}{c} m_1(k, l) = k + l - 1 \\ m_i(i, l) = l, \quad m_i(k, i) = k \end{array}\right\} \qquad (2)$$

obtained from (a) and (b) we can calculate every $m_i(k, l)$.

We obtain e. g. easily

$$m_2(k-1, l-1) = \binom{k+l}{k}. \qquad (3)$$

The function mentioned in the introduction has the form

$$N(k) = m(5, k). \qquad (4)$$

Finally, for the special case $i = 2$, we give a graphotheoretic formulation of Ramsey's theorem and present a very simple proof of it.

THEOREM: *In an arbitrary graph let the maximum number of independent points*[2]) *be* k; *if the number of points is* $N \geqq m(k, l)$ *then there exists in our graph a complete graph*[3]) *of order* l.

---

[2]) Two points are said to be independent if they are not connected; k points are independent if every pair is independent.

[3]) A complete graph is one in which every pair of points is connected.

Proof. For $l=1$, the theorem is trivial for any k, since the maximum number of independent points is k and if the number of points is $(k+1)$, there must be an edge (complete graph of order 1).

Now suppose the theorem proved for $(l-1)$ with any k. Then at least $\frac{N-k}{k}$ edges start from one of the independent points. Hence if

$$\frac{N-k}{k} \geqq m(k, l-1),$$

i. e.,
$$N \geqq k \cdot m(k, l-1) + k, \tag{5}$$

then, out of the end points of these edges we may select, in virtue of our induction hypothesis, a complete graph whose order is at least $(l-1)$. As the points of this graph are connected with the same point, they form together a complete graph of order l.

Second proof.

The foundation of the second proof of our main theorem is formed partly by geometrical and partly by combinatorial considerations. We start from some similar problems and we shall see, that the numerical limits are more accurate then in the previous proof; they are in some respects exact.

Let us consider the first quarter of the plane, whose points are determined by coordinates (x, y). We choose n points with monotonously increasing abscissae [4]).

Theorem: *It is always possible to choose at least* $\sqrt{n}$ *points with increasing abscissae and either monotonously increasing or monotonously decreasing ordinates.* If two ordinates are equal, the case may equally be regarded as increasing or decreasing.

Let us denote by $f(n, n)$ the minimum number of the points out of which we can select n monotonously increasing or decreasing ordinates.

We assert that

$$f(n+1, n+1) = f(n, n) + 2n - 1. \tag{6}$$

Let us select n monotonously increasing or decreasing points out of the $f(n, n)$. Let us replace the last point by one of the $(2n-1)$ new points. Then we shall have once more $f(n, n)$ points, out

[4]) The same problem was considered independently by Richard Rado.

of which we can select as before n monotonous points. Now we replace the last point by one of the new ones and so on. Thus we obtain $2n$ points each an endpoint of a monotonous set. Suppose that among them $(n+1)$ are end points of monotonously increasing sets. Then if $y_l \geqq y_k$ for $l > k$ we add P_l to the monotonously increasing set of P_k and thus, with it, we shall have an increasing set of $(n+1)$ points. If $y_k \geqq y_l$ for every $k < l$, then the $(n+1)$ decreasing end-points themselves give the monotonous set of $(n+1)$ members. If between the $2n$ points there are at least $(n+1)$ end-points of monotonous decreasing sets, the proof will run in just the same way.

But it may happen that, out of the $2n$ points, just n are the end-points of increasing sets, and n the end-points of decreasing sets. Then by the same reasoning, the end-points of the decreasing sets necessarily increase. But after the last end-point P there is no point, for its ordinate would be greater or smaller than that of P. If it is greater, then together with the n end-points it forms a monotonously increasing $(n+1)$ set and if it is smaller, with the n points belonging to P, it forms a decreasing set of $(n+1)$ members. But by the same reasoning the last of the n increasing end-points Q ought to be also an extreme one and that is evidently impossible. Thus we may deduce by induction

$$f(n+1, n+1) = n^2 + 1. \tag{7}$$

Similarly let $f(i, k)$ denote the minimum number of points out of which it is impossible to select either i monotonously increasing or k monotonously decreasing points. We have then

$$f(i, k) = (i-1)(k-1) + 1. \tag{8}$$

The proof is similar to the previous one.

It is not difficult to see, that this limit is exact i. e. we can give $(i-1)(k-1)$ points such that it is impossible to select out of them the desired number of monotonously increasing or decreasing ordinates.

We solve now a similar problem:

$P_1, P_2, \ldots$ are given points on a straight line. Let $f_1(i, k)$ denote the minimum number of points such that proceeding from left to right we shall be able to select either i points so that the distances of two neighbouring points monotonously increase or k points so that the same distances monotonously decrease. We assert that

$$f_1(i, k) = f_1(i-1, k) + f_1(i, k-1) - 1. \tag{9}$$

Let the point C bisect the distance $\overline{AB}$ (A and B being the first and the last points). If the total number of points is $f_1(i-1, k) + f_1(i, k-1) - 1$, then either the number of points in the first half is at least $f_1(i-1, k)$, or else there are in the second half at least $f_1(i, k-1)$ points. If in the first half there are $f_1(i-1, k)$ points then either there are among them k points whose distances, from left to right, monotonously decrease and then the equation for $f_1(i, k)$ is fulfilled, or there must be $(i-1)$ points with increasing distances. By adding the point B, we have i points with monotonely increasing distances. If in the second interval there are $f_1(i, k-1)$ points, the proof runs in the same way. (The case, in which two distances are the same, may be classed into either the increasing or the decreasing sets.)

It is possible to prove that this limit is exact. If the limits $f_1(i-1, k)$ and $f_1(i, k-1)$ are exact (i. e. if it is possible to give $[f_1(i-1, k) - 1]$ points so that there are no $(i-1)$ increasing nor k decreasing distances) then the limit $f_1(i, k)$ is exact too. For if we choose e. g. $[f_1(i-1, k) - 1]$ points in the $0 \ldots 1$ interval, and $[f_1(i, k-1) - 1]$ points in the $2 \ldots 3$ interval, then we have $[f_1(i, k) - 1]$ points out of which it is equally impossible to select i points with monotonously increasing and k points with decreasing distances.

We now tackle the problem of the convex n-gon. If there are n given points, there is always a straight line which is neither parallel nor perpendicular to any join of two points. Let this straight line be e. Now we regard the configuration $A_1A_2A_3A_4 \ldots$ as convex, if the gradients of the lines A_1A_2, $A_2A_3, \ldots$ decrease monotonously, and as concave if they increase monotonously. Let $f_2(i, k)$ denote the minimum number of the points such that from them we may pick out either i sided convex or k-sided concave configurations. We assert that

$$f_2(i, k) = f_2(i-1, k) + f_2(i, k-1) - 1. \tag{10}$$

We consider the first $f_2(i-1, k)$ points. If out of them there can be taken a concave configuration of k points then the equation for $f_2(i, k)$ is fulfilled. If not, then there is a convex configuration of $(i-1)$ points. The last point of this convex configuration we replace by another point. Then we have once more either k concave points and then the assertion holds, or $(i-1)$ convex ones. We go on replacing the last point, until we have made use of all points. Thus we obtain $f_2(i, k-1)$ points, each of which is an end-point of a convex configuration of $(i-1)$

elements. Among them, there are either i convex points and then our assertion is proved, or $(k-1)$ concave ones. Let the first of them be A_1, the second A_2. A_1 is the end-point of a convex configuration of $(i-1)$ points. Let the neighbour of A_1 in this configuration be B. If the gradient of BA_1 is greater than that of A_1A_2, then A_2 together with the $(i-1)$ points form a convex configuration; if the gradient is smaller, then B together with $A_1A_2\ldots$ form a concave k-configuration. This proves our assertion.

The deduction of the recurrence formula may start from the statement: $f_2(3, n) = f_2(n, 3) = n$ (by definition). Thus we easily obtain

$$f_2(k, k) = \binom{2k-4}{k-2} + 1. \tag{11}$$

As before we may easily prove that the limit given by (11) is exact, i.e. it is possible to give $\binom{2k-4}{k-2}$ points such, that they contain neither convex nor concave k points.

Since by connection of the first and last points, every set of k convex or concave points determines a convex k-gon it is evident that $\left[\binom{2k-4}{k-2} + 1\right]$ points always contain a convex k-gon.

And as in every convex $(2k-1)$ polygon there is always either a convex or a concave configuration of k points, it is evident that it is possible to give $\binom{2k-4}{k-2}$ points, so that out of them no convex $(2k-1)$ polygon can be selected. Thus the limit is also estimated from below.

Professor D. König's lemma [5]) of infinity also gives a proof of the theorem that if k is a definite number and n sufficiently great, the n points always contain a convex k-gon. But we thus obtain a pure existence-proof, which allows no estimation of the number n. The proof depends on the statement that if M is an infinite set of points we may select out of it another convex infinite set of points.

(Received December 7th, 1934.)

[5]) D. König, Über eine Schlußweise aus dem Endlichen ins Unendliche [Acta Szeged 3 (1927), 121—130].

Reprinted from
Compositio Math. 2 (1935), 463–470

ON REPRESENTATIVES OF SUBSETS

P. HALL†.

1. Let a set S of mn things be divided into m classes of n things each in two distinct ways, (a) and (b); so that there are m (a)-classes and m (b)-classes. Then it is always possible to find a set R of m things of S which is at one and the same time a C.S.R. (= complete system of representatives) for the (a)-classes, and also a C.S.R. for the (b)-classes.

This remarkable result was originally obtained (in the form of a theorem about graphs) by D. König‡.

In the present note we are concerned with a slightly different problem, viz. with the problem of the existence of a C.D.R. (= complete system of *distinct* representatives) for a finite collection of (arbitrarily overlapping) subsets of any given set of things. The solution, Theorem 1, is very simple. From it may be deduced a general criterion, viz. Theorem 3, for the existence of a *common* C.S.R. for two distinct classifications of a given set; where it is not assumed, as in König's theorem, that all the classes have the same number of terms. König's theorem follows as an immediate corollary.

2. Given any set S and any finite system of subsets of S:

$$T_1, T_2, \ldots, T_m; \tag{1}$$

we are concerned with the question of the existence of a *complete set of distinct representatives* for the system (1); for short, a C.D.R. of (1).

By this we mean a set of m *distinct* elements of S:

$$a_1, a_2, \ldots, a_m, \tag{2}$$

such that

$$a_i \in T_i \tag{3}$$

(a_i belongs to T_i) for each $i = 1, 2, \ldots, m$. We may say, a_i *represents* T_i.

It is not necessary that the sets T_i shall be finite, nor that they should be distinct from one another. Accordingly, when we speak of a system of

† Received 23 April, 1934; read 26 April, 1934.

‡ D. König, "Über Graphen und ihre Anwendungen", *Math. Annalen*, 77 (1916), 453. For the theorem in the form stated above, *cf.* B. L. van der Waerden, "Ein Satz über Klasseneinteilungen von endlichen Mengen", *Abhandlungen Hamburg*, 5 (1927), 185; also E. Sperner, *ibid.*, 232, for an extremely elegant proof.

k of the sets (1), it is understood that k *formally* distinct sets are meant, not necessarily k actually distinct sets.

It is obvious that, if a C.D.R. of (1) does exist, then any k of the sets (1) must contain between them at least k elements of S. For otherwise it would be impossible to find distinct representatives for those k sets.

Our main result is to show that this obviously necessary condition is also sufficient. That is

THEOREM 1. *In order that a C.D.R. of* (1) *shall exist, it is sufficient that, for each* $k = 1, 2, \ldots, m$, *any selection of* k *of the sets* (1) *shall contain between them at least* k *elements of* S.

If $A, B, \ldots$ are any subsets of S, then their *meet* (the set of all elements common to $A, B, \ldots$) will be written

$$A \wedge B \wedge \ldots.$$

Their *join* (the set of all elements which lie in at least one of $A, B, \ldots$) will be written

$$A \vee B \vee \ldots.$$

To prove Theorem 1, we need the following

LEMMA. *If* (2) *is any C.D.R. of* (1), *and if the meet of all the C.D.R. of* (1) *is the set* $R = a_1, a_2, \ldots, a_\rho$ (ρ *can be* 0, *i.e.* R *the null set*), *then the* ρ *sets*

$$T_1, T_2, \ldots, T_\rho$$

contain between them exactly ρ *elements, viz. the elements of* R.

R is, by definition, the set of all elements of S which occur as representatives of some T_i in every C.D.R. of (1).

To prove the lemma, let R' be the set of all elements a of S with the following property: there exists a sequence of suffixes

$$i, j, k, \ldots, l', l$$

such that

$$a \in T_i,$$

$$a_i \in T_j,$$

$$a_j \in T_k,$$

$$\ldots$$

$$a_{l'} \in T_l,$$

and, further,

$$l \leq \rho.$$

First, we shall show that every element a of R' belongs to (2). For, if not, replace, in (2),

$$a_i, a_j, a_k, \dots, a_l$$

by

$$a, a_i, a_j, \dots, a_{l'},$$

respectively; we obtain a new C.D.R. of (1) which does not contain a_l. Hence a_l does not belong to R, which contradicts $l \leqslant \rho$.

There will be no loss of generality in assuming that

$$R' = a_1, a_2, \dots, a_\omega.$$

For it is clear that R' contains R.

Next, it is clear that if a is any element of T_i, where $i \leqslant \omega$, then

$$a \in R'.$$

For then $a_i \in R'$, and hence $j, k, \dots, l$ can be found with $l \leqslant \rho$ and such that

$$a_i \in T_j,$$

$$\dots$$

$$a_{l'} \in T_l.$$

And $a \in T_i$ then shows that $a \in R'$ also. Hence every element of T_i $(i \leqslant \omega)$ belongs to R'. In other words, the ω sets

$$T_1, T_2, \dots, T_\omega$$

contain between them exactly ω elements, viz. the elements of R'. In *every* C.D.R. of (1), therefore, these ω T_i's are necessarily represented by these same ω elements. This shows that R' is contained in R. Hence

$$R' = R,$$

and

$$\rho = \omega,$$

and

$$R = T_1 \vee T_2 \vee \dots \vee T_\rho.$$

This is the assertion of the lemma.

The proof of Theorem 1 now follows by induction over m. The case $m = 1$ is trivial.

We assume then that any k of the sets (1) contain between them at least k elements of S, and also that the theorem is true for $m-1$ sets. We may therefore apply the theorem to the $m-1$ sets

(4) $$T_1, T_2, \dots, T_{m-1}.$$

These have, accordingly, at least one C.D.R. Hence (1) will also have at least one C.D.R., provided only that T_m is not contained in *all* the C.D.R. of (4).

But if (without loss of generality)

$$R^* = a_1, a_2, \ldots, a_\rho \quad (\rho \geqq 0)$$

is the meet of *all* the C.D.R. of (4), and if T_m is contained in R^*, then, by the lemma, the $k = \rho + 1$ sets

$$T_1, T_2, \ldots, T_\rho, T_m$$

contain between them only ρ elements, viz. those of R^*. This being contrary to hypothesis, T_m is not contained in R^*; and so, if a_m is any element of T_m not in R^*, there exists a C.D.R. of (4) in which a_m does not occur. This C.D.R. of (4) together with a_m constitutes the desired C.D.R. of (1).

An elementary transformation of Theorem 1 gives

THEOREM 2. *If S is divided into any number of classes (e.g. by means of some equivalence relation),*

$$S = S_1 \vee S_2 \vee S_3 \vee \ldots,$$

and $S_i \wedge S_j$ is the null set, for $i \neq j$, then there always exists a set of m elements

$$a_1, a_2, \ldots, a_m,$$

no two of which belong to the same class, such that

$$a_i \epsilon T_i \quad (i = 1, 2, \ldots, m),$$

provided only that, for each $k = 1, 2, \ldots, m$, any k of the sets T_i contain between them elements from at least k classes.

Proof. Denote by t_i the set of all classes S_j for which the meet

$$S_j \wedge T_i$$

is not null. The condition to be satisfied by the sets T_i may then be expressed thus: any k of the t_i's contain between them at least k members. Applying Theorem 1, it follows that there exists a set of m *distinct* classes, for simplicity

$$S_1, S_2, \ldots, S_m,$$

such that, for $i = 1, 2, \ldots, m$, the set

$$S_i \wedge T_i = M$$

is not null. Choosing for a_i an arbitrary element from M_i, the result follows.

A particular case of some interest is

THEOREM 3. *If the set S is divided into m classes in two different ways,*

$$S = S_1 \vee S_2 \vee \ldots \vee S_m,$$

$$S = S_1' \vee S_2' \vee \ldots \vee S_m';$$

$S_i \wedge S_j = S_i' \wedge S_j' =$ null set, for $i \neq j$, then, provided that, for each $k = 1, 2, \ldots, m$, any k of the classes S_j' always contain between them elements from at least k of the classes S_i, it will always be possible to find m elements of S,

$$a_1, a_2, \ldots, a_m,$$

such that (possibly after permuting the suffixes of the S_i')

$$a_i \in S_i \wedge S_i' \quad (i = 1, 2, \ldots, m).$$

The case in which all the classes have the same (finite) number of elements clearly fulfils the proviso. Theorem 3 then becomes the well-known theorem of König, referred to above.

The generalization of König's theorem due to R. Rado† may also be deduced without difficulty from Theorem 3.

King's College,
Cambridge.

Reprinted from
J. London Math. Soc. **10** (1935), 26–30

ON THE ABSTRACT PROPERTIES OF LINEAR DEPENDENCE.[1]

By Hassler Whitney.

1. Introduction. Let $C_1, C_2, \cdots, C_n$ be the columns of a matrix $\boldsymbol{M}$. Any subset of these columns is either linearly independent or linearly dependent; the subsets thus fall into two classes. These classes are not arbitrary; for instance, the two following theorems must hold:

(a) Any subset of an independent set is independent.

(b) If $\boldsymbol{N}_p$ and $\boldsymbol{N}_{p+1}$ are independent sets of p and $p+1$ columns respectively, then $\boldsymbol{N}_p$ together with some column of $\boldsymbol{N}_{p+1}$ forms an independent set of $p+1$ columns.

There are other theorems not deducible from these; for in § 16 we give an example of a system satisfying these two theorems but not representing any matrix. Further theorems seem, however, to be quite difficult to find. Let us call a system obeying (a) and (b) a "matroid." The present paper is devoted to a study of the elementary properties of matroids. The fundamental question of completely characterizing systems which represent matrices is left unsolved. In place of the columns of a matrix we may equally well consider points or vectors in a Euclidean space, or polynomials, etc.

This paper has a close connection with a paper by the author on linear graphs;[2] we say a subgraph of a graph is independent if it contains no circuit. Although graphs are, abstractly, a very small subclass of the class of matroids, (see the appendix), many of the simpler theorems on graphs, especially on non-separable and dual graphs, apply also to matroids. For this reason, we carry over various terms in the theory of graphs to the present theory. Remarkably enough, for matroids representing matrices, dual matroids have a simple geometrical interpretation quite different from that in the case of graphs (see § 13).

The contents of the paper are as follows: In Part I, definitions of matroids in terms of the concepts rank, independence, bases, and circuits are considered, and their equivalence shown. Some common theorems are deduced (for instance Theorem 8). Non-separable and dual matroids are studied in

[1] Presented to the American Mathematical Society, September, 1934.

[2] "Non-separable and planar graphs," *Transactions of the American Mathematical Society*, vol. 34 (1932), pp. 339-362. We refer to this paper as G.

Part II; this section might replace much of the author's paper G. The subject of Part III is the relation between matroids and matrices. In the appendix, we completely solve the problem of characterizing matrices of integers modulo 2, of interest in topology.

I. Matroids.

2. Definitions in terms of rank. Let a set M of elements $e_1, e_2, \cdots, e_n$ be given. Corresponding to each subset N of these elements let there be a number $r(N)$, the *rank* of N. If the three following postulates are satisfied, we shall call this system a *matroid.*

(R_1) *The rank of the null subset is zero.*

(R_2) *For any subset N and any element e not in N,*

$$r(N+e)=r(N)+k, \qquad (k=0 \text{ or } 1).$$

(R_3) *For any subset N and elements e_1, e_2 not in N, if $r(N+e_1)=r(N+e_2)=r(N)$, then $r(N+e_1+e_2)=r(N)$.*

Evidently *any subset of a matroid is a matroid.* In what follows, M is a fixed matroid. We make the following definitions:

$$\rho(N) = \text{number of elements in } N.$$

$$n(N)=\rho(N)-r(N) = \textit{nullity} \text{ of } N.$$

N is *independent*, or, the elements of N are independent, if $n(N)=0$; otherwise, N, and its set of elements, are *dependent.*

Lemma 1. *For any N, $r(N)\geqq 0$ and $n(N)\geqq 0$. If $N\subset M$, then $r(N)\leqq r(M)$, $n(N)\leqq n(M)$.*

Lemma 2. *Any subset of an independent set is independent.*

e is *dependent on* N if $r(N+e)=r(N)$; otherwise e is *independent of* N.

A *base* is a maximal independent submatroid of M, i. e. a matroid B in M such that $n(B)=0$, while $B\subset N$, $B\neq N$ implies $n(N)>0$. See also Theorem 7. A *base complement* $A=M-B$ is the complement in M of a base B. A *circuit* is a minimal dependent matroid, i. e. a matroid P such that $n(P)>0$, while $N\subset P$, $N\neq P$ implies $n(N)=0$.[3]

Theorem 1. *N is independent if and only if it is contained in a base, or, if and only if it contains no circuit.*

[3] Compare G, Theorem 9.

THEOREM 2. *A circuit is a minimal submatroid contained in no base, i. e. containing at least one element from each base complement. A base is a maximal submatroid containing no circuit. A base complement is a minimal submatroid containing at least one element from each circuit.*

The above facts follow at once from the definitions. Note the reciprocal relationship between circuits and base complements. Note also that the definitions of independence and of being a circuit depend only on the given subset, while the property of being a base depends on the relationship of the subset to M.

3. Properties of rank. Our object here is to prove Theorem 3. The following definition will be useful:

$$\Delta(M, N) = r(M + N) - r(M). \tag{3.1}$$

LEMMA 3. $\Delta(M + e_2, e_1) \leqq \Delta(M, e_1)$.

Suppose first $r(M + e_1) = r(M) + 1$; then $r(M + e_1 + e_2) = r(M) + k$, $k = 1$ or 2. If $k = 2$, then $r(M + e_2) = r(M) + 1$, on account of (R_2), and the inequality holds; if $k = 1$, $r(M + e_2) = r(M) + l$, $l = 0$ or 1, and it holds again. If $r(M + e_2) = r(M) + 1$, the same reasoning applies. If finally $r(M + e_1) = r(M + e_2) = r(M)$, the inequality follows from (R_3).

LEMMA 4. $\Delta(M + N, e) \leqq \Delta(M, e)$.

If $N = e_1 + \cdots + e_p$, the last lemma gives

$$\Delta(M + N, e) \leqq \Delta(M + e_1 + \cdots + e_{p-1}, e) \leqq \cdots \leqq \Delta(M, e).$$

THEOREM 3. $\Delta(M + N_2, N_1) \leqq \Delta(M, N_1)$, *or,*

$$r(M + N_1 + N_2) \leqq r(M + N_1) + r(M + N_2) - r(M). \tag{3.2}$$

This is true if N_1 contains but a single element. For the general case, we apply the last lemma and induction, setting $N_1 = N' + e$:

$$\begin{aligned}\Delta(M + N_2, N_1) &= \Delta(M + N_2 + e, N') + \Delta(M + N_2, e)\\ &\leqq \Delta(M + e, N') + \Delta(M, e) = \Delta(M, N_1).\end{aligned}$$

(3.2) is evidently equivalent to:

$$r(M_1 + M_2) \leqq r(M_1) + r(M_2) - r(M_1 M_2). \tag{3.3}$$

4. Deduction of (I_1), (I_2) from (R_1), (R_2), (R_3). The first postulate

4

on independent sets below obviously holds if (R_1) and (R_2) hold. To prove (I_2), take N, N' as given there; then

$$r(N) = p, \qquad r(N') = p + 1.$$

We must show that for some i, $\Delta(N, e'_i) = 1$. (Then e'_i does not lie in N.) If this is not so, then on using Lemma 4 we find

$$\begin{aligned} 1 = r(N') - r(N) &\leqq \Delta(N, N') \\ &= \Delta(N, e'_1) + \Delta(N + e'_1, e'_2) + \cdots + \Delta(N + e'_1 + \cdots + e'_p, e'_{p+1}) \\ &\leqq \Delta(N, e'_1) + \Delta(N, e'_2) + \cdots + \Delta(N, e'_{p+1}) = 0, \end{aligned}$$

a contradiction.

5. Deduction of (C_1), (C_2) **from** (R_1), (R_2), (R_3). We shall need here a theorem showing how the nullity (or rank) of a matroid may be determined when we know what circuits it contains.

LEMMA 5. *Each element of a circuit is dependent on the rest of the circuit.*

If e is an element of the circuit P, then $n(P) = 1$, $n(P - e) = 0$; hence $r(P) = \rho(P) - 1 = \rho(P - e) = r(P - e)$.

LEMMA 6. *If e is dependent on P_1 but on no proper subset of P_1, then $P = P_1 + e$ is a circuit.*

As $\Delta(P_1, e) = 0$, $r(P) = r(P_1) \leqq \rho(P_1) < \rho(P)$, $n(P) > 0$, and P contains a circuit P'. If P' does not contain e, take e' in P'; then

$$\Delta(P_1 - e', e') \leqq \Delta(P' - e', e') = 0,$$

hence $r(P_1 - e') = r(P_1)$, and

$$\begin{aligned} \Delta(P_1 - e', e) &= r(P_1 - e' + e) - r(P_1 - e') \\ &\leqq r(P_1 + e) - r(P_1) = \Delta(P_1, e) = 0, \end{aligned}$$

and e is dependent on the proper subset $P_1 - e'$ of P_1, a contradiction. Therefore P' contains e. As P' is a circuit, e is dependent on the rest of P'; hence $P' = P$.

THEOREM 4. *If e is not in N, there is a circuit in $N + e$ which contains e if and only if e is dependent on N.*

Suppose $P_1 + e = P$ is a circuit, $P_1 \subset N$. Then

$$\Delta(N, e) \leqq \Delta(P_1, e) = 0,$$

and e is dependent on N. Suppose, conversely, $\Delta(N, e) = 0$. Let P_1 be a smallest subset of N on which e is dependent; then by the last lemma, $P = P_1 + e$ is a circuit. (It may be that $P = e$.)

THEOREM 5. *If N is formed element by element, then $n(N)$ is just the number of times that adding an element increases the number of circuits present.*

Say $N = e_1 + \cdots + e_p$. Then if O is the null set,

$$r(N) = \Delta(O, e_1) + \Delta(e_1, e_2) + \cdots + \Delta(e_1 + \cdots + e_{p-1}, e_p).$$

Each $\Delta(e_1 + \cdots + e_{i-1}, e_i) = 0$ or 1, and $= 0$ if and only if e_i is dependent on $e_1 + \cdots + e_{i-1}$, i. e. if and only if there is a circuit in $e_1 + \cdots + e_i$ containing e_i. The number of terms is $p = \rho(N)$, and the theorem follows.

We turn now to the proof of (C_1) and (C_2). The first is obvious. To prove the second, take P_1, P_2, e_1, e_2 as given. As

$$\Delta(P_1 - e_2, e_2) = \Delta(P_2 - e_1, e_1) = 0,$$

we have

$$\Delta(P_1 + P_2 - e_2, e_2) = \Delta(P_1 + P_2 - e_1 - e_2, e_1) = 0.$$

These equations give

$$r(P_1 + P_2 - e_1 - e_2) = r(P_1 + P_2 - e_2) = r(P_1 + P_2).$$

Using (R_2) gives

$$r(P_1 + P_2 - e_1) = r(P_1 + P_2 - e_1 - e_2);$$

hence the required circuit P_3 exists, by Theorem 4.

6. Postulates for independent sets. Let M be a set of elements. Let any subset N of M be either "independent" or "dependent." Let the two following postulates be satisfied:

(I_1) *Any subset of an independent set is independent.*

(I_2) *If $N = e_1 + \cdots + e_p$ and $N' = e'_1 + \cdots + e'_{p+1}$ are independent, then for some i such that e'_i is not in N, $N + e'_i$ is independent.*

The resulting system is equivalent to a matroid, as we now show. Given any subset N of M, we let $r(N)$ be the number of elements in a largest independent subset of N. Obviously Postulates (R_1) and (R_2) are satisfied; we must prove (R_3). Say

$$r(N + e_1) = r(N + e_2) = r(N) = r.$$

Then $r(N + e_1 + e_2) = r$ or $r + 1$. If it equals $r + 1$, there is an independent set $N' = e'_1 + \cdots + e'_{r+1}$ in $N + e_1 + e_2$. Let $N'' = e_1'' + \cdots + e_r''$ be an independent set in N. By (I_2) there is an i such that $N'' + e'_i$ is an independent set of $r + 1$ elements. But $N'' + e'_i$ lies in $N + e_1$ or in $N + e_2$, and hence $r(N + e_1)$ or $r(N + e_2) \geqq r + 1$, a contradiction. Therefore $r(N + e_1 + e_2) = r$, as required.

We have shown how to deduce either set of postulates (R) or (I) from the other. Moreover the definitions of the rank and the independence or dependence of any subset of M agree under the two systems, and hence they are equivalent.

7. Postulates for bases. Let M be a set of elements, and let each subset either be or not be a " base." We assume

(B_1) *No proper subset of a base is a base.*

(B_2) *If B and B' are bases and e is an element of B, then for some element e' in B', $B - e + e'$ is a base.*

We shall prove the equivalence of this system with the preceding one. We write here $e_1 e_2 \cdots$ instead of $e_1 + e_2 + \cdots$ for short.

THEOREM 6. *All bases contain the same number of elements.*

For suppose

$$B = e_1 \cdots e_p e_{p+1} \cdots e_q e_{q+1} \cdots e_r,$$
$$B' = e_1 \cdots e_p e'_{p+1} \cdots e'_q$$

are bases, with exactly $e_1, \cdots, e_p$ in common, and $r > q$. We might have $p = 0$. $q > p$, on account of (B_1). By (B_2), we can replace e_{p+1} in B by an element e' of B', giving a base B_1. $e' = e'_{i_1}$ is one of the elements $e'_{p+1}, \cdots, e'_q$, for otherwise B_1 would be a proper subset of B. Hence

$$B_1 = e_1 \cdots e_p e'_{i_1} e_{p+2} \cdots e_q e_{q+1} \cdots e_r.$$

If $q > p + 1$, we replace e_{p+2} in B_1 by an element e'_{i_2} of B', giving a base B_2. Continuing in this manner, we obtain finally the base

$$B_{q-p} = e_1 \cdots e_p e'_{p+1} \cdots e'_q e_{q+1} \cdots e_r.$$

But this contains B' as a proper subset, contradicting (B_1).

We shall say a subset of M is independent if it is contained in a base. (I_1) obviously holds; we shall prove (I_2). Let N, N' be independent sets in the bases B, B'. Say

$$B = e_1 \cdots e_p e_{p+1} \cdots e_q e_{q+1} \cdots e_r e_{r+1} \cdots e_s,$$
$$B' = e_1 \cdots e_p e'_{p+1} \cdots e'_q e'_{q+1} \cdots e'_r e_{r+1} \cdots e_s,$$
$$N = e_1 \cdots e_p e_{p+1} \cdots e_q, \qquad N' = e_1 \cdots e_p e'_{p+1} \cdots e'_q e'_{q+1}.$$

Then N and N' have just $e_1, \cdots, e_p$ in common, and B and B' have just these elements and $e_{r+1}, \cdots, e_s$ in common. By (B_2), there is an element e'_{i_1} of B' such that

$$B_1 = B - e_{q+1} + e'_{i_1}$$

is a base. (This element cannot be any of $e_1, \cdots, e_p, e_{r+1}, \cdots, e_s$, by (B_1).) If i_1 is one of the numbers $p+1, p+2, \cdots, q+1$, then $N + e'_{i_1}$ is in a base B_1, as required. Suppose not; then there is a base

$$B_2 = B_1 - e_{q+2} + e'_{i_2},$$

with $i_2 \neq i_1$. If $p + 1 \leqq i_2 \leqq q + 1$, $N + e'_{i_2}$ is in a base B_2. If not, we find a base B_3, etc. We can drop out each of the $r - q$ elements $e_{q+1}, \cdots, e_r$ in turn; as there are only $r - q - 1$ elements e'_i with $i > q + 1$, we find at some point a base containing $e_1, \cdots, e_q, e'_j$ with $p + 1 \leqq j \leqq q + 1$. Then e'_j is in N', and $N + e'_j$ is in a base and is thus independent, as required.

The definitions of base and independent sets in the two systems (I) and (B) are easily seen to agree. Suppose (I_1) and (I_2) hold. (B_1) obviously holds; using (I_2), we prove that all bases contain the same number of elements; (B_2) now follows at once from (I_2). Hence the two systems are equivalent.

THEOREM 7. *B is a base in M if and only if*

$$r(B) = r(M), \qquad n(B) = 0.$$

Evidently B is a base under the given conditions. To prove the converse, we note first that there exists a base with $r(M)$ elements, as $r(M)$ is the maximum number of independent elements in M (see § 6). By Theorem 6, all bases have this many elements, and the equations follow.

THEOREM 8. *If B is a base and N is independent, then for some N' in B, $N + N'$ is a base.*

This follows from repeated application of Postulate (I_2) and the last theorem.

8. Postulates for circuits. Let M be a set of elements, and let each subset either be or not be a "circuit." We assume:

(C_1) *No proper subset of a circuit is a circuit.*

(C_2) *If P_1 and P_2 are circuits, e_1 is in both P_1 and P_2, and e_2 is in P_1 but not in P_2, then there is a circuit P_3 in $P_1 + P_2$ containing e_2 but not e_1.*

(C_2) may be phrased as follows: If the circuits P_1 and P_2 have the common element e, then $P_1 + P_2 - e$ is the union of a set of circuits.

We shall define the rank of any subset of M, and shall then show that the postulates for rank are satisfied. Let $e_1, \cdots, e_p$ be any ordered set of elements of M. Set $\Gamma_i = 0$ if there is a circuit in $e_1 + \cdots + e_i$ containing e_i, and set $\Gamma_i = 1$ otherwise (compare Theorem 5). Let the "rank" of $(e_1, \cdots, e_p)$ be

$$r(e_1, \cdots, e_p) = \sum_{i=1}^{p} \Gamma_i.$$

LEMMA 7. $r(e_1, \cdots, e_{q-2}, e_{q-1}, e_q) = r(e_1, \cdots, e_{q-2}, e_q, e_{q-1})$.

To prove this, let N be the ordered set $e_1, \cdots, e_{q-2}$, and set

$$r(N) = r, \qquad r(N, e_{q-1}) = r_1, \qquad r(N, e_q) = r_2,$$
$$r(N, e_{q-1}, e_q) = r_{12}, \qquad r(N, e_q, e_{q-1}) = r_{21}.$$

CASE 1. There is no circuit in $N + e_{q-1}$ containing e_{q-1}, and none in $N + e_q$ containing e_q. Then

$$r_1 = r_2 = r + 1.$$

If there is a circuit in $N + e_{q-1} + e_q$ containing e_{q-1} and e_q, then

$$r_{12} = r_1 = r_2 = r_{21};$$

otherwise,

$$r_{12} = r_1 + 1 = r_2 + 1 = r_{21}.$$

CASE 2. There is a circuit P_2 in $N + e_{q-1}$ containing e_{q-1}, and a circuit P_1 in $N + e_{q-1} + e_q$ containing e_{q-1} and e_q. Then, by (C_2), there is a circuit P_3 in $N + e_q$ containing e_q. Hence

$$r_{12} = r_1 = r = r_2 = r_{21}.$$

CASE 3. There is a circuit P_2 as above, but no circuit P_1 as above. If there is a circuit P_3 as above, the last set of equations hold. Otherwise,

$$r_{12} = r_1 + 1 = r + 1 = r_2 = r_{21}.$$

CASE 4. There is a circuit in $N + e_q$ containing e_q. This case overlaps the two preceding ones; the proof above applies here also.

LEMMA 8. *The rank of any subset N is independent of the ordering of the elements of N.*

We saw above that interchanging the last two elements of any subset does not alter the rank; hence, evidently, interchanging any two adjacent elements leaves the rank unchanged. Any ordering of M may be obtained from any other by a number of interchanges of adjacent elements; the rank remains unchanged at each step, proving the lemma.

Postulates (R_1) and (R_2) are obviously satisfied. To prove (R_3), suppose $r(N + e_1) = r(N + e_2) = r(N)$. Then there is a circuit in $N + e_1$ containing e_1 and one in $N + e_2$ containing e_2; hence $r(N + e_1 + e_2) = r(N)$.

The definitions of rank and of circuits under the two systems (R), (C) agree, and hence the systems are equivalent.

9. Fundamental sets of circuits. The circuits $P_1, \cdots, P_q$ of a matroid M form a *fundamental set of circuits* if $q = n(M)$ and the elements $e_1, \cdots, e_n$ of M can be ordered so that P_i contains e_{n-q+i} but no e_{n-q+j} $(j > i)$. The set is *strict* if P_i contains e_{n-q+i} but no e_{n-q+j} $(0 < j < i$ or $j > i)$. These sets may be called sets *with respect to* $e_{n-q+1}, \cdots, e_n$.

THEOREM 9. *If $B = e_1 + \cdots + e_{n-q}$ is a base in $M = e_1 + \cdots + e_n$, then there is a strict fundamental set of circuits with respect to $e_{n-q+1}, \cdots, e_n$; these circuits are uniquely determined.*

As $r(B) = r(M)$, $\Delta(B, e_i) = 0$ $(i = n - q + 1, \cdots, n)$. Hence, by Theorem 4, there is a circuit P_i containing e_i and elements (possibly) of B. $P_{n-q+1}, \cdots, P_n$ is the required set. Suppose, for a given i, there were also a circuit $P'_i \neq P_i$. Then Postulate (C_2) applied to P_i and P'_i would give us a circuit P in B, which is impossible.

This theorem corresponds to the theorem that if a square submatrix N of a matrix M is non-singular, then N can be turned into the unit matrix by a linear transformation on the rows of M.

THEOREM 10. *If $P_1, \cdots, P_q$ form a fundamental set of circuits with* *respect to* $e_{n-q+1}, \cdots, e_n$, *then there is a unique strict set* $P'_1, \cdots, P'_q$ *with respect to* $e_{n-q+1}, \cdots, e_n$.

Set $B = M - (e_{n-q+1} + \cdots + e_n)$. The existence of $P_1, \cdots, P_q$ shows that $r(M) = r(M - e_n) = \cdots = r(B)$. Hence $\rho(B) = n - q = r(M) = r(B)$, and B is a base, by Theorem 7. Theorem 9 now applies.

Note that a matroid is not uniquely determined by a fundamental set of circuits (but see the appendix). This is shown by the following two matroids, in each of which the first two circuits form a strict fundamental set:

M, with circuits 1234, 1256, 3456;
M', with circuits 1234, 1256, 13456, 23456.

II. Separability, Dual Matroids.

10. Separable matroids. If $M = M_1 + M_2$, then $r(M) \leqq r(M_1) + r(M_2)$, on account of (3. 3). If it is possible to divide the elements of M into two groups, M_1 and M_2, each containing at least one element, such that

$$r(M) = r(M_1) + r(M_2), \tag{10. 1}$$

or, which is equivalent (as M_1 and M_2 have no common elements),

$$n(M) = n(M_1) + n(M_2), \tag{10. 2}$$

we shall say M is *separable*; otherwise, M is *non-separable*.[4] Any single element forms a non-separable matroid. Any maximal non-separable part of M is a *component* of M.[5]

Theorem 11. *If*

$$M = M_1 + M_2, \qquad r(M) = r(M_1) + r(M_2),$$
$$M'_1 \subset M_1, \qquad M'_2 \subset M_2, \qquad M' = M'_1 + M'_2,$$

then

$$r(M') = r(M'_1) + r(M'_2).$$

Set $M_1'' = M_1 - M_1'$, $M_2'' = M_2 - M_2'$. The relations (see Theorem 3)

$$\begin{aligned} r(M) &= \Delta(M_1 + M_2', M_2'') + \Delta(M', M_1'') + r(M') \\ &\leqq \Delta(M_2', M_2'') + \Delta(M_1', M_1'') + r(M') \\ &= r(M_2) - r(M_2') + r(M_1) - r(M_1') + r(M') \end{aligned}$$

[4] Compare G, Theorem 15.
[5] See G, § 4.

together with the fact that $r(M) = r(M_1) + r(M_2)$ show that $r(M') \geqq r(M'_1) + r(M'_2)$ and hence $r(M') = r(M'_1) + r(M'_2)$.

THEOREM 12.[6] *If $M = M_1 + M_2$, $r(M) = r(M_1) + r(M_2)$, M' is non-separable, and $M' \subset M$, then either $M' \subset M_1$ or $M' \subset M_2$.*

For suppose $M' = M'_1 + M'_2$, $M'_1 \subset M_1$, $M'_2 \subset M_2$, and M'_1 and M'_2 each contain an element. By the last theorem, $r(M') = r(M'_1) + r(M'_2)$, which cannot be.

THEOREM 13. *If M_1 and M_2 are non-separable matroids with a common element e, then $M = M_1 + M_2$ is non-separable.*

For suppose $M = M'_1 + M'_2$, $r(M) = r(M'_1) + r(M'_2)$. By the last theorem, $M_1 \subset M'_1$ or $M_1 \subset M'_2$, and $M_2 \subset M'_1$ or $M_2 \subset M'_1$; this shows that either M'_1 or M'_2 is void.

THEOREM 14. *No two distinct components of M have common elements.*

This is a consequence of the last theorem. From this follows:

THEOREM 15.[7] *Any matroid may be expressed as a sum of components in a unique manner.*

THEOREM 16.[8] *A non-separable matroid M of nullity 1 is a circuit, and conversely.*

If M_1 is a proper non-null subset of the non-separable matroid M, and $M_2 = M - M_1$, then $r(M) < r(M_1) + r(M_2)$. Hence

$$1 = n(M) > n(M_1) + n(M_2),$$

and $n(M_1) = 0$, proving that M is a circuit.

Conversely, if $M = M_1 + M_2$ is a circuit, and M_1 and M_2 each contain elements, then

$$\begin{aligned} r(M_1) + r(M_2) &= \rho(M_1) + \rho(M_2) - n(M_1) - n(M_2) \\ &= \rho(M) > r(M), \end{aligned}$$

showing that M is non-separable.

---

[6] Compare G, Lemma, p. 344.

[7] Compare G, Theorem 12.

[8] Compare G, Theorem 10.

LEMMA 9. *Let $M = M_1 + M_2$ be non-separable, and let M_1 and M_2 each contain elements but have no common elements. Then there is a circuit P in M containing elements of both M_1 and M_2.*

Suppose there were no such circuit. Say $M_2 = e_1 + \cdots + e_s$. Using Theorem 4, we see that

$$\Delta(M_1 + e_1 + \cdots + e_{i-1}, e_i) = \Delta(e_1 + \cdots + e_{i-1}, e_i) \quad (i = 1, \cdots, s),$$

and hence $r(M) = r(M_1) + r(M_2)$, a contradiction.

THEOREM 17.[9] *Any non-separable matroid M of nullity $n > 0$ can be built up in the following manner: Take a circuit M_1; add a set of elements which forms a circuit with one or more elements of M_1, forming a non-separable matroid M_2 of nullity 2 (if $n(M) > 1$); repeat this process till we have $M_n = M$.*

As $n > 0$, M contains a circuit M_1. If $n > 1$, we use the preceding lemma $n - 1$ times. The matroid at each step is non-separable, by Theorems 16 and 13.

THEOREM 18.[10] *Let $M = M_1 + \cdots + M_p$, and let $M_1, \cdots, M_p$ be non-separable. Then the following statements are equivalent:*

(1) *$M_1, \cdots, M_p$ are the components of M.*

(2) *No two of the matroids $M_1, \cdots, M_p$ have common elements, and there is no circuit in M containing elements of more than one of them.*

(3) $r(M) = r(M_1) + \cdots + r(M_p)$.

We cannot replace rank by nullity in (3); see G, p. 347.

(2) follows from (1) on application of Theorems 13 and 16.

To prove (1) from (2), take any M_i. If it is not a component of M, there is a larger non-separable submatroid M'_i of M containing it. By Lemma 9, there is a circuit P in M'_i containing elements of M_i and elements not in M_i; P must contain elements of some other M_j, a contradiction.

Next we prove (3) from (1). If $p > 1$, M is separable; say $M = M'_1 + M'_2$, $r(M) = r(M'_1) + r(M'_2)$. By Theorem 12, each M_i is in either M'_1 or M'_2; hence M'_1 and M'_2 are each a sum of components of M. If one of these

[9] See G, Theorem 19; also Whitney, "2-isomorphic graphs," *American Journal of Mathematics*, vol. 55 (1933), p. 247, footnote.

[10] Compare G, Theorem 17.

contains more than one component, we separate it similarly, etc. (3) now follows easily.

Finally we prove (1) from (3). Let M' be a component of M, and suppose it has an element in M_i. As

$$r(M) = r(M_i) + \sum_{j \neq i} r(M_j),$$

M' is contained in M_i, by Theorem 12; as M_i is non-separable, $M' = M_i$.

THEOREM 19.[11] *The elements e_1 and e_2 are in the same component of M if and only if they are contained in a circuit P.*

If e_1 and e_2 are both in P, they are part of a non-separable matroid, which lies in a single component of M. Suppose now e_1 and e_2 are in the same component M_0 of M, and suppose there is no circuit containing them both. Let M_1 be e_1 plus all elements which are contained in a circuit containing e_1. By Lemma 9, there is a subset M^* of $M_0 - M_1$ which forms with part of M_1 a circuit P_3. P_3 does not contain e_1. If e'_4 is an element of P_3 in M_1, there is a circuit P_1 in M_1 containing e_1 and e'_4. Let e_3 be an element of M^*. Then in $M_1 + M^*$ there are circuits P_1 and P_3 which contain e_1 and e_3 respectively, and have a common element.

Let M' be a smallest subset of M_0 which contains circuits P'_1 and P'_3 such that one contains e_1, the other contains e_3, and they have common elements. Then P'_1 and P'_3 are distinct, and $M' = P'_1 + P'_3$. Let e_4 be a common element. By Postulate (C_2), there is a circuit P_1 in $M' - e_4$ containing e_1, and a circuit P_3 in $M' - e_4$ containing e_3. By the definition of M', P_1 and P_3 have no common elements. By Postulate (C_1), P_1 is not contained in P'_1; hence it contains an element e_5 of $M' - P'_1$. P_3 does not contain e_5. As P_3 is not contained in P'_3, it contains an element e_6 of P'_1. But now P'_1 contains e_1, P_3 contains e_3, $P'_1 + P_3$ have a common element e_6, and $P'_1 + P_3$ does not contain e_5 and is thus a proper subset of M', a contradiction. This proves the theorem.

11. Dual matroids. Suppose there is a $1-1$ correspondence between the elements of the matroids M and M', such that if N is any submatroid of M and N' is the complement of the corresponding matroid of M', then

$$r(N') = r(M') - n(N). \tag{11.1}$$

[11] Compare D. König, *Acta Litterarum ac Scientiarum Szeged*, vol. 6, pp. 155-179, 4. (p. 159). The present theorem shows that a "glied" is the same as a component.

We say then that M' is a *dual* of M.[12]

THEOREM 20. *If M' is a dual of M, then*

$$r(M') = n(M), \qquad n(M') = r(M).$$

Set $N = M$; then $n(N) = n(M)$. In this case N' is the null matroid, and $r(N') = 0$. (11.1) now gives $r(M') = n(M)$. Also

$$n(M') = \rho(M') - r(M') = \rho(M) - n(M) = r(M).$$

THEOREM 21. *If M' is a dual of M, then M is a dual of M'.*

Take any N and corresponding N' as before. The equations

$$r(N') = r(M') - n(N), \qquad r(M') = n(M),$$
$$\rho(N) + \rho(N') = \rho(M)$$

give

$$\begin{aligned} r(N) &= \rho(N) - n(N) = \rho(N) - [r(M') - r(N')] \\ &= \rho(N) - n(M) + [\rho(N') - n(N')] \\ &= \rho(M) - n(M) - n(N') = r(M) - n(N'), \end{aligned}$$

as required.

THEOREM 22. *Every matroid has a dual.*

This is in marked contrast to the case of graphs, for only a planar graph has a dual graph (see G, Theorem 29).

Let M' be a set of elements in 1 — correspondence with elements of M. If N' is any subset of M', let N be the complement of the corresponding subset of M, and set $r(N') = n(M) - n(N)$. (R_1), (R_2), (R_3) are easily seen to hold in M', as they hold in M; hence M' is a matroid. Obviously $r(M') = n(M)$, and M' is a dual of M.

THEOREM 23. *M and M' are duals if and only if there is a 1—1 correspondence between their elements such that bases in one correspond to base complements in the other.*

Suppose first M and M' are duals. Let B be a base in either matroid, say in M, and let B' be the complement of the corresponding submatroid of the other matroid, M'. Then

[12] Compare G, § 8. Theorems 20, 21, 24, 25 correspond to Theorems 20, 21, 23, 25 in G. Note that *two duals of the same matroid are isomorphic*, that is, there is a 1—1 correspondence between their elements such that corresponding subsets have the same rank. Such a statement cannot be made about graphs. Compare H. Whitney, "2-isomorphic graphs," *American Journal of Mathematics*, vol. 55 (1933), pp. 245-254.

$$r(B') = r(M') - n(B) = r(M'),$$
$$n(B') = r(M) - r(B) = 0,$$

and B' is a base in M', by Theorem 7.

Suppose, conversely, that bases in one correspond to base complements in the other. Let N be a submatroid of M and let N' be the complement of the corresponding submatroid of M'. There is a base B' in M' with $r(N')$ elements in N', by Theorem 8. The complement in M of the submatroid corresponding to B' in M' is a base B in M with $\rho(N') - r(N') = n(N')$ elements in $M - N$, and hence with $r(M) - n(N')$ elements in N. This shows that

$$r(N) = r(M) - n(N') + k, \qquad k \geqq 0.$$

In a similar fashion we see that

$$r(N') = r(M') - n(N) + k', \qquad k' \geqq 0.$$

As B contains $r(M)$ elements and B' contains $r(M')$ elements, $r(M) + r(M') = \rho(M)$. Hence, adding the above equations,

$$\begin{aligned} k + k' &= r(N) + r(N') + n(N) + n(N') - r(M) - r(M') \\ &= \rho(N) + \rho(N') - \rho(M) = 0. \end{aligned}$$

Hence $k = 0$, and the first equation above shows that M and M' are duals.

There are various other ways of stating conditions on certain submatroids of M and M' which will ensure these matroids being duals.[13]

THEOREM 24. *Let $M_1, \cdots, M_p$ and $M'_1, \cdots, M'_p$ be the components of M and M' respectively, and let M'_i be a dual of M_i $(i = 1, \cdots, p)$. Then M' is a dual of M.*

Let N be any submatroid of M, and let the parts of N in $M_1, \cdots, M_p$ be $N_1, \cdots, N_p$. Let N'_i be the complement in M'_i of the submatroid corresponding to N_i; then $N' = N'_i + \cdots + N'_p$ is the complement in M' of the submatroid corresponding to N. By Theorems 18 and 11 we have

$$r(N') = r(N'_1) + \cdots + r(N'_p), \qquad n(N) = n(N_1) + \cdots + n(N_p).$$

Also

$$r(M') = r(M'_1) + \cdots + r(M'_p), \qquad r(N'_i) = r(M'_i) - n(N_i);$$

adding the last set of equations gives $r(N') = r(M') - n(N)$, as required.

[13] See for instance a paper by the author "Planar graphs," *Fundamenta Mathematicae*, vol. 21 (1933), pp. 73-84, Theorem 2. Cut sets may of course be defined in terms of rank.

THEOREM 25. *Let M and M' be duals, and let $M_1, \cdots, M_p$ be the components of M. Let $M'_1, \cdots, M'_p$ be the corresponding submatroids of M'. Then $M'_1, \cdots, M'_p$ are the components of M', and M'_i is a dual of M_i $(i = 1, \cdots, p)$.*

The complement in M of the submatroid corresponding to M'_i in M' is $\sum_{j \neq i} M_j$. Hence, as M and M' are duals and the M_j $(j \neq i)$ are the components of $\sum_{j \neq i} M_j$ (see Theorem 18),

$$r(M'_i) = r(M') - n(\sum_{j \neq i} M_j) = r(M') - \sum_{j \neq i} n(M_j).$$

Adding gives

$$\sum_i r(M'_i) = pr(M') - (p-1) \sum_j n(M_j) = pr(M') - (p-1)n(M)$$
$$= pr(M') - (p-1)r(M') = r(M').$$

Therefore, by Theorem 12, each component of M' is contained in some M'_i. In the same way we see that each component of M is contained in a matroid corresponding to a component of M'; hence the components of one matroid correspond exactly to the components of the other.

Let N_i be any submatroid of M_i, and let N' and N'_i be the complements in M' and M'_i of the submatroid corresponding to N_i. The equations

$$r(M') = \sum_j r(M'_j), \qquad r(N') = r(N'_i) + \sum_{j \neq i} r(M'_j),$$
$$r(N') = r(M') - n(N_i),$$

give

$$r(N'_i) = r(M'_i) - n(N_i),$$

which shows that M'_i is a dual of M_i.

THEOREM 26. *A dual of a non-separable matroid is non-separable.*

This is a consequence of the last theorem.

III. MATRICES AND MATROIDS.

12. Matrices, matroids, and hyperplanes. Consider the matrix

$$M = \left\| \begin{matrix} a_{11} & \cdots & a_{1n} \\ \cdot & \cdot & \cdot \\ a_{m1} & \cdots & a_{mn} \end{matrix} \right\|;$$

let its columns be $C_1, \cdots, C_n$. Any subset N of these columns forms a matrix, and this matrix has a rank, $r(N)$. If we consider the columns as abstract elements, we have a matroid M. The proof of this is simple if we consider the rank of a matrix as the number of linearly independent columns in it. (R_1) and (R_2) are then obvious. To prove (R_3), suppose $r(N + C_1) = r(N + C_2) = r(N)$; then C_1 and C_2 can each be expressed as a linear combination of the other columns of N, and hence $r(N + C_1 + C_2) = r(N)$. The terms independent and base carry over to matrices and agree with the ordinary definitions; a base in M is a minimal set of columns in terms of which all remaining columns of M may be expressed.

We may interpret M geometrically in two different ways; the second is the more interesting for our purposes:

(a) Let E_m be Euclidean space of m dimensions. Corresponding to each column C_i of M there is a point X_i in E_m with coördinates $a_{1i}, \cdots, a_{mi}$. The subset $C_{i_1}, \cdots, C_{i_p}$ of M is linearly independent if and only if the points $O = (0, \cdots, 0)$, $X_{i_1}, \cdots, X_{i_p}$ are linearly independent in E_m, i. e. if and only if these $p + 1$ points determine a hyperplane in E_m of dimension p. A base in M corresponds to a minimal set of points $X_{i_1}, \cdots, X_{i_p}$ in E_m such that each X_j of M lies in the hyperplane determined by $O, X_{i_1}, \cdots, X_{i_p}$. Then p is the rank of M.

(b) Let E_n be Euclidean space of n dimensions. Let $R_1, \cdots, R_m$ be the rows of M. If $Y_1, \cdots, Y_m$ are the corresponding points of E_n: $Y_i = (a_{i1}, \cdots, a_{in})$, then the points $O, Y_1, \cdots, Y_m$ determine a hyperplane $H = H(M)$, which we shall call the *hyperplane associated with* M. The dimension $d(H)$ of H is $r(M)$. Let $N = C_{i_1} + \cdots + C_{i_p}$ be a subset of M, and let E' be the p-dimensional coördinate subspace of E_n containing the x_{i_1} and ... and the x_{i_p} axes. The j-th row of N corresponds to the point Y'_j in E' with coördinates $(a_{ji_1}, \cdots, a_{ji_p})$; this is just the projection of Y_j onto E'. If H' is the hyperplane in E' determined by the points $O, Y'_1, \cdots, Y'_m$, then H' is exactly the projection of H onto E', and

$$d(H') = r(N). \tag{12.1}$$

Let $N = (C_{i_1}, \cdots, C_{i_p})$ be any subset of M, and let E', H' correspond to N. Then N *is independent if and only if*

$$d(H') = p,$$

and is a base if and only if

$$d(H') = d(H) = p.$$

THEOREM 27. *There is a unique matroid M associated with any hyperplane H through the origin in E_n.*

Let M contain the elements $e_1, \cdots, e_n$, one corresponding to each coördinate of E_n. Given any subset $e_{i_1}, \cdots, e_{i_p}$, we let its rank be the dimension of the projection of H onto the corresponding coördinate hyperplane E' of E_n. It was seen above that if $\boldsymbol{M}$ is any matrix determining H, then M is the matroid associated with $\boldsymbol{M}$.

13. Orthogonal hyperplanes and dual matroids. We prove the following theorem:

THEOREM 28. *Let H be a hyperplane through the origin in E_n, of dimension r, and let H' be the orthogonal hyperplane through the origin, of dimension $n - r$. Let M and M' be the associated matroids. Then M and M' are duals.*

We shall show that bases in one matroid correspond to base complements in the other; Theorem 23 then applies. Let

$$\boldsymbol{M} = \left\| \begin{matrix} a_{11} \cdots a_{1n} \\ \cdot \quad \cdot \quad \cdot \quad \cdot \\ a_{r1} \cdots a_{rn} \end{matrix} \right\|, \qquad \boldsymbol{M}' = \left\| \begin{matrix} b_{11} & \cdots b_{1n} \\ \cdot \quad \cdot & \cdot \quad \cdot \quad \cdot \\ b_{n-r,1} & \cdots b_{n-r,n} \end{matrix} \right\|$$

be matrices determining H and H' respectively. Say the first r columns of $\boldsymbol{M}$ form a base in $\boldsymbol{M}$, i. e. the corresponding determinant A is $\neq 0$. As H and H' are orthogonal, we have for each i and j

$$a_{i1}b_{j1} + a_{i2}b_{j2} + \cdots + a_{in}b_{jn} = 0.$$

Keeping j fixed, we have a set of r linear equations in the b_{jk}. Transpose the last $n - r$ terms in each equation to the other side, and solve for b_{jk}. We find

$$b_{jk} = \frac{-1}{A} \sum_{l=r+1}^{n} b_{jl} \left| \begin{matrix} a_{11} \cdots a_{1l} \cdots a_{1r} \\ \cdot \quad \cdot \quad \cdot \quad \cdot \quad \cdot \quad \cdot \quad \cdot \\ a_{r1} \cdots a_{rl} \cdots a_{rr} \end{matrix} \right| = \sum_{l=r+1}^{n} c_{kl} b_{jl} \qquad (k = 1, \cdots, r).$$

This is true for each $j = 1, \cdots, n - r$, and the c_{kl} are independent of j. Thus the k-th column of $\boldsymbol{M}'$ is expressed in terms of the last $n - r$ columns. As this is true for $k = 1, \cdots, r$, the last $n - r$ columns form a base in $\boldsymbol{M}'$, as required.

14. The circuit matrix of a given matrix. Consider the matrix $\boldsymbol{M}$ of § 12. Suppose the columns $C_{i_1}, \cdots, C_{i_p}$ form a circuit, i. e. the corresponding elements of the corresponding matroid form a circuit. Then these columns are linearly dependent, and there are numbers $b_1, \cdots, b_n$ such that

$$
(14.1) \qquad \begin{array}{ll} a_{i1}b_1 + \cdots + a_{in}b_n = 0 & (i = 1, \cdots, m), \\ b_j = 0 \quad (j \neq i_1, \cdots, i_p), & b_j \neq 0 \quad (j = i_1, \cdots, i_p). \end{array}
$$

The b_j are all $\neq 0$ $(j = i_1, \cdots, i_p)$, for otherwise a proper subset of the columns would be dependent, contrary to the definition of a circuit. (They are uniquely determined except for a constant factor; see Lemma 11.) Suppose the circuits of $\boldsymbol{M}$ are $P_1, \cdots, P_s$. Then there are corresponding sets of numbers $b_{i1}, \cdots, b_{in}$ $(i = 1, \cdots, s)$, forming a matrix

$$
\boldsymbol{M}' = \left\| \begin{array}{ccc} b_{11} & \cdots & b_{1n} \\ \cdot & \cdot & \cdot \\ b_{s1} & \cdots & b_{sn} \end{array} \right\|,
$$

the *circuit matrix* of the matrix $\boldsymbol{M}$.

THEOREM 29. *Let $P_1, \cdots, P_q$ be a fundamental set of circuits in $\boldsymbol{M}$* (see § 9). *Then the corresponding rows of the circuit matrix $\boldsymbol{M}'$ form a base for the rows of $\boldsymbol{M}'$. Hence $r(\boldsymbol{M}') = q = n(\boldsymbol{M})$.*

Suppose the columns of $\boldsymbol{M}$ are ordered so that P_i contains C_{n-q+i} but no column C_{n-q+j} $(j > i)$. Then if the corresponding row of $\boldsymbol{M}'$ is $R'_i = (b_{i1}, \cdots, b_{in})$, we have $b_{i,n-q+i} \neq 0$ and $b_{i,n-q+j} = 0$ $(j > i)$. Hence the rows $R'_1, \cdots, R'_q$ of $\boldsymbol{M}'$ are linearly independent, and $r(\boldsymbol{M}') \geqq q$. Hence $r(\boldsymbol{M}') = n(\boldsymbol{M}) = q$, and each row of $\boldsymbol{M}'$ may be expressed in terms of $R'_1, \cdots, R'_q$.

THEOREM 30. *If $\boldsymbol{M}'$ is the circuit matrix of $\boldsymbol{M}$ and H', H are the corresponding hyperplanes, then H' is the hyperplane of maximum dimension orthogonal to H.*

This is a consequence of (14.1) and the last theorem.

THEOREM 31. *The matroids corresponding to a matrix and its circuit matrix are duals.*

This follows from the last theorem and Theorem 28.

15. On the structure of a circuit matrix. Let M be any matroid, and M', its dual. If there exists a matrix $\boldsymbol{M}$ corresponding to M, it is perhaps most easily constructed by considering it as the circuit matrix of a matrix $\boldsymbol{M}'$

5

corresponding to M'. Let H and H' be the hyperplanes corresponding to $\mathbf{M}$ and $\mathbf{M}'$. We shall say the set of numbers $(a_1, \cdots, a_n)$ is in $Z_{i_1 \ldots i_p}$ if

$$a_j \neq 0 \quad (j = i_1, \cdots, i_p), \qquad a_j = 0 \quad (j \neq i_1, \cdots, i_p).$$

If $(a_1, \cdots, a_n)$ is in H and in $Z_{i_1 \ldots i_p}$, then the columns $C_{i_1}, \cdots, C_{i_p}$ of $\mathbf{M}'$ are dependent, evidently.

LEMMA 10. *Let $(b_1, \cdots, b_n)$ be a point of H. If it is in $Z_{i_1 \ldots i_p}$, then the matroid $N' = e_{i_1} + \cdots + e_{i_p}$ is the union of a set of circuits in M'.*

Here e_i in M' corresponds to C_i in $\mathbf{M}$. We need merely show that for each i_s there is a circuit P in N' containing e_{i_s}. Let $k_1 = i_s, k_2, \cdots, k_q$ be a minimal set of numbers from $(i_1, \cdots, i_p)$ containing i_s such that there is a point $(c_1, \cdots, c_n)$ of H in $Z_{k_1 \ldots k_q}$; then $e_{k_1} + \cdots + e_{k_q}$ is the required circuit. For if it were not a circuit, there would be a proper subset $(l_1, \cdots, l_r)$ of $(k_1, \cdots, k_q)$ and a point $(d_1, \cdots, d_n)$ of H in $Z_{l_1 \ldots l_r}$. No $l_i = k_1$, on account of the minimal property of $(k_1, \cdots, k_q)$. Say $l_1 = k_t$, and set

$$a_i = d_{k_t} c_i - c_{k_t} d_i \qquad (i = 1, \cdots, n).$$

Then $(a_1, \cdots, a_n)$ is in H and in $Z_{m_1 \ldots m_u}$ with $(m_1, \cdots, m_u)$ a proper subset of $(k_1, \cdots, k_q)$ containing k_1, again a contradiction.

LEMMA 11. *If $P = e_{i_1} + \cdots + e_{i_p}$ is a circuit of M' and $(b_1, \cdots, b_n)$ and $(b'_1, \cdots, b'_n)$ are in H and in $Z_{i_1 \ldots i_p}$, then these two sets are proportional.*

For otherwise, $(c_1, \cdots, c_n)$ with $c_i = b'_{i_1} b_i - b_{i_1} b'_i$ would be a point of H in some $Z_{k_1 \ldots k_q}$ with $(k_1, \cdots, k_q)$ a proper subset of $(i_1, \cdots, i_p)$, and P would not be a circuit.

It is instructive to show directly that Postulate (C_2) holds for matrices: P_1 and P_2 are represented by rows $(b_1, \cdots, b_n)$ and $(b'_1, \cdots, b'_n)$ of $\mathbf{M}$, lying in $Z_{12 i_1 \ldots i_p}$ and $Z_{1 k_1 \ldots k_q}$ respectively, where $k_1, \cdots, k_q \neq 2$. Set $c_i = b'_1 b_i - b_1 b'_i$; then $(c_1, \cdots, c_n)$ is in H and in $Z_{2 l_1 \ldots l_r}$, with $(l_1, \cdots, l_r)$ a subset of $(i_1, \cdots, i_p, k_1, \cdots, k_q)$; the existence of P_3 now follows from Lemma 10.

THEOREM 32. *Let $\mathbf{M}$ be the circuit matrix of $\mathbf{M}'$. Let $P_1, \cdots, P_q$ form a strict fundamental set of circuits in M' with respect to $e_{n-q+1}, \cdots, e_n$, and let the first q rows in $\mathbf{M}$ correspond to $P_1, \cdots, P_q$. Let $(i_1, \cdots, i_s)$ be any set of numbers from $(1, \cdots, q)$, let $(j_1, \cdots, j_s)$ be any set from $(1, \cdots, n-q)$, and let $(i'_1, \cdots, i'_{q-s})$ be the set complementary to $(i_1, \cdots, i_s)$ in $(1, \cdots, q)$.*

Then the determinant D in $\boldsymbol{M}$ with rows $i_1, \cdots, i_s$ and columns $j_1, \cdots, j_s$ equals zero if and only if the determinant D' with rows $1, \cdots, q$ and columns $j_1, \cdots, j_s$, $n - q + i'_1, \cdots, n - q + i'_{q-s}$ equals zero, or, if and only if there exists a circuit P in $\boldsymbol{M'}$ containing none of the columns $e_{j_1}, \cdots, e_{j_s}$, $e_{n-q+i'_1}, \cdots, e_{n-q+i'_{q-s}}$.

In the matrix of the last $q = r(\boldsymbol{M})$ columns of $\boldsymbol{M}$, the terms along the main diagonal and only those are $\neq 0$. If we expand D' by Laplace's expansion in terms of the columns $n - q + i'_1, \cdots, n - q + i'_{q-s}$, we see at once that $D' = 0$ if and only if $D = 0$.

Suppose $D = 0$. Then there is a set of numbers $(\alpha_1, \cdots, \alpha_q)$, not all zero, with $\alpha_i = 0$ $(i \neq i_1, \cdots, i_s)$, such that

$$b_k = \alpha_1 b_{1k} + \cdots + \alpha_q b_{qk} = 0 \qquad (k = j_1, \cdots, j_s),$$

$(b_{i1}, \cdots, b_{in})$ being the i-th row of $\boldsymbol{M}$., $b_k = 0$ also for $k = n - q + i'_1, \cdots$, $n - q + i'_{q-s}$, as each term is zero for such k. The point $(b_1, \cdots, b_n)$ is in H. Any circuit given by Lemma 10 is the required circuit P.

Suppose the circuit P exists. Then it is represented by a row $(b_1, \cdots, b_n)$ in $\boldsymbol{M}$. As the first q rows of $\boldsymbol{M}$ are of rank $q = r(\boldsymbol{M})$, $(b_1, \cdots, b_n)$ can be expressed in terms of them; say $b_k = \Sigma \alpha_i b_{ik}$. As $b_k = 0$ $(k = n - q + i'_1, \cdots$, $n - q + i'_{q-s})$, certainly $\alpha_k = 0$ $(k = i'_1, \cdots, i'_{q-s})$. $D = 0$ now follows from the fact that $b_k = 0$ $(k = j_1, \cdots, j_s)$.

16. A matroid with no corresponding matrix.[14] The matroid M' has seven elements, which we name $1, \cdots, 7$. The bases consist of all sets of three elements except

(16.1) 124, 135, 167, 236, 257, 347, 456.

Defining rank in terms of bases, we have: Each set of k elements is of rank k if $k \leqq 2$ and of rank 3 if $k \geqq 4$; a set of three elements is of rank 2 if the set is in (16.1) and is of rank 3 otherwise. It is easy to see that the postulates for rank are satisfied. (R_3) in the case that N contains two elements is satisfied vacuously. For suppose $r(N + e_1) = r(N + e_2) = r(N) = 2$. Then $N + e_1$ and $N + e_2$ are both in (16.1); but any two of these sets have but a single element in common.

[14] After the author had noted that M' satisfies (C^*) and corresponds to no linear graph, and had discovered a matroid with nine elements corresponding to no matrix, Saunders MacLane found that M' corresponds to no matrix, and is a well known example of a finite projective geometry (see O. Veblen and J. W. Young, *Projective* Geometry, pp. 3-5).

If there exists a matrix $\boldsymbol{M}'$, corresponding to M', then let $\boldsymbol{M}$ be its circuit matrix. 123 is a base in M', and hence

(16.2) $$124, \quad 135, \quad 236, \quad 1237$$

form a fundamental set of circuits in M'. Let R_1, R_2, R_3, R_4 be the corresponding rows of $\boldsymbol{M}$. By multiplying in succession row 1, column 2, rows 2, 3, 4, and columns 4, 5, 6, 7 by suitable constants $\neq 0$, we bring $\boldsymbol{M}$ into the following form:

(16.3) $$\boldsymbol{M} = \left\| \begin{array}{ccc|cccc} 1 & 1 & 0 & 1 & 0 & 0 & 0 \\ 1 & 0 & a & 0 & 1 & 0 & 0 \\ 0 & 1 & b & 0 & 0 & 1 & 0 \\ 1 & c & d & 0 & 0 & 0 & 1 \\ \cdot & \cdot & \cdot & \cdot & \cdot & \cdot & \cdot \end{array} \right\| ;$$

a, b, c and d are $\neq 0$. We now apply Theorem 32 with

$$(i_1, \cdots, i_s;\ j_1, \cdots, j_s) = (1, 4;\ 1, 2), \quad (2, 4;\ 1, 3), \quad (3, 4;\ 2, 3),$$

i. e. using the circuits 347, 257, 167. This gives

$$\begin{vmatrix} 1 & 1 \\ 1 & c \end{vmatrix} = \begin{vmatrix} 1 & a \\ 1 & d \end{vmatrix} = \begin{vmatrix} 1 & b \\ c & d \end{vmatrix} = 0,$$

and hence $c = 1$, $a = d = b$. Using the circuit 456, with sets $(1, 2, 3;\ 1, 2, 3)$ gives $2a = 0$, $a = 0$, a contradiction.

In regard to this example, see the end of the paper.

APPENDIX.

Matrices of Integers mod 2.

We wish to characterize those matroids M corresponding to matrices $\boldsymbol{M}$ of integers mod 2,[15] i. e. matrices whose elements are all 0 or 1, where rank etc. is defined mod 2. We shall consider linear combinations, *chains:*

(A. 1) $$\alpha_1 e_1 + \cdots + \alpha_n e_n \qquad (\alpha\text{'s integers mod } 2)$$

in the elements of M. The α's may be taken as 0 or 1; (A. 1) may then be interpreted as the submatroid N whose elements have the coefficient 1. Conversely, any $N \subset M$ may be written as a chain. Submatroids are added

[15] See O. Veblen, "Analysis situs," 2nd ed., *American Mathematical Society Colloquium Publications*, Ch. I and Appendix 2.

(mod 2) by adding the corresponding chains (mod 2). For instance, $(e_1 + e_2) + (e_2 + e_3) \equiv e_1 + e_3 \pmod 2$.

Any sum (mod 2) of circuits in M we shall call a *cycle* in M. N is the *true sum* of $N_1, \cdots, N_s$ if these latter have no common elements and $N = N_1 + \cdots + N_s$. We consider matroids which satisfy the following postulate:

(C*) *Each cycle is a true sum of circuits.*

Postulate (C_2) is a consequence of (C*). For the cycle $P_1 + P_2$ is a submatroid containing e_2 but not e_1; The existence of P_3 now follows from (C*).

A simple example of a matroid not satisfying (C*) is given by the matroid M' at the end of § 9.

THEOREM 33. *A circuit is a minimal non-null cycle, and conversely.*

This is proved with the aid of Postulates (C_1) and (C*).

THEOREM 34. *Let $P_1, \cdots, P_q$ be a strict fundamental set of circuits in M with respect to $e_{n-q+1}, \cdots, e_n$. Then there are exactly 2^q cycles in M, formed by taking all sums* (mod 2) *of $P_1, \cdots, P_q$.*

First, each sum $P_{i_1} + \cdots + P_{i_s}$ (mod 2) is a cycle, containing $e_{n-q+i_1}, \cdots, e_{n-q+i_s}$ and elements (perhaps) from $B = e_1, \cdots, e_{n-q}$; obviously distinct sums give distinct cycles. Now let Q be any cycle in M; say Q contains $e_{n-q+k_1}, \cdots, e_{n-q+k_r}$ and elements (perhaps) from B. Set $Q' = P_{k_1} + \cdots + P_{k_r}$; then $Q + Q'$ is a cycle containing elements from B alone. But B is a base (see the proof of Theorem 10), and hence contains no circuits. Consequently $Q + Q'$ is the null cycle, and $Q = Q'$.

THEOREM 35. *As soon as the circuits of a strict fundamental set are known, all the circuits may be determined.*

This is a consequence of the last two theorems. It is to be contrasted with the final remark of § 9.

Remark. The word "strict" may be omitted in the last two theorems.

THEOREM 36. *Let $e_1, \cdots, e_n$ be a set of elements, and let $P_1, \cdots, P_q$ be any subsets such that P_i contains e_{n-q+i} and possibly elements from $e_1, \cdots, e_{n-q}$ alone. Then there is a unique matroid M satisfying* (C*), *with $P_1, \cdots, P_q$ as a strict fundamental set of circuits.*

We form the 2^q cycles of Theorem 34. Those cycles which contain no other non-null cycle as a proper subset we call circuits; in particular, $P_1, \cdots, P_q$ are circuits. To prove (C*), let Q be a non-null cycle. If it is not a circuit, it contains a circuit P as a proper subset. Q and P are sums (mod 2) from $P_1, \cdots, P_q$, hence the same is true of $Q - P$, and $Q - P$ is one of the 2^q cycles. If it is not a circuit, we again extract a circuit, etc.

This theorem furnishes a simple method of constructing all matroids satisfying (C*).

We turn now to the study of matrices of integers (mod 2)

$$\boldsymbol{M} = \begin{Vmatrix} a_{11} & \cdots & a_{1n} \\ \cdot & \cdot\ \cdot & \cdot \\ a_{m1} & \cdots & a_{mn} \end{Vmatrix} \qquad \text{(each } a_{ij} = 0 \text{ or } 1\text{)}.$$

Any linear combination (mod 2) of the columns

$$\text{(A. 2)} \qquad \alpha_1 C_1 + \cdots + \alpha_n C_n \qquad (\alpha\text{'s integers mod 2})$$

is a set of numbers $(\Sigma\alpha_i a_{1i}, \cdots, \Sigma\alpha_i a_{mi})$, which we call a *chain* (mod 2) in $\boldsymbol{M}$. As before, we may take each coefficient as 0 or 1, and we may consider any chain merely as a submatrix of $\boldsymbol{M}$. The chain is a *cycle* if each of the corresponding numbers is $\equiv 0$ (mod 2). The columns $C_{i_1}, \cdots, C_{i_p}$ are *independent* (mod 2) if there exists no set of integers $\alpha_1, \cdots, \alpha_n$ not all $\equiv 0$ (mod 2), with $\alpha_i = 0$ $(i \neq i_1, \cdots, i_p)$, such that $\Sigma\alpha_i C_i$ is a cycle, i. e. if no non-null subset of $C_{i_1}, \cdots, C_{i_p}$ is a cycle. Using this definition, the terms base, circuit, rank, nullity etc. (mod 2) can be defined as in Part I.

Let M be a set of elements $e_1, \cdots, e_n$ corresponding to $C_1, \cdots, C_n$ in $\boldsymbol{M}$, and let $e_{i_1} + \cdots + e_{i_p}$ be a circuit in M if and only if $C_{i_1}, \cdots, C_{i_p}$ is a circuit in $\boldsymbol{M}$. We shall show that *M is a matroid satisfying* (C*) *and the definitions of cycle in M and $\boldsymbol{M}$ agree.*

We show first that each circuit is a cycle in $\boldsymbol{M}$. If $C_{i_1}, \cdots, C_{i_p}$ is a circuit, then these columns are dependent; hence $\Sigma\alpha_i C_i$ is a cycle, with $\alpha_i = 0$ $(i \neq i_1, \cdots, i_p)$. Moreover $\alpha_i = 1$ $(i = i_1, \cdots, i_p)$, for otherwise a proper subset of $C_{i_1}, \cdots, C_{i_p}$ would be dependent. Hence $C_{i_1} + \cdots + C_{i_p}$ is a cycle. Next, any sum (mod 2) of circuits is a cycle, evidently. Next we prove (C*). Suppose $Q = C_{i_1} + \cdots + C_{i_p}$ is a cycle. Let $(k_1, \cdots, k_q)$ be a minimal subset of $(i_1, \cdots, i_p)$ such that $P = C_{k_1} + \cdots + C_{k_q}$ is a cycle; then P is a circuit. $Q - P$ is a cycle; from it we extract a circuit, just as above, etc. It follows from (C*) that the definitions of cycle in M and $\boldsymbol{M}$ agree. Theorems 33, 34 and 35 now apply to $\boldsymbol{M}$ also.

We are now ready to prove the final theorem:

THEOREM 37. *Let M be any matroid satisfying* (C*). *Suppose $\rho(M) = n$, and $e_1 + \cdots + e_{n-q}$ is a base. Then if $\mathbf{M}_1$ is any matrix of integers* (mod 2) *with $n - q$ columns which are independent* (mod 2), *columns $C_{n-q+1}, \cdots, C_n$ can be adjoined in a unique manner to $\mathbf{M}_1$, forming a matrix $\mathbf{M}$ of which the corresponding matroid is M.*

Let $P_1, \cdots, P_q$ be a strict fundamental set of circuits in M with respect to $e_{n-q+1}, \cdots, e_n$ (Theorem 9). Say $P_1 = e_{i_1} + \cdots + e_{i_p} + e_{n-q+1}$. Set $C_{n-q+1} \equiv C_{i_1} + \cdots + C_{i_p}$ (mod 2); this determines C_{n-q+1} as a column of 0's and 1's so that $P'_1 = C_{i_1} + \cdots + C_{i_p} + C_{n-q+1}$ is a circuit. (P'_1 is a cycle; (C*) shows that it is a single circuit, as $C_1 + \cdots + C_{n-q}$ contains no circuit.) C_{n-q+1} evidently must be chosen in this manner. We choose the remaining columns of $\mathbf{M}$ similarly. Let M' be the matroid corresponding to $\mathbf{M}$. Then $P'_1, \cdots, P'_q$ is a strict set of circuits in M'. These same sets form a strict set in M; hence, by Theorem 35, the circuits in M' correspond to those in M. Consequently $M' = M$, completing the proof.

We end by noting that the matroid M' of § 16 satisfies Postulate (C*) but corresponds to no linear graph. For letting 123 be a base and (16. 2) a fundamental set of circuits and determining the matroid as in Theorem 36, we come out with exactly M'. A corresponding matrix of integers mod 2 is constructed from (16. 3) with $a = b = c = d = 1$; we interchange rows and columns in the left-hand portion, leave out the last row and column of the right-hand portion, and interchange these two parts. (The relation $2a = 0$ is of course true mod 2.)

On the other hand, it is easily seen that if the element 7 is left out, there is a corresponding graph, which must be of the following sort: It has four vertices a, b, c, d, and the arcs corresponding to the elements $1, \cdots, 6$ are

$$ab, \quad ac, \quad ad, \quad bc, \quad bd, \quad cd.$$

There is no way of adding the required seventh arc.

The problem of characterizing linear graphs from this point of view is the same as that of characterizing matroids which correspond to matrices (mod 2) with exactly two ones in each column.

HARVARD UNIVERSITY.

Reprinted from
Amer. J. Math. **57** (1935), 509–533

Reprinted from Duke Mathematical Journal
Vol. 7, December, 1940

THE DISSECTION OF RECTANGLES INTO SQUARES

By R. L. Brooks, C. A. B. Smith, A. H. Stone and W. T. Tutte

Introduction. We consider the problem of dividing a rectangle into a finite number of non-overlapping squares, no two of which are equal. A dissection of a rectangle R into a finite number n of non-overlapping squares is called a *squaring* of R of *order* n; and the n squares are the *elements* of the dissection. The term "elements" is also used for the lengths of the sides of the elements. If there is more than one element and the elements are all unequal, the squaring is called *perfect*, and R is a *perfect rectangle*. (We use R to denote both a rectangle and a particular squaring of it.) Examples of perfect rectangles have been published in the literature.[1]

Our main results are:

Every squared rectangle has commensurable sides and elements.[2] (This is (2.14) below.)

Conversely, every rectangle with commensurable sides is perfectible in an infinity of essentially different ways. (This is (9.45) below.) (**Added in proof.** Another proof of this theorem has since been published by R. Sprague: Journal für Mathematik, vol. 182(1940), pp. 60–64; Mathematische Zeitschrift, vol. 46(1940), pp. 460–471.)

In particular, we give in §8.3 a perfect dissection of a square into 26 elements.[3]

There are no perfect rectangles of order less than 9, and exactly two of order 9.[4] (This is (5.23) below.)

The first theorem mentioned is due to Dehn, who remarked[5] that the difficulty of the problem is the semi-topological one of characterizing how the elements fit together. This is overcome here in §1 by associating a certain linear graph (the "normal polar net") with each "oriented" squared rectangle. The metrical properties of the squared rectangle are found to be determined by a certain flow of electric current through this network. Accordingly, in §2 we collect the relevant results from the theory of electrical networks. In particular, the elements of the squared rectangle can be calculated from determinants formed from the incidence matrix of the network. In §3, the elements are expressed in a different way, in terms of the subtrees of the network. This leads

Received May 7, 1940. We are indebted to Dr. B. McMillan, of Princeton University, for help with the diagrams.

[1] A bibliography is given at the end of this paper. Numbers in square brackets refer to this bibliography.

[2] Cf. [6], p. 319.

[3] This disproves a conjecture of Lusin; cf. [10], p. 272. For an independent example of a perfect square (published while this paper was in preparation) see [13].

[4] Partly confirming and partly disproving a conjecture of Toepken (see [18]).

[5] [12], p. 402.

to some relations between determinants and the subtrees of a network, and to some duality theorems. In §4, these duality theorems are applied to prove the converse of §1: that to any "polar net" corresponds a squared rectangle; and moreover, it is shown that (roughly speaking) the networks which correspond to the same squared rectangle in its two orientations are dual. In §5, the polar net is used to determine all the squared rectangles of a given order; in particular, the "simple" perfect rectangles of orders <12 are tabulated. §6 contains some theorems on the factorization properties of the elements of a squared rectangle, as determined in §2; as corollaries, we have some sufficient conditions for a squared rectangle to be perfect ((6.20), (6.21)). In §7, we give "non-uniqueness" constructions—in §7.1, of rectangles which can be dissected into the same elements in essentially different ways, and, in §7.2, of pairs of squared rectangles having the same shape but different elements. These constructions depend mostly on considerations of symmetry or duality in the corresponding networks. In §8, the results of §7.2 are used to give "perfect" squares; and in §9, a whole family of "totally different" perfect squares is worked out, and this leads to the result that every rectangle whose sides are commensurable is perfectible.

We conclude (§10) by outlining some generalizations—notably "rectangled rectangles", squared cylinders and tori, "triangulated" equilateral triangles, and "cubed cubes". We prove in particular that no "perfect" dissection of a rectangular parallelopiped into cubes is possible.[6]

1. The net associated with a squared rectangle

1.1. In any squaring of a rectangle R,[7] the sides of all the elements and of R will clearly be parallel to two perpendicular lines. We *orient* R by choosing one of these lines to be "horizontal" (i.e., parallel to the x-axis). The distinction between this configuration, and its reflections in the coördinate axes, is unimportant; but it is convenient to distinguish it from R in the other orientation (obtained by rotating R through an angle of $\frac{1}{2}\pi$), called the *conjugate* of R.

Consider the point-set formed by the horizontal sides of the elements of R. Its connected components will be horizontal line-segments (each consisting of a set of horizontal sides of elements of R); enumerate them as $p_1, \cdots, p_N$, say, where p_1, p_N are the upper and lower edges of R. Take N points $P_1, \cdots, P_N$ in the plane. Let E be an element of R; its upper edge will lie in some one of $p_1, \cdots, p_N$, say p_i : similarly, its lower edge will lie in p_j $(i \neq j)$. Join the points P_i, P_j by a line (simple arc) e. By taking all elements E of R, we get a network (linear graph) on $P_1, \cdots, P_N$ as vertices and the e's as 1-cells. Figure 1 provides an example.

The points P_1, P_N are the *poles* of the network. We can arrange the joins e in such a way that

[6] Answering a question raised by Chowla in [5].

[7] Throughout, all squares are supposed to have positive sides; thus zero elements are excluded.

(1.11) the network is realizable in a plane with no two 1-cells intersecting (except at a vertex).

(1.12) No circuit encloses a pole.

For we can realize the network as follows. Take P_i to be the mid-point of p_i; and take $\epsilon > 0$ sufficiently small. For each element E, take the vertical segment which bisects E, and cut off a length ϵ from each end, leaving a segment AB, say. Join the upper end of AB, A, to the P_i corresponding to the upper boundary of E, by a straight line-segment, and similarly join B to P_j corresponding to the lower edge of E. The path P_iABP_j is defined to be e. It is now easily verified that (1.11) and (1.12) hold.

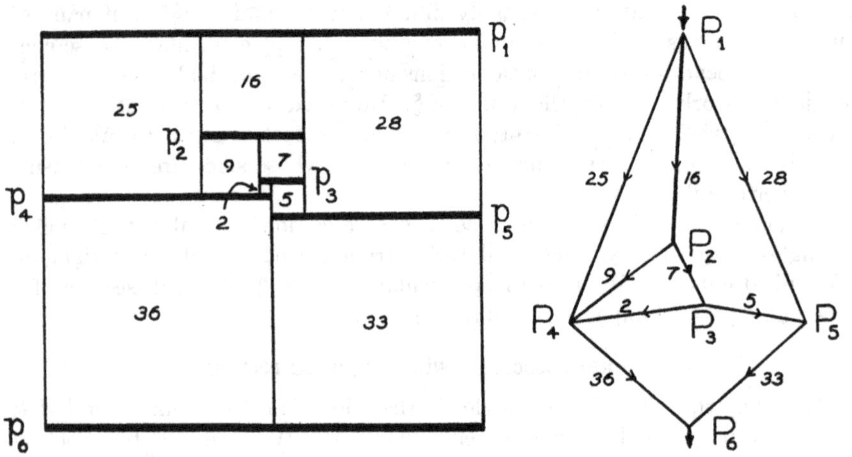

FIG. 1

Also we have clearly

(1.13) The network is connected.

Remark. In general there may be several 1-cells joining two vertices, though not if the squaring is perfect.

(1.14) DEFINITIONS. A network with more than one vertex, satisfying (1.11) and (1.13), is called a *net*. If two of the vertices of a net are assigned as "poles", and (1.12) is satisfied, the net is a *polar net* (p-net). The network constructed above is the *normal polar net* of the squared rectangle.

1.2. **Kirchhoff's laws.** With each 1-cell $e = P_iP_j$ of our normal p-net, associate the length of the side of the corresponding element E, directed from the "upper" point (P_i) to the "lower" point (P_j); call this the *current* in e. Then

(1.21) Except at the poles, the total current flowing into P_i is zero.

(For current flowing in = length of p_i = current flowing out.)

(1.22) The algebraic sum of the currents round any circuit is zero.
(For the current in a "wire" $e = P_iP_j$ is the vertical height of p_i above p_j.)

(1.23) The sum of the currents flowing into P_1 = length of horizontal side of R = sum of the currents flowing out of P_N.

(1.21) and (1.22) are the usual Kirchhoff laws for a flow of electric current in the net from P_1 to P_N, it being assumed that each 1-cell is a wire of unit conductance.

["Rectangulations" of rectangles can be dealt with similarly; the conductance of e will then be the ratio of the sides of E.]

Equations (1.21) and (1.22) can be interpreted differently. Consider the cellular 2-complex formed by embedding our p-net in a 2-sphere. We have on it a *Kirchhoff chain* (K-chain), viz., the 1-chain Σ (current in e)$\cdot e$. Then

(1.24) The K-chain is a cycle modulo its poles. (This $\leftrightarrow$ (1.21).)

(1.25) The K-chain is an absolute cocycle. (This $\leftrightarrow$ (1.22).)

2. Some results from the electrical theory of networks

2.1. In the previous section, we reduced the study of squared rectangles to the study of certain flows of electricity in networks. Here we collect the results on electrical networks in general which will be useful later.

Let $\mathfrak{N}$ be a connected network whose vertices are $P_1, \cdots, P_N$ ($N \geqq 2$). The 1-cells are called *wires*; there may be more than one wire joining two vertices, and there may be wires whose two ends coincide. With each wire is associated a positive real number, its *conductance*. We define a matrix $\{c_{rs}\}$ as follows:

(2.11) If $r \neq s$,

$$-c_{rs} = \begin{cases} \text{sum of conductances of all wires joining } P_r, P_s, \\ 0 \text{ if there are no such wires;} \end{cases}$$

$$c_{rr} = \text{sum of conductances of all wires joining } P_r \text{ to } \textit{other} \text{ vertices.}$$

Thus

(2.12)
$$c_{rs} = c_{sr}, \qquad \sum_r c_{rs} = 0.$$

We make the convention that if $\mathfrak{N}$ is explicitly called a *net*, all its conductances are 1. (The matrix $\{-c_{rs}\}$ is then the product of the usual incidence matrix of the oriented network, with its transpose.)

Let us return to the general case; from (2.12) we can readily show that all first cofactors of $\{c_{rs}\}$ are equal. We call their common value the *complexity* of the network, and denote it by C. It is known that $C > 0$. (An independent proof is given below; see (3.14).)

The second cofactor obtained by taking the cofactor of the component c_{su} in the cofactor of c_{rt} ($r \neq s$, $t \neq u$) is denoted by $[rs, tu]$. (If $N = 2$, $[12, 12] = 1 = -[21, 12]$.) We put $[rr, tu] = 0 = [rs, tt]$. The $[rs, tu]$'s are called the *transpedances* (generalized transfer impedances) of $\mathfrak{N}$.

Consider a flow of current from P_x to P_y (the *poles*). The currents in the wires satisfy (1.21); the potential differences (P.D.'s) satisfy the analogue of (1.22); and the total current I is given by (1.23). It is known[8] that these conditions (with Ohm's Law) determine the flow *uniquely* when I is given, and that

(2.13) P.D. from P_r to P_s when current I enters at P_x and leaves at P_y is $[xy, rs] \cdot I/C$.

It is convenient to take $I = C$, thus fixing the values of the currents and P.D.'s of the network. The flow with $I = C$ is called the *full flow*; and we speak of the "full currents", etc.

Applying this to the normal p-net of a squared rectangle, where all conductances are 1, so that all the transpedances are integers, we see from (1.21)–(1.23) and (2.13) that

(2.14) *Every squared rectangle has commensurable sides and elements.*

The H.C.F. of the full currents of a p-net is the *reduction* ρ of the p-net. Notice that ρ is also the H.C.F. of all the full P.D.'s of the p-net. The flow with $I = C/\rho$ is the *reduced* flow.

2.2. **Properties of the transpedances.** We have

(2.21) $$[rs, tu] = [tu, rs] = -[sr, tu],$$

(2.22) $$\sum_x c_{tx} \cdot [rs, tx] = C \cdot (\delta_{ts} - \delta_{tr}),$$

(2.23) $$[rs, tu] + [rs, uv] = [rs, tv].$$

(2.22) and (2.23) verify that (2.13) does in fact provide a solution of the Kirchhoff equations, and that the current at each pole is C.

We call $[rs, rs]$ the *impedance* of r, s, and write it $V(rs)$. Then

(2.24) $$V(rs) = V(sr), \qquad V(rr) = 0,$$

(2.25) $$2 \cdot [rs, tu] = V(ru) + V(st) - V(su) - V(rt) \qquad \text{(from (2.23))},$$

(2.26) $$[rs, tu] + [tr, su] + [st, ru] = 0.$$

2.3. **Alterations to the network.** For later use, we need to know the effect on the transpedances of making certain alterations to the network $\mathfrak{N}$.

I. Introduce a new wire joining a vertex P_m of $\mathfrak{N}$ to a new vertex P_0. Let the new wire have conductance c; then, in the new network $\mathfrak{N}_1$,

$$C_1 = cC, \qquad V_1(m0) = C;$$

(2.31) $$[ab, xy]_1 = c \cdot [ab, xy] \quad \text{if } 0 \neq a, b, x, y; \qquad [ab, m0]_1 = 0;$$

$$V_1(x0) = V_1(xm) + V_1(m0) = c \cdot V(xm) + C.$$

[8] [8], pp. 324–331.

These results are immediate from the definitions.

II. Identify two points P_x, P_y and ignore any wire that may have joined them. In the new network $\mathfrak{N}_2$,

$$C_2 = [xy, xy] = V(xy) \qquad \text{(from the definitions)}, \tag{2.32}$$

$$[rs, tu]_2 = \frac{[rs, tu] \cdot V(xy) - [rs, xy] \cdot [tu, xy]}{C} \tag{2.33}$$

(for these expressions satisfy Kirchhoff's laws for $\mathfrak{N}_2$, and agree with (2.32)). In particular,

$$V_2(rs) = \frac{V(rs) \cdot V(xy) - [rs, xy]^2}{C}. \tag{2.34}$$

((2.33) may be generalized as follows: C^n *divides the* $(n + 1)$*-th order determinants formed as minors of the matrix of transpedances.* This is an extension of the Cauchy-Sylvester identity.[9])

III. Introduce a new wire of conductance c in $\mathfrak{N}$, joining P_x and P_y. In the new network $\mathfrak{N}_3$ we have, from their definitions as determinants,

$$C_3 = C + c \cdot V(xy) = C + c \cdot C_2 \qquad \text{(from (2.32))}; \tag{2.35}$$

$$[rs, xy]_3 = [rs, xy]; \qquad \text{in particular, } V_3(xy) = V(xy). \tag{2.36}$$

Also

$$[rs, tu]_3 = [rs, tu] + c \cdot [rs, tu]_2 ; \tag{2.37}$$

for III is a combination of I and II. We introduce a new vertex P_0, join it to P_x by a wire of conductance c, and identify P_y and P_0. This enables us to verify (2.37).

3. Subtrees of a network: duality

We shall now characterize the complexity (and hence the transpedances) of a network more topologically, in terms of the "subtrees" of the network. This enables us to prove some duality theorems which will be useful later (§4) and are of interest in themselves.

3.1. As in the previous section, let $\mathfrak{N}$ be a connected network with conductances. By a *subnetwork* $\mathfrak{M}$ of $\mathfrak{N}$, we mean a network consisting of *all* the vertices of $\mathfrak{N}$ and some (or all) of the wires of $\mathfrak{N}$. A *subtree* of $\mathfrak{N}$ is a subnetwork which is a "tree"; i.e., is connected and has no circuits. Enumerate all the subtrees of $\mathfrak{N}$; let M_r be the product of the conductances of the wires of the r-th tree. Define H by:

$$H = \sum_r M_r. \tag{3.11}$$

[9] [19], p. 87.

When a new wire of conductance c is inserted joining P_x, P_y, let "H" for the new network be H_3; and when P_x, P_y are identified (as in §2.3, II), let "H" become H_2. Clearly,

$$H_3 = H + c \cdot H_2 . \tag{3.12}$$

But this is the relation which holds between the complexities of these networks (2.35).

Also, for a connected network with only two vertices, C = sum of conductances of the wires joining P_1 to $P_2 = H$. Hence, by induction on the numbers of vertices and wires in $\mathfrak{N}$, we have:[10]

(3.13) THEOREM. *For any connected network with more than one vertex, having conductances assigned to the* 1*-cells*, $C = H$.

If the conductances are all positive, we clearly have $H > 0$. This proves

$$C > 0. \tag{3.14}$$

This interpretation of complexity in terms of trees enables us, if (2.32) is used, to express $V(xy)$ in terms of the trees of networks formed from $\mathfrak{N}$ by identifying certain pairs of its vertices, and hence in terms of the "tree-pairs" of $\mathfrak{N}$ (formed by omitting one wire from a subtree). Hence, using (2.25), we can get similar interpretations for all the transpedances.

In the case of a net, all conductances are 1, so H = number of subtrees of $\mathfrak{N}$; thus (3.13) gives an explicit formula for the number of subtrees of any connected network, in terms of the incidences of the network.

3.2. **Duality relations.** Now suppose that $\mathfrak{N}$ can be imbedded in a 2-sphere, and let $\mathfrak{N}^*$ be its dual on the sphere. The conductivity of a wire of $\mathfrak{N}^*$ is defined to be the reciprocal of that of the dual wire of $\mathfrak{N}$. Thus $\mathfrak{N}^{**} = \mathfrak{N}$, and the dual of a net is a net. The *codual* of a subnetwork $\mathfrak{M}$ of $\mathfrak{N}$ is the subnetwork $\mathfrak{M}^c$ of $\mathfrak{N}^*$ whose 1-cells are those *not* dual to any wire of $\mathfrak{M}$. Clearly $\mathfrak{M}^{cc} = \mathfrak{M}$.

It can be shown that

(3.21) A subnetwork $\mathfrak{M}$ of $\mathfrak{N}$ is a tree if and only if both $\mathfrak{M}$ and $\mathfrak{M}^c$ are connected.

Hence

(3.22) If $\mathfrak{M}$ is a subtree of $\mathfrak{N}$, then $\mathfrak{M}^c$ is a subtree of $\mathfrak{N}^*$; and conversely.

Let M_r^* equal the product of conductances of wires in the subtree (of $\mathfrak{N}^*$) which is codual to the r-th subtree of $\mathfrak{N}$. Let ω equal the product of conductances of all wires of $\mathfrak{N}$. Then, clearly,

$$M_r = \omega \cdot M_r^* . \tag{3.23}$$

[10] This result is due in principle to Kirchhoff ([9], p. 497). Cf. also [3].

Hence, using (3.22), (3.11), and (3.13), we have

(3.24) If C^* is the complexity of the dual of $\mathfrak{N}$, $\omega \cdot C^* = C$.

In particular, we have proved

(3.25) THEOREM. *Dual nets have equal complexities.*

3.3. **Polar duality.** Let $\mathcal{P}$ be a p-net. By (1.12), we can join the poles of $\mathcal{P}$ by an extra wire e_0, without violating (1.11). The resulting net $\mathcal{C}$ is called the *completed* net (c-net) of $\mathcal{P}$. Let $\mathcal{C}$ be imbedded in a 2-sphere, and let $\mathcal{C}^*$ be the dual of $\mathcal{C}$. From $\mathcal{C}^*$ omit e_0^*, the dual of e_0, and take the ends of e_0^* as poles. We get a p-net $\mathcal{P}'$, the *polar dual* of $\mathcal{P}$.[11]

Clearly $\mathcal{P}'' = \mathcal{P}$.

(The importance of polar duality arises from the fact that, as we shall show in §4.3, polar dual p-nets correspond to the same squared rectangle in its two "orientations" (§1.1).)

The p-dual (polar dual) of any 1-chain on $\mathcal{P}$ is defined in the obvious way (as having the same multiplicity on e_i^* as the given chain has on e_i).

(3.31) THEOREM. *The p-dual of the full Kirchhoff chain on a p-net $\mathcal{P}$ is the full Kirchhoff chain on the p-dual p-net $\mathcal{P}'$.*

Proof. We use $\mathfrak{S}\mathfrak{N}$ to denote the cellular 2-complex formed by a network $\mathfrak{N}$ imbedded in a 2-sphere. F, δ are (as usual) boundary and coboundary operators, and * denotes duality with respect to the 2-sphere.

By (1.24), (1.25), the full K-chain $\mathcal{K}$ on $\mathcal{P}$ is a cycle relative to P_1, P_N (the poles of $\mathcal{P}$), and an absolute cocycle on $\mathfrak{S}\mathcal{P}$. Hence, in $\mathfrak{S}\mathcal{C}$ (where $\mathcal{C}$ is the completed net of $\mathcal{P}$) $\mathcal{K}$ is

(i) a relative cycle mod P_1, P_N, and

(ii) a relative cocycle mod the two 2-cells, say σ_1, σ_2, which have incidence with e_0, the "extra" join.

Dualizing, in $\mathfrak{S}\mathcal{C}^*$, we see that $\mathcal{K}^*$ is

(i) a relative cocycle mod the 2-cells P_1^*, P_N^* and

(ii) a relative cycle mod σ_1^* and σ_2^*, the poles of $\mathcal{P}'$.

But $\mathcal{K}^*$ has zero multiplicity on e_0^*, for $\mathcal{K}$ has zero multiplicity on e_0. Hence $\mathcal{K}^*$ is (from (i)) a cycle on $\mathcal{P}'$ mod its poles, and (from (ii)) a cocycle on $\mathfrak{S}\mathcal{P}'$ mod the 2-cell consisting of P_1^* and P_N^* together. But a single 2-cell cannot be a coboundary; for, dualizing, this would require a single vertex to be a boundary. Hence $\mathcal{K}^*$ is an absolute cocycle on $\mathfrak{S}\mathcal{P}'$, besides being a cycle mod its poles. So $\mathcal{K}^*$ is a K-chain on $\mathcal{P}'$.

Let $\mathcal{K}'$ be the full K-chain on $\mathcal{P}'$; thus $\mathcal{K}^* = k \cdot \mathcal{K}'$, for some k.

[11] There may be several ways of placing e_0 on the sphere, and consequently several polar duals of $\mathcal{P}$ (differing, however, only trivially). We suppose that one of these is chosen arbitrarily. In the open plane, a convention will be introduced to make $\mathcal{P}'$ unique; cf. §§4.2, 4.3.

Let $\mathcal{P}$ have complexity C, and $V(1N) = V$. Let the corresponding numbers for $\mathcal{P}'$ be C', V'. Using (2.22) (with $c_{tx} = 1$), we have, in $\mathcal{P}$,

$$F(\mathcal{K}) = C \cdot (P_1 - P_N).$$

Therefore, in $S\mathcal{P}'$, $\delta(\mathcal{K}^*) = C \cdot (P_1^* - P_N^*)$ (these cells being oriented suitably). So

$$C = \begin{cases} \text{sum of currents around } F(P_1^*) \text{ in } \mathcal{K}^*, \\ \text{sum of currents along a path joining the end-points of } e_0^*, \\ \text{total P.D. between the poles of } \mathcal{P}', \text{ in the flow } \mathcal{K}^*. \end{cases}$$

Thus

(3.32) $$C = k \cdot V'.$$

Similarly,

(3.33) $$C' = (1/k) \cdot V.$$

Now, by (2.35), the complexity of $\mathcal{C}$ is $C + V$. Similarly, the complexity of $\mathcal{C}^*$ is $C' + V' = k \cdot (C + V)$, by (3.32), (3.33). But by (3.25) these complexities are equal. Hence $k = 1$, and $\mathcal{K}^*$ is the *full* K-chain on $\mathcal{P}'$.

4. The correspondence between p-nets and squared rectangles

4.1. We now sketch a proof showing that to each p-net corresponds a squared rectangle. This correspondence is many-one and is clarified by introducing the "normal form" of a p-net (§4.2). We can then set up a 1-1 correspondence between classes of p-nets (having the same normal form) and "oriented" squared rectangles, and can prove that p-dual p-nets correspond to "conjugate" squared rectangles. (Cf. §1.1.)

(4.10) LEMMA. *For a K-chain in a p-net $\mathcal{P}$, whose poles are P_1, P_N (suitably numbered),*

(4.11) *the potential of each vertex lies between the potentials of the poles;*

(4.12) *no currents go into P_1, or out of P_N;*

(4.13) *at a vertex P_i, there is an angle (in the plane) containing all ingoing currents, whose reflex contains all outgoing currents;*

(4.14) *on the boundary of a 2-cell of $S\mathcal{P}$, there are two vertices P_i, P_j such that no current round this boundary goes from P_j towards P_i.*

(We make the convention that zero currents do not go in or out.)

Proof. Let P_i be any vertex, and suppose a current goes into P_i. Then a current goes out of P_i along at least one wire, ending at P_j, say; and so on, until we reach a pole P_N (say). All this time the potential has been falling, so P_N is eventually reached; and the potential of P_i is thus not less than that of P_N. If all the currents at P_i are zero, we can connect P_i to a vertex P_k at which not all currents are zero, by a path of zero currents; and P_i, P_k have the same potential. Thus in all cases the potential of P_i is not less than that of P_N; and similarly it is not greater than that of P_1. This proves (4.11).

(4.12) follows at once from (4.11).

(4.13) has been proved for the poles; so let $i \neq 1, N$, and suppose that two outgoing currents at P_i separate (in the plane) two ingoing ones. As in the proof of (4.11), we can continue each of the first two wires into a path down to P_N, along which the current falls; and similarly we can extend the other two wires into paths of rising potential up to P_1. Hence one of the two former paths must intersect one of the latter again, say in P_j $(i \neq j)$. The potential of P_j is both less than and greater than the potential of P_i. This is a contradiction, and so (4.13) is proved.

(4.14) follows from (4.13) and (4.12) by dualizing, if we use (3.31).

4.2. **Normal form of a p-net.** Let $\mathcal{P}$ be a p-net imbedded in the open plane in such a way that its poles, P_1, P_N, can be joined in the "outside region" of $\mathcal{SP}$. (That is, $\mathcal{P}$ is first imbedded in the closed 2-sphere, an extra join e_0 of the poles is inserted, and the "point at infinity" is then taken to be in the 2-cell of $\mathcal{SP}$ which contains e_0.) We define the *normal form* of $\mathcal{P}$, as so placed in the plane, as follows:

Consider any (not identically zero) K-chain $\mathcal{K}$ on $\mathcal{P}$. Some currents may be zero; delete the corresponding wires, and delete all vertices at which all currents are zero. Since $C > 0$, we are left with a p-net still, having P_1, P_N as poles. Using (2.31), (2.37), (2.36) (with $c = 1$), we see that $\mathcal{K}$ is a K-chain for the new p-net $\mathfrak{N}$. Next take each *finite* 2-cell of $\mathcal{S}\mathfrak{N}$, and consider the vertices on its boundary. By (4.14), the 2-cell with its boundary is homeomorphic to a convex polygon which has one highest point and one lowest point, and in which the potentials of the vertices increase with their heights. Moreover, they increase *strictly*; for now no currents are zero. Hence equipotential vertices on this boundary occur at most in pairs, which can all be respectively identified by a deformation across the 2-cell. Making all these identifications for all the finite 2-cells, we end with a p-net $\mathfrak{N}_0$, on P_1, P_N as poles, on which $\mathcal{K}$ is still a K-chain (by (2.33)). And there are now no two vertices at the same potential which can be joined without crossing some wire of $\mathfrak{N}_0$, or separating the poles in the "outside" region. In particular, there are no zero currents. $\mathfrak{N}_0$ is called the *normal form* of $\mathcal{P}$, in its given imbedding in the plane.

Notice that, while we have proved that $\mathcal{P}$, $\mathfrak{N}_0$ have the same *reduced* K-chains, they need not have the same full K-chains.

It is easily seen that the normal p-net of a squared rectangle is its own normal form.

4.3. We next prove

(4.31) Theorem. *To every p-net $\mathcal{P}$ in the open plane corresponds a squared rectangle R, whose normal p-net is the normal form of $\mathcal{P}$. Polar dual p-nets correspond to conjugate squared rectangles.*

(The polar dual of a p-net $\mathcal{P}$ in the open plane is itself put in the open plane in the obvious way—e_0^* is taken to be in the "outside".)

Proof. Consider the full K-chain $\mathcal{K}$ on $\mathcal{P}$ and its dual, the full K-chain on the p-dual net $\mathcal{P}'$. (By (3.31).) Let $\mathcal{P}$ have complexity C, and let the P.D. between its poles be V ($= V(xy)$). Thus ((3.32), (3.33)) the analogous numbers for $\mathcal{P}'$ are V and C respectively. We can take the lowest potentials in $\mathcal{P}$ and $\mathcal{P}'$ to be zero. Suppose a wire e in $\mathcal{P}$ has its end-points at potentials V_1, V_2, and its dual e^* has its end-points at potentials V_1', V_2'. If μ is a number such that $V_1 < \mu < V_2$, we say that *e comprises* $(\ , \mu)$; and if λ is such that $V_1' < \lambda < V_2'$, then *e comprises* $(\lambda, \ )$. If both relations are true, we say that *e comprises* (λ, μ).

Now, observing that $V_2 - V_1 =$ current in $e =$ current in $e^* = V_2' - V_1'$, we construct a squared rectangle R as follows: In a rectangle of height V and base C, we take, for each wire e of $\mathcal{P}$, the (closed) *square* E whose horizontal sides are at a height V_1, V_2 above the base (x-axis) and whose vertical sides are at a distance V_1', V_2' to the right of the left-hand vertical side (y-axis). If the current in e is zero, this square reduces to a single point, and is omitted.

Let $\lambda \neq$ any potential of a vertex of $\mathcal{P}'$, and $\mu \neq$ any potential of a vertex of $\mathcal{P}$. Then, if $0 < \lambda < C$, and $0 < \mu < V$, we have the following:

The wires (of $\mathcal{P}$) comprising $(\lambda, \ )$ form a single path from pole to pole, along which the direction of the current is constant. For, by (4.12) and duality, there is just one such wire terminating at each pole; and from (4.14), if one such wire carries current to a vertex, then just one such wire carries current from that vertex, and no more such wires terminate at that vertex.

Along this path, the potential increases steadily from pole to pole; also, by choice of λ, the currents along the path are non-zero. Hence just one wire in it comprises $(\ , \mu)$. So just one wire of $\mathcal{P}$ comprises (λ, μ). Thus the point of coördinates (λ, μ) belongs to just one of the squares E. It follows that the whole rectangle is filled completely and without overlap (except of boundaries of squares).

It is easy to see that the normal p-net of the squared rectangle so constructed is—to within reflection in the axes (which we always disregard)—the normal form of $\mathcal{P}$. Also, it is clear from the construction that the squared rectangle assigned to $\mathcal{P}$ differs from that assigned to $\mathcal{P}'$ only by interchange of horizontal and vertical; i.e., the two squared rectangles are conjugate.

In this way, we have a 1-1 correspondence between classes of p-nets in the plane having the same normal form, and "oriented" squared rectangles.

Definitions. As suggested by (4.31), the complexity of a p-net is called its (full) *horizontal side* (often written H instead of C); and the full P.D. between its poles is its *vertical side* (V). The "full elements" and "full sides" of a squared rectangle refer to those of its normal p-net. The "reduced elements" will be the same for all corresponding p-nets.

4.4. Defining a *cross* as a point of a squared rectangle which is common to four elements, and an "uncrossed" squared rectangle as one which has no crosses, we have:

The normal p-nets of uncrossed conjugate squared rectangles are p-duals.

For let $\mathcal{P}$ be the normal p-net of the squared rectangle R; and let $\mathcal{P}'$ be the p-dual of $\mathcal{P}$. Let $\mathcal{Q}$ be the normal p-net of the conjugate R' of R; thus, from

§4.3, $\mathcal{Q}$ is the normal form of $\mathcal{P}'$. Now, in deriving the normal form of $\mathcal{P}$ (as in §4.2) there are no zero currents to suppress; and there are no identifications of vertices possible, as otherwise R', and hence R, would have a cross. So $\mathcal{P}' = \mathcal{Q}$. That is, $\mathcal{P}$ and $\mathcal{Q}$ are p-duals.

(This result could be extended to crossed squared rectangles by making a suitable convention modifying the normal p-net when crosses are present; e.g., by regarding a cross as an "element of side zero".)

5. Enumeration of squared rectangles

5.1. **Computation.** To find all the squared rectangles of a given order n, we have only to make a list of all p-nets having n wires. There is no difficulty in this, if n is not too large. We can save some labor by noting that p-dual nets give essentially the same rectangles; also we can assume that no part of a net, not containing a pole, is joined to the rest only at one vertex. (For the currents in this part would all be zero, whereas we can restrict ourselves to "normal forms".) A convenient way of carrying out the calculations is to consider the c-nets. From each net of $n + 1$ wires, we remove one wire and take its end-points as poles in the remaining net (if it is a net; i.e., is connected). Dual c-nets give rise to pairs of polar dual p-nets; so we need consider only half the c-nets. The working can be simplified by a proper use of §2.2. In practice, the Kirchhoff equations are best solved directly (without using determinants); a single determinant then gives the *full* elements for all the p-nets derived from one c-net.

It follows from §2.3 that all p-nets derived as above from the same c-net will have the same (full) semiperimeter, viz., the horizontal side of the c-net; and that two p-nets which differ only in the choice of poles, and their (non-polar) duals, all have the same (full) horizontal sides, viz., the complexity of the nets. (By (3.25).) Thus a number which appears in the $(n + 1)$-th order as a side appears (several times) in the n-th order as a semiperimeter. These facts are illustrated in the table below (§5.3).

5.2. **The perfect rectangles of least order. "Simple" perfect rectangles**

(5.21) A squared rectangle which contains a smaller squared rectangle (and any p-net corresponding to it) is called *compound*; all other squared rectangles and p-nets are *simple*. A p-net $\mathcal{P}$, without zero currents, which has a part $\mathcal{Q}$ such that $\mathcal{Q}$ contains more than one wire, $\neq \mathcal{P}$, is joined to the rest at only two vertices Q_1, Q_2, and contains no pole (except perhaps for Q_1 or Q_2) is compound. For $\mathcal{Q}$ must be connected; and the squared rectangle corresponding to $\mathcal{P}$ will contain the smaller squared rectangle which corresponds to $\mathcal{Q}$ (with Q_1, Q_2 as poles).

(5.22) "*Trivial*" *imperfection.* If a p-net has two equal non-zero currents, it is *imperfect*, and these currents constitute an "imperfection". (This is equivalent to saying that the corresponding squared rectangle is not perfect.) If a p-net has a part, not containing a pole, joined to the rest by only two wires, or if it has a pair of vertices joined by two (or more) wires, these two wires will clearly have equal currents. If these currents are non-zero, the resulting imperfection is said to be *trivial*. A p-net which has a non-trivial imperfection is called *non-trivially imperfect*. A non-trivially imperfect p-net may or may not have a trivial imperfection.

We now have the theorem:

(5.23) *The c-net derived from a simple perfect rectangle has no part (consisting of more than one wire and of less than all but one wire) joined to the rest at less than three vertices; and the same is true of its dual.*

For the normal p-net of the simple perfect squared rectangle (or of the conjugate squared rectangle) will otherwise have a zero current, or a trivial imperfection, or be compound.

A perfect rectangle of the smallest possible order must evidently be simple. Applying (5.23) to the method of §5.1, we readily find that

There are no perfect rectangles of order less than 9, *and exactly two perfect rectangles of order* 9.

Of the latter, one is well known;[12] the other is, we believe, new and has been drawn in Figure 1.

Below, we give a list of the simple perfect rectangles of orders 9–11. The compound perfect rectangles of these orders follow trivially.

5.3. Table of simple perfect rectangles.

Order	Full Sides	Semi-perimeter	Description of Polar Net (current from P_a to P_b = ab)	Reduction
9	66, 64	130	$ab = 30, ac = 36, bd = 14, cd = 8, be = 16, de = 2, ef = 18, df = 20, cf = 28.$	2
	69, 61	130	$ac = 25, ab = 16, ae = 28, bc = 9, bd = 7, dc = 2, de = 5, cf = 36, ef = 33.$	1
10	114, 110	224	$ab = 60, ac = 54, cb = 6, ce = 22, cd = 26, be = 16, ed = 4, bf = 50, ef = 34, df = 30.$	2
	130, 94	224	$ab = 44, ac = 38, ae = 48, cb = 6, ce = 10, cd = 22, ed = 12, bf = 50, df = 34, ef = 46.$	2
	104, 105	209	$ab = 60, ac = 44, cb = 16, cd = 28, bd = 12, be = 19, de = 7, bf = 45, ef = 26, df = 33.$	1
	111, 98	209	$ab = 44, ad = 26, ae = 41, dc = 11, de = 15, ce = 4, cb = 7, eb = 3, bf = 54, ef = 57.$	1
	115, 94	209	$ab = 34, ac = 19, ad = 23, ae = 39, cb = 15, cd = 4, de = 16, db = 11, bf = 60, ef = 55.$	1
	130; 79	209	$ab = 34, ac = 23, ad = 35, ae = 38, cb = 11, cd = 12, de = 3, bf = 45, df = 44, ef = 41.$	1

[12] First found, apparently, by Moroń [11]. See also [10], p. 272; [2], p. 93; [14], p. 8; and [4].

The full sides and semiperimeters of the simple perfect rectangles of the 11-th order are:

Order	Semi-perimeter	Sides
11	336	127, 209; 151, 185
	353	144, 209; 159, 194; 162, 191; 166, 187; 168, 185; 176, 177
	368	159, 209; 169, 199; 172, 196; 177, 191; 183, 185
	377	168, 209; 178, 199; 183, 194
	386	162, 224; 177, 209; 181, 205; 190, 196; 191, 195; 192, 194

Four of these are reducible, with reduction = 2; these are the rectangles whose sides are both even.

Of the 67 simple perfect rectangles of the 12-th order, eleven have reduction 2, eight have reduction 3, and one has reduction 4.

6. Theorems on reduction

In perfect rectangles of higher orders, much larger reductions occur; for example, a 19-th order rectangle with reduced sides 144 and 155 has $\rho = 80$. Its reduced elements are: $ab = 46$, $ad = 40$, $af = 28$, $ag = 41$, $bc = 10$, $bi = 36$, $ci = 26$, $dc = 16$, $de = 3$, $dh = 21$, $eh = 18$, $fe = 15$, $fg = 13$, $gk = 54$, $hl = 39$, $ij = 62$, $kj = 49$, $kl = 5$, $lj = 44$.

6.1. The following theorems on reduction are of interest.

(6.11) THEOREM. *If one of the currents in a p-net is zero, the net is reducible.*

Let the poles be P_r, P_s, and the zero current be in a wire joining P_x, P_y. Then the transpedance $[rs, xy]$ is zero. On removing the wire in question (use (2.37) with $c = -1$, and (2.33)), the new value for $[rs, tu]$ is

$$[rs, tu]' = [rs, tu] - \frac{[rs, tu]\cdot V(xy) - [rs, xy]\cdot[tu, xy]}{C}$$

$$= [rs, tu]\cdot\frac{C - V(xy)}{C}.$$

Now, $C > C - V(xy) = C'$ (by (2.35)) > 0 (by (3.14)). Hence the H.C.F. of the $[rs, tu]$'s must be at least $C/(C - V(xy)) > 1$.

DEFINITION. Let a positive integer $n = m\cdot k^2$, where m is square-free. Then k is called the *lower square root* of n, and mk is the *upper square root.*

(6.12) THEOREM. *Let the full sides of a p-net be H, V. Then the reduction ρ is a multiple of the upper square root of the H.C.F. of H and V.*

By (2.34), remembering that $V_2(rs)$ is an integer, we have

$$C \text{ divides } V(rs)\cdot V(xy) - [rs, xy]^2.$$

Since $C = H$, and $V(rs) = V$ (taking P_r, P_s as poles), it follows that the H.C.F. of H, V divides $[rs, xy]^2$; whence the result.

(6.13) COROLLARY. *If the reduced sides of a squared rectangle have H.C.F. σ, then the reduction of any corresponding p-net is divisible by σ.*

(For, by (6.12), σ is a factor of the lower square root of the H.C.F. of the full sides and hence—since the lower square root divides the upper—of ρ.)

(An example is the rectangle 96×99 given in §7.1.)

(6.14) COROLLARY. *Any p-net of a squared square has for reduction a multiple of its reduced side.*

(6.15) COROLLARY. *A necessary and sufficient condition that a p-net be irreducible is that its two full sides be coprime.*

(6.16) THEOREM. *All non-trivially imperfect p-nets are reducible.*

(6.17) LEMMA. *If H, V, k are positive integers such that, for each positive integer n, $(H + nV, k) > 1$, then H, V, k all have a common factor greater than 1.*

Proof. Let N_0 be the product of all the primes which divide k but not H. (Empty product $= 1$.) Let p_0 be a prime factor of $(H + N_0V, k)$. Suppose $p_0 \nmid H$. Then $p_0 \mid N_0$. Hence, since $p_0 \mid (H + N_0V)$, we have $p_0 \mid H$, and this is a contradiction. So $p_0 \mid H$. Therefore p_0 divides N_0V but not N_0; so that p_0 divides V as well as H and k.

Proof of (6.16). Now let $\mathcal{P}$ be a p-net with full sides H ($= C$) and V ($= V(1N)$); and let a non-trivial imperfection be $[1N, ab] = [1N, pq] = k$, say. Thus $k > 0$, and we do not have both $a = p$ and $b = q$. (Else the imperfection is trivial.)

Join P_1, P_N to produce the completed net $\mathcal{C}$. Let $\mathcal{Q}$ be the p-net formed from $\mathcal{C}$ by taking P_a, P_b as poles, and omitting one wire joining P_a, P_b. (Of course, there is such a wire; there may be several. It is easy to see, from considerations of "triviality", that $\mathcal{Q}$ is connected, and therefore a p-net.) Applying (2.33) to $\mathcal{C}$, and using (2.35), (2.36), (2.37), we have

$$(6.18) \qquad (H + V) \mid k \cdot (V(ab) - [ab, pq]),$$

where $V(ab)$, $[ab, pq]$ refer to the p-net formed by $\mathcal{C}$ with P_a, P_b as poles, and hence (2.36) refer equally well to $\mathcal{Q}$.

Now, we have $0 < V(ab) - [ab, pq] \leq$ semiperimeter of $\mathcal{Q}$, with equality only if the current $[ab, qp]$ equals the total current of $\mathcal{Q}$. In this case, $\mathcal{C}$ must consist of two parts, joined only by the two wires P_aP_b and P_pP_q. Further, P_1, P_N, being joined in $\mathcal{C}$ by a wire not P_aP_b or P_pP_q, must lie in the same part. Hence the imperfection in $\mathcal{P}$ with which we started was trivial.

Hence (6.18) gives (since semiperimeter of $\mathcal{Q}$ = complexity of $\mathcal{C} = H + V$)

$$(6.19) \qquad (H + V, k) > 1.$$

Now let n be any positive integer. Join P_1, P_N by $n - 1$ extra wires (of unit conductance). The new p-net will have the same non-trivial imperfection (by (2.36)), so, applying (6.19) to the new net, and using (2.35), (2.32) repeatedly, we have

$$(H + nV, k) > 1.$$

The lemma (6.17) now shows that $(H, V) > 1$. Hence, by (6.15), $\mathcal{P}$ is reducible.

(6.20) COROLLARY. *All irreducible p-nets having no trivial imperfections give perfect squared rectangles.*

(6.21) COROLLARY. *If the complexity of a c-net is prime, all the squared rectangles derived from it (as in §5.1) will be perfect.*

These results are sometimes useful as tests for perfection.

For the reduced elements, we can prove (using the Euler polyhedron formula, and some consideration of the various cases)

(6.22) At least three of the reduced elements of any perfect rectangle are even. (Three is the best number possible.)

7. Construction of some special squared rectangles

7.1. **Conformal rectangles.** Two squared rectangles (or p-nets in this plane) which have the same shape (that is, have proportional sides) but are not merely rigid displacements of each other (in the case of p-nets, have not the same normal form) are called *conformal.* An example of a conformal pair is provided by the 9-th order rectangle 64×66 and a 12-th order rectangle of reduced sides 96, 99, whose (reduced) net is specified by: $ga = 31$, $ge = 21$, $gc = 44$, $ea = 10$, $ed = 11$, $ad = 1$, $dc = 12$, $ac = 13$, $ab = 27$, $cb = 14$, $bf = 41$, $cf = 15$.

Two conformal rectangles need not have the same full sides or reduction; for example, the rectangle 96×99 has reduction 3 (cf. (6.21)).

We now show how to construct conformal pairs having the *same* reduced elements (but differently arranged).

Suppose that a p-net $\mathcal{P}$ has a part $\mathcal{Q}$ joined to the rest only at vertices $A_1, \cdots, A_m$, say, and containing no pole different from an A_i. If $\mathcal{Q}$ has rotational symmetry about a vertex P, in which the A's are a set of corresponding points, then a simple symmetry argument shows that the potential of P (in $\mathcal{P}$) will be the mean of the potentials of $A_1, \cdots, A_m$. Hence if this is also true for another vertex P', P and P' will have equal potentials.

Coalesce P and P', forming (if this can be done in the plane) the p-net $\mathcal{P}_2$. If C is the complexity of $\mathcal{P}$, we see from (2.33) that, if $[ab, 1N]$ and $[ab, 1N]_2$ are corresponding elements in $\mathcal{P}$ and $\mathcal{P}_2$ (with 1, N referring to the poles),

$$[ab, 1N]_2 = \frac{V(PP')}{C}[ab, 1N].$$

Hence the elements of $\mathcal{P}_2$ are proportional to those of $\mathcal{P}$; $\mathcal{P}$ and $\mathcal{P}_2$ have the same reduced elements and sides. Their reductions are clearly in the ratio $C : V(PP')$. This construction enables conformal p-nets with the same elements to be written down.

A simple example is shown in Figure 2. Here A_1 and A_2 are poles. The rectangles are perfect and simple, and have reductions 5 and 6, and reduced sides 75 and 112.

In a more complicated example, illustrating a variation on the method, we make the potentials of three points P_1, P_2, P_3 equal. Although the network we start with is not planar, it becomes so when either P_1P_2 or P_1P_3 coincide. Such a network is specified below. It gives conformal simple perfect rectangles of the 28-th order, with reductions 96 and 120, reduced sides 6834 and 14065, and reduced elements: $A_1a = 3288$, $A_1P_1 = 3480$, $A_1b = 2512$, $A_1d = 2247$, $A_1i = 2538$, $aP_3 = 192$, $aA_3 = 3096$, $bP_3 = 968$, $bA_2 = 1544$, $P_1A_2 = 576$, $P_1A_3 = 2904$, $P_3c = 1160$, $A_2c = 584$, $cA_3 = 1744$, $de = 1014$, $dP_2 = 1233$, $eA_2 = 795$,

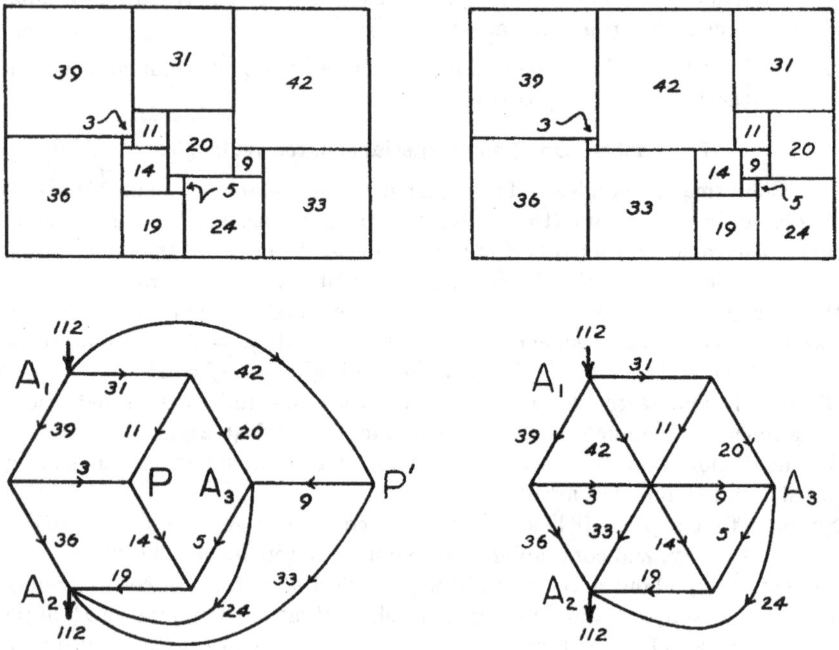

FIG. 2

$eP_2 = 219$, $iP_2 = 942$, $ih = 1596$, $P_2h = 654$, $P_2f = 579$, $P_2g = 1161$, $hA_3 = 2250$, $A_2f = 3$, $fg = 582$, $gA_3 = 1743$, $A_2A_3 = 2328$. (The poles are A_1 and A_3.)

These examples show that, even when the sides and elements of a simple perfect rectangle are given, the configuration is far from uniquely determined.

We now turn to the opposite problem of constructing conformal pairs of squared rectangles having *different* sets of elements. Again, symmetry considerations enable us to do this. We are led to pairs of rectangles (and p-nets) which are not merely conformal but have the same full sides. Such pairs are said to be *equivalent.*

7.2. Symmetry method. Let a p-net $\mathcal{P}$ have a part $\mathcal{Q}$ joined to the rest only at vertices $A_1, \cdots, A_m$, and containing no pole different from an A_i. Suppose that $\mathcal{Q}$ has rotational symmetry in which the A's are a set of corresponding points, and that $\mathcal{Q}$ is not identical with its mirror-image. $\mathcal{Q}$ is the *rotor*, and the wires of $\mathcal{P} - \mathcal{Q}$ form the *stator*. In $\mathcal{P}$, replace $\mathcal{Q}$ by its mirror-image. It is easy to see that the full currents in the stator will be entirely unaffected, though (in general) the rotor currents will change. (This can be proved, e.g., by induction over the number of wires in the stator, if we use §2.) So we have, in general, a pair of equivalent rectangles, with different (though overlapping) sets of elements.

One of the simplest examples of this method is shown in Figure 3. This gives equivalent simple perfect rectangles of order 16, reduction 5, and reduced sides 671 and 504.[13]

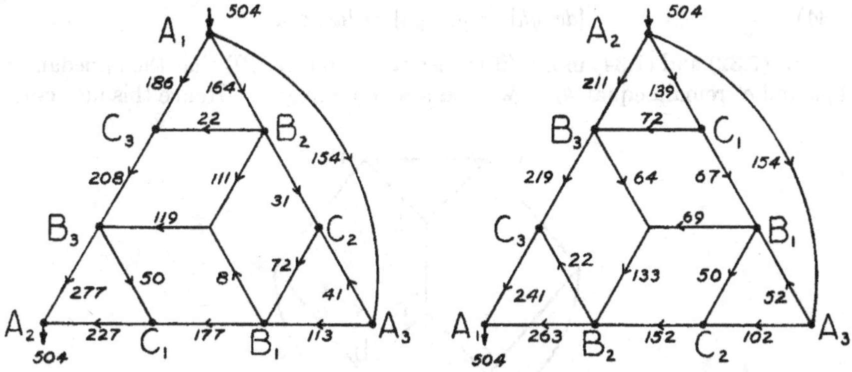

FIG. 3

We may generalize this method by noting that it remains effective when some of the A's are coincided (corresponding to the introduction of "wires of infinite conductance" in the stator). Or, again, we may take the stator to be itself a rotor, with $A_1, \cdots, A_m$ as its set of corresponding points (with possible coincidences). By reflecting both parts we can get pairs of equivalent rectangles having no elements in common.

7.3. Special methods. The preceding methods (and similar ones based on duality instead of symmetry) are useful for existence theorems, as in the next section; but other devices are more suitable for producing equivalent rectangles of small orders.

If, in a c-net $\mathcal{C}$, we can find two wires whose end-points—say P_a, P_b and P_x, P_y, respectively—satisfy

$$V(ab) = V(xy) \qquad \text{(in } \mathcal{C}\text{)}, \tag{7.31}$$

[13] The rotor of Figure 3 has a remarkable property. If currents I_1, I_2, I_3 (summing to zero) enter the *rotor* (considered as a net) at A_1, A_2, A_3, then the currents in B_3C_1, B_1C_2, B_2C_3 will be $I_1/7$, $I_2/7$, $I_3/7$, respectively. This explains the "extra" equalities of the currents in Figure 3. Other rotors of 15 wires (having the same type of symmetry) behave in a similar way. This phenomenon is not yet fully explained.

then the corresponding p-nets (obtained from $\mathcal{C}$ by omitting each of the two wires in turn, and taking its ends as poles) will be equivalent, if not identical. For they have the same semiperimeter in any case, viz., the complexity of $\mathcal{C}$.

By using the properties of symmetrical or self-dual networks, we can often demonstrate an equality like (7.31). For example, in Figure 4, it is clear that

$$V(gh) = V(cb) \tag{7.32}$$

and

$$[da, gh] = 0. \tag{7.33}$$

Hence (by (7.33) and (2.23) and symmetry)

$$[de, gh] = [ae, gh] = [de, cb]. \tag{7.34}$$

Now, (7.32) and (7.34) imply (if we use (2.37) and (2.33)) that the impedances of *gh* and *cb* remain equal when we add a wire joining *de*. Hence this new c-net

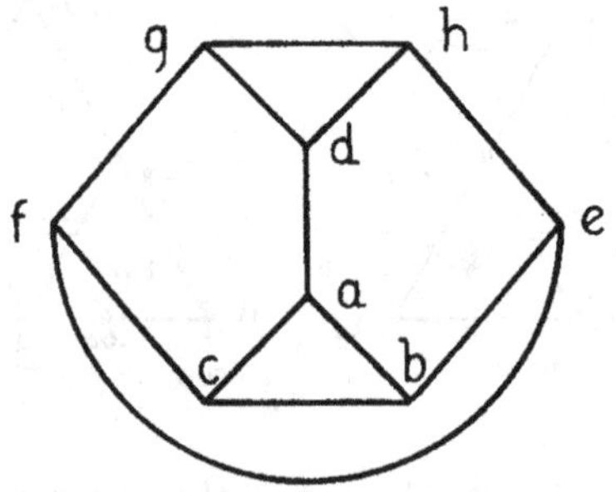

FIG. 4

satisfies (7.31), and so we get a pair of equivalent squared rectangles of the 12-th order. These rectangles are perfect, and provide the simplest example of equivalence among perfect rectangles. They both have reduction 2 and reduced sides 142 and 162. Their (reduced) specifications are respectively:

$gf = 57$, $gd = 85$, $dh = 77$, $de = 12$, $ad = 4$, $fe = 40$, $be = 13$, $eh = 65$, $ab = 3$, $ca = 7$, $cb = 10$, $fc = 17$; and $cf = 59$, $ca = 83$, $fe = 40$, $fg = 19$, $gh = 10$, $he = 11$, $gd = 9$, $dh = 1$, $ad = 4$, $de = 12$, $eb = 63$, $ab = 79$.

8. Construction of perfect squares

8.1. **Definition.** Two conformal rectangles are said to be *totally different* if C_2 times an element of the first is never equal to C_1 times an element of the second, where C_1, C_2 are their respective (corresponding) horizontal sides.

For equivalent rectangles this is equivalent to: No element of the first equals an element of the second.

A pair of totally different simple perfect squared rectangles gives us a perfect square at once; we have only to place them as in Figure 5, and add two corner squares. This idea, though often in modified form, underlies all the constructions for perfect squares in this paper.

(8.11) It is easy to show (by the use of determinants) that if H, V and H', V' are the full sides of the rectangles used in this construction, then the resulting square will have full side $(H + V)\cdot(H' + V')$. In particular, if the rectangles are *equivalent*, the full side of the square is the square of an integer.

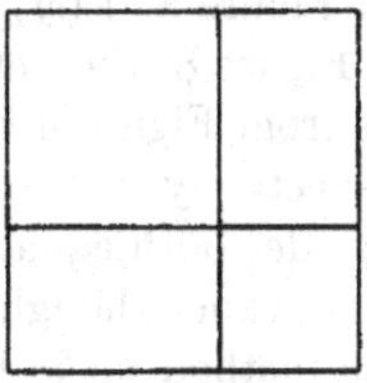

FIG. 5

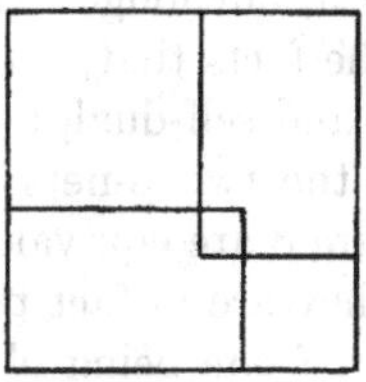

FIG. 6

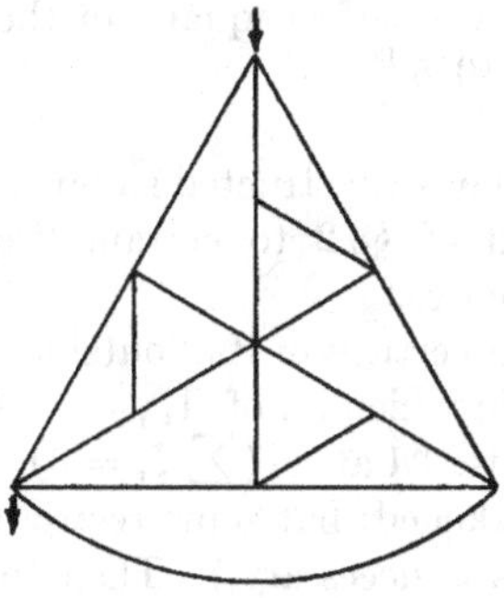

FIG. 7

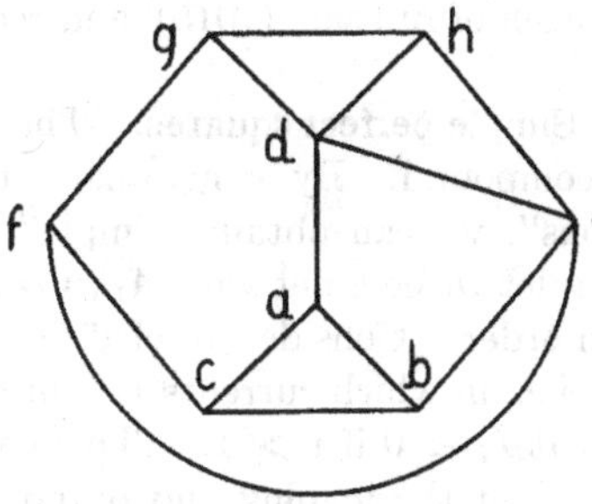

FIG. 8

8.2. **Symmetry method.** Equivalent perfect rectangles constructed as in §7.2 can be used to give us a perfect square. The stator is taken to be a single wire A_iA_j (drawn outside the rotor), one of whose end-points is a pole. The equivalent rectangles so obtained will have, *in general*,[14] just one element in common, the element corresponding to this stator. As this element is placed at a *corner* in both rectangles, we may "overlap" the rectangles as in Figure 6 to get a square.

One of the simplest perfect squares formed in this way is based on the rotor and stator shown in Figure 7. The square is of the 39-th order.

(8.21) It can bé shown that, if H, V are the full sides of the equivalent rectangles used in this construction (§8.2), and E is the common element, then the full

[14] The "exceptional case", in which two elements from the following set: the rotor, its reflection, and the stator-element, are equal, seems in practice to be rare. It does occur, however, if the rotor has trivial imperfections, or if it has too much symmetry, or if it has triad symmetry and only 15 wires (cf. the previous footnote).

side of the resulting squared square is $(H + V - E)^2$, the square of an integer. In the case of *triad* symmetry ($m = 3$ in §7.2), we can show that $E \cdot (2H + 2V - E) = HV$, so that the full side of the squared square is, in this case, $H^2 + HV + V^2$.

8.3. **Perfect squares of smaller orders.** A perfect square of much smaller order is given by an elaboration of §7.3. We can show by an argument similar to that in §7.3, but longer, that in the net shown in Figure 8, $V(cf) = V(ge)$. (We use the facts that, if g and f are coalesced in Figure 8, the net becomes symmetrical and self-dual, and that Figure 8 results from Figure 4 by joining de.) Hence the two p-nets obtained by taking respectively c, f and g, e as poles in Figure 8 are equivalent (for their horizontal sides both equal the complexity). They are in fact perfect and totally different; and, though not both simple (the c, f one being obviously compound), the method of §8.1 is easily modified to give a perfect square, which is drawn in Figure 9. It is of the 26-th order. (The least possible order of a perfect square is unknown.)

We have also constructed, in a similar way, two perfect squares of the 28-th order, each of full side $(1015)^2$ and reduced side 1015.[15]

8.4. **Simple perfect squares.** The perfect squares constructed so far have all been compound. By generalizing the method of §8.2 to certain "squared polygons", we can obtain "simple" perfect squares.

First, let $\mathfrak{N}$ be a net with $A_1, \cdots, A_m$ as the vertices of its "outside" polygon, in order. Consider an electric flow in $\mathfrak{N}$ in which all of $A_1, \cdots, A_m$ are poles—i.e., in which currents I_i (not all zero) enter $\mathfrak{N}$ at A_i $(\sum I_i = 0)$. Suppose that $I_i \geqq 0$ if $i > 1$. (This could be weakened; but some restriction on the order of the ingoing and outgoing currents is necessary.) Then the flow in $\mathfrak{N}$ corresponds to a squared polygon, of angles $\frac{1}{2}\pi$ and $\frac{3}{2}\pi$.

Proof. We reduce the number of poles of $\mathfrak{N}$ as follows: Suppose A_i is at potential V_i. Suppose there is more than one i for which $I_i > 0$; let $1'$, $2'$ be the least and second least such i's. If $V_{1'} = V_{2'}$, coalesce $A_{1'}$ and $A_{2'}$ (by joining them by a line outside the polygon $A_1 \cdots A_m$ and shrinking the line to a point); and let current $I_{1'} + I_{2'}$ enter there, the other currents being as before. The currents in $\mathfrak{N}$ will be unaltered, and there is now one fewer positive current entering the network. If $V_{1'} \neq V_{2'}$, we can suppose $V_{1'} > V_{2'}$. Join $A_{1'}$, $A_{2'}$ by a wire of conductance $I_{2'}/(V_{1'} - V_{2'})$ (passing outside the polygon $A_1 \cdots A_m$) and take currents $(I_{1'} + I_{2'})$ at $A_{1'}$, 0 at $A_{2'}$ and I_i at A_i for the other i's. Again, the currents in $\mathfrak{N}$ will be unaltered, and one fewer positive current enters the system. Repeating this process till there is only one positive external current left, we have the flow in $\mathfrak{N}$ "imbedded" in a flow with only two poles; in fact, in a p-net flow (except that some of the extra wires may have conductances different from 1). This corresponds to a "rectangled rectangle" R.

[15] See [16].

Stripping off the elements of R which correspond to the extra wires, we are left with a squared polygon, corresponding to $\mathfrak{N}$.

Since the currents I_i are (apart from sign) at our disposal, the shape of the squared polygon can be controlled. (It has $m - 2$ degrees of freedom.)

Now take for $\mathfrak{N}$ a pure rotor—i.e., a network having skew symmetry; and suppose that the points $A_1, \cdots, A_m$ are a set of corresponding points in $\mathfrak{N}$.

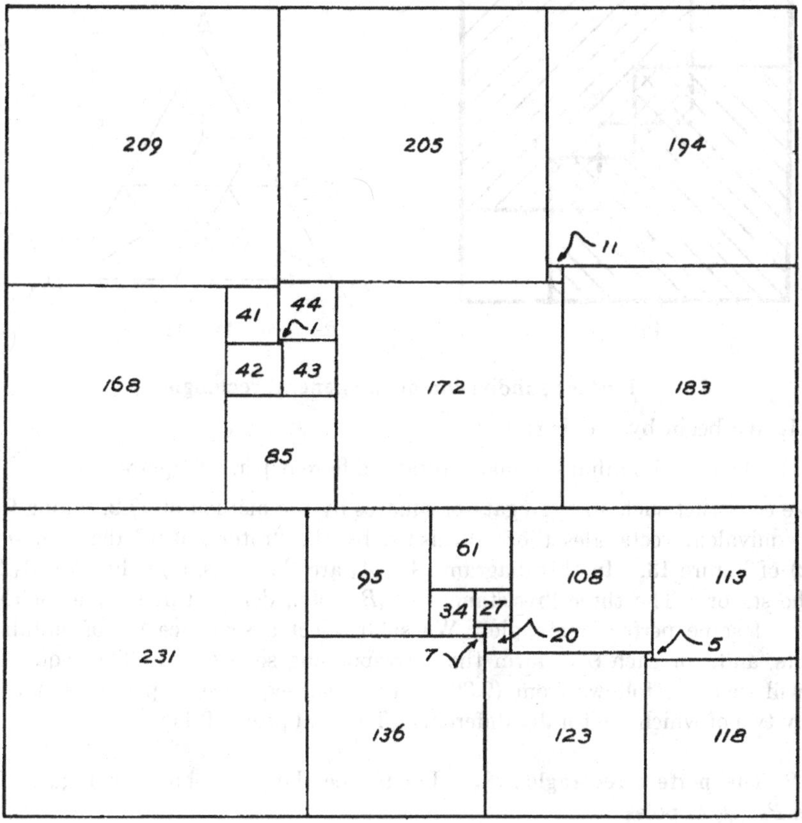

FIG. 9

If $\mathfrak{N}$ is replaced by its reflection (leaving the currents I_i invariant), the new squared polygon will have the same shape as the old—in fact, the two squared polygons will be "equivalent". For, as in §7.2, the rectangled rectangle R will be replaced by an equivalent one, in which the "extra" elements are the same as before.

By combining such a pair of equivalent polygons, as in Figure 10, and arranging their shape so that the overlapped portions coincide with elements (which are then removed), and inserting three extra squares (in the center and at the corners), we can obtain a "simple" perfect square.

For instance, the rotor shown in Figure 11 gives rise to a simple "uncrossed" perfect square of order 55, which, when drawn out, disguises its symmetrical origin very skillfully.

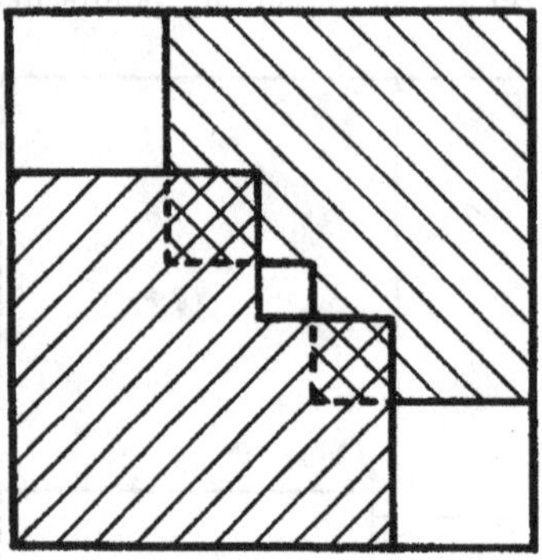

FIG. 10

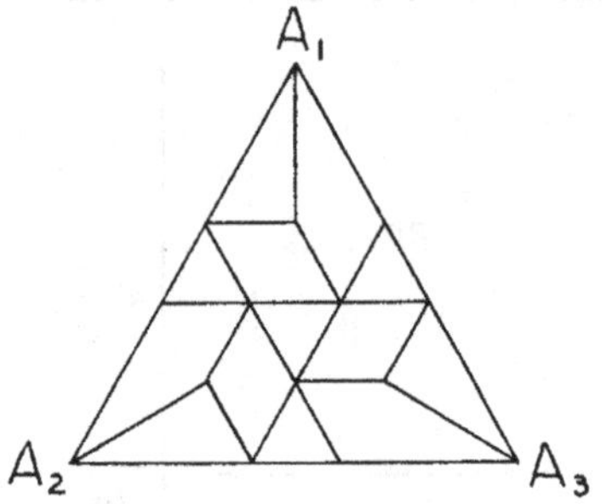

FIG. 11

9. Perfect subdivision of the general rectangle

9.1. We begin by proving:

(9.11) There exist infinitely many totally different perfect squares.

We construct such an aggregate of squares by the method of §8.2, taking for our equivalent rectangles those furnished by the "rotor-stator" diagram (cf. §7.2) of Figure 13. In this diagram, A_1, A_2 are the poles, and the wire A_1A_3 is the stator. The three "resistances" A_1B_2, etc., denote three copies of the p-net of some perfect rectangle. We shall select a sequence $\mathfrak{R}_n$ of suitable p-nets, and, for each $\mathfrak{R}_n$, form the corresponding square $\mathfrak{S}_n$. The sequence $\mathfrak{S}_n$ will then (as follows from (9.39)) have a subsequence of perfect squares, every two of which are totally different. This will prove (9.11).

9.2. **The perfect rectangles $\mathfrak{R}_n$.** Let $\mathfrak{R}_n$ be the p-net shown in Figure 12, with P_0, Q_0 as poles.

Write $\phi_r = [(2 + \sqrt{3})^r - (2 - \sqrt{3})^r]/2\sqrt{3}$. Thus

(9.21) ϕ_r is an integer; $\phi_0 = 0$; and $\phi_{r+1} - 4\phi_r + \phi_{r-1} = 0$.

It will readily be verified that a solution of Kirchhoff's equations is given by:

(9.22) Current in P_0P_r (from P_0 to P_r) is a_r, where

$$a_r = \tfrac{1}{2}\cdot[5\phi_n + \phi_{n-1} + 3\phi_r - 3\phi_{r-1}] \qquad \text{if } 0 < r < n,$$

$$a_{n+1} = 3\phi_n .$$

Current in P_rQ_0 is b_r, where

$$b_r = \tfrac{1}{2}\cdot[5\phi_n + \phi_{n-1} - 3\phi_r - 3\phi_{r-1}] \qquad \text{if } 0 < r < n + 1,$$

$$b_{n+1} = 2\phi_n + \phi_{n-1} .$$

Current in P_rP_{r+1} is c_r, where

$$c_r = 3\phi_r \qquad \text{if } 0 < r < n,$$

$$-c_n = \phi_n - \phi_{n-1}.$$

(This solution is in fact the full flow.)

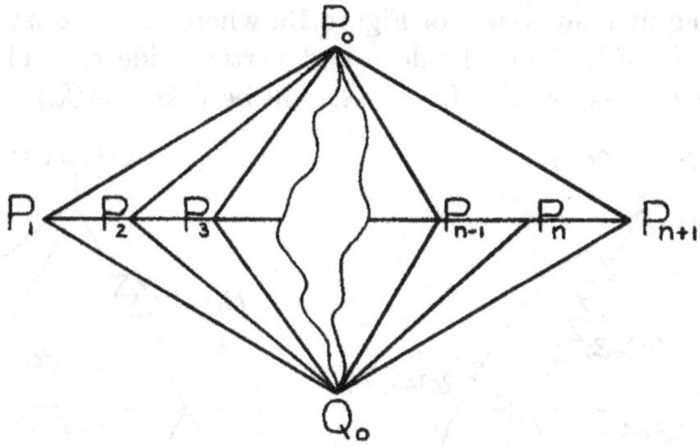

FIG. 12

Also the total current p_n (the horizontal side of $\mathcal{R}_n$), and the total P.D. q_n (the vertical side) are given by:

(9.23) $$p_n = \tfrac{1}{2}\cdot[(5n+1)\phi_n + (n+2)\phi_{n-1}]; \qquad q_n = 5\phi_n + \phi_{n-1}.$$

Now, if $n > 2$, we see that

$$0 < c_1 < c_2 < \cdots < c_{n-2} < (-c_n) < c_{n-1} < b_n < b_{n+1} < b_{n-1} < b_{n-2}$$
$$< \cdots < b_1 < a_1 < a_2 < \cdots < a_{n-1} < a_{n+1}.$$

Hence

(9.24) If $n > 2$, $\mathcal{R}_n$ is perfect.

From (9.23), we have

(9.25) q_n and $p_n/q_n \to \infty$ with n.

For later use, we note that

(9.26) $$(p_n, q_n) \mid 9.$$

Proof. From (9.23),

$$(n+2)q_n - 2p_n = 9\phi_n \qquad \text{and} \qquad (5n+1)q_n - 10p_n = 9\phi_{n-1}.$$

Now, we can prove by induction (using (9.21)) that $(\phi_n, \phi_{n-1}) = 1$. Thus (9.26) follows.

[(9.24) can be generalized: If in Figure 12 the wire P_0P_n is inserted and the wire P_0P_r removed, where $1 < r \leq \frac{1}{2}n$, the resulting p-net $\mathcal{R}_{nr}$ is perfect. $\mathcal{R}_{n2}$

is essentially the same as $\mathfrak{R}_n$. The reduction ρ_r of $\mathfrak{R}_{nr}$ can be calculated; for instance, it can be shown that ρ_r is a factor of $(\phi_r - \phi_{r-1})$; and that $\rho_r = (\phi_r - \phi_{r-1})$ if and only if $n \equiv 0 \pmod{2r-1}$.]

9.3. We next prove

(9.31) THEOREM. *For all large n, the squared square $\mathfrak{S}_n$ is perfect.*

Consider the equivalent p-nets of Figure 13, where each "resistance" denotes a certain p-net $\mathfrak{R}$, of horizontal side p and vertical side q. (The other wires have conductance 1, as usual. Later, $\mathfrak{R}_n$ will be taken as $\mathfrak{R}$.)

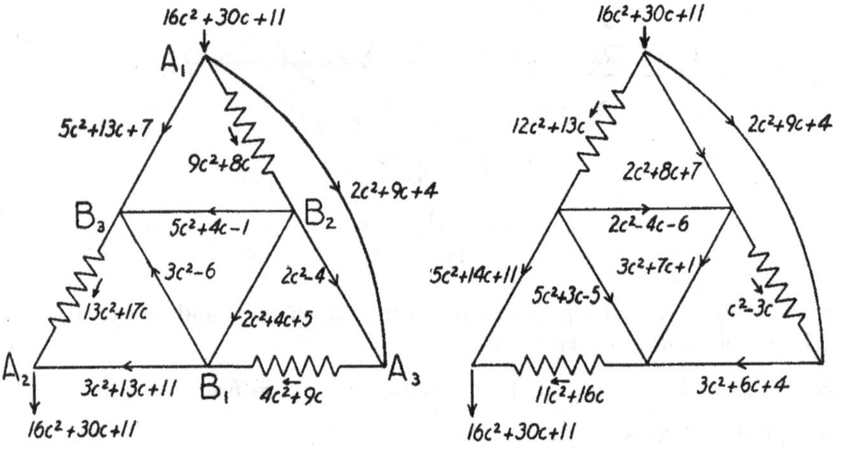

FIG. 13

Setting $c = p/q$ = effective conductance of these resistors, we find that the flows are as indicated in the diagram. (The quantities shown are currents.)

Hence, multiplying through by q^2, and adjoining the extra elements required in forming the squared square $\mathfrak{S}$ as in §8.2, we find that the elements of $\mathfrak{S}$ (some integral multiple of the reduced elements) are:

$$(9.32)\quad \begin{cases} \text{(A)} \quad 14p^2 + 21pq + 7q^2,\ 5p^2 + 14pq + 11q^2,\ 5p^2 + 13pq + 7q^2, \\ \qquad 5p^2 + 4pq - q^2,\ 5p^2 + 3pq - 5q^2, \\ \qquad 3p^2 + 17pq + 20q^2,\ 3p^2 + 13pq + 11q^2,\ 3p^2 + 7pq + q^2, \\ \qquad 3p^2 + 6pq + 4q^2,\ 3p^2 - 6q^2, \\ \qquad 2p^2 + 9pq + 4q^2,\ 2p^2 + 8pq + 7q^2,\ 2p^2 + 4pq + 5q^2, \\ \qquad 2p^2 - 4q^2,\ 2p^2 - 4pq - 6q^2. \\ \text{(B)} \quad \text{Multiples of the elements of } \mathfrak{R}, \text{ the multipliers being respectively} \\ \qquad 13p + 17q,\ 12p + 13q,\ 11p + 16q,\ 9p + 8q,\ 4p + 9q,\ p - 3q. \end{cases}$$

We also find that

(9.33) The side of $\mathfrak{S}$ is $19p^2 + 47pq + 31q^2$.

Now take $\mathcal{R}$ to be $\mathcal{R}_n$, so that $p = p_n$ and $q = q_n$; and let n be so large that (in virtue of (9.25))

(9.34) $$p_n > 180q_n .$$

We prove that, under this condition, $\mathfrak{S} = \mathfrak{S}_n$ is perfect.

(9.35) *The elements* (A) *are all different, and no element* (A) *equals an element* (B).

For the elements (A) in the above list are in strictly decreasing order; so no two of them are equal. Also the least element (A) is $2p^2 - 4pq - 6q^2$ which $> (13p + 17q)q$, which is greater than any element (B). Thus (9.35) follows.

(9.36) *No two elements* (B) *are equal.*

For suppose that two such elements are equal:

(9.37) $$\xi(\alpha p + \beta q) = \eta(\gamma p + \delta q),$$

where ξ, η are elements of $\mathcal{R}_n$, and $\alpha p + \beta q$, $\gamma p + \delta q$ are two multipliers of (9.32). They are *different* multipliers; for $\mathcal{R}_n$ is perfect by (9.24). Hence, by inspection of (9.32), $\alpha\delta - \beta\gamma \neq 0$. But, from (9.37), $(\alpha\xi - \gamma\eta)p = (\delta\eta - \beta\xi)q$. Hence

(9.38) $$p \mid (\delta\eta - \beta\xi)\cdot(p, q).$$

Now, if $\delta\eta - \beta\xi = 0$, we have (since $p \neq 0$) $\alpha\xi - \gamma\eta = 0$, and hence, if we eliminate ξ, η (which are not zero), it follows that $\alpha\delta - \beta\gamma = 0$. So we have $0 < |\delta\eta - \beta\xi| < 20q$ (by inspection of (9.32), since $0 < \xi, \eta < q$).

Hence, if we use (9.26), (9.38) gives $p < 180q$. This contradicts (9.34). And (9.31) now follows from (9.35) and (9.36); the squares $\mathfrak{S}_n$ are perfect, for large enough n.

(9.39) THEOREM. *Given any large enough n, then for all large enough N, $\mathfrak{S}_n$ and $\mathfrak{S}_N$ are totally different.*

Write $p_n = p$, $q_n = q$, $p_N = P$, $q_N = Q$. We bring $\mathfrak{S}_n$ and $\mathfrak{S}_N$ to the same size by multiplying the elements of $\mathfrak{S}_n$ (as given by (9.32)) by $19P^2 + 47PQ + 31Q^2$ and those of $\mathfrak{S}_N$ by $19p^2 + 47pq + 31q^2$. (This follows from (9.33).)

(9.40) *Each element* (B) *of $\mathfrak{S}_N$ is less than every element of $\mathfrak{S}_n$.*

For a typical element (B) of $\mathfrak{S}_N$ is

$$e = (\alpha P + \beta Q)\cdot(19p^2 + 47pq + 31q^2), \quad \text{where } |\alpha|, |\beta| \leqq 17.$$

If n and N are large, this gives $e < 360Pp^2$. (This follows from (9.25).) But each element of $\mathfrak{S}_n$ is at least as large as P^2p (times some non-zero constant). Hence if $n >$ some n_0, and if then $N >$ some $N_0(n)$, so that P is large compared with p (see (9.25)), we have $e <$ each element of $\mathfrak{S}_n$.

(9.41) *Each element* (A) *of* $\mathfrak{S}_N$ *is greater than every element* (B) *of* $\mathfrak{S}_n$.

For any element (A) of $\mathfrak{S}_N$ is at least as large as P^2p^2 (times some non-zero constant), whereas an element (B) of $\mathfrak{S}_n$ is less than $360P^2p$.

(9.42) *No element* (A) *of* $\mathfrak{S}_N$ *can equal any element* (A) *of* $\mathfrak{S}_n$.

Otherwise we have

$$(aP^2 + bPQ + cQ^2)\cdot(19p^2 + 47pq + 31q^2) = (a'p^2 + b'pq + c'q^2)\cdot(19P^2 + 47PQ + 31Q^2),$$

where by (9.32) a, a', etc., are integers numerically less than 22. Hence

$$P^2\cdot[(19a - 19a')p^2 + (47a - 19b')pq + (31a - 19c')q^2] = \text{similar terms in } PQ \text{ and } Q^2. \tag{9.43}$$

Now, $47a - 19b' \neq 0$; for otherwise $19 \mid a$, whereas $0 < a < 19$ (from (9.32)). Hence the left side of (9.43) is numerically at least as large as P^2pq (times some non-zero constant); in fact, if $a \neq a'$, it is as large as P^2p^2. But the right side of (9.43) is at most PQp^2 (times a constant). Hence, if N is taken large enough, so that P dominates both p and Q (this is possible, by (9.25)), (9.43) is impossible.

(9.40), (9.41), and (9.42) imply (9.39).

(9.44) Corollary. *There is a sequence* $\{\mathfrak{T}_n\}$ *of perfect squares, every two of which are totally different.*

This is immediate from (9.31) and (9.39) and proves (9.11).

A rough calculation shows that we may take $\mathfrak{T}_r = \mathfrak{S}_{10^{3(r+1)}}$. This could probably be greatly improved.

(9.45) Theorem. *Any rectangle whose sides are commensurable can be squared perfectly in an infinity of totally different ways.*

Magnifying the rectangle suitably, we may suppose that its sides are integers h, k. Divide it into hk squares of side 1, by lines parallel to its sides. Take any positive integer n, and replace the i-th of these unit squares by $\mathfrak{T}_{nhk+i}$ (suitably contracted). By (9.44), this gives, for each n, a perfect subdivision of the given rectangle; and these subdivisions for any two values of n are "totally different".

Using the theorem of (2.14), we see that a rectangle can be squared *perfectly* if it can be squared at all.

It is plausible that any commensurable-sided rectangle can be squared perfectly and *simply*; possibly this can be proved in a similar way if we use some extension of §8.4; but this seems to involve laborious calculations.

10. Some generalizations

We mention briefly some of the extensions of the methods and results of this paper. A fuller discussion may perhaps appear later.

10.1. **Rectangled rectangles.** An immediate and natural generalization (as pointed out in §1.2) is to the problem of a rectangle dissected into a finite number of rectangles. The wires of the p-net merely have general (not necessarily equal) conductances.

There is also (cf. §8.4) a rather trivial extension in which the dissection is of a polygon (of angles $\frac{1}{2}\pi$ and $\frac{3}{2}\pi$). A more natural generalization, however, is given in the following section.

10.2. **Squared cylinders and tori.** We may regard a squared rectangle, after identification of its left and right sides, as a "trivial" example of a squared cylinder. The squared cylinders are found to correspond exactly to the relaxation of the condition (1.12) that no circuit of the p-net may enclose a pole. A second step brings us to the "squared torus". Using the existence theorem of (9.45), we can easily construct such figures. It is also possible to construct a *simple* non-trivial perfect torus; but this is not so easy.

Of course, the word "squared" may be replaced by "rectangled".

10.3. **Triangulations of a triangle.** In a rather different direction, we may consider dissections of a triangle into a finite number of triangles; particularly when all the triangles considered are *equilateral.* It is easily proved that *there is no perfect equilateral triangle*; i.e., that in any such dissection of an equilateral triangle into equilateral triangles, two of the latter are equal. Apart from this, the theory extends fairly completely. Duality relations, for example, are replaced by "triality" relations. We could also consider dissections into a mixture of equilateral triangles and regular hexagons, no two of these elements having equal sides; essentially this amounts to agglomerating the imperfections of an "equilateral triangled triangle" together by sixes. There is no difficulty in constructing such figures empirically, or in finding "perfect isosceles right-angled triangles"; however, it can be done by using the theory.

10.4. **Three dimensions.** We have seen that the "p-net" and its generalizations are satisfactory for plane dissections. As yet, however, there is no satisfactory analogue in three dimensions. The problem is less urgent, because *there is no perfect cube* (*or parallelopiped*). That is, in any dissection of a rectangular parallelopiped into a finite number of cubes ("elements"), two of the latter are equal.

Proof. It is easily seen that in any perfect rectangle, the smallest element is not on the boundary of the rectangle. Suppose we have a "perfect" cubed parallelopiped P. Let R_1 be its base. The elements of P which rest on R_1 "induce" a dissection of R_1 into a perfect rectangle. (We can clearly assume that more than one cube rests on R_1.) Let s_1 be the smallest element of R_1. Let c_1 be the corresponding element of P. Then c_1 is surrounded by *larger*, and therefore *higher*, cubes on all four sides; for, as remarked above, s_1 is surrounded by larger squares. Hence the upper face of c_1 is divided into a perfect rectangle R_2 by the elements of P which rest on it; let s_2 be the smallest element of R_2; and so on. In this way, we get an infinite sequence of elements c_n of P, all different (for $c_{n+1} < c_n$). This is a contradiction.

This proof excludes generalizations of "perfect cylinders" to three (or more, a fortiori) dimensions; but it does not exclude the possibility of a *perfect three-dimensional torus* (product of three circles). It is not known whether such a thing can exist.

Bibliography

1. M. Abe, *On the problem to cover simply and without gap the inside of a square with a finite number of squares which are all different from one another*, Proceedings of the Physico-Mathematical Society of Japan, (3), vol. 14(1932), pp. 385–387.
2. W. W. Rouse Ball, *Mathematical Recreations*, 11th ed., New York, 1939.
3. C. W. Borchardt, *Ueber eine der Interpolation entsprechende Darstellung der Eliminations-Resultante*, Journal für Mathematik, vol. 57(1860), pp. 111–121.
4. S. Chowla, *Division of a rectangle into unequal squares*, Mathematics Student, vol. 7 (1939), p. 69.
5. S. Chowla, ibid., Question 1779.
6. M. Dehn, *Zerlegung von Rechtecke in Rechtecken*, Mathematische Annalen, vol. 57 (1903), pp. 314–332.
7. Jarenkiewycz, Zeitschrift für die mathematische und naturwissenschaftliche Unterricht, vol. 66(1935), p. 251, Aufgabe 1242; and solution, op. cit., vol. 68(1937), p. 43. (Also solutions by Mahrenholz and Sprague.)
8. J. H. Jeans, *The Mathematical Theory of Electricity and Magnetism*, Cambridge, 1908.
9. G. Kirchhoff, *Ueber die Auflösung der Gleichungen, auf welche man bei der Untersuchung der linearen Vertheilung galvanischer Ströme geführt wird*, Annalen d. Physik und Chemie, vol. 72(1847), p. 497.
10. M. Kraitchik, *La Mathématique des Jeux*, Brussels, 1930.
11. Z. Moroń, Przegląd Mat. Fiz., vol. 3(1925), pp. 152, 153.
12. A. Schoenflies-M. Dehn, *Einfuehrung in die analytische Geometrie der Ebene und des Raumes*, 2d ed., Berlin, 1931.
13. R. Sprague, Mathematische Zeitschrift, vol. 45 (1939), p. 607.
14. H. Steinhaus, *Mathematical Snapshots*, New York, 1938.
15. A. Stöhr, *Zerlegung von Rechtecken in inkongruente Quadrate*, Thesis, Berlin.
16. A. H. Stone, Question E. 401 and solution, American Mathematical Monthly, vol. 47 (1940).
17. H. Toepken, Aufgabe 242, Jahresberichte der deutschen Mathematiker-Vereinigung, vol. 47(1937), p. 2.
18. H. Toepken, Aufgabe 271, op. cit., vol. 48(1938), p. 73.
19. H. W. Turnbull, *Theory of Matrices, Determinants, and Invariants*, London, 1929.

Trinity College, Cambridge, England, and Princeton University.

[Extracted from the *Proceedings of the Cambridge Philosophical Society*,
Vol. XXXVII. Pt. II.]
PRINTED IN GREAT BRITAIN

ON COLOURING THE NODES OF A NETWORK

By R. L. BROOKS

Communicated by W. T. TUTTE

Received 15 November 1940

The purpose of this note is to prove the following theorem.

Let N be a network (or linear graph) such that at each node not more than n lines meet (where $n > 2$), and no line has both ends at the same node. Suppose also that no connected component of N is an n-simplex. Then it is possible to colour the nodes of N with n colours so that no two nodes of the same colour are joined.

An *n-simplex* is a network with $n+1$ nodes, every pair of which are joined by one line.

N may be infinite, and need not lie in a plane.

A network in which not more than n lines meet at any node is said to be *of degree not greater than n*. The colouring of its nodes with n colours so that no two nodes of the same colour are joined is called an "n-colouring".

Without loss of generality we may suppose that N is connected, for otherwise the theorem can be proved for each connected component; and that it is not a simplex. With these suppositions, N is finite or enumerable, both as regards nodes and lines.

Now we can $(n+1)$-colour N with colours $c_0, c_1, \ldots, c_n$, by giving to each node in turn a colour different from all those already assigned to nodes to which it is directly joined. We can then apply the following operations, in which the colours of directly joined nodes remain distinct.

(1) A node directly joined to not more than $n-1$ colours can be recoloured not-c_0. (In the term "recolouring" we include for convenience the case in which no colour is altered.) In particular, a node directly joined to two nodes of the same colour may be recoloured not-c_0.

(2) If P and Q are directly joined they can be recoloured without altering any other nodes, so that P is not-c_0. For neglecting the join PQ, we may recolour P not-c_0, by (1); and Q can then be recoloured (possibly c_0).

(3) Let $P, P', P'', \ldots, Q$ be a path, i.e. suppose every consecutive pair of nodes directly joined. Then we can recolour $P, P', \ldots, Q$ successively, without altering any other nodes, so that at most Q has finally the colour c_0.

Corollary 1. If N is finite, choose Q arbitrarily in N. Since there is a path joining Q to every node P in N, we can recolour N with at most the node Q coloured c_0.

Corollary 2. If N is infinite, let F be any connected finite part, and Q a node directly joined to, but not in, F. Then we can recolour F and Q so that no node of F is coloured c_0.

Proof of the main theorem for a finite network

Case 1. If any node X meets fewer than n lines, we can n-colour N. For let N be $(n+1)$-coloured, with at most the node X coloured c_0. Then by (1), X also can be recoloured not-c_0.

Case 2. Suppose that if P, Q, A, B are any four distinct nodes, there is a path from P to Q not including A or B.

Since N is not a simplex, we can find nodes P, Q not directly joined. Let N be $(n+1)$-coloured so that only Q is coloured c_0. Then P and all nodes directly joined to it are not-c_0. Either P meets fewer than n lines, when N may be n-coloured by case 1, or there are two nodes A, B, directly joined to P, which have the same colour. But there is a path joining P to Q, not including A or B. Hence, by (3), N can be recoloured, without altering A or B, so that at most P is coloured c_0. Since A and B have the same colour, P can be recoloured not-c_0, by (1). Thus N is n-coloured.

Case 3. Suppose there exist distinct nodes P, Q, A, B, such that every path from P to Q passes through A or B.

Then consider the networks, contained in N, with the following specifications:

N_1. *Nodes*: P, and all nodes joined to P by some path not passing through A or B as an intermediate point.

Lines: all lines connecting the above nodes in N.

N_2. *Nodes*: A, B, and all nodes of N not in N_1.

Lines: all lines connecting the above nodes in N.

Thus N_1 and N_2 are connected non-null networks, together making up the whole network N, and having in common at least one of the nodes A, B, and at most A, B, and any lines AB. Therefore if m_i is the number of lines in N_i containing A, and m_0 is the number of lines AB,

$$m_1 + m_2 \leqslant m_0 + n. \tag{X}$$

Clearly N_1 and N_2 have degree not exceeding n.

There are three subcases, 3·1, 3·2 and 3·3.

Case 3·1. Suppose N_1 and N_2 have only one node, say A, in common. Then in each of them, the node A meets fewer than n lines. Thus by case 1, N_1 and N_2

may be n-coloured; and if we permute the colours of N_2 so that the colours of A in N_1 and N_2 become the same, the whole network N is n-coloured.

Case 3·2. One of N_1 and N_2 (say N_1), is such that when the line AB is added it becomes an n-simplex.

N_1 can be n-coloured by assigning arbitrary colours to the $n-1$ nodes other than A and B, and the remaining colour to A and B. By (X) there is just one line in N_2 meeting A, and just one meeting B or case 3·1 holds. Thus if A and B were identified in N_2, there would still (since $n > 2$) be fewer than n lines meeting A $(=B)$. Hence by case 1 the resulting network can be n-coloured; i.e. N_2 can be n-coloured with A and B the same colour. The colours can be chosen so that A and B have the same as in N_1. N is therefore n-coloured.

Case 3·3. Neither N_1 nor N_2 becomes an n-simplex on adding a join AB.

Suppose they become M_1 and M_2 respectively by this addition. Then M_1 and M_2 each contain fewer nodes than N, and by (X) they are of degree not greater than n. (If not we should have case 3·1.)

If M_1 and M_2 are n-colourable so is N. For since both contain a line AB, in any n-colouring A must have a different colour from B in each network. We can permute the colours of M_1 so that the colours of A and B are the same as in M_2, and then by combination obtain an n-colouring of N.

Thus if N is a finite connected network of degree not exceeding n, and is not an n-simplex, either it is n-colourable, or it is n-colourable if two networks which satisfy the same conditions, but have fewer nodes, are n-colourable. Now it is obvious that the theorem is true for a network with less than four nodes. Therefore, by induction over the number of nodes, N is always n-colourable.

Infinite networks

If F is a network or set of nodes in N, we denote by $N-F$ the network composed of the nodes of N not in F, and the lines of N neither end of which is in F.

Lemma. For each positive r we can find a connected finite network F_r such that

(i) the connected components of $N-F_1-F_2-\dots-F_r$ are infinite for all r,

(ii) every node of N lies in one and only one F_r.

Let the nodes of N be enumerated as $P_1, P_2, P_3, \dots$, and let F_r be defined inductively as follows.

When F_s has been chosen, for $s < r$ (or without any such choice when $r = 1$), let R_r be the first P_m not in any F_s. Suppose further that the connected components of $N-F_1-F_2-\dots-F_{r-1}$ are infinite. Then all the finite connected components of $N-F_1-F_2-\dots-F_{r-1}-R_r$ must contain a node directly joined to R_r in N. Therefore the number of these components cannot exceed n, and at least one infinite component of $N - F_1 - F_2 - \ldots - F_{r-1} - R_r$ contains a node joined by a line to R_r.

Take F_r to be the "logical sum" of R_r, all finite connected components of $N - F_1 - \ldots - F_{r-1} - R_r$, and all lines joining them in N. Thus F_r is a finite connected network, and has no node in common with F_s, for $s < r$. Further, the connected components of $N - F_1 - \ldots - F_r$ are simply the infinite connected components of $N - F_1 - \ldots - F_{r-1} - R_r$. The inductive construction is therefore complete.

By the method of choosing R_r, P_m must lie in some F_s $(m \geqslant s)$.

A node Q_r can be chosen in an infinite connected component of

$$N - F_1 - \ldots - F_{r-1} - R_r,$$

so that Q_r is directly joined in N to R_r, which lies in F_r. Thus Q_r does not lie in F_s, if $s \leqslant r$.

Now N can be $(n+1)$-coloured; and by (3), corollary 2, we can recolour F_r in n colours, altering only F_r and Q_r, i.e. not altering F_s for $s < r$. Thus we can recolour F_1, F_2, ..., in turn in n colours, each recolouring not affecting the nodes already recoloured: that is, we can n-colour N.

TRINITY COLLEGE
CAMBRIDGE

SOLUTION OF THE "PROBLÈME DES MÉNAGES"

IRVING KAPLANSKY

The *problème des ménages* asks for the number of ways of seating n husbands and n wives at a circular table, men alternating with women, so that no husband sits next to his wife. Despite the considerable literature devoted to this problem (cf. the appended bibliography), the following simple solution seems to have been missed.

It is convenient first to solve two preliminary problems, perhaps of some interest in themselves.

LEMMA 1. *The number of ways of selecting k objects, no two consecutive, from n objects arrayed in a row is ${}_{n-k+1}C_k$.*

Let $f(n, k)$ be the desired number. We split the selections into two subsets: those which include the last of the n objects and those which do not. The former are $f(n-2, k-1)$ in number (since further selection of the second last object is forbidden); the latter are $f(n-1, k)$ in number. Hence

$$f(n, k) = f(n-1, k) + f(n-2, k-1),$$

and, combining this with $f(n, 1)=n$, we readily prove by induction that $f(n, k)={}_{n-k+1}C_k$.

LEMMA 2. *The number of ways of selecting k objects, no two consecutive, from n objects arrayed in a circle is ${}_{n-k}C_k n/(n-k)$.*

This differs from the preceding problem only in the imposition of the further restriction that no selection is to include both the first and last objects; and the number of such selections which are otherwise acceptable is $f(n-4, k-2)$. Hence the desired result is $f(n, k) - f(n-4, k-2) = {}_{n-k}C_k n/(n-k)$.

Presented to the Society, September 13, 1943; received by the editors May 4, 1943.

We now restate the *problème des ménages* in the usual fashion by observing that the answer is $2n!u_n$, where u_n is the number of permutations of $1, \cdots, n$ which do not satisfy any of the following $2n$ conditions: 1 is 1st or 2nd, 2 is 2nd or 3rd, $\cdots$, n is nth or 1st. Now let us select a subset of k conditions from the above $2n$ and inquire how many permutations of $1, \cdots, n$ there are which satisfy all k; the answer is $(n-k)!$ or 0 according as the k conditions are compatible or not. If we further denote by v_k the number of ways of selecting k compatible conditions from the $2n$, we have, by the familiar argument of inclusion and exclusion, $u_n = \sum(-1)^k v_k(n-k)!$. It remains to evaluate v_k, for which purpose we note that the $2n$ conditions, when arrayed in a circle, have the property that only consecutive ones are not compatible. It follows from Lemma 2 that $v_k = {}_{2n-k}C_k 2n/(2n-k)$, and hence

$$u_n = n! - \frac{2n}{2n-1}\,{}_{2n-1}C_1(n-1)! + \frac{2n}{2n-2}\,{}_{2n-2}C_2(n-2)! - \cdots.$$

From this result it follows without difficulty that $u_n/n! \to e^{-2}$ as $n \to \infty$.

Bibliography

A. Cayley, *A problem of arrangements*, Proceedings of the Royal Society of Edinburgh vol. 9 (1878) pp. 338–341.

E. Lucas, *Théorie des nombres*, Paris, 1891, pp. 491–495.

P. A. MacMahon, *Combinatory analysis*, vol. 1, Cambridge, 1915, pp. 253–254.

E. Netto, *Lehrbuch der Combinatorik*, Berlin, 1927, pp. 75–80.

J. Touchard, *Sur un problème de permutations*, C. R. Acad. Sci. Paris vol. 198 (1934) pp. 631–633.

Harvard University

Reprinted from
Bull. Amer. Math. Soc. **49** (1943), 784–785

A RING IN GRAPH THEORY

BY W. T. TUTTE

Received 10 April 1946

1. INTRODUCTION

We call a point set in a complex K a 0-*cell* if it contains just one point of K, and a 1-*cell* if it is an open arc. A set L of 0-cells and 1-cells of K is called a *linear graph* on K if

(i) no two members of L intersect,

(ii) the union of all the members of L is K,

(iii) each end-point of a 1-cell of L is a 0-cell of L

and (iv) the number of 0-cells and 1-cells of L is finite and not 0.

Clearly if L is a linear graph on K, then K is either a 0-complex or a 1-complex, and L contains at least one 0-cell.

A 1-cell of L is called a *loop* if its two end-points coincide and a *link* otherwise.

We say that L is connected if K is connected. If not then the subset of L consisting of the 0-cells and 1-cells of L which are in a component K_1 of K constitute a *component* of L. A component of a linear graph is itself a linear graph.

Let the numbers of 0-cells and 1-cells of a linear graph L on a complex K be $\alpha_0(L)$ and $\alpha_1(L)$ respectively. Then if $p_i(L) = p_i(K)$ is the Betti number of dimension i of K we have by elementary homology theory

$$\alpha_1(L) - \alpha_0(L) = p_1(L) - p_0(L). \tag{1}$$

Let L_1, L_2 be linear graphs on K_1, K_2 respectively. Then if there is a homoeomorphism of K_1 on to K_2 which maps each i-cell of L_1 on to an i-cell of L_2 $(i = 0, 1)$ we say that L_1 and L_2 are *isomorphic* and write

$$L_1 \cong L_2. \tag{2}$$

If L_1 and L_2 are two linear graphs whose complexes K_1 and K_2 do not meet, then together they constitute a linear graph L on the union of K_1 and K_2. We call it the product of L_1 and L_2 and write

$$L = L_1 L_2. \tag{3}$$

The set of all the 0-cells of a linear graph L, together with an arbitrary subset of the 1-cells constitutes a linear graph S which we call a *subgraph* of L. We call S a *subtree* of L if $p_0(S) = 1$ and $p_1(S) = 0$.

Let A be a link in a linear graph L on a complex K. By suppressing A we derive from L a linear graph L'_A on a complex K'_A. By identifying all the points of the closure of A in K and taking the resulting point as a 0-cell of the new linear graph we derive from L a linear graph L''_A on a complex K''_A.

Now there exist single-valued functions $W(L)$ on the set of all linear graphs to the ring I of rational integers which obey the general laws

$$W(L_1) = W(L_2) \quad \text{if} \quad L_1 \cong L_2 \tag{4}$$

and

$$W(L) = W(L'_A) + W(L''_A), \tag{5}$$

where A is any link of L. Some of these functions also satisfy

$$W(L_1 L_2) = W(L_1)\, W(L_2), \tag{6}$$

whenever the product $L_1 L_2$ exists.

We give here three examples; all three satisfy (4) and (5) and the last two satisfy (6). Proofs of these statements will emerge later, but the reader may easily verify them at once.

(I) $W(L)$ is the number of subtrees of L. This function is connected with the theory of Kirchhoff's Laws. A summary of its properties and an application of it to dissection problems is given in a paper entitled 'The dissection of rectangles into squares' by Brooks, Smith, Stone and Tutte (*Duke Math. J.* 7 (1940), 312–40). These authors call it the *complexity* of L.

(II) $(-1)^{\alpha_0(L)}\, W(L)$ vanishes whenever L contains a loop, and is otherwise equal to the number of single-valued functions on the set of 0-cells of L to some fixed set H of a finite number n of elements such that for each 1-cell of L the two end-points are associated with different elements of H.

Important papers dealing with such 'colourings of the 0-cells of L in n colours' are 'The coloring of graphs' by Hassler Whitney* (*Ann. Math.* 33 (1932), 688–718) and 'On colouring the nodes of a network' by R. L. Brooks (*Proc. Cambridge Phil. Soc.* 37 (1941), 194–97).

(III) If we orient the 1-cells of L and adopt the convention that the boundary of an oriented loop vanishes, we can define 1-cycles on L with coefficients in a fixed additive Abelian group G of finite order λ. $(-1)^{\alpha_0(L)+\alpha_1(L)}\, W(L)$ is the number of such 1-cycles on L in which no 1-cell has for coefficient the zero element of G.

These examples suggest that a general theory of functions satisfying the laws (4) and (5) should be constructed, and this paper represents an attempt to develop such a theory. For this purpose it is convenient to have the following definitions.

A *W-function* (*V-function*) is a single-valued function on the set of all linear graphs to an additive Abelian group G (commutative ring H) which satisfies equations (4) and (5) (equations (4), (5) and (6)).

In the second section of this paper a ring R is defined such that each linear graph L is associated with a unique element $f(L)$ of R, and it is shown that every W-function to G (V-function to H) can be expressed in the form $hf(L)$ where h is a homomorphism of R considered as an additive group (considered as a ring) into the group G (ring H), and that every such homomorphism is a W-function to G (V-function to H).

In the third section a V-function $Z(L)$ defined in terms of the subgraphs of L is studied; it is used in the next section in the proof of the following theorem.

THEOREM. *Let $(x_0, x_1, x_2, \ldots)$ be an infinite sequence of independent indeterminates over the ring I of rational integers. Then R is isomorphic with the ring of all polynomials over I in the x_i having no constant term.*

* The $p_1(L)$ of this paper is Whitney's 'nullity', and $p_0(L)$ is Whitney's P. The 'components' in this paper are Whitney's 'pieces': he uses the word 'component' with a different meaning. A footnote to Whitney's paper, dealing with some work of R. M. Forster, is particularly interesting with respect to the subject of the present paper.

Further there is a particular isomorphism in which the element of R corresponding to x_r is the element associated with a linear graph having just one 0-cell and just r 1-cells.

In the fifth section those W- and V-functions which are topological invariants of the complexes K are considered, and in the sixth section a particular V-function is applied to some colouring problems.

In the seventh section those 1-complexes K which admit of a simplicial dissection in which each 0-simplex is an end-point of not less than two and not more than three 1-simplexes are studied. A class of topologically invariant functions of these 1-complexes, one member of which is associated with a well-known colouring problem, is investigated, and it is shown that each of these functions has a unique extension as a topologically invariant W-function to all linear graphs.

2. THE RING R

From the definitions of L'_A and L''_A it is evident that

$$\alpha_0(L) = \alpha_0(L'_A) \quad = \alpha_0(L''_A)+1 \tag{7}$$

and

$$\alpha_1(L) = \alpha_1(L''_A)+1 = \alpha_1(L'_A)+1. \tag{8}$$

We say that the link A is an *isthmus* if its suppression increases the number of components of a linear graph. Evidently

$$p_0(L) = p_0(L''_A) = p_0(L'_A) \quad \text{or} \quad p_0(L'_A)-1, \tag{9}$$

according as A is not or is an isthmus. Hence by (1)

$$p_1(L) = p_1(L''_A) = p_1(L'_A) \quad \text{or} \quad p_1(L'_A)+1, \tag{10}$$

according as A is or is not an isthmus.

We call the class of all linear graphs isomorphic with L the isomorphism class L^* of L. We also use clarendon type for isomorphism classes.

If $\mathbf{L}_1$ and $\mathbf{L}_2$ are any two isomorphism classes not necessarily distinct we can find L_1 in $\mathbf{L}_1$ and L_2 in $\mathbf{L}_2$ such that the product $L_1 L_2$ exists. All products formed in this way from $\mathbf{L}_1$ and $\mathbf{L}_2$ are clearly isomorphic. We call their isomorphism class $\mathbf{L}$ the *product* of $\mathbf{L}_1$ and $\mathbf{L}_2$ and write

$$\mathbf{L} = \mathbf{L}_1 \mathbf{L}_2. \tag{11}$$

A *graphic form* is a linear form in the isomorphism classes $\mathbf{L}$ with integer coefficients of which only a finite number may be non-zero. We do not distinguish between an isomorphism class $\mathbf{L}$ and the graphic form in which the coefficient of $\mathbf{L}$ is unity and all the other coefficients are zero.

We define addition and multiplication for graphic forms by

$$\sum_i \lambda_i \mathbf{L}_i + \sum_i \mu_i \mathbf{L}_i = \sum_i (\lambda_i + \mu_i)\,\mathbf{L}_i \tag{12}$$

and

$$(\sum_i \lambda_i \mathbf{L}_i)\,(\sum_j \mu_j \mathbf{L}_j) = \sum_{i,j} (\lambda_i \mu_j)\,\mathbf{L}_i \mathbf{L}_j, \tag{13}$$

where the $\mathbf{L}_i$ are isomorphism classes and the λ_i are rational integers.

With these definitions the graphic forms are the elements of a commutative ring B. For the commutative, associative and distributive laws are evidently satisfied; and if $\mathbf{X} = \sum \lambda_i \mathbf{L}_i$ and $\mathbf{Y} = \sum_i \mu_i \mathbf{L}_i$ are any two graphic forms there is a unique graphic form $\mathbf{Z} = \sum (\lambda_i - \mu_i)\,\mathbf{L}_i$ such that $\mathbf{Y} + \mathbf{Z} = \mathbf{X}$. We write $\mathbf{Z} = \mathbf{X} - \mathbf{Y}$.

If $\mathbf{X} = \sum_i \lambda_i \mathbf{L}_i$ is any graphic form and λ an integer, we denote by $\lambda\mathbf{X}$ the graphic form $\sum_i \lambda\lambda_i \mathbf{L}_i$. We also denote by $\mathbf{O}$ the graphic form whose coefficients are all zero.

If A is a link in a linear graph L we say that the graphic form $L^* - (L'_A)^* - (L''_A)^*$ is a *W-form*. Let W denote the set of all linear combinations of a finite number of W-forms taken with integer coefficients. Then W is a modul of B, for with $\mathbf{X}$ and $\mathbf{Y}$ it contains also $\mathbf{X} - \mathbf{Y}$.

Now if L_0 is any linear graph such that the product $L_0 L$ exists we have

$$(L_0 L)'_A = L_0 L'_A \quad \text{and} \quad (L_0 L)''_A = L_0 L''_A.$$

Therefore if $\mathbf{X}$ is any W-form and $\mathbf{L}$ any isomorphism class, then $\mathbf{LX}$ is also a W-form. Hence by (12) and (13) for any $\mathbf{Y} \in W$ and any $\mathbf{Z} \in B$, we have $\mathbf{YZ} \in W$. That is, W is an ideal of the commutative ring B. We denote the difference ring $B - W$ by R.

The elements of R are the cosets mod. W in B. If we denote the coset of $\mathbf{X}$ mod. W by $[\mathbf{X}]$, addition and multiplication in R satisfy

$$[\mathbf{X}] + [\mathbf{Y}] = [\mathbf{X} + \mathbf{Y}], \tag{14}$$

$$[\mathbf{X}][\mathbf{Y}] = [\mathbf{XY}]. \tag{15}$$

THEOREM I. *A single-valued function $W(L)$ on the set of all linear graphs L to an additive Abelian group G (commutative ring H) is a W-function (V-function) if and only if it is of the form $h[L^*]$ where h is a homomorphism of the additive group R (ring R) into G (H).*

Now the functions $W(L)$ which satisfy (4) depend only on the isomorphism classes. For such functions we write $W(\mathbf{L}) = W(L)$ where $\mathbf{L}$ is the isomorphism class of the linear graph L. $W(\mathbf{L})$ can now be extended to all graphic forms by writing

$$W\left(\sum_i \lambda_i \mathbf{L}_i\right) = \sum_i \lambda_i W(\mathbf{L}_i). \tag{16}$$

If $W(L)$ satisfies (6) we have also $W(\mathbf{L}_1 \mathbf{L}_2) = W(\mathbf{L}_1) W(\mathbf{L}_2)$ for any two isomorphism classes $\mathbf{L}_1$ and $\mathbf{L}_2$, and therefore, by (13) and (16),

$$W(\mathbf{X}_1 \mathbf{X}_2) = W(\mathbf{X}_1) W(\mathbf{X}_2), \tag{17}$$

where $\mathbf{X}_1$ and $\mathbf{X}_2$ are any two graphic forms.

By (16) and (17) any single-valued function $W(L)$ satisfying equation (4) (equations (4) and (6)) is of the form $h_0 L^*$ where h_0 is a homomorphism of B considered as an additive group (considered as a ring) into G (H); and conversely it is evident that if h_0 is any such homomorphism, the function $h_0 L^*$ satisfies equation (4) (equations (4) and (6)).

$W(L)$ then satisfies (5) if and only if h_0 maps all W-forms and therefore all elements of W on to the zero element of G (H). This is equivalent to the condition that $h_0 L^*$ shall depend only on the coset $[L^*]$. The theorem now follows from (14) and (15).

Let y_r denote any linear graph having just one 0-cell and just r 1-cells (necessarily loops). Clearly all such linear graphs (for a fixed r) are isomorphic. We denote their isomorphism class by $\mathbf{y}_r$. We call the members of the $\mathbf{y}_r$ *elementary graphs*. Clearly

$$p_0(y_r) = 1, \tag{18}$$

$$p_1(y_r) = r. \tag{19}$$

THEOREM II. *If L is any linear graph then $[L^*]$ can be expressed as a polynomial $P[L^*] = P([L^*]; [\mathbf{y}_0], [\mathbf{y}_1], [\mathbf{y}_2], \ldots)$ in the $[\mathbf{y}_i]$ such that*

(i) *$P[L^*]$ has no constant term,*

(ii) *the coefficients of $P[L^*]$ are non-negative rational integers,*

(iii) *the degree of $P[L^*]$ is $\alpha_0(L)$,*

(iv) *$P[L^*]$ involves no suffix i greater than $p_1(L)$,*

and (v) *if L is connected and has no isthmus A such that for some component L_0 of L'_A, $p_1(L_0) = 0$; then $P[L^*]$ is of the form $[\mathbf{y}_p] + [\mathbf{Q}]$ where $p = p_1(L)$ and $[\mathbf{Q}]$ is a polynomial in those $[\mathbf{y}_i]$ for which i is less than p.*

The proof is by induction. We first observe that if $\alpha_1(L)$ is zero, then L is the product of $\alpha_0(L)$ elementary graphs each isomorphic with y_0. Hence by (15)

$$[L^*] = [\mathbf{y}_0]^{\alpha_0(L)},$$

and so the theorem is true for L.

Assume that the theorem is true for all connected linear graphs having fewer than some finite number n of 1-cells. Let L be any linear graph having just n 1-cells. If L is connected, then either $\alpha_0(L) = 1$, in which case

$$[L^*] = [\mathbf{y}_n],$$

and so the theorem is true for L, or else L contains a link A. In the second case we have

$$(L^* - (L'_A)^* - (L''_A)^*) \in W,$$

and therefore
$$[L^*] = [(L'_A)^*] + [(L''_A)^*]. \qquad (20)$$

By (8) L'_A and L''_A have each fewer 1-cells than L and so by the inductive hypothesis the theorem is true for them. The propositions (i) to (iv) follow immediately for L from (20) with the help of (7) and (10).

Now suppose that L satisfies the conditions of (v). Then L''_A also satisfies these conditions since it is formed from L by identifying all the points and end-points of a link, a process which cannot alter the number of components of L'_B, where B is any link other than A of L. Hence by hypothesis

$$[(L''_A)^*] = [\mathbf{y}_p] + [\mathbf{Q}_p], \qquad (21)$$

where p is $p_1(L)$, and $[\mathbf{Q}_p]$ denotes any polynomial (not always the same polynomial) in those $[\mathbf{y}_i]$ for which $i < p$.

We also have
$$[(L'_A)^*] = [\mathbf{Q}_p]. \qquad (22)$$

This follows at once from (10) and (iv) when A is not an isthmus. If A is an isthmus. L'_A is of the form $L_0 L_1$. Since L satisfies the conditions of (v), $p_1(L_0), p_1(L_1) > 0$, and therefore since $p_1(L_0) + p_1(L_1) = p_1(L'_A)$ we have $p_1(L_0), p_1(L_1) < p_1(L'_A)$. Consequently $[(L'_A)^*] = [(L_0)^*][(L_1)^*] = [\mathbf{Q}_p]$ and (22) is still valid.

By (20), (21) and (22)
$$[L^*] = [\mathbf{y}_p] + [\mathbf{Q}_p].$$

This completes the proof that the theorem for connected linear graphs is true when $\alpha_1(L) = n$ if it is true for $\alpha_1(L) < n$. We have proved it for $\alpha_1(L) = 0$ and therefore it is true in general. If L is not connected we can obtain $P[L^*]$ satisfying the theorem by multiplying together the polynomials of its components.

COROLLARY. *Any element $[\mathbf{X}]$ of R can be expressed as a polynomial in the $[\mathbf{y}_i]$ with rational integer coefficients and no constant term.*

For $\mathbf{X}$ is a finite linear form in the $\mathbf{L}_i$ with integer coefficients.

3. Subgraphs

Let S denote any subgraph of a linear graph L. Let the number of components T of S such that $p_1(T) = r$ be $i_r(S)$. We define a function $Z(L)$ of L by

$$Z(L) = \sum_S \prod_r z_r^{i_r(S)}, \tag{23}$$

where the z_r are independent indeterminates over the ring I of rational integers. Although (23) involves a formal infinite product, yet for a given S only a finite number of the $i_r(S)$ can be non-zero and so, for each L, $Z(L)$ is a polynomial in the z_i.

Theorem III. *$Z(L)$ is a V-function.*

For first it is obvious that $Z(L)$ satisfies (4).

Secondly, if A is any link of L, then the subgraphs of L which do not contain A are simply the subgraphs of L'_A, and the subgraphs S of L which do contain A are in 1-1 correspondence with the subgraphs S''_A of L''_A. For, for such an S, S''_A is a subgraph of L''_A; and if S_1 is any subgraph of L''_A there is one and only one subgraph S of L having the same 1-cells as S_1 with the addition of A and therefore satisfying $S''_A = S$.

Further S''_A differs from S only in that a component T of S is replaced by T''_A; and, by (9) and (10), T''_A is connected and $p_1(T'') = p_1(T)$. Hence $i_r(S''_A) = i_r(S)$ for all r.

Hence by (23) $$Z(L) = \sum_{S(L'_A)} \prod_r z_r^{i_r(S)} + \sum_{S(L''_A)} \prod_r z_r^{i_r(S)},$$

where $S(L'_A)$ for example denotes a subgraph S of L'_A. Therefore

$$Z(L) = Z(L'_A) + Z(L''_A), \tag{24}$$

so that $Z(L)$ satisfies (5).

Thirdly, for any product $L_1 L_2$ the subgraphs of $L_1 L_2$ are simply the products of the subgraphs S_1 of L_1 with the subgraphs S_2 of L_2. It is evident that

$$i_r(S_1 S_2) = i_r(S_1) + i_r(S_2),$$

and therefore $$Z(L_1 L_2) = \sum_{S_1, S_2} \prod_r z_r^{i_r(S_1)+i_r(S_2)}$$

$$= \left(\sum_{S_1} \prod_r z_r^{i_r(S_1)}\right)\left(\sum_{S_2} \prod_r z_r^{i_r(S_2)}\right) = Z(L_1)\,Z(L_2). \tag{25}$$

Thus $Z(L)$ satisfies (4), (5) and (6). That is, it is a V-function.

Theorem IV. $$Z(y_r) = \sum_i \binom{r}{i} z_{r-i}. \tag{26}$$

For each subgraph of y_r has just one 0-cell (§ 2), and therefore just one component. Hence $Z(y_r)$ is a linear form in the z_r. The number of subgraphs S such that $p_1(S) = k$ is the number with $\alpha_1(S) = k$, by (19), and this is the number of ways of choosing k 1-cells out of r.

4. Structure of the ring R

Lemma. $$\sum_{i=0}^{r} (-1)^i \binom{r}{i}\binom{i}{j} = (-1)^r \delta_{rj}. \tag{27}$$

This equality can be obtained by expanding $x^r = ((x-1)+1)^r$ in powers of $(x-1)$, expanding each of the terms in the resulting series in powers of x, and then equating coefficients.

THEOREM V. *R is isomorphic with the ring R_0 of all polynomials in the z_i with integer coefficients and no constant term.*

For by Theorem III $Z(L)$ is a V-function with values in R_0. Hence by Theorem I

$$Z(L) = h[L^*], \tag{28}$$

where h is a homomorphism of R into R_0.

Let $[\mathbf{t}_i]$ be the element of R defined by

$$[\mathbf{t}_i] = \sum_{j=0}^{i} (-1)^{i+j} \binom{i}{j} [\mathbf{y}_j]. \tag{29}$$

Then, by Theorem IV and the lemma,

$$h[\mathbf{t}_i] = \sum_{j=0}^{i} \sum_{s=0}^{j} (-1)^{i+j} \binom{i}{j} \binom{j}{s} z_s = z_i. \tag{30}$$

If we multiply (29) by $\binom{r}{i}$, sum from $i = 0$ to $i = r$, and use the lemma we find

$$[\mathbf{y}_r] = \sum_{i=0}^{r} \binom{r}{i} [\mathbf{t}_i]. \tag{31}$$

Hence by Theorem II, Corollary, any element $[\mathbf{X}]$ of R can be expressed as a polynomial in the $[\mathbf{t}_i]$ with integer coefficients and no constant term. Moreover this expression is unique; otherwise there would be a polynomial relationship between the $[\mathbf{t}_i]$, and therefore by (30) between the z_i, with integer coefficients, and this would contradict the definition of the z_i. It follows that h is an isomorphism of R on to R_0 (for every integer polynomial in the $[\mathbf{t}_i]$ is in R).

THEOREM VI. *Let $x_0, x_1, x_2, \ldots$ be an infinite sequence of connected linear graphs, and $\mathbf{x}_0, \mathbf{x}_1, \mathbf{x}_2, \ldots$ the corresponding isomorphism classes, such that*

(i) $x_0 \cong y_0$,

(ii) $p_1(x_r) = r$,

and (iii) *x_r contains no isthmus A such that for some component L_0 of $(x_r)'_A$, $p_1(L_0) = 0$.*

Then any element $[\mathbf{X}]$ of R has a unique expression as a polynomial in the $[\mathbf{x}_i]$ with integer coefficients and no constant term.

By Theorem II (v) and equation (31) we have, for $r > 0$,

$$[\mathbf{x}_r] = [\mathbf{t}_r] + [\mathbf{S}_r], \tag{32}$$

where $[\mathbf{S}_r]$ is a polynomial in those $[\mathbf{t}_i]$ for which $i < r$. Hence

$$[\mathbf{t}_r] = [\mathbf{x}_r] + [\mathbf{U}_r], \tag{33}$$

where $[\mathbf{U}_r]$ is a polynomial in those $[\mathbf{x}_i]$ for which $i < r$. (If we assume this for $r < n$ it follows for $r = n$ by substitution in (32). Since $[\mathbf{x}_0] = [\mathbf{y}_0] = [\mathbf{t}_0]$ it is true for $r = 0$, and therefore it is true in general.) Clearly $[\mathbf{S}_r]$ and $[\mathbf{U}_r]$ have no constant terms.

By Theorem II, Corollary, and equations (31) and (33), $[\mathbf{X}]$ can be expressed as a polynomial without a constant term in the $[\mathbf{x}_i]$.

Suppose this expression not unique. Then there will be a polynomial relationship

$$P([\mathbf{x}_i]) = 0 \tag{34}$$

between the $[\mathbf{x}_i]$. Of the terms of non-zero coefficient in $P([\mathbf{x}_i])$ pick out the subset M_1 of those which involve the greatest suffix occurring in them raised to the highest power to which it occurs. Of this subset M_1 pick out the subset M_2 of terms involving the second greatest suffix appearing in M_1 raised to the highest power to which it occurs in M_1, and so on. This process must terminate in a subset M_k consisting of a single term

$$A[\mathbf{x}_i]^{a(i)}[\mathbf{x}_j]^{a(j)}\dots.$$

It is evident that if we substitute from (32) in (34), we shall obtain a polynomial relationship

$$Q([\mathbf{t}_i]) = 0$$

between the $[\mathbf{t}_i]$, in which the coefficient of $[\mathbf{t}_i]^{a(i)}[\mathbf{t}_j]^{a(j)}\dots$ is $A \neq 0$. But it was shown in the proof of Theorem V that there is no polynomial relationship between the $[\mathbf{t}_i]$.

This contradiction proves uniqueness and so completes the proof of the theorem.

5. Topologically invariant W-functions

Let A be a 1-cell of a linear graph L on a complex K. Let p be any point of A. We can obtain a new linear graph M on K from L by replacing A by the point p, taken as a 0-cell of M, and the two components of $A-p$ taken as 1-cells of M. We call this operation a subdivision of A by p.

Given any two linear graphs L_1, L_2 on the same K we can find a linear graph L_3 which can be obtained from either by suitable subdivisions. Such a linear graph is evidently obtained by taking as the set V of 0-cells the set of all points of K which are 0-cells either of L_1 or of L_2, and by taking as 1-cells the components of $K-V$.

We seek the condition that a W-function $W(L)$ shall be topologically invariant, i.e. depend only on K. By the above considerations a necessary and sufficient condition for this is that $W(L)$ shall be invariant under subdivision operations. (For then $W(L_1) = W(L_3) = W(L_2)$.)

Suppose therefore that A is any 1-cell of L, possibly a loop, and let M be obtained from L by subdividing A by a point p. Let us denote the new 1-cells by B and C. Then by (5) for any W-function $W(L)$

$$\begin{aligned} W(M) &= W(M'_B) + W(M''_B) \\ &= W((M'_B)'_C) + W((M'_B)''_C) + W(M''_B) \\ &= W(p\,.\,(M'_B)''_C) + W((M'_B)''_C) + W(M''_B). \end{aligned}$$

Here p is used to denote the linear graph which consists solely of the 0-cell p. It is isomorphic to y_0.

By making use of the obvious isomorphisms $M''_B \cong L$ and $(M'_B)''_C \cong L_0$, where L_0 is the linear graph derived from L by suppressing A, we obtain

$$W(M) - W(L) = W(y_0\,.\,L_0) + W(L_0),$$

Therefore

$$W(M) - W(L) = h([\mathbf{y}_0][L_0^*] + [L_0^*]), \tag{35}$$

where h is a homomorphism of R, regarded as an additive group, into an additive Abelian group G (Theorem I).

Let N denote the set of all elements of R which are of the form $[\mathbf{y}_0][\mathbf{X}] + [\mathbf{X}]$. Clearly N is an ideal of R. Let $\{\mathbf{X}\}$ denote that element of the difference ring $R-N$ which contains $[\mathbf{X}]$.

Theorem VII. *A function $W(L)$ on the set of all linear graphs L to the additive Abelian group G (commutative ring H) is a topologically invariant W-function (V-function) if* *and only if it is of the form $k\{L^*\}$, where k is a homomorphism of the additive group $R-N$ (ring $R-N$) into G (H).*

For in (35), by a proper choice of L, we can have any linear graph we please as L_0. It follows that the necessary and sufficient condition for the W-function $W(L)$ to be topologically invariant is that h shall map all elements of R of the form $[\mathbf{y}_0][L^*]+[L^*]$ and therefore all elements of N on to the zero of G. This proves the theorem for W-functions. The same argument applies to V-functions, except that h in (35) is then a homomorphism of R (as a ring) into the ring H.

THEOREM VIII. *Let $x_0, x_1, x_2, \ldots$ be as in the enunciation of Theorem VI.*

Then any element $\{\mathbf{X}\}$ of $R-N$ has a unique expression as a polynomial in the $\{\mathbf{x}_i\}$ $(i>0)$ with integer coefficients.

For we can obtain such an expression for $\{\mathbf{X}\}$ by replacing each $[\mathbf{x}_i]$ by the corresponding $\{\mathbf{x}_i\}$ in the expression for $[\mathbf{X}]$ in terms of the $[\mathbf{x}_i]$ whose existence is asserted in Theorem VI. Now for all $\{\mathbf{X}\}$, $\{\mathbf{X}\}+\{\mathbf{y}_0\}\{\mathbf{X}\}=\{\mathbf{O}\}$, and so $R-N$ has a unity element $-\{\mathbf{y}_0\}=-\{\mathbf{x}_0\}$ which we may denote by 1. Hence $\{\mathbf{x}_0\}$ is not an indeterminate over I. and we can regard our polynomial for $\{\mathbf{X}\}$ as a polynomial in those $\{\mathbf{x}_i\}$ for which $i>0$ (with perhaps a constant term).

If this expression for $\{\mathbf{X}\}$ is not unique then there will be a polynomial $\{\mathbf{P}\}$ in the $\{\mathbf{x}_i\}$ $(i>0)$ without a constant term such that

$$A\{\mathbf{x}_0\}+\{\mathbf{P}\}=\{\mathbf{O}\},$$

where A is some integer. Hence if $[\mathbf{P}]$ is the polynomial of the same form in the $[\mathbf{x}_i]$ we must have

$$A[\mathbf{x}_0]+[\mathbf{P}]+[\mathbf{X}_0]+[\mathbf{x}_0][\mathbf{X}_0]=[\mathbf{O}] \tag{36}$$

for some $[\mathbf{X}_0]$.

Equating coefficients of like powers of $[\mathbf{x}_0]$, as is permissible by Theorem VI, we see that $[\mathbf{X}_0]$ cannot involve $[\mathbf{x}_0]$, and hence that $A=-[\mathbf{X}_0]=[\mathbf{P}]$. Consequently $\{\mathbf{P}\}$ is a constant and therefore, by its definition, the zero polynomial in the $\{\mathbf{x}_i\}$. The theorem follows.

6. SOME COLOURING PROBLEMS

The homomorphism of the ring R_0 (see Theorem V) into the ring of polynomials in two independent indeterminates t and z by the correspondence $z_i \to tz^i$ transforms $Z(L)$ into

$$Q(L;t,z)=\sum_S t^{p_0(S)}z^{p_1(S)} \tag{37}$$

by (23). Since $Z(L)$ is of the form $h[L^*]$ where h is a homomorphism of R into R_0 (Theorems I and III). $Q(L;t,z)$ can be defined by a homomorphism of R into the ring of polynomials in t and z and is therefore a V-function (Theorem I).

The coefficient of $t^a z^b$, for fixed a, b, therefore satisfies (4) and (5) and so is a W-function. Writing $a=1, b=0$ we obtain the function of Example I of the Introduction. This function satisfies $W(L_1L_2)=0$ (by (37) since $p_0(S)$ is always positive) and so it can be regarded as a V-function with values in the ring constructed from the additive group of the rational integers by defining the 'product' of any two elements as 0.

$Q(L;t,z)$ has an interesting property which we call

THEOREM IX. *If L_1 and L_2 are connected dual linear graphs on the sphere then*

$$\frac{1}{t}Q(L_1; t, z) = \frac{1}{z}Q(L_2; z, t). \tag{38}$$

This follows from (37) as a consequence of the fact that there is a 1-1 correspondence $S \to S'$ between the subgraphs S of L_1 and the subgraphs S' of L_2 such that

$$p_0(S) = p_1(S') + 1$$

and

$$p_1(S) = p_0(S') - 1.$$

(S' is that subgraph of L_2 whose 1-cells are precisely those not dual to 1-cells of S.) For a proof of this proposition reference may be made to the paper 'Non-separable and planar graphs' by Hassler Whitney (*Trans. American Math. Soc.* 34 (1932), 339–62).

We go on to consider two kinds of colourings of a linear graph, which we distinguish as *α-colourings* and *β-colourings*. An α-colouring of L of degree λ is a single-valued function on the set of 0-cells of L to a fixed set H the number of whose elements is λ.

If f is an α-colouring let $\phi(f)$ denote the number of 1-cells A of L such that f associates all the end-points of A with the same element of H (e.g. every loop has this property). We say that any subgraph of L all of whose 1-cells have this property for f is *associated* with f. We use the symbol $S(f)$ to denote a subgraph associated with a given f, and $f(S)$ to denote any α-colouring with which a given S is associated.

THEOREM X. *Let $J(L; \lambda, \phi)$ be the number of α-colourings f of L of degree λ for which $\phi(f)$ has the value ϕ. Then the following identity is true.*

$$\sum_{\phi} J(L; \lambda, \phi)\, x^{\phi} = (x-1)^{\alpha_0(L)}\, Q\left(L; \frac{\lambda}{x-1}, x-1\right) \tag{39}$$

where x is an indeterminate over I.

For, by (37) and (1), the right-hand side is

$$(x-1)^{\alpha_0(L)} \sum_S \lambda^{p_0(S)} (x-1)^{p_1(S)-p_0(S)} = \sum_S (x-1)^{\alpha_1(S)} \lambda^{p_0(S)}$$

$$= \sum_S \left((x-1)^{\alpha_1(S)} \sum_{f(S)} 1\right);$$

for the α-colourings associated with S are precisely those which map all the 0-cells in the same component of S on to the same element of H. This last expression equals

$$\sum_f \sum_{S(f)} (x-1)^{\alpha_1(S)} = \sum_f x^{\phi(f)}$$

since the number of subgraphs associated with f and having just $\alpha_1(S)$ 1-cells is the number of ways of choosing $\alpha_1(S)$ 1-cells out of $\phi(f)$. This completes the proof of the theorem.

If we write $x = 0$ in (39) we find that $(-1)^{\alpha_0(L)} J(L; \lambda, 0)$, which is Example II of the Introduction, is the V-function $Q(L; -\lambda, -1)$. We thus obtain the well-known result*

$$J(L; \lambda, 0) = \sum_S (-1)^{\alpha_1(S)} \lambda^{p_0(S)}.$$

* Hassler Whitney, 'A logical expansion in mathematics', *Bull. American Math. Soc.* 38 (1932), 572–9.

If we orient the 1-cells of L and adopt the convention that the boundary of an oriented loop vanishes, we can define 1-cycles on L with coefficients in some fixed additive Abelian group G of finite order λ. The number* of such 1-cycles on L will be $\lambda^{p_1(L)}$. We call them *β-colourings* of L with respect to G.

Let $E(L;\, G, \psi)$ be the number of such 1-cycles for which just ψ of the 1-cells have coefficient zero. Let g_G be any β-colouring with respect to G of L and let $\psi(g_G)$ be the number of its zero coefficients†. We say that a subgraph S of L is associated with g_G if every 1-cell of L not in S is assigned the zero element of G as its coefficient in g_G. We use the symbol $S(g_G)$ to denote a subgraph of L associated with g_G and $g_G(S)$ to denote a β-colouring with which a given subgraph is associated. Clearly the number of β-colourings associated with a given S is the number of β-colourings of S, which is $\lambda^{p_1(S)}$.

THEOREM XI. *If x is an indeterminate over I then*

$$\sum_{\psi} E(L;\, G.\psi)\, x^{\psi} = (x-1)^{\alpha_1(L)-\alpha_0(L)}\, Q\left(L;\, x-1, \frac{\lambda}{x-1}\right). \tag{40}$$

For, by (37) and (1), the right-hand side is

$$(x-1)^{\alpha_1(L)-\alpha_0(L)} \sum_{S} (x-1)^{p_0(S)-p_1(S)} \lambda^{p_1(S)} = \sum_{S} \left((x-1)^{\alpha_1(L)-\alpha_1(S)} \sum_{g_G(S)} 1\right)$$
$$= \sum_{g_G,\, S(g_G)} (x-1)^{\alpha_1(L)-\alpha_1(S)} = \sum_{g_G} x^{\psi(g_G)};$$

for the number of subgraphs of L associated with g_G and having just $\alpha_1(L)-\psi(g_G)+r$ 1-cells is the number of ways of choosing r 1-cells out of the $\psi(g_G)$ which have zero coefficient in g_G.

COROLLARY. *$E(L;\, G, \psi)$ is the same for all additive Abelian groups G of the same order λ.*

If we write $x = 0$ in (40) we find that $(-1)^{\alpha_1(L)-\alpha_0(L)}\, E(L;\, G, 0)$, which is Example III of the Introduction, is the V-function $Q(L;\, -1, -\lambda)$. It takes the value -1 when L is y_0 and therefore corresponds to a homomorphism of R into the ring of rational integers which maps N into 0. It is therefore, by the preceding section, topologically invariant.

If L_1 and L_2 are dual linear graphs on the sphere, the β-colourings of L_1 are closely connected with the α-colourings of L_2. In fact a 1-cycle g_G bounds on the sphere and any 2-chain which it bounds on the map defined by L_1 has a dual 0-chain which is an α-colouring f_λ of L_2 such that $\phi(f_\lambda) = \psi(g_G)$.

There is also a relationship between the α-colourings and the β-colourings of the same linear graph L expressed by the following identity in x

$$(x-1)^{\alpha_1(L)} \sum_{\psi} \left(E(L;\, G, \psi) \left(\frac{\lambda}{x-1}+1\right)^{\psi}\right) = \lambda^{\alpha_1(L)-\alpha_0(L)} \sum_{\phi} J(L;\, \lambda, \phi)\, x^{\phi}. \tag{41}$$

This is obtained by writing $\lambda/(x-1)$ for $(x-1)$ in (40) and then eliminating the function Q by means of (39).

* See Lefschetz, *Algebraic Topology* (Amer. Math. Soc. Colloquium Publications, vol. 27), p. 106.

† It may be mentioned that for graphs on the sphere a β-colouring is essentially equivalent to a colouring of the regions of the map defined by a graph in λ colours. The colours can be represented by elements of G and so the colouring can be represented by a 2-chain on the map with coefficients in G. A β-colouring is simply the boundary of such a 2-chain. The number of 1-cells incident with two regions of the same colour (or incident with only one region) in a given colouring is given by the number $\psi(g_G)$ where g_G is the corresponding β-colouring.

7. Cubical networks

We define a *cubical network* as a 1-complex for which there exists a finite simplicial dissection in which each 0-simplex is incident with not less than two, and not more than three 1-simplexes. Clearly any other simplicial dissection of such a complex will have the same property. The 0-simplexes which are each incident with three 1-simplexes we call *nodes*. The set of nodes is evidently independent of the particular simplicial dissection taken.

A component of a cubical network which does not contain a node is evidently a simple closed curve, and if a component does contain nodes then the remainder of it

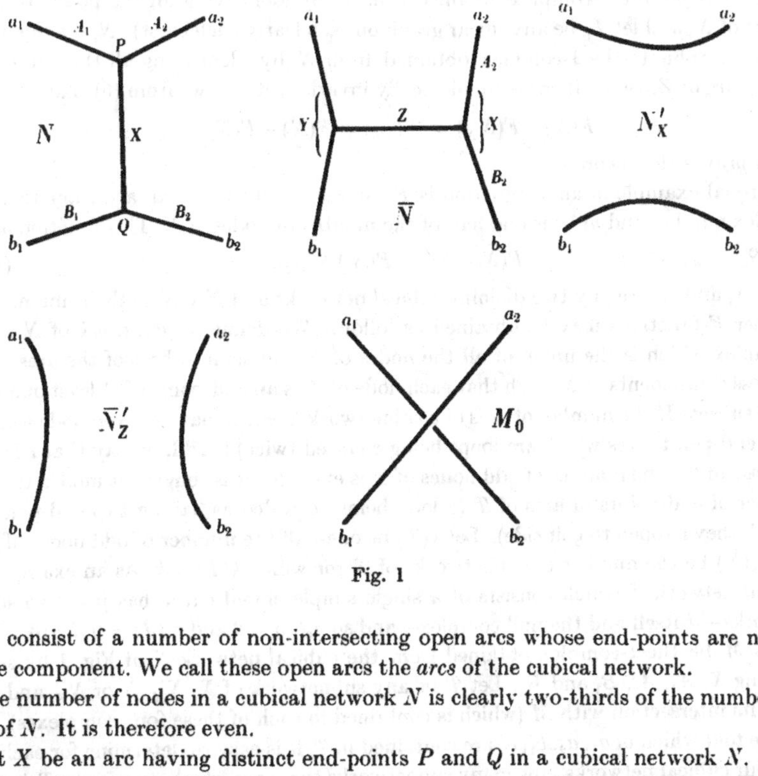

Fig. 1

must consist of a number of non-intersecting open arcs whose end-points are nodes of the component. We call these open arcs the *arcs* of the cubical network.

The number of nodes in a cubical network N is clearly two-thirds of the number of arcs of N. It is therefore even.

Let X be an arc having distinct end-points P and Q in a cubical network N. In a simplicial dissection of N let A_1, A_2 be those 1-simplexes incident with P, and B_1, B_2 those 1-simplexes incident with Q, which are not in X. Let a_1, a_2, b_1, b_2 be the other end-points of A_1, A_2, B_1, B_2 respectively. By suitable subdivisions of a given simplicial dissection we can always arrange that a_1, b_1, a_2, b_2 are distinct points and not nodes of N.

Other cubical networks can be obtained from N by replacing X, A_1, A_2, B_1, B_2, P and Q by other systems of simplexes (see Fig. 1). If for example we suppress A_1 and B_1, introduce a new arc Y joining a_1 to b_1 and then introduce an arc Z joining a point in Y

to a point in X, we obtain $\bar{N}$. We call this process a Λ-*operation* on N. If N_1 can be obtained from N by a finite sequence of Λ-operations we say that N and N_1 are Λ-*equivalent*. In such a case it is clear that N_1 has the same number of nodes as N and that if N is connected, so is N_1.

By suppressing X in N we obtain N'_X, and by suppressing Z in $\bar{N}$ we obtain $\bar{N}'_Z$. We define an *F-function* as a single-valued topologically invariant function on the set of all cubical networks to an additive Abelian group G or commutative ring H which satisfies the general law

$$F(N)-F(N'_X) = F(\bar{N})-F(\bar{N}'_Z). \tag{42}$$

THEOREM XII. *If $W(L)$ is a topologically invariant W-function, and $F(N)$ is the value of $W(L)$ for any linear graph on the cubical network N, then $F(N)$ is an F-function.*

For let N_0 be the 1-complex obtained from N by identifying all the points of the closure of X, and let L_0 be any linear graph on N_0 (clearly such exist). N_0 is evidently homoeomorphic to the 1-complex obtained from $\bar{N}$ by identifying all the points of the closure of Z. Since $W(L)$ is topologically invariant it follows from (5) that

$$F(N)-F(N'_X) = W(L_0) = F(\bar{N})-F(\bar{N}'_Z),$$

which proves the theorem.

A trivial example of an F-function is $F(N) = x^{n(N)}$ where x is an arbitrary real or complex number and $n(N)$ is one-half of the number of nodes of N. This function also satisfies

$$F(N_1 \cup N_2) = F(N_1)\,F(N_2), \tag{43}$$

where N_1 and N_2 are any two disjoint cubical networks and $N_1 \cup N_2$ is their union.

Other F-functions may be obtained as follows. We define a *subnetwork* of N as a 1-complex which is the union of all the nodes of N and some subset of the arcs and nodeless components of N, such that each node of N is an end-point of at least one arc of the subset. If the number of arcs of a subnetwork T which have a given node v of N as an end-point (arcs which are loops being counted twice) is odd, we say that v is an *odd node* of T. The number of odd nodes of T is even, for it is congruent mod. 2 to the number of end-points of arcs of T (a loop being regarded as having two end-points, though they happen to coincide). Let $k(T)$ be one-half the number of odd nodes of T. Let $\pi_k(N)$ be the number of subnetworks of N for which $k(T) = k$. As an example a cubical network J which consists of a single simple closed curve has just two subnetworks—J itself and the null complex—and so $\pi_0(J) = 2$ and $\pi_i(J) = 0$ $(i > 0)$.

Let M be the 1-complex obtained from the cubical network N of Fig. 1 by suppressing X, A_1, A_2, B_1 and B_2. Let T be any subnetwork of N, N'_X, $\bar{N}$ or $\bar{N}'_Z$, and let T_0 be its intersection with M (which is contained in each of these four complexes). If we are told which of a_1, a_2, b_1, b_2 are contained in T_0 it is easy to determine for each of the four cubical networks how many subnetworks there are which agree with T_0 in M, and how many of these have 0 (or 1, or 2) odd nodes outside T_0. A consideration of the possible cases will show

$$\pi_k(N)+\pi_k(N'_X) = \pi_k(\bar{N})+\pi_k(\bar{N}'_Z), \tag{44}$$

whence $(-1)^{n(N)}\pi_k(N)$ satisfies (42) and is thus an F-function.

If therefore we define a polynomial $D(N; x)$ by

$$D(N; x) = \sum_k \pi_k(N)\,x^k$$

then $(-1)^{n(N)}D(N; x)$ will be an F-function. Further, by an argument analogous to the proof of (25) this F-function satisfies (43).

If N has no nodeless component, $\pi_0(N) = D(N; 0)$ is by its definition the number of solutions of Petersen's problem* for N.

We define a *Hamiltonian circuit* of N as a subnetwork of N which is connected and has no odd nodes. It is easily verified that the residue mod. 2 of the number of Hamiltonian circuits of N satisfies (42), so this also is an F-function.

Let γ_{i+1} $(i \geqslant 1)$ be a cubical network with just $2i$ nodes $a_1, a_2, a_3, \ldots, a_{2i}$, having just one arc linking each pair of nodes a_r, a_{r+1} for which r is odd, having just two arcs linking each pair of nodes a_r, a_{r+1} for which r is even, and having two arcs which are loops the end-points of one coinciding in a_1 and those of the other in a_{2i}. The nodes and arcs define a linear graph which we also denote by γ_{i+1}.

THEOREM XIII. *Any connected cubical network N of $2n$ nodes $(n > 0)$ is Λ-equivalent to a homoeomorph of γ_{n+1}.*

For first, if N, not being homoeomorphic to γ_{n+1}, contains a simple closed curve K of $k > 1$ arcs, then N is Λ-equivalent to a cubical network N_1 containing a simple closed curve of $k-1$ arcs. For we can suppose that K contains the arc X (Fig. 1) and also a_1 and b_1. Then $\bar{N}$ clearly has the property desired. It follows that by a sequence of Λ-operations we can convert N into a cubical network having a loop.

Let δ_r be the 1-complex derived from γ_{r+1} $(r > 0)$ by suppressing the loop on a_{2r}. If part of a cubical network M meeting the rest of M only in a single node is homoeomorphic with δ_r, we call it a *frond* of M of *degree* r, and say that the node corresponding to a_{2r} is the *base* of the frond. The above argument showed that N is Λ-equivalent to a cubical network N_2 having a frond f (of degree r say).

Secondly either N_2 contains a simple closed curve passing through the base of f, or it is Λ-equivalent to a cubical network having a frond of degree at least r with a simple closed curve through its base. For if the base c_0 of f is not on such a curve there will be a sequence $c_0, c_1, c_2, c_3, \ldots, c_s$ of minimum length such that consecutive nodes c_i, c_{i+1} are linked by an arc C_i, and such that c_s is on a simple closed curve K_1 in N_2. Otherwise we could extend the sequence $c_0, c_1, c_2, \ldots$ indefinitely in such a way that C_i differed from C_{i+1} for each i without repetitions, which is absurd since N_2 has only a finite number of nodes. By Λ-operations on $C_0, C_1, \ldots$ in turn it is possible to transfer the frond to a base on a simple closed curve without altering its degree.

Now at this stage the simple closed curve through the base of the frond may be a loop, in which case N has been transformed into a γ_i-homoeomorph, and $i = n+1$ since connexion and number of nodes are invariant under Λ-operations; or it may contain just two arcs in which case N_2 has been transformed into a cubical network having a frond of degree exceeding r; or it can be reduced to a curve of just two arcs by a sequence of Λ-operations on those of its arcs not meeting the base of the frond. Hence if N_2 is not homoeomorphic with γ_{n+1} it can be transformed into a cubical network with a frond of degree greater than r. A finite number of such transformations will therefore change it into a homoeomorph of γ_{n+1}.

* Dénes König, *Theorie der Endlichen und unendlichen Graphen* (Leipzig, 1936), p. 186.

Theorem XIV. *Let $F(N)$ be any F-function. Then there is a unique topologically invariant W-function $W(L)$ such that $W(L) = F(N)$ whenever L is a linear graph on N.*

For the linear graphs γ_{i+1} may be taken as the linear graphs x_{i+1} of Theorem VI. If we make the definitions $\gamma_0 = y_0$ and $\gamma_1 = y_1$ then the γ_i clearly satisfy the conditions of Theorem VI, and so by Theorem VIII $\{L^*\}$ has a unique expression as a polynomial in the $\{\gamma_i\}$. Hence there is a unique topologically invariant W-function $W(L)$ which is equal to $F(N)$ whenever N is a product of γ_i and L is on N. By Theorem XII there is a unique F-function $F_1(N)$ such that $W(L) = F_1(N)$ whenever L is on N.

But if the value of an F-function is given for every product of γ_i, then it is determined for all N. For by (42) if it is known for all N such that $n(N) = p$ and for one cubical network M such that $n(M) = p+1$, then it is determined for any cubical network M_1 Λ-equivalent to a homoeomorph of M. By applying Theorem XIII to each component having a node we see that every cubical network is Λ-equivalent to a homoeomorph of a product of γ_i and so the required result follows by induction. Since $F(N) = F_1(N)$ whenever N is a product of γ_i it follows that $F(N) = F_1(N)$ for every cubical network N. This proves the theorem.

Corollary. *For an F-function satisfying* (43) '*W-function*' *can be replaced by* '*V-function*' *in the above argument.*

As an example we mention an application of the above theory to the problem of functions obeying the law

$$f(\bar{N}) = f(N'_X) + f(M_0) \tag{45}$$

(see Fig. 1).

By eliminating $f(M_0)$ from two equations of the form (45) it is easy to show that $f(N)$ is an F-function multiplied by $(-1)^{n(N)}$. Hence it is fixed when its values for the products of the γ_i are given. But by applying (45) to these products we can show that for them $f(N) = 2^{n(N)}A$ where A is a constant. Since $2^{n(N)}A$ is obviously a solution of (45) it follows that it is the general solution.

Trinity College
Cambridge

Reprinted from
Proc. Cambridge Phil. Soc. **43** (1947), 26–40

ANNALS OF MATHEMATICS
Vol. 51, No. 1, January, 1950

A DECOMPOSITION THEOREM FOR PARTIALLY ORDERED SETS

BY R. P. DILWORTH

(Received August 23, 1948)

1. Introduction

Let P be a partially ordered set. Two elements a and b of P are *camparable* if either $a \geqq b$ or $b \geqq a$. Otherwise a and b are *non-comparable*. A subset S of P is *independent* if every two distinct elements of S are non-comparable. S is *dependent* if it contains two distinct elements which are comparable. A subset C of P is a *chain* if every two of its elements are comparable.

This paper will be devoted to the proof of the following theorem and some of its applications.

THEOREM 1.1. *Let every set of $k + 1$ elements of a partially ordered set P be dependent while at least one set of k elements is independent. Then P is a set sum of k disjoint chains.*[1]

It should be noted that the first part of the hypothesis of the theorem is also necessary. For if P is a set sum of k chains and S is any subset containing $k + 1$ elements, then at least one pair must belong to the same chain and hence be comparable.

Theorem 1.1 contains as a very special case the Radó-Hall theorem on representatives of sets (Hall [1]). Indeed, we shall derive from Theorem 1.1 a general theorem on representatives of subsets which contains the Kreweras (Kreweras [2]) generalization of the Radó-Hall theorem.

As a further application, Theorem 1.1 is used to prove the following imbedding theorem for distributive lattices.

THEOREM 1.2. *Let D be a finite distributive lattice. Let $k(a)$ be the number of distinct elements in D which cover a and let k be the largest of the numbers $k(a)$. Then D is a sublattice of a direct union of k chains and k is the smallest number for which such an imbedding holds.*

2. Proof of Theorem 1.1.

We shall prove the theorem first for the case where P is finite. The theorem in the general case will then follow by a transfinite argument. Hence let P be a finite partially ordered set and let k be the maximal number of independent elements. If $k = 1$, then every two elements of P are comparable and P is thus

[1] This theorem has a certain formal resemblance to a theorem of Menger on graphs (D. König, *Theorie der endlichen und unendlichen Graphen*, Leipzig, (1936)). Menger's theorem, however, is concerned with the characterization of the maximal number of disjoint, *complete* chains. Another type of representation of partially ordered sets in terms of chains has been considered by Dushnik and Miller [3] (see also Komm [4]). It can be shown that if n is the maximal number of non-comparable elements, then the dimension of P in the sense of Dushnik and Miller is at most n. Except for this fact, there seems to be little connection between the two representations.

a chain. Hence the theorem is trivial in this case and we may make an argument by induction. Let us assume, then, that the theorem holds for all finite partially ordered sets for which the maximal number of independent elements is less than k. Now it will be sufficient to show that if $C_1, \cdots, C_k$ are k disjoint chains of P and if a is an element belonging to none of the C_i, then $C_1 + \cdots + C_k + a$ is a set sum of k disjoint chains. For beginning with a set $a_1, \cdots, a_k$ of independent elements (which exist by hypothesis) we may add one new element at a time and be sure that at each stage we have a set sum of k disjoint chains. Since P is finite, we finally have P itself represented as a set sum of k chains.

Let, then, $C_1, \cdots, C_k$ be k disjoint chains and let a be an element not belonging to $C_1 + \cdots + C_k$. Let U_i be the set of all elements of C_i which contain a, let L_i be the set of all elements of C_i which are contained in a, and let N_i be the set of all elements of C_i which are non-comparable with a. Finally let

$$U = U_1 + \cdots + U_k$$
$$L = L_1 + \cdots + L_k$$
$$N = N_1 + \cdots + N_k$$
$$C = C_1 + \cdots + C_k.$$

Clearly $U_i + N_i + L_i = C_i$ and $U + N + L = C$.

We show now that for some m the maximal number of independent elements in $N + U - U_m$ is less than k. For suppose that for each j there exists a set S_j consisting of k independent elements of $N + U - U_j$. Since there are k elements in S_j and they belong to $C = C_1 + \cdots + C_k$, there is exactly one element of S_j in each of the chains C_i. Since S_j contains no elements of U_j it follows that S_j contains exactly one element of N_j. Thus $S = S_1 + \cdots + S_k$ contains at least one element of N_i for each i. Now let s_i be the minimal element of S which belongs to C_i. s_i exists since the intersection of S and C_i is a finite chain which we have proved to be non-empty. Furthermore, $s_i \in N_i$ since there is at least one element of N_i which belongs to S and all of the elements of U_i properly contain all of the elements of N_i. Hence $s_1, \cdots, s_k \in N$. Now if $s_i \geqq s_j$ for $i \neq j$, let $s_j \in S_r$. Since S_r contains an element t_i belonging to C_i, we have from the definition of s_i that $t_i \geqq s_i \geqq s_j$ and $t_i \neq s_j$ since $t_i \in C_i$ and $s_j \in C_j$. But this contradicts our assumption that the elements of S_r are independent. Hence we must have $s_j \ngeqq s_j$ for $i \neq j$ and $s_1, \cdots, s_k$ form an independent set. But since s_i belongs to N, s_i is non-comparable with a and hence $a, s_1, \cdots, s_k$ is an independent set containing $k + 1$ elements. But this contradicts the hypothesis of the theorem and hence we conclude that for some m, the maximal number of independent elements in $N + U - U_m$ is less than k.

In an exactly dual manner it follows that for some l, the maximal number of independent elements in $N + L - L_l$ is less than k.

Now let T be an independent subset of $C - U_m - L_l$. If T contains an element x belonging to $U - U_m$ and an element y belonging to $L - L_l$, then $x \geqq a \geqq y$ contrary to the independence of T. Since

$$(N + U - U_m) + (N + L - L_l) = C - U_m - L_l$$

it follows that T is either a subset of $N + U - U_m$ or of $N + L - L_l$. Hence the number of elements in T is less than k and thus the maximal number of independent elements in $C - U_m - L_l$ is less than k. Since $U_m + L_l$ is a chain there is at least one independent set of $k - 1$ elements in $C - U_m - L_l$. Hence by the induction hypothesis $C - U_m - L_l = C_1' + \cdots + C_{k-1}'$ where $C_1', \cdots, C_{k-1}'$ are disjoint chains. Let C_k' be the chain $U_m + a + L_l$. Then

$$C + a = C_1' + \cdots + C_k'$$

and our assertion is proved.

We turn now to the proof of the general case. Again when $k = 1$ the theorem is trivial and we may proceed by induction. Hence let the theorem hold for all partially ordered sets having at most $k - 1$ independent elements and let P satisfy the hypotheses of the theorem. A subset C of P is said to be *strongly dependent* if for every finite subset S of P, there is a representation of S as a set sum of k disjoint chains such that all of the elements of C which belong to S are members of the same chain. Clearly any strongly dependent subset is a chain. Also from the theorem in the finite case it follows that a set consisting of a single element is always strongly dependent. Since strong dependence is a finiteness property it follows from the Maximal Principle that P contains a maximal strongly dependent subset C_1. Suppose that $P - C_1$ contains k independent elements $a_1, \cdots, a_k$. Then from the maximal property of C_1 we conclude that $C_1 + a_i$ is not strongly dependent for each i. Hence there exists a finite subset S_i such that in any representation as a set sum of k chains there are at least two chains which contain elements of $C_1 + a_i$. S_i must clearly contain a_i since C_1 is strongly dependent. Let $S = S_1 + \cdots + S_k$. By the strong dependence of C_1, $S = K_1 + \cdots + K_k$ where $K_1, \cdots, K_k$ are disjoint chains such that for some $n \leqq k$ we have $S \cdot C_1 \subseteqq K_n$. Since S contains $a_1, \cdots, a_k$ which are independent, for some $m \leqq k$ we have $a_m \in K_n$. Let K_i' be the chain $S_m \cdot K_i$. Then $S_m = K_1' + \cdots + K_k'$ and $S_m \cdot C_1 \subseteqq S_m \cdot S \cdot C_1 \subseteqq S_m \cdot K_n = K_n'$. But by definition $a_m \in S_m$ and $a_m \in K_n$. Hence $S_m \cdot (C_1 + a_m) \subseteqq K_n'$ which contradicts the definition of S_m. We conclude that $P - C_1$ contains at most $k - 1$ independent elements. But since C_1 is a chain and P contains a set of k independent elements, it follows that $P - C_1$ contains a set of $k - 1$ independent elements. Thus by the induction hypothesis we have $P - C_1 = C_2 + \cdots + C_k$. Hence

$$P = C_1 + \cdots + C_k$$

and the proof of the theorem is complete.

3. Application to representatives of sets.

G. Kreweras has proved the following extension of the Radó-Hall theorem on representatives of sets:

Let $\mathfrak{A}$ and $\mathfrak{B}$ be two partitions of a set into n parts and let h be the smallest number such that for any r, r parts of $\mathfrak{A}$ contain at most $r + h$ parts of $\mathfrak{B}$. Let k be the smallest number such that $n + k$ elements serve to represent both partitions. Then $h = k$.

To show the power of Theorem 1.1 we shall prove an even more general theorem in which the partition requirement is dropped. Now if $\mathfrak{A}$ is any finite collection of subsets of a set S we shall say that a set of n elements (repetitions being counted) represents $\mathfrak{A}$ if there exists a one-to-one correspondence of the sets of $\mathfrak{A}$ onto a subset of the n elements such that each set contains its corresponding element. For example, the set $\{1, 1, 1\}$ represents the three sets $\{1, 2\}$, $\{1, 3\}$, and $\{1, 4\}$. The theorem can then be stated as follows:

THEOREM 3.1. *Let $\mathfrak{A}$ and $\mathfrak{B}$ be two finite collections of subsets of some set. Let $\mathfrak{A}$ and $\mathfrak{B}$ contain m and n sets respectively. Let h be the smallest number such that for every r, the union of any $r + h$ sets of $\mathfrak{A}$ intersects at least r sets of $\mathfrak{B}$. Let k be the smallest number such that $n + k$ elements serve to represent both collections $\mathfrak{A}$ and $\mathfrak{B}$. Then $h = k$.*

It can be easily verified that if $\mathfrak{A}$ and $\mathfrak{B}$ are partitions of a set, then h as defined in Theorem 2.1 is equivalent to the definition given in the theorem of Kreweras.

For the proof let $\mathfrak{A}$ consist of sets $A_1, \cdots, A_m$ and $\mathfrak{B}$ consist of sets $B_1, \cdots, B_n$. We make the sets $A_1, \cdots, A_m, B_1, \cdots, B_n$ into a partially ordered set P as follows:

$$A_i \geqq A_i \quad i = 1, \cdots, m$$

$$B_j \geqq B_j \quad j = 1, \cdots, n.$$

$$A_i \geqq B_j \text{ if and only if } A_i \text{ and } B_j \text{ intersect.}$$

It is obvious that P is a partially ordered set under this ordering. Now let w be the maximal number of independent elements of P. Since the union of any $r + h$ sets of $\mathfrak{A}$ intersects at least r sets of $\mathfrak{B}$, it follows that any independent subset of P can have at most $r + h + (n - r) = n + h$ elements. Hence $w \leqq n + h$. On the other hand for some r there are $r + h$ sets of $\mathfrak{A}$ whose union intersects precisely r sets of $\mathfrak{B}$. Hence these $r + h$ sets of $\mathfrak{A}$ and the remaining $n - r$ sets of $\mathfrak{B}$ form an independent subset of P containing $n + h$ elements. Thus $w = n + h$. By Theorem 1.1, P is the set sum of w chains $C_1, \cdots, C_w$. Now if a chain C_i contains two sets they have a non-null intersection by definition. Hence for each C_i there is an element a_i common to the sets of C_i. But since $A_1, \cdots, A_m$ are independent in P it follows that they belong to different chains and hence the w elements $a_1, \cdots, a_w$ represent $\mathfrak{A}$. Similarly, $a_1, \cdots, a_w$ represent $\mathfrak{B}$ and thus $n + k \leqq w$. But since P cannot be represented as a set sum of less than w chains, it follows that $n + k = w = n + h$. Hence $h = k$ and the theorem is proved.

4. Proof of Theorem 1.2.

Let us recall that an element q of a finite distributive lattice D is (*union*) *irreducible* if $q = x \cup y$ implies $q = x$ or $q = y$. It can be easily verified that if q is irreducible, then $q \leqq x \cup y$ implies $q \leqq x$ or $q \leqq y$. From the finiteness[2] of S it

[2] L is assumed to be finite for sake of simplicity. The theorem holds without this restriction. In the proof, "elements covered by a" must be replaced by "maximal ideals in a" and "irreducible elements" must be replaced by "prime ideals."

follows that every element of D can be expressed as a union of irreducible elements. From this fact we conclude that if $x > y$, there exists at least one irreducible q such that $x \geqq q$ and $y \not\geqq q$.

Now let P be the partially ordered set of union irreducible elements of D. Let a be such that $k = k(a)$. Then there are k elements $a_1, \cdots, a_k$ which cover a. Let q_i be an irreducible such that $a_i \geqq q_i$ and $a \not\geqq q_i$. Then if $q_i \geqq q_j$ where $i \neq j$ we have $a = a_i \cap a_j \geqq q_i \cap q_j \geqq q_j$ which contradicts $a \not\geqq q_j$. Hence $q_1, \cdots, q_k$ are an independent set of elements of P.

Next let $q_1', \cdots, q_l'$ be an arbitrary independent subset of P. Let $a' = q_1' \cup \cdots \cup q_l'$ and for each i let $p_i' = q_1' \cup \cdots \cup q_{i-1}' \cup q_{i+1}' \cup \cdots \cup q_l'$. Now if $p_i' = a'$ for some i, then

$$q_i' = q_i' \cap a' = q_i' \cap p_i'$$
$$= (q_i' \cap q_1') \cup \cdots \cup (q_i' \cap q_{i-1}') \cup (q_i' \cap q_{i+1}') \cup \cdots \cup \mathrm{M}\, (q_i' \cap q_l')$$

and hence $q_i' = q_i' \cap q_j'$ for some $j \neq i$. But then $q_j' \geqq q_i'$ contrary to independence. Thus $a' > p_i'$ for each i and $p_i' \cup p_j' = a'$ for $i \neq j$. Let $a = p_1' \cap \cdots \cap p_l'$ and for each i let $p_i = p_1' \cap \cdots \cap p_{i-1}' \cap p_{i+1}' \cap \cdots \cap p_l'$. If $p_i = a$, then $p_i' = p_i' \cup a = p_i' \cup p_i = (p_i' \cup p_1') \cap \cdots \cap (p_i' \cup p_{i-1}') \cap (p_i' \cup p_{i+1}') \cap \cdots \cap (p_i' \cup p_l') = a'$ which contradicts $p_i' < a'$. Hence $p_i > a$ and $p_i \cap p_j = a$ for $i \neq j$. Let $p_i \geqq a_i$ where a_i covers a. Then $a \leqq a_i \cap a_j \leqq p_i \cap p_j = a$ for $i \neq j$ and hence $a_i \cap a_j = a$, $i \neq j$. Thus $a_1, \cdots, a_l$ are distinct elements of D covering a. It follows that $l \leqq k$ and hence k is the maximal number of independent elements of P.

Now by Theorem 1.1 P is the set sum of k disjoint chains $C_1, \cdots, C_k$. We adjoin the null element z of D to each of the chains C_i. Then for each $x \in D$, there is a unique maximal element x_i in C_i which is contained in x. Now suppose $x > x_1 \cup \cdots \cup x_k$ in D. Then there exists an irreducible q such that $x \geqq q$ and $x_1 \cup \cdots \cup x_k \not\geqq q$. But $q \in C_i$ for some i and hence $x_1 \cup \cdots \cup x_k \geqq x_i \geqq q$ contrary to the definition of q. Hence $x = x_1 \cup \cdots \cup x_k$. Consider the mapping of D into the direct union of $C_1, \cdots, C_k$ given by

$$x \to \{x_1, \cdots, x_k\}.$$

Now if $x_i = y_i$ for $i = 1, \cdots, k$, then $x = x_1 \cup \cdots \cup x_k = y_1 \cup \cdots \cup y_k = y$ and the mapping is thus one-to-one. Since $x \cup y \geqq x_i \cup y_i$ we have $(x \cup y)_i \geqq x_i \cup y_i$. But since $(x \cup y)_i$ is union irreducible we get $x \cup y \geqq (x \cup y)_i \to x \geqq (x \cup y)_i$ or $y \geqq (x \cup y)_i \to x_i \geqq (x \cup y)_i$ or $y_i \geqq (x \cup y)_i \to x_i \cup y_i \geqq (x \cup y)_i$. Thus $(x \cup y)_i = x_i \cup y_i$ and we have

$$x \cup y \to \{x_1 \cup y_1, \cdots, x_k \cup y_k\}.$$

Similarly $x \cap y \geqq x_i \cap y_i \to (x \cap y)_i \geqq x_i \cap y_i$. But $x \geqq x \cap y \to x_i \geqq (x \cap y)_i$ and $y \geqq x \cap y \to y_i \geqq (x \cap y)_i$. Hence $x_i \cap y_i \geqq (x \cap y)_i$. Thus $(x \cap y)_i = x_i \cap y_i$ and we have

$$x \cap y \to \{x_1 \cap y_1, \cdots, x_k \cap y_k\}.$$

This completes the proof that D is isomorphic to a sublattice of a direct union of k chains.

Now suppose that D is a sublattice of the direct union of l chains $C'_1, \cdots, C'_l$ where $l < k$. Again let a be such that $k(a) = k$ and let $a_1, \cdots, a_k$ be the k distinct elements covering a. Define $a' = a_1 \cup \cdots \cup a_k$ and let $a'_i = a_1 \cup \cdots \cup a_{i-1} \cup a_{i+1} \cup \cdots \cup a_k$ for each i. Now $a'_i = q'_1 \cup \cdots \cup q'_l$ where $q'_i \in C'_i$. And if $q'_i = x' \cup y'$, then $q'_i = x'_i \cup y'_i$ where $x'_i, y'_i \in C'_i$. But then either $q'_i = x'_i \cup y'_i = x'_i$ or $q'_i = x'_i \cup y'_i = y'_i$ and hence either $q'_i = x'$ or $q'_i = y'$. Thus each q'_i is union irreducible. But $a_1 \cup \cdots \cup a_k = a' \geqq q'_i$ for $i = 1, \cdots, l$. Thus for each $i \leqq l$ there is a j such that $a_j \geqq q'_i$. Since $l < k$ there is some r such that $a'_r \geqq q'_i \cup \cdots \cup q'_l = a' \geqq a_r$. But then $a_r = a'_r \cap a_r = a$ which contradicts the fact that a_r covers a. Hence $l \geqq k$ and we conclude that k is the least number of chains whose direct union contains D as a sublattice. This completes the proof of Theorem 1.2.

Yale University
California Institute of Technology

REFERENCES

1. P. Hall. *On representatives of subsets*. J. London Math. Soc. 10 (1935), 26–30.
2. G. Kreweras. *Extension d'un thèorème sur les rèpartitions en classes*. C. R. Acad. Sci. Paris 222 (1946), 431–432.
3. B. Dushnik and E. W. Miller. *Partially ordered sets*. Amer. J. of Math. vol. 63 (1941), 600–610.
4. H. Komm. *On the dimension of partially ordered sets*. Amer. J. of Math. vol. 20 (1948), 507–520.

THE MARRIAGE PROBLEM.*

By Paul R. Halmos and Herbert E. Vaughan.

In a recent issue of this journal Weyl[1] proved a combinatorial lemma which was apparently considered first by P. Hall.[2] Subsequently Everett and Whaples[3] published another proof and a generalization of the same lemma. Their proof of the generalization appears to duplicate the usual proof of Tychonoff's theorem.[4] The purpose of this note is to simplify the presentation by employing the statement rather than the proof of that result. At the same time we present a somewhat simpler proof of the original Hall lemma.

Suppose that each of a (possibly infinite) set of boys is acquainted with a finite set of girls. Under what conditions is it possible for each boy to marry one of his acquaintances? It is clearly necessary that every finite set of k boys be, collectively, acquainted with at least k girls; the Everett-Whaples result is that this condition is also sufficient.

We treat first the case (considered by Hall) in which the number of boys is finite, say n, and proceed by induction. For $n = 1$ the result is trivial. If $n > 1$ and if it happens that every set of k boys, $1 \leqq k < n$, has at least $k + 1$ acquaintances, then an arbitrary one of the boys may marry any one of his acquaintances and refer the others to the induction hypothesis. If, on the other hand, some group of k boys, $1 \leqq k < n$, has exactly k acquaintances, then this set of k may be married off by induction and, we assert, the remaining $n - k$ boys satisfy the necessary condition with respect to the as yet unmarried girls. Indeed if $1 \leqq h \leqq n - k$, and if some set of h bachelors were to know fewer than h spinsters, then this set of h bachelors together with the k married men would have known fewer than $k + h$ girls. An

* Received June 6, 1949.

[1] H. Weyl, "Almost periodic invariant vector sets in a metric vector space," *American Journal of Mathematics*, vol. 71 (1949), pp. 178-205.

[2] P. Hall, "On representation of subsets," *Journal of the London Mathematical Society*, vol. 10 (1935), pp. 26-30.

[3] C. J. Everett and G. Whaples, "Representations of sequences of sets," *American Journal of Mathematics*, vol. 71 (1949), pp. 287-293. Cf. also M. Hall, "Distinct representatives of subsets," *Bulletin of the American Mathematical Society*, vol. 54 (1948), pp. 922-926.

[4] C. Chevalley and O. Frink, Jr., "Bicompactness of Cartesian products," *Bulletin of the American Mathematical Society*, vol. 47 (1941), pp. 612-614.

application of the induction hypothesis to the $n-k$ bachelors concludes the proof in the finite case.

If the set B of boys is infinite, consider for each b in B the set $G(b)$ of his acquaintances, topologized by the discrete topology, so that $G(b)$ is a compact Hausdorff space. Write G for the topological Cartesian product of all $G(b)$; by Tychonoff's theorem G is compact. If $\{b_1,\cdots,b_n\}$ is any finite set of boys, consider the set H of all those elements $g=g(b)$ of G for which $g(b_i)\neq g(b_j)$ whenever $b_i\neq b_j$, $i,j=1,\cdots,n$. The set H is a closed subset of G and, by the result for the finite case, H is not empty. Since a finite union of finite sets is finite, it follows that the class of all sets such as H has the finite intersection property and, consequently, has a non empty intersection. Since an element $g=g(b)$ in this intersection is such that $g(b')\neq g(b'')$ whenever $b'\neq b''$, the proof is complete.

It is perhaps worth remarking that this theorem furnishes the solution of the celebrated problem of the monks.[5] Without entering into the history of this well-known problem, we state it and its solution in the language of the preceding discussion. A necessary and sufficient condition that each boy b may establish a harem consisting of $n(b)$ of his acqaintances, $n(b)=1, 2, 3,\cdots$, is that, for every finite subset B_0 of B, the total number of acquaintances of the members of B_0 be at least equal to $\Sigma n(b)$, where the summation runs over every b in B_0. The proof of this seemingly more general assertion may be based on the device of replacing each b in B by $n(b)$ replicas seeking conventional marriages, with the understanding that each replica of b is acquainted with exactly the same girls as b. Since the stated restriction on the function n implies that the replicas satisfy the Hall condition, an application of the Everett-Whaples theorem yields the desired result.

UNIVERSITY OF CHICAGO
AND
UNIVERSITY OF ILLINOIS.

---

[5] H. Balzac, *Les Cent Contes Drôlatiques*, IV, 9: *Des moines et novices*, Paris (1849).

Reprinted from
Amer. J. Math. 72 (1950), 214–215

CIRCUITS AND TREES IN ORIENTED LINEAR GRAPHS

by T. van Aardenne-Ehrenfest (Dordrecht) and N. G. de Bruijn (Delft)

§ 1. $P_n^{(\sigma)}$-cycles.

In this § we state the problem which gave rise to our investigations about graphs. The further contents of the paper are independent of this § 1.

Consider a set of σ figures 1, 2, . . ., σ, and let n be a natural number. A sequence of n figures will be called an n-tuple. Clearly, there are σ^n different n-tuples.

An oriented circular array, consisting of σ^n figures, will be called a P_n-cycle, whenever it has the property that each n-tuple occurs exactly once as a set of n consecutive figures of the cycle. An example, with $\sigma = 3$, $n = 2$ is the cycle 1 1 2 2 3 3 1 3 2. [1])

The existence of $P_n^{(\sigma)}$-cycles, for arbitrary values of σ and n, was proved by M. H. Martin [3], I. J. Good [2] and D. Rees [4]. One of us showed ([1]) that, for $\sigma = 2$, the number of different $P_n^{(2)}$-cycles equals $2^{f(n)}$, $f(n) = 2^{n-1} - n$.

This result was derived as follows. The number of $P_n^{(2)}$-cycles can be interpreted as the number of circuits in a certain graph N_{n+1} (compare also [2]). The graph N_{n+1} can be obtained by a certain operation from N_n, and by a general theorem on circuits in oriented graphs the number of circuits of N_{n+1} could be expressed in the number of circuits of N_n. This theorem on graphs was proved in [1] only for the case that at any vertex 2 edges point outward and 2 inward. In the present paper we shall deal, among other things, with the general case (theorem 4). This result immediately enables us to determine the number of $P_n^{(\sigma)}$-cycles for arbitrary σ. Referring to [1] for details, we only state the result: The number of different $P_n^{(\sigma)}$-cycles is $\sigma^{-n}(\sigma!)^q$, where $q = \sigma^{n-1}$.

For example, there are 24 $P_2^{(3)}$-cycles. Six of them are

1 1 2 3 3 2 2 1 3	1 1 2 2 3 2 1 3 3
1 1 2 3 2 2 1 3 3	1 1 2 2 3 3 1 3 2
1 1 2 2 3 3 2 1 3	1 1 2 2 3 1 3 3 2

1) It has to be understood that these figures have to be placed around an oriented circle. Therefore, 21 is one of the 2 – tuples occurring in the cycle. Naturally, 112233132 and 331321122 are considered as one and the same cycle, but 112313322, which has the reversed order, is a different one.

Another six are obtained from these by interchanging the figures 2 and 3 everywhere. By reversing the orientation, 12 new cycles arise.

§ 2. Preliminaries about permutation groups.

Let $\mathfrak{S}_m$ be the symmetric group of degree m, that is, the group of all $m!$ permutations of a set E_m of m objects. If $\mathfrak{A}$ is a subset of $\mathfrak{S}_m$ then the number of cyclic permutations in $\mathfrak{A}$ will be denoted by $|\mathfrak{A}|$, and the total number of elements in $\mathfrak{A}$ by $n(\mathfrak{A})$.

A subset $\mathfrak{D}$ of $\mathfrak{S}_m$ will be called a *D-set* (in $\mathfrak{S}_m$), whenever it has the property that $|S\mathfrak{D}|$ has the same value for all $S \varepsilon \mathfrak{S}_m$. It is easily seen that, in that case, we have $|S\mathfrak{D}| = m^{-1} . n(\mathfrak{D})$. For, if C is any cyclic permutation, then there are exactly $n(\mathfrak{D})$ possibilities for S such that $S\mathfrak{D}$ contains C, and it follows that $m! \, |S\mathfrak{D}| = (m-1)! \, n(\mathfrak{D})$.

Furthermore, it may be remarked that $|S\mathfrak{D}| = |\mathfrak{D}S|$, since SBS^{-1} is cyclic whenever B is cyclic. Therefore, if $\mathfrak{D}$ is a D-set and if P is an arbitrary element of $\mathfrak{S}_m$, then $\mathfrak{D}P$ is also a D-set. $\mathfrak{S}_m$ itself clearly is a D-set in $\mathfrak{S}_m$, but theorem 1 will show that non-trivial D-sets exist.

Let E_l be a sub-set of the set of objects E_m, containing l objects. Consider the sub-group $\mathfrak{G} \subset \mathfrak{S}_m$ of all permutations which only permute the elements of E_l, leaving the remaining elements of E_m invariant. If G is any permutation of $\mathfrak{G}$, then $\bar{G}$ denotes the corresponding permutation of the objects of E_l, that is to say, we disregard the objects belonging to $E_m - E_l$ (which are invariant under G). G is defined uniquely by $\bar{G}$, and vice versa. The same notation will be used for sets: if $\mathfrak{A} \subset \mathfrak{G}$, then $\bar{\mathfrak{A}}$ denotes the set of all $\bar{G}$, where $G \varepsilon \mathfrak{A}$.

L e m m a 1. Let $\mathfrak{B}$ be a sub-set of $\mathfrak{G}$ such that $\bar{\mathfrak{B}}$ is a D-set in $\bar{\mathfrak{G}}$, and let $C \varepsilon \mathfrak{S}_m$ be a cyclic permutation. Then we have

$$|\mathfrak{B} C| = l^{-1} . n(\mathfrak{B}).$$

P r o o f. We shall deal with the cyclic representations of the permutations involved. Let $\bar{G}$ be the element of $\bar{\mathfrak{G}}$ whose cyclic representation is obtained by cancelling the objects of $E_m - E_l$ from the cyclic representation of C. Further, let G_1 be an arbitrary permutation of $\mathfrak{G}$. Then it is easily verified that $\bar{G}_1\bar{G}$ (of degree l) shows the same number of cycles as G_1C (of degree m). Hence G_1C is cyclic whenever $\bar{G}_1\bar{G}$ is cyclic. Therefore

$$|\mathfrak{B} C| = |\bar{\mathfrak{B}} \bar{G}| = l^{-1} . n(\bar{\mathfrak{B}}) = l^{-1} . n(\mathfrak{B}).$$

Lemma 2. Let $\mathfrak{B}$ be a subset of $\mathfrak{G}$ such that $\overline{\mathfrak{B}}$ is a D-set in $\overline{\mathfrak{G}}$. Let Q be any arbitrary permutation of $\mathfrak{S}_m$. Then we have

$$\frac{|\mathfrak{B}\, Q|}{n(\mathfrak{B})} = \frac{|\mathfrak{G}\, Q|}{n(\mathfrak{G})}.$$

Proof. If there is no $G \varepsilon \mathfrak{G}$ such that GQ is cyclic, then both sides are equal to zero. Now assume that $G \varepsilon \mathfrak{G}$ is such that GQ is cyclic; put $GQ = C$.

We have $\mathfrak{G}\, Q = \mathfrak{G}\, C$, and $\mathfrak{B}\, Q = (\mathfrak{B}\, G^{-1})\, C$. The set $\overline{\mathfrak{B}}_1 = \overline{\mathfrak{B}}\, \overline{G}^{-1}$ is a D-set in $\overline{\mathfrak{G}}$, since $\overline{\mathfrak{B}}$ was a D-set in $\overline{\mathfrak{G}}$. Now, by lemma 1,

$$|\mathfrak{B}\, Q| = |\mathfrak{B}_1 C| = l^{-1} \,.\, n(\mathfrak{B}_1) = l^{-1} \,.\, n(\mathfrak{B}).$$

Analogously

$$|\mathfrak{G}\, Q| = |\mathfrak{G}\, C| = l^{-1} \,.\, n(\mathfrak{G}),$$

and lemma 2 has been proved.

Let k and n be natural numbers, and take $m = kn$. We consider a set E_m of m objects, divided into k systems, each of them containing n objects. We shall again denote by $\mathfrak{S}_m$ the group of all permutations of E_m. $\mathfrak{H}$ denotes the group consisting of all $k!\,(n!)^k$ permutations H with the property that Ha and Hb belong to the same system whenever a and b belong to the same system. Or, shortly, H transforms systems into systems.

Theorem 1. $\mathfrak{H}$ is a D-set in $\mathfrak{S}_m$.

Proof. If either $k = 1$ or $n = 1$, then we have $\mathfrak{H} = \mathfrak{S}_m$, and the theorem is trivial.

Next we shall deal with the case $k = 2$, $m = 2n$. It has to be shown that $|S\, \mathfrak{H}|$ does not depend on S ($S\, \varepsilon \mathfrak{S}_{2n}$). Let $\mathfrak{H}_1$ be the set of all permutations mapping the first system onto itself, and let $\mathfrak{H}_2$ be the set of those mapping the first system onto the second. Thus $\mathfrak{H} = \mathfrak{H}_1 + \mathfrak{H}_2$.

Let p be the number of objects of the first system mapped into the first system by S, then there are $q = n - p$ objects of the first system which are mapped into the second system. Then we have

$$|S\, \mathfrak{H}_1| = q\{(n-1)!\}^2.$$

This can be seen, for instance, by interpreting $|S\, \mathfrak{H}_1|$ as the number of circuits (see § 3) in the following graph. Take two vertices, A and B, and $2n$ oriented edges: p of them from A to A, p from B to B, q from A to B and q from B to A. The number of circuits can be shown to be $q\{(n-1)!\}^2$. It can be very rapidly determined by theorem 6, for there are exactly q trees with root A.

$\mathfrak{H}_2$ can be written as $S_0\,\mathfrak{H}_1$, where S_0 is an arbitrary element of $\mathfrak{H}_2$. Now $|\,S\,\mathfrak{H}_2\,| = |\,S_1\,\mathfrak{H}_1\,|$, where $S_1 = SS_0$. S_1 has the same nature as S, apart from the fact that p and q changed their roles.

Hence $|\,S\,\mathfrak{H}_2\,| = p\,\{(n-1)!\}^2$, and so

$$|\,S\,\mathfrak{H}\,| = |\,S\,\mathfrak{H}_1\,| + |\,S\,\mathfrak{H}_2\,| = (p+q)\,\{\,(n-1)!\}^2 = (n!)^2/n.$$

This does not depend on S, and so our theorem has been proved in the case $k = 2$.

Next we consider the general case $k > 2$. We have to show that $|\,S_1\,\mathfrak{H}\,| = |\,S_2\,\mathfrak{H}\,|$ for any pair S_1, S_2 $(S_1 \,\varepsilon\, \mathfrak{S}_m,\ S_2 \,\varepsilon\, \mathfrak{S}_m)$. Since any $S \,\varepsilon\, \mathfrak{S}_m$ can be written as a product of transpositions, it is sufficient to prove that $|\,S\,\mathfrak{H}\,| = |\,S\,T\,\mathfrak{H}\,|$ for all S and for any transposition T. Or, what is the same thing, that

$$(2.1) \qquad |\,\mathfrak{H}\,Q\,| = |\,T\,\mathfrak{H}\,Q\,|$$

for any $Q \,\varepsilon\, \mathfrak{S}_m$ and any transposition $T \,\varepsilon\, \mathfrak{S}_m$.

We may assume that T interchanges two symbols belonging to the first and to the second system, respectively (if T interchanges two symbols of the same system, then we have $\mathfrak{H} = T\,\mathfrak{H}$, and (2.1) is trivial). Let $\mathfrak{H}^*$ be the sub-group of $\mathfrak{H}$ consisting of all permutations of $\mathfrak{H}$ which leave all individual elements of the 3rd, 4th, ..., k^{th} system invariant, and let $\mathfrak{G}$ be the group arising from $\mathfrak{S}_m$ in the same manner. We now apply lemma 2, with $l = 2n$ and $\mathfrak{B} = \mathfrak{H}^*$. Since the theorem has been proved for the case $k = 2$, we know that $\overline{\mathfrak{H}^*}$ is a D-set in $\overline{\mathfrak{G}}$.

Therefore

$$\frac{|\,\mathfrak{H}^*Q\,|}{n\,(\mathfrak{H}^*)} = \frac{|\,\mathfrak{G}\,Q\,|}{n\,(\mathfrak{G})}, \qquad \frac{|\,T\,\mathfrak{H}^*Q\,|}{n\,(\mathfrak{H}^*)} = \frac{|\,T\,\mathfrak{G}\,Q\,|}{n\,(\mathfrak{G})}.$$

Evidently $T\,\mathfrak{G} = \mathfrak{G}$, and so $|\,\mathfrak{H}^*Q\,| = |\,T\,\mathfrak{H}^*Q\,|$ for all $Q \,\varepsilon\, \mathfrak{S}_m$.

Since $\mathfrak{H}^*$ is a sub-group of $\mathfrak{H}$, we can split $\mathfrak{H}$ into classes, $\mathfrak{H} = \Sigma\,\mathfrak{H}^*Q_i$, and now (2.1) follows immediately. The order of the group $\mathfrak{H}$ is $k!(n!)^k$, and therefore

$$(2.2) \quad |\,\mathfrak{H}\,| = m^{-1}\,n(\mathfrak{H}) = m^{-1}\,k!\,(n!)^k = n^{-1}\,(k-1)!\,(n!)^k.$$

Let $\mathfrak{K}$ be the set of all permutations K with the property that the n objects of each system are transformed into objects of n different systems. In other words, K is such that, if a and b belong to the same system, then Ka and Kb belong to different systems. Clearly $\mathfrak{K}$ is empty if $k < n$.

It is not difficult to show that $\mathfrak{K}$ *is a D-set*. For, if H is an arbitrary permutation of $\mathfrak{H}$, we have $\mathfrak{K} = \mathfrak{K}\,H$. It follows that $\mathfrak{K}$ is the sum of a number of left-classes mod $\mathfrak{H}$: $\mathfrak{K} = \Sigma K_i \mathfrak{H}$. Each component $K_i \mathfrak{H}$ is a D-set, by theorem 1. Hence $\mathfrak{K}$ is a D-set.

It is easily seen that in the special case $k = n$ the number of elements in $\mathfrak{K}$ is $(n!)^{2n}$, and so we have

$$| \mathfrak{K} | = (n!)^{2n} n^{-2} \qquad (k = n). \tag{2.3}$$

§ 3. T-Graphs.

In §§ 3, 4, 5, 6 we shall be mainly concerned with a special type of finite oriented linear graphs, called *T-graphs* [1]). These have the property that, at each vertex P, the number σ_i of oriented edges pointing to P_i equals the number of edges pointing away from P_i. For simplicity we assume $\sigma_i > 0$ for all i. If this number happens to be the same for all vertices ($\sigma_i = \sigma$ for all i) then we shall call the graph a $T^{(\sigma)}$ [2]).

We do not exclude the possibility that, in a T-graph, several different edges point from P_i to P_j, and we neither exclude edges pointing from P_i to P_i itself (closed loops).

Therefore, a T-graph can be interpreted as a pair of mappings of a finite set of edges $\{e_1, \ldots, e_m\}$ onto a finite set of vertices $\{P_1, \ldots, P_N\}$ such that each vertex is the image of the same number of edges in both mappings. The first mapping maps every edge onto the point where it starts from, and the second one onto the point where it terminates. We shall call these vertices the *tail* and the *head* of the edge, respectively. If the head of e_i coincides with the tail of e_j, then e_i and e_j will be called *consecutive* (which does not imply that e_j and e_i are consecutive).

By a complete circuit (a *circuit* for short) is meant any cyclic arrangement of the set of edges in such a manner that the head of each edge coincides with the tail of the next one in the circuit. Or, in other words, such that consecutive edges in the circuit are consecutive in the graph.

Naturally, two circuits are considered as identical whenever the first one is a cyclic permutation of the second. It has to be understood that the order of the edges counts, and not only the order of the heads. So, for instance, if $m = 3$, $N = 1$, then P_1 is the head as well as the tail of all edges. There are two different circuits, viz. (e_1, e_2, e_3) and (e_1, e_3, e_2).

---

[1]) Tutte [5] calls them *simple oriented networks*.

[2]) In the paper [1] the name "T-net" denoted the same thing as $T^{(2)}$ does in our present notation.

The number of circuits of a graph T will be denoted by $|T|$ [1]).

A permutation P of the set of edges $e_1, \ldots, e_m$ will be called *conservative* (with respect to T), whenever $Pe_i = e_j$ always implies that the head of e_i coincides with the tail of e_j. We choose one special conservative permutation A_0, arbitrary, but fixed in the sequel. The set of all conservative permutations of T can be represented as $\mathfrak{G} A_0$, where $\mathfrak{G}$ is the group of all permutations which leave the tails of all edges invariant.

Evidently, any circuit determines a cyclic conservative permutation, and vice versa.

Therefore,

$$|T| = |\mathfrak{G} A_0|.$$

This simple relation between the number of circuits in a graph and the number of cyclic permutations in a set explains why we choose the same notation $|\quad|$ for both.

Consider a vertex P_i where σ_i edges start and σ_i edges terminate. By the *local symmetric group* $\mathfrak{G}_i$ we shall denote the group of all $\sigma_i!$ permutations which permute the σ_i edges whose tail is P_i, but which leave invariant all edges whose tail is not P_i.

Clearly $\mathfrak{G}$ is the direct product of $\mathfrak{G}_1, \ldots, \mathfrak{G}_N$.

§ 4. **Traffic regulations.**

We shall also consider circuits described under certain restrictive conditions, called traffic regulations.

Let $\mathfrak{B}_1, \ldots, \mathfrak{B}_n$ be sub-sets of $\mathfrak{G}_1, \ldots, \mathfrak{G}_n$, respectively, and construct the set

$$\mathfrak{B} = \mathfrak{B}_1 \times \mathfrak{B}_2 \times \ldots \times \mathfrak{B}_N, \tag{4.1}$$

defined in the same way as the direct product

$$\mathfrak{G} = \mathfrak{G}_1 \times \mathfrak{G}_2 \times \ldots \times \mathfrak{G}_N. \tag{4.2}$$

Now a circuit described under the traffic regulation $\mathfrak{B}$ is defined as a circuit corresponding to a permutation BA_0, where $B \in \mathfrak{B}$, and A_0 is the fixed permutation chosen in § 3.

Denoting the number of circuits described under the traffic regulation $\mathfrak{B}$ by $|T|_{\mathfrak{B}}$, we have

$$|T| = |T|_{\mathfrak{G}}, \quad |T|_{\mathfrak{B}} = |\mathfrak{B} A_0|. \tag{4.3}$$

---

[1]) We have $|T| > 0$ if and only if T is connected (see [2]). For non-connected graphs our theorems are trivial. Nevertheless, all our proofs are valid for that case also.

The traffic regulation (4.1) will be called *regular* if, for each i, $\overline{\mathfrak{B}}_i$ is a D-set in $\overline{\mathfrak{G}}_i$ [1]).

Theorem 2. [2]) If $\mathfrak{B}$ is regular, then we have

$$(4.4)\qquad \frac{1}{n(\mathfrak{B})} \mid T \mid_{\mathfrak{B}} = \frac{1}{n(\mathfrak{G})} \mid T \mid_{\mathfrak{G}},$$

where $n(\mathfrak{B})$ and $n(\mathfrak{G})$ denote the number of elements of $\mathfrak{B}$ and $\mathfrak{G}$, respectively.

Proof. Since $\mathfrak{G}_i$ itself satisfies the condition imposed on $\mathfrak{B}_i$, it is sufficient to show that the value of the left-hand-side of (4.4) does not change if some $\mathfrak{B}_i$ is replaced by the corresponding $\mathfrak{G}_i$. If this has been proved, we can replace all $\mathfrak{B}_i$'s by $\mathfrak{G}_i$'s one after the other, and (4.4) follows.

To this end we consider $\mathfrak{B}$, defined by (4.1) and $\mathfrak{B}^*$, defined by

$$(4.5)\qquad \mathfrak{B}^* = \mathfrak{G}_1 \times \mathfrak{B}_2 \times \mathfrak{B}_3 \times \ldots \times \mathfrak{B}_N,$$

and we have to show that

$$(4.6)\qquad \mid T \mid_{\mathfrak{B}} : \mid T \mid_{\mathfrak{B}^*} = n(\mathfrak{B}_1) : n(\mathfrak{G}_1),$$

for the latter ratio equals $n(\mathfrak{B}) : n(\mathfrak{B}^*)$.

Referring to (4.3) we write

$$(4.7)\qquad \mid T \mid_{\mathfrak{B}} = \Sigma \mid \mathfrak{B}_1 B_2 \ldots B_N A_0 \mid,$$

$$\mid T \mid_{\mathfrak{B}^*} = \Sigma \mid \mathfrak{G}_1 B_2 \ldots B_N A_0 \mid,$$

where, in both sums, $B_2, \ldots, B_N$ run independently through the elements of $\mathfrak{B}_2, \ldots, \mathfrak{B}_N$, respectively. If we put $B_2 \ldots B_N A_0 = Q$, then we have, by lemma 2,

$$\mid \mathfrak{B}_1 Q \mid : \mid \mathfrak{G}_1 Q \mid = n(\mathfrak{B}_1) : n(\mathfrak{G}_1).$$

Applying this to each pair of corresponding terms of the sums in (4.7), we obtain (4.6).

§ 5. Special traffic regulations.

Let T be a T-graph with N vertices and m edges. Again, the numbers of edges pointing towards $P_1, \ldots, P_N$ are denoted by $\sigma_1, \ldots, \sigma_N$, respectively, and so $m = \sigma_1 + \ldots + \sigma_N$.

Let λ be a positive integer. Then by T^λ [3]) we denote the graph which arises from T if we replace any edge P_iP_j of T by λ edges P_iP_j, with the same orientation. Hence T^λ has N vertices and λm edges. The edges of T^λ arising from one and the same edge of T are said to form a *bundle*.

---

1) As in § 2, the bar indicates that the permutations are considered as permutations of the σ_i edges whose tail is P_i, whereas the other edges are disregarded.

2) This theorem was used implicitly in [1].

3) The notations T^σ and $T^{(\sigma)}$ (for the latter see § 3) must not be confused.

In T^λ we shall consider several possible traffic regulations. We first choose a fixed conservative permutation A_0 which transforms bundles into bundles.

A traffic regulation will be obtained by choosing, at each vertex P_i, a set $\mathfrak{B}_i$ of permutations of the $\lambda\sigma_i$ edges starting from that vertex. We shall consider three possibilities, all regular in the sense of § 4.

1°. $\mathfrak{B}_i = \mathfrak{G}_i$, where $\mathfrak{G}_i$ is the local symmetric group (of order $(\lambda\sigma_i)!$).

2°. $\mathfrak{B}_i = \mathfrak{H}_i$. Here $\mathfrak{H}_i$ is the sub-set of $\mathfrak{G}_i$ which transforms bundles into bundles. In other words, as to the edges whose tail is P_i it acts like the group $\mathfrak{H}$ of theorem 1, where the systems are given by the bundles. Thus $n = \lambda$, $k = \sigma_i$.

3°. $\mathfrak{B}_i = \mathfrak{K}_i$. Here $\mathfrak{K}_i$ is the sub-set of $\mathfrak{H}_i$ consisting of the permutations which transform the edges of each outgoing bundle at P_i into sets of edges belonging to λ different bundles (see the end of § 2).

We have, by theorem 2,

$$(5.1) \qquad \frac{|T^\lambda|_{\mathfrak{G}}}{\prod_{i=1}^{N} (\lambda\sigma_i)!} = \frac{|T^\lambda|_{\mathfrak{H}}}{\prod_{i=1}^{N} \sigma_i!\,(\lambda!)^{\sigma_i}} = \frac{|T^\lambda|_{\mathfrak{K}}}{\prod_{i=1}^{N} \varphi(\lambda, \sigma_i)},$$

where $\varphi(\lambda, \sigma_i)$ is the number of elements of $\mathfrak{K}_i$.

As stated in § 4, we have $|T^\lambda|_{\mathfrak{G}} = |T^\lambda|$.

The number $|T^\lambda|_{\mathfrak{H}}$ can be connected with the number of circuits in T itself. To this end we consider a circuit of T^λ described according to the traffic regulation $\mathfrak{H}$. At any stage, the bundle to which an edge belongs only depends on the bundle containing the preceding edge. Therefore, the sequence of bundles described by the circuit is periodic mod m, and any bundle is used exactly λ times. It follows that each circuit under consideration defines a circuit of T. Conversely, it is easily seen that each circuit of T arises from $\lambda^{-1}(\lambda!)^m$ different circuits of T^λ in this manner. Hence

$$(5.2) \qquad |T^\lambda|_{\mathfrak{H}} = \lambda^{-1}(\lambda!)^m \,.\, |T|,$$

and so we obtain from (5.1)

Theorem 3. $\qquad |T^\lambda| = \lambda^{-1} \,.\, |T| \,.\, \prod_{i=1}^{N} \frac{(\lambda\sigma_i)!}{\sigma_i!}.$

We shall now make the restriction that T is a $T^{(\sigma)}$, and that $\lambda = \sigma$, that is to say

$$\sigma_1 = \ldots = \sigma_N = \sigma = \lambda, \qquad m = N\sigma.$$

Then we have (see (2.3))

$$\varphi(\lambda, \sigma_i) = (\sigma!)^{2\sigma}$$

and now (5.1) and (5.2) lead to

$$(5.3)\quad |T^\sigma|_{\mathfrak{R}} = |T| \,.\, \sigma^{-1} (\sigma!)^m \,.\, \left(\frac{(\sigma!)^{2\sigma}}{\sigma!\,(\sigma!)^\sigma}\right)^N = |T| \,.\, \sigma^{-1} (\sigma!)^{N(2\sigma-1)}.$$

If T is a $T^{(\sigma)}$, with N vertices and $m = \sigma N$ edges, then by T^* we denote the graph defined as follows. T^* has m vertices $E_1, \ldots, E_m$. Two vertices E_i, E_j are connected in T^* by an edge from E_i to E_j if and only if e_i, e_j are consecutive in T. This process was considered in [1] for the case $\sigma = 2$ only.

Theorem 4. $|T^*| = \sigma^{-1} (\sigma!)^{N(\sigma-1)} \,.\, |T|$.

Proof. By "*σ-cycle in T*" is meant a circular array containing each edge of T exactly σ times, such that two edges are consecutive in the array if and only if they are consecutive in T. A σ-cycle will be called *restricted* if it is such that any pair of consecutive edges of T occurs just once as a pair of consecutive elements in the array. It will be clear that any restricted σ-cycle in T defines uniquely a circuit in T^*, and vice versa.

The restricted σ-cycles in T are closely related to the circuits in T^σ described under the traffic regulation $\mathfrak{R}$. Actually, if we identify the edges of each bundle in T^σ a $\mathfrak{R}$-circuit in T^σ becomes a restricted σ-cycle, owing to the definition of $\mathfrak{R}$. Conversely, any restricted σ-cycle gives rise to a large number of $\mathfrak{R}$-circuits. Any bundle occurs σ times in the cycle, and each time an arbitrary edge of the bundle can be chosen.

So we see that $(\sigma!)^m$ different $\mathfrak{R}$-circuits arise from one restricted σ-cycle. Now the theorem follows from (5.3), since $m = N\sigma$.

In a T-graph which is not necessarily a $T^{(\sigma)}$ we can still consider (unrestricted) σ-cycles [1]). The number of different σ-cycles can be determined from theorem 3. A difficulty lies in the fact that a σ-cycle may be periodical with a period md, where d is a proper divisor of σ, which could not happen with a restricted σ-cycle.

If $c(\varrho)$ denotes the number of those ϱ-cycles in T whose period is exactly $m\varrho$, then we have obviously

$$|T^\rho| = \sum_{d/\rho} \frac{d}{\varrho}\, c(d)\, (\varrho!)^m.$$

[1]) And, if T is a $T^{(\sigma)}$, we can consider (unrestricted) ϱ-cycles, for arbitrary values of ϱ.

Hence we obtain from Möbius' inversion formula,

$$c(\varrho) = \sum_{d/\rho} \frac{d}{\varrho} \mu\left(\frac{\varrho}{d}\right) . (d!)^{-m} . | T^d |,$$

and so the number of unrestricted ϱ-cycles equals

$$\sum_{d/\rho} c(d) = \frac{1}{\varrho} \sum_{d/\varrho} \varphi\left(\frac{\varrho}{d}\right) (d!)^{-m} d . | T^d | ,$$

where φ is Euler's indicator. $| T^d |$ can be evaluated by theorem 3.

Especially, is T is a $T^{(\sigma)}$, then the number of unrestricted ϱ-cycles is

$$\frac{1}{\varrho} \sum_{d/\rho} \varphi\left(\frac{\varrho}{d}\right)\left(\frac{(\sigma d)!}{(d!)^\sigma \sigma!}\right)^N . | T | .$$

§ 6. **Trees in T-graphs.**

Let T be a T-graph with N vertices and m edges. The number of edges whose tail is P_i is again denoted by σ_i. Choose an arbitrary vertex; for convenience of notations we take it to be P_1. We shall define the notion: (oriented) tree with root P.

A tree with root P_1 is a sub-set Λ of the set of edges of T, with the following properties.

a. Any vertex $\neq P_1$ is the tail of just one element of Λ.

b. No element of Λ has its tail in P_1.

c. Any vertex can be connected with P_1 by a set of consecutive edges, all belonging to Λ.

It is easily seen that *c* can be replaced by

*c**. Λ contains no closed oriented cycles.

There is a striking relation between trees and circuits. Choose a fixed edge e_1 whose tail is P_1, and consider an arbitrary circuit of T. We traverse it, starting with e_1. Running through the circuit, each vertex P_i will be visited σ_i times. The edge by which we leave P_i after having visited it for the σ_i-th time will be called the *last exit* of P_i.

Theorem 5a. *The set* Λ *consisting of the last exits of* $P_2, \ldots, P_N$ *is a tree with root* P_1.

Proof. The properties *a* and *b* are trivial. We shall verify *c**.

We can number the edges of T according to the order in the circuit, with indices $1, \ldots, m$; e_1 gets the index 1.

If e_i and e_j both belong to Λ, and if e_i and e_j are consecutive in T, then we have $i < j$. For, e_{i+1} has the same tail as e_j, and j is the maximal value of the indices of all the edges with this tail.

Consequently, Λ does not contain any closed cycle; the indices in such a cycle would increase indefinitely.

Theorem 5b. If a tree Λ with root P_1 is given, and if e_1 is given, then there are exactly

$$(6.1) \qquad \prod_{i=1}^{N} (\sigma_i - 1)!$$

circuits of T whose set of last exits coincides with Λ.

Proof. At any vertex we number the outgoing edges [1]), with the following restrictions: At P_1 the edge e_1 gets the number 1; at P_i $(i > 1)$ the edge belonging to Λ gets the highest possible number, that is σ_i. The number of ways in which this can be arranged is expressed by (6.1). It remains to be shown that, for each numbering of this type, there exists a circuit (and not more than one circuit) corresponding with this numbering.

First thing it will be clear that we have no choice at all if we try to traverse a circuit according to this numbering. Starting with e_1, we arrive at a vertex P_2, say. It is prescribed which outgoing edge we have to take first, etc. If we meet a vertex for a second time, we are forced to leave it by the edge bearing the number 2, and so on. The process has to stop somewhere, the graph being finite. The only reason why it should stop is, that we arrive at a vertex where all outgoing edges have already been taken before. This must be P_1, for all other vertices have been entered at least as often as they have been left.

We can show that at this moment all edges of T have been used, each exactly once of course, which means that a circuit has been described. Assume that a certain edge is *vacant*, that means that it has not yet been used. Considering its head, there is a vacant entry and hence there is a vacant exit. Especially, the exit belonging to Λ has to be vacant, since it has the highest number. This vacant edge of Λ leads into another vacant edge of Λ, and so on. By *c*, we eventually arrive at P_1, and we find that there is a vacant outgoing edge. This contradicts the fact that the process stopped.

From Theorem 5a and 5b we immediately obtain.

Theorem 6. *The number of trees in T with a given root is*

$$| T | . \{ \prod_{i=1}^{N} (\sigma_i - 1)! \}^{-1},$$

---

[1]) This way of numbering is different from the one considered in the proof of theorem 5*a*.

which does not depend on the vertex chosen as the root. As before, $|T|$ denotes the number of circuits of T.

Theorem 6 furnishes a new proof of theorem 3. For, there is a simple relation between the number of trees in T and in T^λ. Any tree in T^λ gives rise to a tree in T, by the mapping $T^\lambda \to T$ which maps entire bundles of T^λ into the corresponding edges of T. Conversely, in any bundle of T^λ an edge can be chosen in λ ways. Any tree in T contains $N—1$ edges, and so we have

$$t(T^\lambda) = t(T) \,.\, \lambda^{N-1}, \tag{6.2}$$

where $t(T)$ and $t(T^\lambda)$ denote the number of trees with a given root, in T and T^λ, respectively. By theorem 6 we have

$$|T^\lambda| = t(T^\lambda) \,.\prod_{i=1}^{N} (\lambda\sigma_i - 1)\,!, \tag{6.3}$$

$$|T| = t(T) \,.\prod_{i=1}^{N} (\sigma_i - 1)\,!\,. \tag{6.4}$$

Theorem 3 follows from (6.2), (6.3) and (6.4).

§ 7. **Trees in arbitrary oriented graphs.**

We consider an oriented graph G, with N vertices $P_1, \ldots, P_N$. We no longer require that it is a T-graph, that is to say, the number of edges starting from P_i need not be the same as the number of edges pointing towards P_i.

Again, we can consider (oriented) trees, with a given root. Tutte [**5**] showed, that the number of trees in T with a given root can be interpreted as the value of a certain determinant. Since his result is in several ways connected with the results of the present paper, we give a full account of his theorem, with a new proof.

Let (a_{ij}) $(i, j = 1, \ldots, N)$ be the following matrix. If $i \neq j$, then $a_{ij} = - b_{ij}$, where b_{ij} denotes the number of oriented edges from P_i to P_j (P_i is the tail and P_j is the head of these edges) Further a_{ii} is such that $\sum_{j=1}^{N} a_{ij} = 0$.

T h e o r e m 7 (Tutte). The number of trees with the given root P_i equals the minor of a_{ii} in the matrix (a_{ij}).

P r o o f. For simplicity of notation we take $i = 1$.

We first consider a special graph, where each vertex $\neq P_1$ is the tail of just one edge, and where no edge leaves P_1. This graph is either a tree or it is not; the possibility of constructing more than one tree in this graph does not exist. We shall show that the minor of a_{11} is 1 or 0 according to whether the graph is or is not a tree.

First assume that the graph is a tree. We shall apply induction with respect to N; for $N = 2$ the result is trivial. Take $N > 2$. There is at least one vertex which is not the head of an edge. This is the case with P_2, say. Then the second column of the matrix reads 0, 1, 0, ..., 0. Hence the value of the minor of a_{11} is not altered if the second row and the second column are both cancelled. The new matrix corresponds to the graph which results by cancelling P_2 and the edge starting from P_2. This new graph is still a tree, and the induction is completed.

Next assume that the graph is not a tree. Then it shows somewhere a cycle of edges not containing P_1. For example, let the cycle consist of the edges P_2P_3, P_3P_4, P_4P_3. Then, in the matrix, the 2nd, 3rd, and 4th row are linearly dependent, for their sum vanishes. It follows that the minor of a_{11} equals zero. This completes the proof of the theorem for our special graph.

The general case is easily reduced to this one by repeated application of the following operation. Divide the set of edges starting from a certain edge, P_2, say, into two groups. Now construct two graphs; the first one arises from the original graph by cancelling the edges of the first group, the second one by cancelling the edges of the second group. The matrices of the graphs are such that the second row of the original matrix equals the sum of the corresponding rows in the new matrices; all other rows are identical in the three matrices. Therefore, the minor of a_{11} in the original matrix is the sum of the minors of a_{11} in the new matrices. On the other hand, the number of trees in the original graph is the sum of the numbers of trees in both graphs. This proves the theorem.

Theorem 6 shows that in a T-graph the number of trees does not depend on the choice of the root. Tutte deduced the same fact from theorem 7. We repeat his argument. Assume that the graph considered in theorem 7 is a T-graph. Then we have $a_{ii} = \sigma_i - \varrho_i$ where ϱ_i is the number of edges from P_i to P_i. Therefore, we also find that the sum of the elements in each column of the matrix is equal to zero. It is a well-known fact that if in a square matrix the sum of the elements in each row and in each column vanishes, then the cofactors of all elements have the same value. Especially, the minor of a_{ii} does not depend on i.

We again consider an arbitrary graph G, which need not be a T-graph. Let $P_1, \ldots, P_n$ be its vertices, and let σ_i be the number of edges starting from P_i, and τ_i the number of edges pointing towards P_i. Furthermore, b_{ij} denotes the number of oriented edges from P_i to P_j. Hence $\sigma_i = \sum_j b_{ij}$, $\tau_j = \sum_i b_{ij}$.

Next we consider a permutation S of the N objects 1, 2, ..., N. Let G_S be the graph arising from G in the following manner: replace each edge $P_i\, P_j$ of G_S by an edge $P_i\, P_{Sj}$, where Sj is the result of S applied to the object j. Therefore, if the analogues of σ_i, τ_j, b_{ij} for the graph G_S are denoted by $\sigma_i{}^{(S)}$, $\tau_i{}^{(S)}$, $b_{ij}{}^{(S)}$, respectively, then we have

$$(7.1) \qquad \sigma_i{}^{(S)} = \sigma_i \; , \; \tau_{Sj}{}^{(S)} = \tau_j \; , \; b_{i,Sj}{}^{(S)} = b_{ij}.$$

Let $t_i\,(G_S)$ denote the number of oriented trees in G_S whose root is P_i, and let $\mathfrak{S}_N$ denote the group of all $N!$ permutations of the objects $1, \ldots, N$. Then we have

Theorem 8. $\sum_{S \in \mathfrak{S}_N} t_i\,(G_S) = (N-1)\,!\prod_{k \neq i} \sigma_k.$

Proof. We may and do assume $i = 1$. We shall apply theorem 7. To this end, we consider the matrix

$$M_S = (\lambda_i \delta_{ij} - b_{ij}{}^{(S)}) \qquad (i, j = 1, \ldots, N)$$

where δ_{ij} is Kronecker's symbol. Its determinant det M_S is a multilinear polynomial in the variables $\lambda_1, \ldots, \lambda_N$:

$$\det M_S = f_S\,(\lambda_1, \ldots, \lambda_N),$$

and it will be clear from theorem 7 that

$$(7.2) \qquad t_1\,(G_S) = \frac{\partial}{\partial \lambda_1} f_S\,(\lambda_1, \sigma_2, \sigma_3, \ldots, \sigma_N).$$

We put

$$(7.3) \qquad P\,(\lambda_1, \ldots, \lambda_N) = \sum_{S \in \mathfrak{S}_N} f_S\,(\lambda_1, \ldots, \lambda_N).$$

In the first place we can show that $P\,(\lambda_1, \ldots, \lambda_N)$ does not contain terms of degree $< N-1$. For instance, consider the term with $\lambda_3\,\lambda_4 \ldots \lambda_N$, which does not contain either λ_1 or λ_2. Let T be the transposition of the objects 1 and 2. Then the coefficient of $\lambda_3 \ldots \lambda_N$ in det M_{TS} is easily seen to be the opposite of the coefficient of $\lambda_3 \ldots \lambda_N$ in det M_S. If S runs through $\mathfrak{S}_N$, then TS does the same, and so the coefficient of $\lambda_3 \ldots \lambda_N$ in $P\,(\lambda_1, \ldots, \lambda_N)$ turns out to be zero.

We next deal with the terms of degree $N-1$, and therefore

we consider $\lambda_1 \lambda_2 \ldots \lambda_{i-1} \lambda_{i+1} \ldots \lambda_N$. Its coefficient in f_S equals $-b_{ii}^{(S)}$. Consequently, its coefficient in $P(\lambda_1, \ldots, \lambda_N)$ is

$$-\sum_{S\in\mathfrak{S}_N} b_{ii}^{(S)} = -\sum_{S\in\mathfrak{S}_N} b_{i,S^{-1}i} = -(N-1)!\sum_j b_{ij} = -(N-1)!\sigma_i$$

Finally, the coefficient of $\lambda_1 \ldots \lambda_N$ in f_S equals 1, and in $P(\lambda_1, \ldots, \lambda_N)$ it is $N!$ Thus we have proved that

$$P(\lambda_1, \ldots, \lambda_N) = \lambda_1 \ldots \lambda_N \,.\, (N-1)!\,\{N - \sum_i \sigma_i/\lambda_i\}.$$

From (7.2) and (7.3) we now deduce

$$\sum_{S\in\mathfrak{S}_N} t_1(G_S) = \frac{\partial}{\partial\lambda} P(\lambda_1, \sigma_2, \ldots, \sigma_N) = \sigma_2 \ldots \sigma_N \,.\, (N-1)!$$

Theorem 8 is, in some sense, a generalization of theorem 1 For, if we apply theorem 8 to a graph which is a $T^{(\sigma)}$, then we have, by theorem 6,

$$(7.4) \qquad \sum_{S\in\mathfrak{S}_N} |G_S| = (N-1)!\,\sigma^{N-1}\{(\sigma-1)!\}^N = (N-1)!\,\frac{1}{\sigma}\,.\,(\sigma!)^N$$

It is not difficult to see that (7.4) is equivalent with theorem 1 (take $n = \sigma$, $k = N$).

Note added in proof. By theorems 6 and 7 the number of circuits in a T-graph can be expressed as a determinant. For the special case that T is a $T^{(2)}$, this result was announced by W. T. Tutte and C. A. B. Smith (On unicursal paths in a network of degree 4, Amer. Math. Monthy **48**, 233—237 (1941)).

REFERENCES

1. N. G. de Bruijn. A combinatorial problem. Nederl. Akad. Wetensch., Proc. **49**, 758—764 (1946) = Indagationes Math. **8**, 461—467 (1946).
2. I. J. Good. Normal recurring decimals, J. London Math. Soc. **21**, 167—169 (1947).
3. M. H. Martin. A problem in arrangements. Bull. Amer. Math. Soc. **40**, 859—864 (1934).
4. D. Rees. Note on a paper by I. J. Good. J. London Math. Soc. **21**, 169—172 (1947).
5. W. T. Tutte. The dissection of equilateral triangles into equilateral triangles. Proc. Cambridge Phil. Soc. **44**, 463—482 (1948).

Reprinted from
Simon Stevin **28** (1951), 203–217

THE FACTORS OF GRAPHS

W. T. TUTTE

1. Introduction. A *graph* G consists of a non-null set V of objects called *vertices* together with a set E of objects called *edges*, the two sets having no common element. With each edge there are associated just two vertices, called its *ends*. Two or more edges may have the same pair of ends.

G is *finite* if both V and E are finite, and *infinite* otherwise.

The *degree* $d_G(a)$ of a vertex a of G is the number of edges of G which have a as an end. G is *locally finite* if the degree of each vertex of G is finite. Thus the locally finite graphs include the finite graphs as special cases.

A *subgraph* H of G is a graph contained in G. That is, the vertices and edges of H are vertices and edges of G, and an edge of H has the same ends in H as in G. A *restriction* of G is a subgraph of G which includes all the vertices of G.

A graph is said to be *regular of order n* if the degree of each of its vertices is n. An *n-factor* of a graph G is a restriction of G which is regular of order n.

The problem of finding conditions for the existence of an n-factor of a given graph has been studied by various authors [**3**; **4**; **5**]. It has been solved, in part, by Petersen for the case in which the given graph is regular. The author has given a necessary and sufficient condition that a given locally finite graph shall have a 1-factor [**6**; **7**]. In this paper we establish a necessary and sufficient condition that a given locally finite graph shall have an n-factor, where n is any positive integer. Actually we obtain a more general result. We suppose given a function f which associates with each vertex a of a given locally finite graph G a positive integer $f(a)$, and obtain a necessary and sufficient condition that G shall have a restriction H such that $d_H(a) = f(a)$ for each vertex a of G. The discussion is based on the method of alternating paths introduced by Petersen [**4**].

We also consider the problem of associating a non-negative integer with each edge of G so that for each vertex c of G the numbers assigned to the edges having c as an end sum to $f(c)$. We obtain a necessary and sufficient condition for the solubility of this problem.

My attention has been drawn to two other papers in which similar theories of factorization have been put forward. In one of these papers, Gallai [**2**] gives a valuable unified theory of factors and gives some new results on the factorization of regular graphs. He also claims to have obtained a necessary and sufficient condition for the existence of a 2-factor in a general locally finite graph, but leaves the discussion of this for another occasion. In the other paper Belck [**1**] establishes a necessary and sufficient condition for the existence of an n-factor in a general finite graph, where n is any positive integer. Prominent

Received February 20, 1951.

in his theory is the *hyper-n-prime* graph, a generalization of the *hyperprime* graph introduced in [**6**].

2. **Recalcitrance.** A *path* in a graph G is a finite sequence

$$P = (a_1, A_1, a_2, A_2, \ldots, A_{r-1}, a_r) \tag{1}$$

satisfying the following conditions:

(i) The members of P are alternately vertices and edges of G, the terms $a_1, a_2, \ldots, a_r$ being vertices.

(ii) If $1 \leqslant i < r$, then a_i and a_{i+1} are the two ends of A_i.

We say that P is a path *from* a_1 *to* a_r, and that its *length* is $r - 1$. We note that the terms of P need not be all distinct. We admit the case in which P has length 0. Then P has just one term, a vertex of G.

The vertices x and y of G are *connected in* G if a path from x to y in G exists. If this is so for each pair $\{x, y\}$ of vertices of G, then G is *connected*. The relation of being connected in G is evidently an equivalence relation. It therefore partitions G into a set $\{G_\alpha\}$ of connected graphs such that each edge or vertex of G belongs to some G_α and no two of the G_α have any edge or vertex in common. We call the graphs G_α the *components* of G.

If S is any proper subset of the set of vertices of a given graph G, we denote by $G(S)$ the subgraph of G obtained by suppressing the members of S and all edges of G having one or both ends in S.

Suppose now that G is locally finite and that S is a finite set of vertices of G.

If S does not include all the vertices of G the graph $G(S)$ is defined. Then if H is any finite component of $G(S)$ we denote the number of edges which have one end in S and the other a vertex of H by $v(H)$. We have

$$v(H) + \sum_{c \in H} d_G(c) \equiv 0 \pmod 2, \tag{2}$$

for the expression on the left is equal to twice the number of edges of G having an end which is a vertex of H. (We have used the symbol $c \in H$ to denote that c is a vertex of H.)

We denote by $K(G, S)$ the set of all finite components H of $G(S)$ which satisfy

$$v(H) + \sum_{c \in H} f(c) \equiv 1 \pmod 2. \tag{3}$$

If $K(G, S)$ is finite we denote the number of its elements by $k(G, S)$. If S includes all the vertices of G we write $k(G, S) = 0$. In either case we write

$$r(G, S) = k(G, S) + \sum_{c \in S} (f(c) - d_G(c)). \tag{4}$$

We call $r(G, S)$ the *recalcitrance* of G with respect to S. If $K(G, S)$ is infinite we say that $r(G, S)$ is infinite.

THEOREM I. *If G is finite, $r(G, S)$ is even or odd according as*

$$\sum_{c \in G} f(c)$$

is even or odd.

Proof. By (2), (3), and (4),

$$r(G, S) \equiv \sum_{c \in G} d_G(c) + \sum_{c \in G} f(c) \pmod 2.$$

But the sum of the degrees of the vertices of G is even, since it is twice the number of edges of G. The theorem follows.

The locally finite graph G is *constricted* with respect to f if there exist disjoint finite sets S and T of vertices of G such that

$$(5) \qquad \sum_{c \in T} f(c) < r(G(T), S).$$

As an example, G is constricted if it has a vertex a such that $d_G(a) < f(a)$. In this case (5) is satisfied if T is null and S has the single element a. Again, G is constricted if $r(G, S) > 0$ for any set S of vertices of G, for then (5) is satisfied with T null. So by Theorem I a finite graph G is constricted if the sum of the numbers $f(c)$, for all the vertices c of G, is odd. In this case (5) is satisfied if S and T are both null.

We define an *f-factor* of the given locally finite graph G as a restriction F of G such that $d_F(c) = f(c)$ for each vertex c of G. Similarly, a restriction F of a subgraph X of G is an f-factor of X if $d_F(c) = f(c)$ for each vertex c of X. A restriction F of a subgraph X of G is an *incomplete* f-factor of X if $d_F(c) \leqslant f(c)$ for each vertex c of X, and $d_F(c) = f(c)$ for all but a finite number of the vertices of X. The *deficiency* of such an incomplete f-factor is the sum

$$\sum (f(c) - d_F(c)),$$

taken over all vertices c of X for which $d_F(c) < f(c)$.

Our object in this paper is to show that G has no f-factor if and only if G is constricted with respect to f.

THEOREM II. *Let F be an incomplete f-factor of G, and let S be any finite set of vertices of G. Then the deficiency of F is not less than $r(G, S)$.*

Proof. If H is any member of $K(G, S)$, let $w(H)$ be the number of edges of F which have one end in S and the other a vertex of H. Analogously with (2) we have

$$(6) \qquad \sum_{c \in H} d_F(c) \equiv w(H) \pmod 2.$$

Let P be the set of all elements H of $K(G, S)$ such that $d_F(c) = f(c)$ for each vertex c of H. Let Q be the set of all other members of $K(G, S)$. Let the numbers of members of P and Q be p and q respectively; q must be finite.

The sum of the numbers $f(c) - d_F(c)$ taken over all vertices of G not in S which satisfy $d_F(c) < f(c)$ is at least q.

If $H \in P$, then by (3) and (6), $v(H) \not\equiv w(H)$. Hence at least p of the edges of G having just one end in S are not edges of F. It follows that

$$\sum_{c\epsilon S} d_F(c) \leqslant \sum_{c\epsilon S} d_G(c) - p,$$

$$\sum_{c\epsilon S} (f(c) - d_F(c)) \geqslant p + \sum_{c\epsilon S} (f(c) - d_G(c)).$$

Hence if D is the deficiency of F we have

$$D \geqslant p + q + \sum_{c\epsilon S} (f(c) - d_G(c)) = r(G, S).$$

THEOREM III. *If G is constricted with respect to f, it has no f-factor.*

Proof. Suppose G is constricted. Then there are disjoint finite subsets S and T of the set of vertices of G such that (5) is satisfied. Assume G has an f-factor F. Then $F(T)$ is an incomplete f-factor of $G(T)$. Its deficiency D is equal to the number n of edges of F having one end in T and the other not in T. Hence, by Theorem II,

$$\sum_{c\epsilon T} f(c) \geqslant n = D \geqslant r(G(T), S).$$

This contradicts the definition of S and T.

3. Alternating paths. An *f-subgraph* of G is a restriction J of G having the following properties:

(i) The number of edges of J is finite.

(ii) $d_J(c) \leqslant f(c)$ for each vertex c of G.

A vertex c of G is *deficient* in J if $d_J(c) < f(c)$.

Let us suppose that we are given an f-subgraph J of G and that a is a vertex of G which is deficient in J. Following a long-established tradition we refer to an edge of G as *blue* or *red* according as it is or is not an edge of J.

An *alternating path based on a* is a path P in G which satisfies the following conditions:

(i) The first term of P is a.

(ii) No edge of G occurs twice as a term of P.

(iii) If P has more than one term the edges of G which occur in P are alternately red and blue, the first one being red.

If P includes the subsequence (c, C, d) where C is an edge of G, we say that P *passes through c and then C*, or *P passes through C and then d*.

Let $\Pi(a)$ be the set of alternating paths based on a; $\Pi(a)$ is not null since it has one member whose only term is a.

Let C be an edge of G, with ends c and d. If no member of $\Pi(a)$ has C as a term, C is *acursal*. If some member of $\Pi(a)$ passes through C and then d, C is *describable to d* or *from c*. If C is describable to d but not to c, C is *unicursal to d* or *from c*. If C is describable both to c and to d, C is *bicursal*.

A vertex of G is *accessible from a* if it is a term of some member of $\Pi(a)$.

The vertex a is *singular* if no deficient vertex of G, other than a itself, is accessible from a.

THEOREM IV. *Only a finite number of vertices of G are accessible from a.*

If b is a vertex of G accessible from a, then either $b = a$, or b is an end of a blue edge, or b is an end of an edge B whose other end is either a or an end of a blue edge. Since the number of blue edges and the degree of each vertex of G are finite, the theorem follows.

THEOREM V. *Let A and B be edges of G which are of different colours and have a common end x. Suppose A is unicursal to x. Then B is describable from x.*

There is a member P of $\Pi(a)$ which passes through A and then x. If B is not a term of P preceding A there is evidently a member of $\Pi(a)$ which agrees with P as far as A and continues $(x, B, \ldots)$. Then B is describable from x.

If B precedes A in P, either the theorem is satisfied or P passes through B and then x. In the latter case there is a member of $\Pi(a)$ which agrees with P as far as B and continues $(x, A, \ldots)$. Then A is not unicursal to x, contrary to hypothesis.

4. Bicursal components. Let us suppose that G has at least one bicursal edge. The bicursal edges of G, with their ends, define a subgraph of G. We refer to the components of this subgraph as the *bicursal components*.

THEOREM VI. *The bicursal components are finite graphs.*

This follows from Theorem IV, since the vertices of a bicursal component are all accessible from a and G is locally finite.

Let L denote any bicursal component. An *entrant* of L is any member of $\Pi(a)$ which has a vertex of L as a term. If P is an entrant of L we denote by $e(P)$ the vertex of L which occurs first as a term of P. We then say that P *enters* L *at* $e(P)$. A vertex of L at which some entrant of L enters L is an *entrance* of L.

Let P be an entrant of L. Let A be the first edge of G in P after the first occurrence of $e(P)$ which is not in L, if such an edge exists. The *section of P by L* is defined as follows. If the edge A exists, the section is the part of P extending from the first occurrence of $e(P)$ to the term immediately preceding A. Otherwise, the section is the part of P extending from the first occurrence of $e(P)$ to the last term of P. In either case the section is an alternating path based on $e(P)$ and having only edges and vertices of L as terms (except that its first edge may be blue).

If e is any entrance of L we denote by $\Delta(e)$ the set of sections by L of those members of $\Pi(a)$ which enter L at e.

Since the edges of L are not acursal, L has at least one entrance. If a is a vertex of L then a is an entrance of L.

In the following series of theorems (VII—XI)) we suppose that some entrance e of L is specified, with the proviso that e is a if a is a vertex of L.

THEOREM VII. *There exists an edge of L which is a term of some member of $\Delta(e)$.*

Proof. Suppose first that e is a. Any red edge of L having a as an end is clearly a term of a member of $\Delta(a)$. Suppose therefore that the edges of L

having a as an end are all blue. Each of these is describable from a, and no one is the first edge of a member of $\Pi(a)$. Hence some red edge C having a as an end is describable to a. But all red edges having a as an end are describable from a. Hence C is bicursal and therefore an edge of L, contrary to supposition.

Now consider the case in which a is not a vertex of L. Let P be an entrant of L such that $e(P) = e$. Let C be the edge of G which immediately precedes the first occurrence of e in P. Then C is unicursal to e. Any edge of L having e as an end and differing in colour from C is clearly a term of a member of $\Delta(e)$. Suppose therefore that the edges of L having e as an end all have the same colour as C. Since they are all describable from e, some edge E of G differing in colour from C is describable to e. But E is describable from e, by Theorem V. Hence E is bicursal and therefore an edge of L, contrary to supposition.

THEOREM VIII. *If A is an edge of L with ends x and y, and if some member P' of $\Delta(e)$ passes through x and then A, then some other member of $\Delta(e)$ passes through y and then A.*

Proof. Since A is bicursal there exists a member Q of $\Pi(a)$ which passes through y and then A. It may happen that every term of Q which precedes A is an edge or vertex of L. Then a is a vertex of L and therefore $e = a$ by the definition of e. Hence the section of Q by L is a member of $\Delta(e)$ which passes through y and then A.

In the remaining case, let B be the last term of Q preceding A which is an edge of G but not an edge of L. Let b be the immediately succeeding term of Q. Then b is a vertex of L. Let C be the first edge of G in P' which succeeds B in Q but does not succeed A in Q. Such an edge exists since A is an edge both of P' and of Q. Let the ends of C be r and s. We may suppose that P' passes through r and then C.

Suppose Q passes through r and then C. Then there is a member of $\Delta(e)$ which agrees with P' as far as C and then continues with the terms of Q from C to A. This member of $\Delta(e)$ passes through y and then A.

Alternatively, suppose Q passes through s and then C. There is a member Q_1 of $\Pi(a)$ which enters L at e, then agrees with P' as far as C, and continues with the terms of Q in reverse order from C to b. Let D be the edge of Q_1 immediately preceding the first occurrence of e. If $B \neq D$ it follows that B is describable from b. But Q passes through B and then b. So B is bicursal and therefore an edge of L, contrary to its definition. We conclude that $B = D$ and therefore $b = e$. Hence there is a member of $\Pi(a)$ which agrees with Q_1 as far as B and agrees with Q from B to A. The section of this path by L is a member of $\Delta(e)$ which passes through y and then A.

THEOREM IX. *Let A be an edge of L which is a term of some member of $\Delta(e)$. Let x be an end of A distinct from e. Then there is an edge B of L which differs in colour from A, which has x as an end, and which is a term of some member of $\Delta(e)$.*

Proof. By Theorem VIII there is a member of $\Delta(e)$ which passes through x and then A. The last edge preceding A in this member of $\Delta(e)$ has the required properties.

THEOREM X. *If A is any edge of L and x is any end of A, then there is a member of $\Delta(e)$ which passes through x and then A.*

Proof. Let U be the set of all edges of L occurring as terms in the members of $\Delta(e)$; U is non-null, by Theorem VII. Let V be the set of all other edges of L.

Assume that V is non-null. Since L is connected there is a vertex z of L which is an end of a member B of U and a member C of V. If z is not e we may suppose that B and C differ in colour, by Theorem IX. By Theorem VIII there is a member of $\Delta(e)$ which passes through B and then z. C is not a term of this member of $\Delta(e)$. Hence there is a member of $\Delta(e)$ which agrees with this one as far as B and then continues with z and C. This contradicts the definition of C.

Suppose now that z is e. If B and C differ in colour we obtain a contradiction as before. We deduce that all the edges of L having e as an end have the same colour. If $e = a$ it follows from Theorem VII that e is an end of some red edge of L. Then C is red. Hence there is a member of $\Delta(e)$ which has C as its first edge, contrary to assumption. If e is not a it follows from Theorem VII that there is a member P of $\Pi(a)$ entering L at e in which the first occurrence of e is immediately succeeded by an edge of L. We may take this edge to be B. Since B and C have the same colour there is a member of $\Pi(a)$ which agrees with P as far as the first occurrence of e and then continues with C. Hence C is a member of U, contrary to assumption.

We conclude that V is null. The theorem now follows from Theorem VIII.

Let G_1 denote any subgraph of G. An edge A of G is said to *touch* G_1 if A is not an edge of G_1 and just one end, say x, of A is a vertex of G_1. Such an edge A is *unicursal to* or *from* G_1 if it is unicursal to or from x respectively.

THEOREM XI. *If a is a vertex of L then all edges of G which touch L are unicursal from L. If a is not a vertex of L then there exists just one edge of G which touches L and is unicursal to L, and all other edges of G which touch L are unicursal from L.*

Proof. Let A be an edge of G which touches L. Let x be the end of A which is a vertex of L. Assume that A is not unicursal from x. We recall that $a = e$ if a is a vertex of L.

If x is not e there is an edge C of L differing in colour from A and having x as an end, by Theorem IX. This is true also if $x = e = a$. For then A is blue since it is not unicursal from a and not bicursal, and Theorem VII shows that some red edge of L has a as an end. In either of these cases it follows from Theorem X that there is a member of $\Pi(a)$ which enters L at e, whose section by L passes through C and then x, and which continues from C with the terms x and A. But A is not bicursal since it is not an edge of L. Hence A is unicursal from x, contrary to assumption.

Now suppose that $x = e$ and e is not a. By Theorem VII, there is a member P of $\Pi(a)$ which enters L at e and in which the first occurrence of e is immediately succeeded by an edge C of L. Let the edge of G which immediately precedes the first occurrence of e in P be B. Clearly B touches L and is unicursal to L.

Suppose that A and B are distinct. If A differs in colour from B it is describable from $x = e$, by Theorem V. If A and B have the same colour this differs from that of C. By Theorem X there is a member Q of $\Pi(a)$ which enters L at e, and whose section by L passes through C and then e. It is clear that A and B cannot both precede C in Q. Hence there is a member Q' of $\Pi(a)$ which agrees with Q as far as C and then continues with e and one of the edges A and B. Actually, it continues with e and A since B is unicursal to e. Hence if A and B are distinct, A is unicursal from x.

This completes the proof of the theorem.

5. Bicursal units. Let T be the set of all vertices of G which are ends of bicursal edges. Let T' be the set of all edges of G having both ends in T. Then T and T' define a subgraph G' of G. We refer to the components of G' as *bicursal units*. Evidently a bicursal component having a given vertex b is a subgraph of the bicursal unit having the vertex b. By Theorem IV the bicursal units are finite graphs.

THEOREM XII. *Let M be any bicursal unit. If a is a vertex of M then all edges of G which touch M are unicursal from M. If a is not a vertex of M then there exists just one edge of G which touches M and is unicursal to M, and all other edges of G which touch M are unicursal from M.*

Proof. Since some edges of M are bicursal there exists a member P of $\Pi(a)$ having a vertex of M as a term. Let e be the first vertex of M to occur in P. If a is not a vertex of M there is an edge E of G which immediately precedes the first occurrence of e in P. Then E touches M and is unicursal to e and M. We denote the bicursal component of which e is a vertex by L. If instead a is a vertex of M we denote the bicursal component of which a is a vertex by L.

A subgraph L' of M which is a bicursal component distinct from L is *supplied from* L if there exists a sequence $(L_1, L_2, \ldots, L_t)$ of bicursal components and a sequence $(A_1, A_2, \ldots, A_{t-1})$ of edges of M such that

(i) $L_1 = L$ and $L_t = L'$,

(ii) the L_i are subgraphs of M,

(iii) for each integer i in the range $1 \leqslant i < t$, A_i is unicursal from L_i and to L_{i+1}.

We can show that any subgraph of M which is a bicursal component distinct from L is supplied from L. For suppose it is not. Then since M is connected there is an edge B of M with ends b and c belonging to bicursal components L' amd L'', where L' is L or is supplied from L, and L'' is not L and is not supplied from L. Now B is not bicursal by the definition of a bicursal component, and is not acursal, by Theorem XI. It is not unicursal to L'', since L'' is not supplied from

L. Hence B is unicursal to L'. But this is contrary to Theorem XI since L' is either L or is supplied from L.

The Theorem now follows by the application of Theorem XI to each of the bicursal components which are subgraphs of M.

If a is not a vertex of the bicursal unit M, we call the edge of G which touches M and is unicursal to M the *entrance-edge* of M. We classify such bicursal units as *red-entrant* and *blue-entrant* according as their entrance-edges are red or blue. A bicursal unit having a as a vertex is *a-entrant*.

6. Singular vertices. In this section we suppose that a is a singular vertex. We denote the numbers of red-entrant and blue-entrant bicursal units by k_r and k_b respectively. These numbers are finite, by Theorem IV.

Let U denote the set of all vertices of G which are not vertices of G'. Thus no bicursal edge has an end in U. Let V be the set of all members of U to which some red edge is unicursal. Let W be the set of all members of U from which some red edge is unicursal or to which some blue edge is unicursal. Clearly,

$$a \notin V. \tag{9}$$

Suppose $c \in V$. Any blue edge of G having c as an end is unicursal from c, by Theorem V. Hence, by (9), no red edge of G can be unicursal from c. There are just $f(c)$ blue edges of G which have c as an end and are therefore unicursal from c since c is accessible from, but distinct from, the singular vertex a.

Now suppose $i \in W$. If some red edge is unicursal from i then either $a = i$ or there is a blue edge unicursal to i. If $a = i$ or there is a blue edge unicursal to i, then each red edge having i as an end is unicursal from i, by Theorem V and the definition of $\Pi(a)$. Hence any red edge having i as an end is unicursal from i. Consequently no blue edge of G can be unicursal from i.

It is clear from these results that V and W are disjoint sets. By Theorem IV they are finite sets.

If $i \in W$, let $y(i)$ be the number of red edges of G unicursal from i which are entrance-edges of red-entrant bicursal units. Let $z(i)$ be the number of blue edges of G which are unicursal to i from members of V.

Let H denote the graph $G(V)$. If $i \in W$, any red edge unicursal from i is unicursal to a vertex p distinct from a. For no red edge is unicursal to a. So by Theorem V, p is either a vertex of G' or a member of V. Hence in the graph H, the number of edges having i as an end is $y(i) + (d_J(i) - z(i))$. Thus we have

$$z(i) = y(i) + (d_J(i) - d_H(i)). \tag{10}$$

By the definition of a bicursal unit the entrance-edge of any bicursal unit which is not a-entrant is unicursal either from a member of V or from a member of W.

Let λ be the number of blue edges of G unicursal from a member of V to a member of W. It is equal to the total number of blue edges unicursal from members of V less the number of the entrance-edges of the blue-entrant bicursal units. The latter number is k_b, by Theorem XII. But λ is also equal to the sum of the numbers $z(i)$ taken over all $i \in W$. The corresponding sum of the $y(i)$ is k_r, by Theorem XII. Hence we have

$$\sum_{c \in V} f(c) = k_b + k_r + \sum_{i \in W} (d_J(i) - d_H(i)). \tag{11}$$

The bicursal units, if any, are connected finite graphs. By Theorem XII, they are components of $(G(V))(W) = H(W)$.

Let M be any bicursal unit. Write $q(M) = 0$ or 1 according as M is or is not a-entrant. Let $u(M)$ be the number of blue edges of G which touch M and let $v(M)$ be the number of edges of G which touch M and have an end in W.

Using Theorem XII we readily obtain the following results: if M is blue-entrant $q(M) = 1$ and $u(M) = v(M) + 1$, if M is red-entrant $q(M) = 1$ and $u(M) = v(M) - 1$, and if M is a-entrant $q(M) = 0$ and $u(M) = v(M)$. In each case we have

$$u(M) \equiv v(M) + q(M) \pmod 2. \tag{12}$$

The sum of $u(M)$ and the degrees in J of the vertices of M is even, since it is twice the number of blue edges of G having vertices of M as ends. Moreover, if c is a vertex of M, we have $d_J(c) = f(c)$ unless $c = a$; and a is a vertex of M if and only if $q(M) = 0$. It follows from (12) that

$$v(M) + \sum_{c \in M} f(c) + (q(M) + 1)(d_J(a) - f(a)) + q(M) \equiv 0 \pmod 2. \tag{13}$$

Referring to the definitions of §2 we see that M is a member of $K(G(V), W)$ if and only if $(q(M) + 1)\,(d_J(a) - f(a)) + q(M) \equiv 1 \pmod 2$. Hence M is not a member of $K(G(V), W)$ if and only if M is a-entrant $(q(M) = 0)$ and the deficiency $f(a) - d_J(a)$ of a in J is even.

THEOREM XIII. *If G is not constricted there exists an a-entrant bicursal unit, and the deficiency of a in J is even.*

Proof. Suppose, first, that a is a member of U but not of W. Then no red edge of G has a as an end. Hence $d_G(a) = d_J(a) < f(a)$. But then G is constricted, contrary to hypothesis, for (5) is satisfied if we take T to be null and S to have the single element a. We conclude that either $a \in W$ or there exists an a-entrant bicursal unit.

Applying formula (4) we have

$$r(G(V), W) = k(G(V), W) + \sum_{i \in W} (f(i) - d_H(i)). \tag{14}$$

If $a \in W$ then $k(G(V), W) \geqslant k_b + k_r$, and $f(a) > d_J(a)$. If there exists an a-entrant bicursal unit and the deficiency of a in J is odd we have

$$k(G(V), W) \geqslant k_b + k_r + 1.$$

Here we have used the results proved above concerning the membership of bicursal units in $K(G(V), W)$. In each of these cases it follows that the expression on the right of (11) is less than $r(G(V), W)$. Then G is constricted, contrary to hypothesis. The theorem follows.

7. Augmentation. In this section we no longer assume that the deficient vertex a is singular.

Suppose P is a member of $\Pi(a)$ which has more than one term, and whose last term is a vertex i of G deficient in J. To *transform J by P* is to replace J by a restriction K of G, defined as follows. The edges of K consist of the blue edges of G which are not terms of P, together with the red edges of G which are terms of P.

We say the f-subgraph J is *augmentable* at the deficient vertex a if there is an f-subgraph K of G satisfying the following conditions:

(i) $d_K(a) > d_J(a)$.

(ii) If $d_J(c) = f(c)$, then $d_K(c) = f(c)$.

Suppose a is not singular. Then there is a member P of $\Pi(a)$ whose last term is a deficient vertex i of G distinct from a. Let K be the restriction of G obtained by transforming J by P. By the definition of $\Pi(a)$ we have

$$d_K(a) = d_J(a) + 1, \quad d_K(i) = d_J(i) \pm 1,$$

and $d_K(c) = d_J(c)$ if c is not a or i. Hence K is an f-subgraph of G, and J is augmentable at a.

Suppose next that a is singular and that G is not constricted. The deficiency of a in J is at least 2, by Theorem XIII. Also by Theorem XIII, a is the entrance of a bicursal unit M_0. By Theorem VII there is a red edge A of M_0 having a as an end. Since A is bicursal there is a member P of $\Pi(a)$ including at least two edges, whose last term is a and whose last edge is A. Let K be the restriction of G obtained by transforming J by P. By the definition of $\Pi(a)$ we have

$$d_K(a) = d_J(a) + 2 \leqslant f(a),$$

and $d_K(c) = d_J(c)$ if c is not a. Hence K is an f-subgraph of G, and J is augmentable at a.

Thus we have the following

THEOREM XIV. *Let J be any f-subgraph of G, and let a be any vertex of G which is deficient in J. Then either G is constricted with respect to f or J is augmentable at a.*

8. Condition for an f-factor.

THEOREM XV. *G has no f-factor if and only if it is constricted with respect to f.*

Proof. Suppose first that the locally finite graph G is constricted with respect to f. Then G has no f-factor, by Theorem III.

Suppose next that G is not constricted with respect to f. Let J_0 be the restriction of G which has no edges. Then J_0 is an f-subgraph of G.

If a vertex a of G is deficient in a given f-subgraph J of G, we can, by Theorem XIV, replace J by an f-subgraph K in which the degree of a is increased and no vertex of G which is not deficient in J is deficient in K. By repeating this process sufficiently often we can obtain an f-subgraph K' of G in which a and those vertices of G not deficient in J are not deficient.

It follows that if S is any finite set of vertices of G, we can, by the above process, build from J_0 an f-subgraph J of G in which no member of S is deficient.

The theorem follows at once in the case in which G is finite. Then we can take S to be the set of all vertices of G, and the corresponding f-subgraph J must be an f-factor of G.

If G is infinite and connected we use the following non-constructive argument. (I have replaced my original proof by a shorter one for which I am indebted to the referee.)

Let x be any vertex of G. The number of paths in G whose first term is x and which have just $2n + 1$ terms, where n is any given non-negative integer, is finite since G is locally finite. Hence the set of paths in G having x as first term is denumerable. Since G is connected it follows that the set of vertices of G is denumerable, say $\{a_1, a_2, \ldots\}$. By the foregoing argument, to every positive integer n there is an f-subgraph J_n such that

$$d_{J_n}(a_r) = f(a_r), \qquad r \leqslant n.$$

The set of edges of G is at most denumerable, say equal to $\{A_1, A_2, \ldots\}$. Put $F_n(s) = 1$ if A_s is an edge of J_n and $F_n(s) = 0$ otherwise. Then by the diagonal process, there is an increasing sequence $n_1, n_2, \ldots$ such that

$$\lim_{k \to \infty} F_{n_k}(s) = F(s)$$

exists for all s. Let J be the restriction of G whose edges are those A_s for which $F(s) = 1$. Then $d_J(a_r) = f(a_r)$ for all r, and J is an f-factor of G.

Lastly, we must consider the case in which G is infinite and not connected. We can show that no component of G is constricted with respect to f. For if this is not so, there is a component G_a of G such that for some disjoint finite subsets S and T of the set of vertices of G,

$$\text{(15)} \qquad \sum_{c \in T} f(c) < k(G_a(T), S) + \sum_{c \in S} (f(c) - d_{G_a(T)}(c)).$$

Clearly each component of $G_a(T)$ is a component of $G(T)$. Hence (15) holds with $G_a(T)$ replaced by $G(T)$, so that

$$\sum_{c \in T} f(c) < r(G(T), S).$$

Thus G is constricted with respect to f, contrary to hypothesis.

Since the theorem has been proved for connected graphs it follows that each component of G has an f-factor. Hence (assuming the multiplicative axiom) there is a set Z of f-factors of components of G which contains just one f-factor of each component of G. The restriction of G whose edges are the edges of the members of Z is an f-factor of G.

9. n-factors. A necessary and sufficient condition for the existence of an n-factor of G, where n is a given positive integer, can be obtained by applying Theorem XV to the special case in which the value of $f(c)$ is n for each vertex c of G.

It is convenient to denote the number of elements of a finite set U by $\alpha(U)$. We then obtain the following

THEOREM XVI. *G has no n-factor if and only if there exist disjoint finite sets S and T of vertices of G such that*

$$n\alpha(T) < k(G(T), S) - \sum_{c \in S} (d_{G(T)}(c) - n). \tag{16}$$

Here $k(G(T),S)$ is the number of finite components H of $(G(T))(S) = G(S \cup T)$ for which n times the number of vertices differs in parity from the number of edges of G which have one end in S and the other end a vertex of H.

A necessary and sufficient condition for the existence of a 1-factor of a given locally finite graph G has been given in previous papers [**6**; **7**]. It is simpler in form than the expression obtained by writing $n = 1$ in (16). In the next section, this simpler formula is deduced from Theorem XV. The argument suggests no analogous simplification in the case $n > 1$.

10. An allied problem. Suppose that we are given a locally finite graph G, and a function f which associates with each vertex c of G a positive integer $f(c)$. We consider the problem of associating with each edge A of G a non-negative integer $h(A)$ so that for each vertex c of G the sum of the numbers $h(A)$, taken over all edges A of G having c as an end, is $f(c)$. If such a set of non-negative integers $h(A)$ exists we say that G is *f-soluble*.

We note that if $f(c) = 1$ for each vertex of G, then G is f-soluble if and only if it has a 1-factor.

Let T be any finite set of vertices of G. We denote by $S(T)$ the set of all vertices c of G having the following properties:

(i) c is not an element of T.

(ii) Each edge of G having c as an end has its other end in T.

If T does not include every vertex of G we denote by $k(T)$ the number of finite components H of $G(T)$ having the following properties:

(i) H has more than one vertex.

(ii) The sum of the numbers $f(a)$, taken over all vertices of H, is odd.

If T is the set of all vertices of G we write $k(T) = 0$.

THEOREM XVII. *G is not f-soluble if and only if there exists a finite set T of vertices of G such that*

$$\sum_{c \in T} f(c) < k(T) + \sum_{c \in S(T)} f(c). \tag{17}$$

Proof. By adjoining new edges to G we can obtain a graph G' having the following properties:

(i) The vertices of G' are the vertices of G.

(ii) Two vertices are joined by an edge in G' if and only if they are joined by an edge in G.

(iii) If two vertices a and b are joined by an edge in G', the number of distinct edges of G' which join them is finite and not less than $d_G(a) + f(a)$.

Clearly G' is locally finite.

If S and T are disjoint finite sets of vertices of G such that S is contained in $S(T)$ it follows from the definition of G' that

$$k(G'(T), S) = k(G(T), S). \tag{18}$$

It is clear that G is f-soluble if and only if G' has an f-factor. Hence, by Theorem XV, G is not f-soluble if and only if there exist disjoint finite sets S and T of vertices of G such that

$$\sum_{c\epsilon T} f(c) < k(G'(T), S) - \sum_{c\epsilon S} (d_{G'(T)}(c) - f(c)). \tag{19}$$

Suppose first that (17) is satisfied for some finite T. If $S(T)$ is not finite then all but a finite number of its elements have degree 0 in G, since G is locally finite. Hence G is not f-soluble since it has a vertex of degree 0. If $S(T)$ is finite it follows from (17) and (18) that (19) will be satisfied if we put $S = S(T)$. Hence G is not f-soluble.

Conversely, suppose that G is not f-soluble. Then (19) is satisfied for some disjoint finite sets S and T. If possible let a be any member of S not in $S(T)$. Consider the effect of replacing S by $S' = S - \{a\}$. Clearly the replacement diminishes

$$\sum_{c\epsilon S} (d_{G'(T)}(c) - f(c))$$

by $d_{G'(T)}(a) - f(a)$, that is, by at least $d_G(a)$, from (iii).

The replacement diminishes $k(G'(T), S)$ by not more than $d_{G(T)}(a)$, the maximum number of finite components of $G'(S \cup T)$ joined to a by an edge of G'. But $d_{G(T)}(a) \leqslant d_G(a)$. Hence, if a is not an element of $S(T)$, formula (19) remains valid when S is replaced by S'.

If S' has an element not in $S(T)$ we repeat the argument with S' replacing S, and so on. Since S is finite we find, eventually,

$$\sum_{c\epsilon T} f(c) < k(G'(T), U) + \sum_{c\epsilon U} f(c), \tag{20}$$

where U is the intersection of S and $S(T)$. But, by (18), $k(G'(T), U)$ is equal to $k(T)$ plus the number of components of $G(T \cup U)$ which consist of a single vertex, the value of f for this vertex being odd. Hence

$$k(G'(T), U) \leqslant k(T) + \sum_{c\epsilon S(T)-U} f(c). \tag{21}$$

Now (20) and (21) imply (17). This completes the proof of the theorem.

If $f(c) = 1$ for each vertex c of G it is clear that G is f-soluble if and only if it has a 1-factor. Applying Theorem XVII to this case we find that G has no 1-factor if and only if there exists a finite set T of vertices of G such that

$$\alpha(T) < h_u(T),$$

where $h_u(T)$ is the number of finite components of $G(T)$ having an odd number of vertices. This is the simple criterion for the existence of a 1-factor mentioned in §9.

References

1. H. B. Belck, *Reguläre Faktoren von Graphen*, J. Reine Angew. Math., vol. 188 (1950), 228-252.
2. T. Gallai, *On factorization of graphs*, Acta Mathematica Academiae Scientarum Hungaricae, vol. 1 (1950), 133-153.
3. P. Hall, *On representation of subsets*, J. London Math. Soc., vol. 10 (1934), 26-30.
4. J. Petersen, *Die Theorie der regulären Graphs*, Acta Math., vol. 15 (1891), 193-220.
5. R. Rado, *Factorization of even graphs*, Quarterly J. Math., vol. 20 (1949), 95-104.
6. W. T. Tutte, *The factorization of linear graphs*, J. London Math. Soc., vol. 22 (1947), 107-111.
7. ———, *The factorization of locally finite graphs*, Can. J. Math., vol. 1 (1950), 44-49.

The University of Toronto

Reprinted from
Canad. J. Math. **4** (1952), 314–328

A PARTITION CALCULUS IN SET THEORY

P. ERDÖS AND R. RADO

1. Introduction. Dedekind's pigeon-hole principle, also known as the box argument or the chest of drawers argument (Schubfachprinzip) can be described, rather vaguely, as follows. *If sufficiently many objects are distributed over not too many classes, then at least one class contains many of these objects.* In 1930 F. P. Ramsey [12] discovered a remarkable extension of this principle which, in its simplest form, can be stated as follows. *Let S be the set of all positive integers and suppose that all unordered pairs of distinct elements of S are distributed over two classes. Then there exists an infinite subset A of S such that all pairs of elements of A belong to the same class.* As is well known, Dedekind's principle is the central step in many investigations. Similarly, Ramsey's theorem has proved itself a useful and versatile tool in mathematical arguments of most diverse character. The object of the present paper is to investigate a number of analogues and extensions of Ramsey's theorem. We shall replace the sets S and A by sets of a more general kind and the unordered pairs, as is the case already in the theorem proved by Ramsey, by systems of any fixed number r of elements of S. Instead of an unordered set S we consider an ordered set of a given order type, and we stipulate that the set A is to be of a prescribed order type. Instead of two classes we admit any finite or infinite number of classes. Further extension will be explained in §§2, 8 and 9.

The investigation centres round what we call *partition relations* connecting given cardinal numbers or order types and in each given case the problem arises of deciding whether a particular partition relation is true or false. It appears that a large number of seemingly unrelated arguments in set theory are, in fact, concerned with just such a problem. It might therefore be of interest to study such relations for their own sake and to build up a *partition calculus* which might serve as a new and unifying principle in set theory.

In some cases we have been able to find best possible partition relations, in one sense or another. In other cases the methods available to the authors do not seem to lead anywhere near the ultimate

---

Part of this paper was material from an address delivered by P. Erdös under the title *Combinatorial problems in set theory* before the New York meeting of the Society on October 24, 1953, by invitation of the Committee to Select Hour Speakers for Eastern Sectional Meetings; received by the editors May 17, 1955.

truth. The actual description of results must be deferred until the notation and terminology have been given in detail. The most concrete results are perhaps those given in Theorems 25, 31, 39 and 43. Of the unsolved problems in this field we only mention the following question. *Is the relation* $\lambda \rightarrow (\omega_0 2, \omega_0 2)^2$ *true or false?* Here, λ denotes the order type of the linear continuum.

The classical, Cantorian, set theory will be employed throughout. In some arguments it will be advantageous to assume the continuum hypothesis $2^{\aleph_0} = \aleph_1$ or to make some even more general assumption. In every such case these assumptions will be stated explicitly.

The authors wish to thank the referee for many valuable suggestions and for having pointed out some inaccuracies.

2. **Notation and definitions.** Capital letters, except Δ, denote sets, small Greek letters, except possibly π, order types, briefly: *types*, and $k, l, m, n, \kappa, \lambda, \mu, \nu$ denote ordinal numbers (*ordinals*). The letters r, s denote non-negative integers, and a, b, d cardinal numbers (*cardinals*). No distinction will be made between finite ordinals and the corresponding finite cardinals. Union and intersection of A and B are $A+B$ and AB respectively, and $A \subset B$ denotes inclusion, in the wide sense. For any A and B, $A-B$ is the set of all $x \in A$ such that $x \notin B$. No confusion will arise from our using 0 to denote both zero and the empty set. If $p(x)$ is a proposition involving the general element x of a set A then $\{x : p(x)\}$ is the set of all $x \in A$ such that $p(x)$ is true.

η and λ are the types, under order by magnitude, of the set of all rational and of all real numbers respectively. λ will also be used freely as a variable ordinal in places where no confusion can arise. The relation $\alpha \leqq \beta$ means that every set, ordered according to β, contains a subset of type α, and $\alpha \nleqq \beta$ is the negation of $\alpha \leqq \beta$. To every type α there belongs the *converse type* α^* obtained from α by replacing every order relation $x<y$ by the corresponding relation $x>y$. We put

$$[m, n] = \{\nu : m \leqq \nu < n\} \qquad \text{for } m \leqq n.$$

The symbol

$$\{x_0, x_1, \cdots\}_<$$

denotes the set $\{x_0, x_1, \cdots\}$ and, at the same time, expresses the fact that $x_0 < x_1 < \cdots$. Brackets $\{\ \}$ are only used in order to define sets by means of a list of their elements. For typographical convenience we write

$$\sum [x \in A] f(x)$$

instead of $\sum_{x\in A} f(x)$, and we proceed similarly in the case of products etc. or when the condition $x\in A$ is replaced by some other type of condition.

The cardinal of S is $|S|$, and the cardinal of α is $|\alpha|$. For every cardinal a, the symbol a^+ denotes the next larger cardinal. If $a=b^+$ for some b, then we put $a^-=b$, and if a is not of the form b^+, i.e. if a is zero or a limit cardinal, then we put $a^-=a$. Similarly, we put $k^-=l$, if $k=l+1$, and $k^-=k$, if $k=0$ or if k is a limit ordinal. If S is ordered by means of the order relation $x<y$, then the type of S is denoted by $\overline{S}_<$ and, if no confusion can arise, by $\overline{S}$. For any $a>1$ we denote by a' the least cardinal $|n|$ such that a can be represented in the form $\sum[\nu<n]a_\nu$ where $a_\nu<a$ for all $\nu<n$. This cardinal a', the *cofinality cardinal* belonging to a, is closely related to the cofinality ordinal $cf(\beta)$ of an ordinal β introduced by Tarski [17]. A *regular* cardinal is a cardinal a such that $a'=a$. The least ordinal of a given cardinal a is the *initial ordinal* belonging to a. Initial ordinals are the finite ordinals and the infinite ordinals ω_ν of cardinal $\aleph_\nu$. We put

$$[S]^a = \{X:X\subset S;\ |X| = a\}.$$

In particular, $[S]^a=0$ if $|S|<a$. The relation

$$A = \sum{}'[\nu < k]A_\nu = A_0 +' A_1 +' \cdots$$

means, by definition, that $A=\sum[\nu<k]A_\nu$ and, also,

$$A_\mu A_\nu = 0 \qquad (\mu < \nu < k).$$

Fundamental throughout this paper is the *partition relation*

$$a \to (b, d)^2$$

introduced in [6]. More generally, for any a, b_ν, k, r the relation

$$a \to (b_0, b_1, \cdots)_k^r \tag{1}$$

is said to hold if, and only if, the following statement is true. The cardinals b_ν are defined for $\nu<k$. Whenever

$$|S| = a; \qquad [S]^r = \sum[\nu < k]K_\nu,$$

then there are $B\subset S$; $\nu<k$ such that

$$|B| = b_\nu; \qquad [B]^r \subset K_\nu.$$

For $k<\omega_0$ we also write (1) in the form

$$a \to (b_0, b_1, \cdots, b_{k-1})^r,$$

and if k is arbitrary, and $b_\nu = b$ for all $\nu < k$, then we may write (1) in the form

$$a \rightarrow (b)_k^r.$$

We also introduce partition relations between types. By definition, the relation

$$\alpha \rightarrow (\beta_0, \beta_1, \cdots)_k^r \tag{2}$$

holds if, and only if, β_ν is defined for $\nu < k$ and if, whenever a set S is ordered and

$$\bar{S} = \alpha; \qquad [S]^r = \sum [\nu < k] K_\nu,$$

there are $B \subset S$; $\nu < k$ such that

$$\bar{B} = \beta_\nu; \qquad [B]^r \subset K_\nu.$$

If $k < \omega_0$, or if all β_ν are equal to each other, we use an alternative notation for (2) analogous to that relating to (1). The negation of (1), and similarly in the case of (2), is denoted by

$$a \nrightarrow (b_0, b_1, \cdots)_k^r.$$

We mention in passing that the gulf between (1) and (2) can be bridged by the introduction of more general partition relations referring to partial orders. These will, however, not be considered here.

If $a \geq \aleph_0$ then, clearly, a' is the least cardinal $\aleph_n$ such that $a \nrightarrow (a)_{\omega_n}^1$. Also, $\aleph_m$ is regular if, and only if, $\aleph_m \rightarrow (\aleph_m)_{\omega_n}^1$ for all $n < m$. Finally, the relation $a \rightarrow (b_0^+, b_1^+, \cdots)_k^1$ is equivalent to $\sum [\nu < k] b_\nu < a$.

We now introduce some abbreviations. Let S be ordered. Then, for $x \in S$,

$$L(x) = \{y : \{y, x\}_< \subset S\}; \qquad R(x) = \{y : \{x, y\}_< \subset S\}.$$

If, in addition, $[S]^r = \sum [\nu < k] K_\nu$, then, for $B \subset S$; $\nu < k$,

$$F_\nu(B) = \{\bar{A} : A \subset B;\ [A]^r \subset K_\nu\},$$
$$[K_\nu] = F_\nu(S).$$

In the special case $r = 2$, we put, for $x \in S$; $\nu < k$, $L_\nu(x) = \{y : \{y, x\}_< \in K_\nu\}$; $R_\nu(x) = \{y : \{x, y\}_< \in K_\nu\}$; $U_\nu = L_\nu + R_\nu$. U_ν is independent of the order of S. If $0 \neq A \subset S$, and if $W(x)$ is any one of the functions $L, R, L_\nu, R_\nu, U_\nu$, then we put

$$W(A) = \prod [x \in A] W(x).$$

Also, $W(0) = S$. If $n < \omega_0$, then we write $W(x_0, \cdots, x_{n-1})$ instead of

$W(\{x_0, \cdots, x_{n-1}\})$. It will always be clear from the context to which ordered set S and to which partition of $[S]^r$ these functions refer. We shall occasionally make use of the notation and the calculus of partitions (distributions) summarized in [**5**, p. 419]. The meaning of *canonical partition relations*

$$\alpha \to *(\beta)^r$$

and that of *polarized partition relations*

$$\begin{pmatrix} a_0 \\ \cdot \\ \cdot \\ \cdot \\ a_{t-1} \end{pmatrix} \to \begin{pmatrix} b_{00} & b_{01} & \cdots \\ \cdot & \cdot & \cdot \\ b_{t-1,0} & & \cdots \end{pmatrix}^{r_0,\cdots,r_{t-1}}_k$$

will be given in §§8 and 9 respectively. The relation $\alpha \to (\beta)^r_d$ will be defined in §4.

3. Previous results.

THEOREM 1. *If* $k < \omega_0$ *then* $\aleph_0 \to (\aleph_0)^r_k$ [12, *Theorem* A].

THEOREM 2. *If* $k, n < \omega_0$, *then, for some* $f = f(k, n, r) < \aleph_0$,

$$f \to (n)^r_k$$

[12, *Theorem* B].

THEOREM 3. (i) *If* $a \geqq \aleph_0$, *then* $a \to (\aleph_0, a)^2$.
(ii) $\aleph_{\omega_0} \not\to (\aleph_1, \aleph_{\omega_0})^2$.

(i) is proved in [**2**, 5.22]. This formula will be restated and proved as Theorem 44.
(ii) is in [**3**, p. 366] and will follow from Theorem 36 (iv).

THEOREM 4. (i) *If* $a \geqq \aleph_0$, *then* $(a^a)^+ \to (a^+)^2_a$.
(ii) *If* $a \geqq \aleph_0$, *then* $a^a \not\to (3)^2_a$.
(iii) *If* $2^{\aleph_n} = \aleph_{n+1}$, *then* $\aleph_{n+2} \to (\aleph_{n+1}, \aleph_{n+2})^2$.

(i) is given in [**3**] and will be deduced as a corollary of Theorem 39.
(ii) is in [**3**, p. 364], and (iii) is [**3**, Theorem II] and follows from Theorem 7(i).[1]

THEOREM 5. *If* $\phi \leqq \lambda$; $|\phi| > \aleph_0$, *then, for* $\alpha < \omega_0 2$; $\beta < \omega_0^2$; $\gamma < \omega_1$,
(i) $\phi \to (\omega_0, \gamma)^2$.
(ii) $\phi \to (\alpha, \beta)^2$.

---

[1] The partition relations occurring in (i) and (ii) are to be interpreted in the obvious way. Their formal definition is given in §4.

(i) is [5, Theorem 5], and (ii) is [5, Theorem 7]. Both results will follow from Theorem 31.

THEOREM 6. $\eta \rightarrow (\aleph_0, \eta)^2$.

This relation, a cross between (1) and (2), has, by definition, the following meaning. If $\bar{S} = \eta$; $[S]^2 = K_0 + K_1$, then there is $A \subset S$ such that either

$$|A| = \aleph_0; \; [A]^2 \subset K_0$$

or

$$\bar{A} = \eta; \; [A]^2 \subset K_1.$$

This result is [5, Theorem 4].

THEOREM 7. *If $a \geqq \aleph_0$, and if b is minimal such that $a^b > a$, then*
(i) $a^+ \rightarrow (b, a^+)^2$
(ii) $a^b \nrightarrow (b^+, a^+)^2$.

These results are contained in [6, p. 437]. (i) will follow from Theorem 34.[2]

THEOREM 8. *If $2^{\aleph_\nu} = \aleph_{\nu+1}$ for all ν, and if a is a regular limit number*[3] *then, for every $b < a$, $a \rightarrow (b, a)^2$.*

This result is [6, Lemma 3], and will follow from Theorem 34.

THEOREM 9. *If $\phi \leqq \lambda$; $|\phi| = |\lambda|$, then $\lambda \nrightarrow (\phi, \phi)^1$.*

This result is due to Sierpiński who kindly communicated it to one of us. It will follow from Theorem 29. Our proof of Theorem 29 uses some of Sierpiński's ideas.

THEOREM 10. *For any a, $a \nrightarrow (\aleph_0, \aleph_0)^{\aleph_0}$.*

This is in [5, p. 434]. The last result justifies our restriction to the case of finite "exponents" r.

4. Simple properties of partition relations.

THEOREM 11. *The two relations*

$$\text{(i)} \; \alpha \rightarrow (\beta_0, \beta_1, \cdots)_k^r \qquad\qquad \text{(ii)} \; \alpha^* \rightarrow (\beta_0^*, \beta_1^*, \cdots)_k^r$$

are equivalent.

---

[2] By methods similar to those used in [17] one can show that (i) $b \leqq a'$ for all $a > 1$, (ii) $b = a'$ for those $a > 1$ for which $d < a$ implies $2^d \leqq a$.

[3] It is not known if regular limit numbers $> \aleph_0$ exist or not. Cf. [13, p. 224].

PROOF. Let (i) hold; $\bar{S}_{<}=\alpha^*$; $[S]^r=\sum[\nu<k]K_\nu$. Then $\bar{S}_{>}=\alpha$. Hence, by hypothesis, there are $A\subset S$; $\nu<k$ such that $\bar{A}_{>}=\beta_\nu$; $[A]^r\subset K_\nu$. Then $\bar{A}_{<}=\beta_\nu^*$. This proves (ii), and the theorem follows by reasons of symmetry.

THEOREM 12. *Let* $\alpha\rightarrow(\beta_0, \beta_1, \cdots)_k^r$; $\alpha\leqq\alpha^{(1)}$; $k\geqq k^{(1)}$;

$$\beta_\nu \geqq \beta_\nu^{(1)} \qquad (\nu < k^{(1)}),$$
$$\beta_\nu \geqq r \qquad (k^{(1)} \leqq \nu < k).$$

Then

$$\alpha^{(1)} \rightarrow (\beta_0^{(1)}, \beta_1^{(1)}, \cdots)_{k^{(1)}}^r.$$

An analogous result holds when the types α, β_ν are replaced by cardinals.

PROOF. It suffices to consider the case of types. Let

$$\bar{S}^{(1)} = \alpha^{(1)}; \qquad [S^{(1)}]^r = \sum[\nu < k^{(1)}]K_\nu^{(1)}.$$

Then there is $S\subset S^{(1)}$ such that $\bar{S}=\alpha$. Then

$$[S]^r = \sum [\nu < k]K_\nu,$$

where $K_\nu=K_\nu^{(1)}[S]^r$ for $\nu<k^{(1)}$, and $K_\nu=0$ otherwise. By hypothesis, there are $A\subset S$; $\nu<k$ such that

$$\bar{A} = \beta_\nu; \qquad [A]^r \subset K_\nu.$$

If $\nu\geqq k^{(1)}$, then $|A|=|\beta_\nu|\geqq r$; $0\neq[A]^r\subset K_\nu$ which is a contradiction. Hence $\nu<k^{(1)}$. There is $A^{(1)}\subset A$ such that $\bar{A}^{(1)}=\beta_\nu^{(1)}$. Then $[A^{(1)}]^r \subset[A]^r\subset K_\nu\subset K_\nu^{(1)}$, and the assertion follows.

THEOREM 13. *If* $\alpha\rightarrow(\beta_0, \beta_1, \cdots)_k^r$ *then*

$$|\alpha| \rightarrow (|\beta_0|, |\beta_1|, \cdots)_k^r.$$

PROOF. Let $|S|=|\alpha|$; $[S]^r=\sum[\nu<k]K_\nu$. We order S so that $\bar{S}=\alpha$. Then there are $A\subset S$; $\nu<k$ such that $\bar{A}=\beta_\nu$; $[A]^r\subset K_\nu$. Then $|A|=|\beta_\nu|$, and the theorem follows.

THEOREM 14. *If β_ν is an initial ordinal, for all $\nu<k$, then the two relations*

$$m \rightarrow (\beta_0, \beta_1, \cdots)_k^r, \tag{3}$$
$$|m| \rightarrow (|\beta_0|, |\beta_1|, \cdots)_k^r \tag{4}$$

are equivalent.

PROOF. By Theorem 13, (3) implies (4). Now suppose that (4)

holds. Let $\overline{S}=m$; $[S]^r=\sum[\nu<k]K_\nu$. Then $|S|=|m|$, and hence, by (4), there are $A\subset S$; $\nu<k$ such that $|A|=|\beta_\nu|$; $[A]^r\subset K_\nu$. Then, as β_ν is an initial ordinal, $\overline{A}\geqq\beta_\nu$, and there is $B\subset A$ such that $\overline{B}=\beta_\nu$. This proves (3).

THEOREM 15. *If* $1+\alpha\rightarrow(1+\beta_0, 1+\beta_1, \cdots)_k^{1+r}$, *then*

$$\alpha\rightarrow(\beta_0, \beta_1, \cdots)_k^r.$$

In this proposition, $1+\alpha$ *and* $1+\beta_\nu$ *may be replaced by* $\alpha+1$ *and* $\beta_\nu+1$ *respectively. Also, the types* α, β_ν *may be replaced by cardinals.*

PROOF. Let $\overline{S}=\alpha$. Let x_0 be an object which is not an element of S, and put $S_0=S+\{x_0\}$. The order of S is extended to an order of S_0 by stipulating that $x_0\in L(S)$. Then $\overline{S}_0=1+\alpha$. Now let $[S]^r=\sum[\nu<k]K_\nu$. Then $[S_0]^{1+r}=\sum[\nu<k]K_{0\nu}$, where $K_{0\nu}=\{\{y_0, \cdots, y_r\}_<: \{y_1, \cdots, y_r\}\in K_\nu\}$. If $1+\alpha\rightarrow(1+\beta_0, 1+\beta_1, \cdots)_k^{1+r}$, then there are $A_0\subset S_0$; $\nu<k$ such that $\overline{A}_0=1+\beta_\nu$; $[A_0]^{1+r}\subset K_{0\nu}$. Then $A_0=\{y_0\}+A$; $y_0\in L(A)$;

$$\overline{A}=\beta_\nu; \qquad [A]^r\subset K_\nu.$$

This proves the first assertion.

Next, if $\alpha+1\rightarrow(\beta_0+1, \cdots)_k^{r+1}$, then, by Theorem 11 and the result just obtained, we conclude that

$$1+\alpha^*\rightarrow(1+\beta_0^*, 1+\beta_1^*, \cdots)_k^{1+r},$$

$$\alpha^*\rightarrow(\beta_0^*, \cdots)_k^r; \qquad \alpha\rightarrow(\beta_0, \cdots)_k^r.$$

Finally, let $1+a\rightarrow(1+b_0, 1+b_1, \cdots)_k^{1+r}$. Let α and β_ν be the initial ordinals belonging to a and b_ν respectively. Then, by Theorems 14 and 13, $1+\alpha\rightarrow(1+\beta_0, \cdots)_k^{1+r}$,

$$\alpha\rightarrow(\beta_0, \cdots)_k^r; \qquad a\rightarrow(b_0, \cdots)_k^r.$$

THEOREM 16. *If* $\alpha\rightarrow(\beta_0, \beta_1, \cdots)_{1+k}^r$; $\beta_0\rightarrow(\gamma_0, \gamma_1, \cdots)_l^r$, *then*

$$\alpha\rightarrow(\gamma_0, \gamma_1, \cdots, \beta_1, \beta_2, \cdots)_{l+k}^r.$$

In this proposition the types α, β_μ, γ_ν *may be replaced by cardinals.*

In formulating the last theorem we use an obvious extension of the symbol (2).

PROOF. We consider the case of types. Let $\overline{S}=\alpha$,

$$[S]^r=\sum[\lambda<l]K_{0\lambda}+\sum[0<\nu<1+k]K_\nu.$$

Put $K_0=\sum[\lambda<l]K_{0\lambda}$. Then, by hypothesis, there are $A\subset S$; $\nu<1+k$

such that $\bar{A}=\beta_\nu$; $[A]^r\subset K_\nu$. If $\nu>0$, then this is the desired conclusion. If $\nu=0$, then $\bar{A}=\beta_0$; $[A]^r\subset\sum[\lambda<l]K_{0\lambda}$ and so, by hypothesis, there are $B\subset A$; $\lambda<l$ such that $\bar{B}=\gamma_\lambda$; $[B]^r\subset K_{0\lambda}$ which, again, is a conclusion of the desired kind. This proves the theorem.

It is clear that, instead of replacing in the relation $\alpha\rightarrow(\beta_0, \beta_1, \cdots)^r_{1+k}$ a single type β_0 by a well-ordered system of types $\gamma_0, \gamma_1, \cdots$, we could have replaced simultaneously every type β_ρ by a system of types and in this way obtained a more general form of the transitive property of the partition relation than that given in Theorem 16.

THEOREM 17. *If* $\alpha\rightarrow(\beta_0, \beta_1, \cdots)^r_k$; $\gamma_\lambda=\beta_{\rho_\lambda}$ $(\lambda<l)$, *where* $\lambda\rightarrow\rho_\lambda$ *is a one-one mapping of* $[0, l]$ *into* $[0, k]$ *such that* $\beta_\nu\geqq r$ *for* $\nu\in[0, k]-\{\rho_\lambda:\lambda<l\}$, *then*

$$\alpha\rightarrow(\gamma_0, \gamma_1, \cdots)^r_l.$$

In particular, the condition on the mapping $\lambda\rightarrow\rho_\lambda$ is satisfied whenever this mapping is on $[0, k]$.

The types α, β_ν may be replaced by cardinals.

PROOF. Let $N=\{\rho_\lambda:\lambda<l\}$, and let $\nu\rightarrow\sigma_\nu$ be the mapping of N on $[0, l]$ which is inverse to the given mapping $\lambda\rightarrow\rho_\lambda$. Now let $\bar{S}=\alpha$; $[S]^r=\sum[\lambda<l]K_\lambda$. Then

$$[S]^r = \sum[\nu\in N]K_{\sigma_\nu} + \sum[\nu\in[0, k]-N]0.$$

By hypothesis, there are $A\subset S$; $\nu<k$ such that $\bar{A}=\beta_\nu$ and either

(i) $$\nu\in N; \qquad [A]^r\subset K_{\sigma_\nu}$$

or

(ii) $$\nu\notin N; \qquad [A]^r = 0.$$

In case (ii), $r\not\leqq\bar{A}=\beta_\nu$ which contradicts the hypothesis. Hence (i) holds, $\bar{A}=\gamma_\lambda$; $[A]^r\subset K_\lambda$, where $\lambda=\sigma_\nu<l$, and the result follows.

We note that the hypothesis relating to $\beta_\nu\geqq r$ cannot be omitted, as is seen from the pair of the obviously correct relations $4\rightarrow(1, 3, 3)^2$; $4\nrightarrow(3, 3)^2$.

COROLLARY. *If* $\alpha\rightarrow(\beta)^r_k$; $|k|=|l|$, *then* $\alpha\rightarrow(\beta)^r_l$.

This shows that, as far as k is concerned, the truth of the relation $\alpha\rightarrow(\beta)^r_k$ depends only on $|k|$. We are therefore able to introduce the relation

$$\alpha\rightarrow(\beta)^r_d$$

which, by definition, holds if, and only if,

$$\alpha \to (\beta)_k^r$$

for some, and hence for all, k such that $|k| = d$. A similar remark applies to the relation

$$a \to (b)_d^r.$$

THEOREM 18. *Let* $k < \omega_0$; $\alpha \to (\beta_0, \beta_1, \cdots)_k^r$;

$$\overline{S} = \alpha; \qquad [S]^r = K_0 + \cdots + K_{k-1}.$$

Then there are sets $M, N \subset [0, k]$ *such that* $|M| + |N| > k$,

$$\beta_\mu \in [K_\nu] \qquad \textit{for } \mu \in M; \nu \in N.$$

In the special case $k = 2$ *we have either*

(i) $\beta_0 \in [K_0][K_1]$ *or* (ii) $\beta_1 \in [K_0][K_1]$

or

(iii) $\beta_0, \beta_1 \in [K_0]$ *or* (iv) $\beta_0, \beta_1 \in [K_1]$.

PROOF. Let

$$P_\nu = \{\mu : \mu < k; \beta_\mu \in [K_\nu]\},$$
$$Q_\nu = [0, k] - P_\nu \qquad (\nu < k).$$

We have to find a set $N \subset [0, k]$ such that $|\prod[\nu \in N]P_\nu| > k - |N|$ or, what is equivalent, $|\sum[\nu \in N]Q_\nu| < |N|$. If no such N exists, i.e. if $|\sum[\nu \in N]Q_\nu| \geqq |N|$ for all $N \subset [0, k]$, then, by a theorem of P. Hall [**8**], it is possible to choose numbers $\rho_\nu \in Q_\nu$ such that $\rho_\mu \neq \rho_\nu$ $(\mu < \nu < k)$. Then $\beta_{\rho_\nu} \notin [K_\nu]$ $(\nu < k)$. On the other hand, by Theorem 17 and the hypothesis, $\alpha \to (\beta_{\rho_0}, \beta_{\rho_1}, \cdots)_k^r$. This is the required contradiction.

THEOREM 19. *Let* $\alpha \to (\beta, \gamma)^2$, *and suppose that* m *is the initial ordinal belonging to* $|\alpha|$. *Then at least one of the following four statements holds.*[4]

(i) $\beta < \omega_0$ (ii) $\gamma < \omega_0$ (iii) $\beta, \gamma \leqq \alpha, m$ (iv) $\beta, \gamma \leqq \alpha, m^*$.

PROOF. Let S be a set ordered by means of the relation $x < y$ and also by means of the relation $x \ll y$, and let the corresponding order types of S be $\overline{S} = \alpha$; $\overline{S}_{<<} = m$. Then $[S]^2 = K_0 +' K_1$ where $K_0 = \{\{x, y\}_< : \{x, y\}_{<<} \subset S\}$. Then, by Theorem 18, we have at least one of the following four cases.

Case 1. $\beta \in [K_0]_<[K_1]_<$. Then there are sets $A, B \subset S$ such that $\overline{A}_< = \overline{A}_{<<} = \beta$; $\overline{B}_< = \overline{B}_{>>} = \beta$, and hence $\beta = \overline{A}_{<<} \leqq \overline{S}_{<<} = m$. Then β is

[4] (iii) means that $\beta \leqq \alpha$; $\beta \leqq m$; $\gamma \leqq \alpha$; $\gamma \leqq m$.

an ordinal. If $\beta \geqq \omega_0$, then the contradiction $\omega_0^* \leqq \beta^* = \overline{B}_{<<} \leqq \overline{S}_{<<} = m$ follows. Hence $\beta < \omega_0$.

Case 2. $\gamma \in [K_0]_<[K_1]_<$. Then, by symmetry, $\gamma < \omega_0$.

Case 3. $\beta, \gamma \in [K_0]_<$. Then, for some sets $A, B \subset S$, $\overline{A}_< = \overline{A}_{<<} = \beta$; $\overline{B}_< = \overline{B}_{<<} = \gamma$, and $\beta, \gamma \leqq \alpha, m$.

Case 4. $\beta, \gamma \in [K_1]_<$. Then, similarly, $\overline{A}_< = \overline{A}_{>>} = \beta$; $\overline{B}_< = \overline{B}_{>>} = \gamma$; $\beta, \gamma \leqq \alpha, m^*$. This proves the theorem.

COROLLARY. *For every* α,

$$(r-2) + \alpha \nrightarrow (\omega_0, (r-2) + \omega_0^*)^r \qquad (r \geqq 2). \tag{5}$$

For none of the relations (i)–(iv) of Theorem 19 holds if $\beta = \omega_0$; $\gamma = \omega_0^*$. Hence $\alpha \nrightarrow (\omega_0, \omega_0^*)^2$, and Theorem 15 yields (5).

The method employed in the proof of Theorem 19, i.e. the definition of a partition of $[S]^2$ from two given orders of S, seems to have been first used by Sierpiński [15]. In that note Sierpiński proves $\aleph_1 \nrightarrow (\aleph_1, \aleph_1)^2$. Cf. Theorem 30.

THEOREM 20. (i) *If* $\beta_0 \leqq \alpha$; $|\beta_0| < r$, *then* $\alpha \to (\beta_0, \beta_1, \cdots)_k^r$ *holds for any* $k, \beta_1, \beta_2, \cdots$.

(ii) *If* $\beta_\nu = r$ *for* $\nu < k$, *then the two relations*

$$\alpha \to (\beta_0, \beta_1, \cdots, \gamma_0, \gamma_1, \cdots)_{k+l}^r, \tag{6}$$

$$\alpha \to (\gamma_0, \gamma_1, \cdots)_l^r \tag{7}$$

are equivalent.

PROOF OF (i). If $\overline{S} = \alpha$; $[S]^r = \sum [\nu < k] K_\nu$, then there is $A \subset S$ such that $\overline{A} = \beta_0$. Then $[A]^r = 0 \subset K_0$.

PROOF OF (ii). By Theorem 12, (6) implies (7). Now suppose that (7) holds. Let $\overline{S} = \alpha$; $[S]^r = \sum [\nu < k+l] K_\nu$. If there is $\nu < k$ such that $K_\nu \neq 0$, then we can choose $A \in K_\nu$, and we shall have $A \subset S$; $\overline{A} = \beta_\nu$; $[A]^r \subset K_\nu$. If, on the other hand, $K_\nu = 0$ for all $\nu < k$, then $[S]^r = \sum [\lambda < l] K_{k+\lambda}$, and there exist, by (7), $B \subset S$; $\lambda < l$ such that $\overline{B} = \gamma_\lambda$; $[B]^r \subset K_{k+\lambda}$. This proves (6).

THEOREM 21. *Let*

$$\alpha \to (\beta_0, \beta_1, \cdots)_k^r. \tag{8}$$

Then either (i) *there is* $\nu_0 < k$ *such that* $\beta_{\nu_0} \leqq \alpha$; $|\beta_{\nu_0}| < r$, *or* (ii) $\beta_\nu \leqq \alpha$ *for all* $\nu < k$.

REMARK. If (i) holds then (8) is true trivially. For, let $\overline{S} = \alpha$; $[S]^r = \sum [\nu < k] K_\nu$. Then we can choose $B \subset S$ such that $\overline{B} = \beta_{\nu_0}$. Then $[B]^r = 0 \subset K_{\nu_0}$. We see, therefore, that the relation (8) need only be studied in the case in which $\beta_\nu \leqq \alpha$ for all $\nu < k$. In particular, if α is an an ordinal, then we may assume, if we wish, that every β_ν is an ordinal.

PROOF. Suppose that (ii) is false. Then there is $\nu_1 < k$ such that $\beta_{\nu_1} \nleqq \alpha$. Let $\bar{S} = \alpha$, and put $[S]^r = \sum' [\nu < k] K_\nu$, where $K_{\nu_1} = [S]^r$. Then, by (8), there are $B \subset S$; $\nu_0 < k$ such that $\bar{B} = \beta_{\nu_0}$; $[B]^r \subset K_{\nu_0}$. Then $\beta_{\nu_0} = \bar{B} \leqq \bar{S} = \alpha$; $\nu_0 \neq \nu_1$; $[B]^r = 0$; $|\beta_{\nu_0}| = |B| < r$. Hence (i) holds.

THEOREM 22. *The following two tables give information about a number of cases in which the truth or otherwise of any of the relations*

$$\alpha \to (\beta_0, \beta_1, \cdots)_k^r, \tag{9}$$

$$a \to (b_0, b_1, \cdots)_k^r \tag{10}$$

can be decided trivially.

$k = 0$:

$r \leqq \|\alpha\|$ $r \leqq a$	+
$r > \|\alpha\|$ $r > a$	−

$k > 0$:

	$\beta_\nu \leqq \alpha$ $b_\nu \leqq a$	$\beta_\nu = \alpha$ $b_\nu = a$	$\beta_\nu \geqq \alpha$; $\beta_0 \nleqq \alpha$ $b_\nu \geqq a$; $b_0 > a$	$\beta_\nu \nleqq \alpha$ $b_\nu > a$	$\beta_0 \ngeqq \alpha$ $b_0 < a$	$\beta_0 \leqq \alpha$ $b_0 \leqq a$	$\beta_0 \nleqq \alpha$ $b_0 > a$	$\beta_0 \ngeqq r$ $b_0 < r$
$r = 0$	+						−	
$0 < r < \|\alpha\|$ $0 < r < a$				−		±		+
$r = \|\alpha\| > 0$ $r = a > 0$		+	−		+			
$r > \|\alpha\|$ $r > a$				−		+		

The proofs may be omitted. When a row or column is headed by two lines of conditions the first line refers to (9) and the second line to (10). Every condition involving the suffix ν is meant to hold for every $\nu<k$. An entry $+$ means that both, (9) and (10) are true, and an entry $-$ means that both, (9) and (10) are false. The one entry $\pm$ marks the only case worth studying, i.e. the case in which (9) or (10) can be either true or false, and this for nontrivial reasons. In each row the entries are chosen in such a way that all possibilities are covered. In the column headings we may, of course, replace β_0 and b_0 by β_{ν_0} and b_{ν_0} respectively, for any choice of $\nu_0<k$. The case $k=0$ has, obviously, only curiosity value but is included for the sake of completeness.

5. **Denumerable order types.**

THEOREM 23. *If* $n<\omega_0$; $\alpha<\omega_0 2$, *then*

(11) $$\omega_0 n \rightarrow (n, \alpha)^2,$$

(12) $$\omega_0 n \not\rightarrow (n+1, \omega_0+1)^2.$$

PROOF. We may assume $n>0$.

(a) In order to prove (12), consider the set $S=\{(\nu, \lambda):\nu<n; \lambda<\omega_0\}$, ordered alphabetically: $(\nu, \lambda)<(\nu_1, \lambda_1)$ if either (i) $\nu<\nu_1$ or (ii) $\nu=\nu_1$; $\lambda<\lambda_1$. Then $[S]^2=K_0+'K_1$, where K_1 is the set of all sets $\{(\nu, \lambda), (\nu, \lambda_1)\}_<\subset S$. Then, clearly, $\overline{S}=\omega_0 n$; $n+1\notin[K_0]$; $\omega_0+1 \notin[K_1]$, and (12) follows.

(b) We now prove (11). Let the set $A=\sum[\nu<n]A_\nu$ be ordered, $\overline{A}_\nu=\omega_0$ for $\nu<n$, and $A_\nu\subset L(A_{\nu+1})$ for $\nu+1<n$. Suppose that $[A]^2 =K_0+K_1$; $n\notin[K_0]$; $\alpha\notin[K_1]$. We want to deduce a contradiction.

By Theorem 1, there is, for every $\nu<n$, a set $B_\nu\in[A_\nu]^{\aleph_0}$ such that $[B_\nu]^2\subset K_{\rho_\nu}$, for some $\rho_\nu<2$. Since $n\notin[K_0]$, we have $\rho_\nu=1$. Let $\lambda<\mu<n$. We define an operator $p_{\lambda\mu}$ as follows. There is at least one set $B\subset B_\lambda+B_\mu$ such that $|BB_\lambda|=\aleph_0$; $[B]^2\subset K_1$. For instance, B_λ is such a set B. Since $\alpha\notin[K_1]$, we have, for every such B, $\omega_0\leqq\overline{B}<\alpha <\omega_0 2$. Hence we can choose B such that $\overline{B}$ is maximal. We fix such a B by any suitable convention and put

$$P_{\lambda\mu}(B_0, B_1, \cdots, B_{n-1}) = (C_0, C_1, \cdots, C_{n-1}),$$

where $C_\lambda=BB_\lambda$; $C_\mu=B_\mu-B$, and $C_\nu=B_\nu$ for $\nu\neq\lambda, \mu$. Then $\overline{C}_\nu=\omega_0$; $|C_\lambda L_1(x)|<\aleph_0$ for $\nu<n$; $x\in C_\mu$. We now apply, in turn, all

$$\binom{n}{2}$$

operators $p_{\lambda\mu}$, corresponding to all choices of λ, μ, to the system $(B_0, \cdots, B_{n-1})$, applying each one of the operators, from the second onwards, to the system obtained by the preceding operator, and obtain, as end product, the system $(D_0, \cdots, D_{n-1})$. Then $D_\nu \subset A_\nu$; $\overline{D}_\nu = \omega_0$ $(\nu < n)$; $|D_\nu L_1(x)| < \aleph_0$ for $\nu < n$; $x \in D_{\nu+1} + \cdots + D_{n-1}$. Hence it is possible to choose, in this order, elements $x_{n-1}, x_{n-2}, \cdots, x_0$ such that

$$x_\nu \in D_\nu L_0(x_{\nu+1}, x_{\nu+2}, \cdots, x_{n-1}) \qquad (\nu < n).$$

Then, putting $D = \{x_\nu : \nu < n\}$, we have $\overline{D} = n$; $[D]^2 \subset K_0$ and therefore $n \in [K_0]$ which is the required contradiction.

THEOREM 24. *If* $\alpha < \omega_0 4$, *then*

$$\alpha \nrightarrow (3, \omega_0 2)^2, \tag{13}$$

$$\omega_0 4 \rightarrow (3, \omega_0 2)^2. \tag{14}$$

This theorem is a special case of the following theorem.

THEOREM 25. *Let* $2 \leq m, n < \omega_0$, *and denote by* $l_0 = l_0(m, n)$ *the least finite number* l *possessing the following property.*[5]

Property P_{mn}. *Whenever* $\rho(\lambda, \mu) < 2$ *for* $\{\lambda, \mu\}_{\neq} \subset [0, l]$, *then there is either* $\{\lambda_0, \cdots, \lambda_{m-1}\}_{\neq} \subset [0, l]$ *such that*

$$\rho(\lambda_\alpha, \lambda_\beta) = 0 \qquad \text{for } \alpha < \beta < m,$$

or there is $\{\lambda_0, \cdots, \lambda_{n-1}\}_{\neq} \subset [0, l]$ *such that*

$$\rho(\lambda_\alpha, \lambda_\beta) = 1 \qquad \text{for } \{\alpha, \beta\}_{\neq} \subset [0, n].$$

Then

$$\omega_0 l_0 \rightarrow (m, \omega_0 n)^2, \tag{15}$$

$$\gamma \nrightarrow (m, \omega_0 n)^2 \qquad \text{for } \gamma < \omega_0 l_0. \tag{16}$$

Moreover, if $l_1 \rightarrow (m, m, n)^2$, *then* $l_0 \leq l_1$.

Deduction of Theorem 24 *from Theorem* 25. We have to prove that $l_0(3, 2) = 4$. (i) By considering the function ρ defined by

$$\rho(0, 1) = \rho(1, 2) = \rho(2, 0) = 0; \quad \rho(2, 1) = \rho(1, 0) = \rho(0, 2) = 1,$$

we deduce that 3 does not possess the property P_{32}. (ii) Let us assume that 4 does not possess P_{32}. Then there is $\rho(\lambda, \mu)$ such that the condition stipulated for P_{32} does not hold, with $l = 4$. If

[5] The existence of such a number l follows from Theorem 2. It will follow from Theorem 39 that we may take $l = (1 + 3^{2m+n-5})/2$.

$$\{\alpha, \beta, \gamma\}_{\neq} \subset [0, 4]; \qquad \rho(\alpha, \beta) = \rho(\alpha, \gamma) = 0,$$

then the assumption $\rho(\beta, \gamma) = 0$ would lead to

$$\rho(\alpha, \beta) = \rho(\beta, \gamma) = \rho(\alpha, \gamma) = 0,$$

i.e. to a contradiction. Hence $\rho(\beta, \gamma) = 1$ and, by symmetry, $\rho(\gamma, \beta) = 1$. This, again, is a contradiction. This argument proves that

(17) $$\text{if } \{\alpha, \beta, \gamma\}_{\neq} \subset [0, 4]; \quad \rho(\alpha, \beta) = 0, \quad \text{then } \rho(\alpha, \gamma) = 1.$$

Since at least one of the numbers $\rho(0, 1)$, $\rho(1, 0)$ is zero, there is a permutation $\alpha, \beta, \gamma, \delta$ of 0, 1, 2, 3 such that $\rho(\alpha, \beta) = 0$. Then repeated application of (17) yields $\rho(\alpha, \gamma) = \rho(\alpha, \delta) = 1$; $\rho(\gamma, \alpha) = 0$; $\rho(\gamma, \delta) = 1$; $\rho(\delta, \alpha) = \rho(\delta, \gamma) = 0$, which contradicts (17). This proves $l_0(3, 2) \leq 4$ and, in conjunction with (i), $l_0(3, 2) = 4$.

PROOF OF THEOREM 25. 1. We begin by proving the last clause. Let

(18) $$l_1 \to (m, m, n)^2.$$

Suppose that $\rho(\lambda, \mu) < 2$ for $\{\lambda, \mu\}_{\neq} \subset [0, l_1]$. Then

$$[S]^2 = K_0 + K_1 + K_2,$$

where $S = [0, l_1]$, and K_ν is the set of all $\{\lambda, \mu\}_< \subset [0, l_1]$ such that

$$\rho(\lambda, \mu) = 0 \qquad (\nu = 0),$$

$$\rho(\lambda, \mu) > \rho(\mu, \lambda) \qquad (\nu = 1),$$

$$\rho(\lambda, \mu) = \rho(\mu, \lambda) = 1 \qquad (\nu = 2).$$

By (18), there is $S_1 = \{\lambda_0, \cdots, \lambda_{k-1}\}_< \subset S$ such that one of the following three statements holds.

(19) $$k = m; \qquad [S_1]^2 \subset K_0,$$

(20) $$k = m; \qquad [S_1]^2 \subset K_1,$$

(21) $$k = n; \qquad [S_1]^2 \subset K_2.$$

(19) implies that $\rho(\lambda_\alpha, \lambda_\beta) = 0$ for $\alpha < \beta < m$;
(20) implies that $\rho(\lambda_{m-1-\alpha}, \lambda_{m-1-\beta}) = 0$ for $\alpha < \beta < m$;
(21) implies that $\rho(\lambda_\alpha, \lambda_\beta) = 1$ for $\{\alpha, \beta\}_{\neq} \subset [0, n]$.

This shows that $l_0(m, n) \leq l_1$.

2. We now prove (15). Let $l = l_0(m, n)$; $A = [0, \omega_0 l]$; $N = [0, \omega_0]$;

$$[A]^2 = K_0 +' K_1 \qquad \text{(partition } \Delta\text{)}.$$

We use the notation of the partition calculus given in detail in [4, p. 419] which can be summarized as follows. If Δ is an equivalence relation on a set M or a partition of M into disjoint classes then $|\Delta|$ denotes the cardinal of the set of nonempty classes, and the relation

$$x \equiv y(\cdot \Delta)$$

expresses the fact that x and y belong to M and lie in the same class of Δ. If, for $\rho \in R$, Δ_ρ is a partition of M, and if $t \rightarrow f_\rho(t)$ is a mapping of a set T into M, then the formula

$$\Delta'(t) = \prod [\rho \in R] \Delta_\rho(f_\rho(t)) \qquad (t \in T)$$

defines that partition Δ' of T for which

$$s \equiv t(\cdot \Delta')$$

if, and only if,

$$f_\rho(s) \equiv f_\rho(t)(\cdot \Delta) \qquad \text{for } \rho \in R.$$

We continue the proof of (15) by putting

$$\Delta'(\{\sigma, \tau\}) = \prod [\lambda, \mu < l] \Delta(\{\omega_0\lambda + \sigma, \omega_0\mu + \tau\}) \qquad (\sigma < \tau < \omega_0).$$

By Theorem 1 there is $N' \in [N]^{\aleph_0}$ such that $|\Delta'| = 1$ in $[N']^2$. Then, by definition of Δ', there is $\rho(\lambda, \mu) < 2$ such that

$$\{\omega_0\lambda + \sigma, \omega_0\mu + \tau\} \in K_{\rho(\lambda,\mu)} \qquad \text{for } \lambda, \mu < l;\ \{\sigma, \tau\}_< \subset N'.$$

By definition of l this implies that there is a set $\{\lambda_0, \cdots, \lambda_{k-1}\}_{\neq} \subset [0, l]$ such that either

$$(22) \qquad k = m; \qquad \rho(\lambda_\alpha, \lambda_\beta) = 0 \qquad \text{for } \alpha < \beta < m$$

or

$$(23) \qquad k = n; \qquad \rho(\lambda_\alpha, \lambda_\beta) = 1 \qquad \text{for } \{\alpha, \beta\}_{\neq} \subset [0, n].$$

If (22) holds, then we put

$$A' = \{\omega_0\lambda_\alpha + \sigma_\alpha : \alpha < m\},$$

where σ_α is chosen such that $\{\sigma_0, \sigma_1, \cdots, \sigma_{m-1}\}_< \subset N'$. Then $[A']^2 \subset K_0$, so that the desired conclusion $m \in [K_0]$ is reached.

If we now assume that $m \notin [K_0]$, then $\rho(\lambda, \lambda) = 1$ for $\lambda < l$, and, furthermore, (23) holds. Then we put $N' = \{\sigma_0, \sigma_1, \cdots\}_<$; $A'' = \{\omega_0\lambda_\alpha + \sigma_{\alpha+tn} : \alpha < n;\ t < \omega_0\}$. We find that $[A'']^2 < K_1$; $\omega_0 n = A'' \in [K_1]$. This proves (15).

3. Finally, let $\gamma < \omega_0 l_0$. Then there is $l < l_0$ such that $\omega_0 l \leqq \gamma < \omega_0(l+1)$. Then, by definition of l_0, there is $\rho(\lambda, \mu) < 2$ for $\{\lambda, \mu\}_{\neq} \subset [0, l]$ such that, whenever $\{\lambda_0, \cdots, \lambda_{m-1}\}_{\neq} \subset [0, l]$, then

$$(24) \qquad \rho(\lambda_\alpha, \lambda_\beta) \neq 0$$

for some $\{\alpha, \beta\}_< \subset [0, m]$, and, whenever $\{\lambda_0, \cdots, \lambda_{n-1}\}_{\neq} \subset [0, l]$, then

$$\rho(\lambda_\alpha, \lambda_\beta) \neq 1 \tag{25}$$

for some $\{\alpha, \beta\}_{\neq} \subset [0, n]$. Then, if $A = [0, \gamma]$, we have $[A]^2 = K_0 +' K_1$, where K_0 is the set of all $\{\omega_0\lambda + \sigma, \omega_0\mu + \tau\}$ such that $\{\lambda, \mu\}_{\neq} \subset [0, l]$; $\sigma < \tau < \omega_0$; $\rho(\lambda, \mu) = 0$. If, now, $A' \in [A]^m$; $[A']^2 \subset K_0$, then

$$A' = \{\omega_0\lambda_\alpha + \sigma_\alpha : \alpha < m\}; \qquad \sigma_0 < \cdots < \sigma_{m-1} < \omega_0;$$

$\{\lambda_0, \cdots, \lambda_{m-1}\}_{\neq} \subset [0, l]$;

$$\rho(\lambda_\alpha, \lambda_\beta) = 0 \qquad \text{for } \alpha < \beta < m,$$

which contradicts (24). If, on the other hand,

$$A'' \subset A; \qquad \overline{A''} = \omega_0 n; \qquad [A'']^2 \subset K_1,$$

then there is $\{\lambda_0, \cdots, \lambda_{n-1}\}_< \subset [0, l]$ such that $\overline{B}_\alpha = \omega_0$ for $\alpha < n$, where $B_\alpha = A''[\omega_0\lambda_\alpha, \omega_0(\lambda_\alpha + 1))$ $(\alpha < n)$. Then $\rho(\lambda_\alpha, \lambda_\beta) = 1$ for $\{\alpha, \beta\}_{\neq} \subset [0, n]$, which contradicts (25). Hence neither A' nor A'' exist, with the properties stated, so that (16) follows. This completes the proof of Theorems 24 and 25.

6. **The linear continuum.** Our object is to investigate relations of the form

$$\lambda \rightarrow (\alpha_0, \alpha_1, \cdots)_k^r$$

and their negatives. It turns out[6] that every positive relation we were able to prove holds not only for the particular type λ of the set of all real numbers but for every type ϕ such that

$$|\phi| > \aleph_0; \quad \omega_1, \omega_1^* \not\leq \phi. \tag{26}$$

This fact seems to suggest that, given any type ϕ satisfying (26), there always exists λ_1 such that

$$\lambda_1 \leqq \lambda, \phi; \qquad |\lambda_1| > \aleph_0,$$

i.e., that every nondenumerable type which does not "contain" ω_1 or ω_1^* contains a nondenumerable type which is embeddable in the real continuum. This conjecture has, as far as the authors are aware, neither been proved nor disproved.[7]

Throughout this section S denotes the set of all real numbers x such that $0 < x < 1$, ordered by magnitude. The letters x, y, z denote elements of S, and $\lambda = \overline{S}$.

---

[6] Cf. Theorems 31, 32.

[7] Since this paper was submitted E. Specker has disproved this conjecture.

THEOREM 26.

(i) $\lambda \nrightarrow (\omega_1)^r_k$ *for* $r \geqq 0; k > 0.$

(ii) $\lambda \nrightarrow (r+1)^r_{\omega_0}$ *for* $r \geqq 2.$

PROOF. (i) is trivial, in view of $\omega_1 \nleqq \lambda$. In order to prove (ii) it suffices, by Theorem 15, to consider the case $r=2$. Let $\{x_\nu : \nu < \omega_0\}$ be the set of all rational numbers in S, and denote, for $n < \omega_0$, by K_n the set of all $\{x, y\}_<$ such that the least ν satisfying $x < x_\nu < y$ is $\nu = n$. Then $[S]^2 = \sum [\nu < \omega_0] K_\nu$. Also, if $[\{x, y, z\}_<]^2 \subset K_n$, then the contradiction $x < x_n < y < x_n < z$ follows. Hence $3 \notin [K_n]$, and Theorem 26 is proved.

THEOREM 27. $\lambda \nrightarrow (\omega_0, \omega_0+2)^r$ *for* $r \geqq 3$.

PROOF. By Theorem 15, we need only consider the case $r=3$. We have $[S]^3 = K_0 +' K_1$, where $K_0 = \{\{x, y, z\}_< : y - x < z - y\}$.

ASSUMPTION 1. Let $[\{x_0, x_1, \cdots\}_<]^3 \subset K_0$. Then $\lim x_\nu = u$ as $\nu \to \infty$, and we have, for $0 < m < \omega_0$,

$$\{x_0, x_m, x_{m+1}\} \in K_0; \qquad x_m - x_0 < x_{m+1} - x_m.$$

If $m \to \infty$, then the contradiction $u - x_0 \leqq u - u$ follows.

ASSUMPTION 2. Let $A \subset S$; $\bar{A} = \omega_0 + 2$; $[A]^3 \subset K_1$. Then $A = B + \{y, z\}_<$; $B = \{x_0, x_1, \cdots\}_< \subset L(y)$; $\lim x_\nu = u$ as $\nu \to \infty$, and we have, for $m < \omega_0$, $\{x_m, x_{m+1}, z\} \in K_1$; $x_{m+1} - x_m \geqq z - x_{m+1}$. If $m \to \infty$, then the contradiction $u - u \geqq z - u$ follows. This proves Theorem 27.

THEOREM 28. $\lambda \nrightarrow (r+1, \omega_0+2)^r$ *for* $r \geqq 4$.

PROOF. It suffices to consider the case $r=4$. We have $[S]^4 = K_0 +' K_1$, where $K_0 = \{\{x_0, x_1, x_2, x_3\}_< : x_2 - x_1 < x_3 - x_2, x_1 - x_0\}$.

ASSUMPTION 1. Let $[\{x_0, x_1, x_2, x_3, x_4\}_<]^4 \subset K_0$. Then $\{x_0, x_1, x_2, x_3\} \in K_0$, and hence $x_2 - x_1 < x_3 - x_2$. Also, $\{x_1, x_2, x_3, x_4\} \in K_0$, and hence $x_3 - x_2 < x_2 - x_1$. This is a contradiction.

ASSUMPTION 2. Let $A \subset S$; $\bar{A} = \omega_0 + 2$; $[A]^4 \subset K_1$. We define B, y, z, x_ν, u as in the proof of Theorem 27. Then there is $m_0 < \omega_0$ such that, for $m_0 \leqq m < \omega_0$, $u - x_m < x_m - x_0$. Then, for $m_0 \leqq m < \omega_0$, $\{x_0, x_m, x_{m+1}, z\} \in K_1$; $x_{m+1} - x_m < u - x_m < x_m - x_0$; $x_{m+1} - x_m \geqq z - x_{m+1}$, and if $m \to \infty$, then the contradiction $u - u \geqq z - u$ follows. This proves Theorem 28.

The next two theorems are extensions of results due to Sierpiński.

THEOREM 29. *If* $2 \leqq |k| \leqq |\lambda|$; $|\alpha_\nu| \geqq |\lambda|$ $(\nu < k)$, *then*

$$\lambda \nrightarrow (\alpha_0, \alpha_1, \cdots)^1_k.$$

Sierpiński proved that $\lambda \nrightarrow (\alpha, \alpha)^1$ if $\alpha \leqq \lambda$; $|\alpha| = |\lambda|$ (Theorem 9).

PROOF.

Case 1. There is $\mu<k$ such that $\alpha_\mu \not\leqq \lambda$. We consider the partition $S=\sum' [\nu<k]K_\nu$, where $K_\mu=S$. We have $\alpha_\mu \not\leqq \lambda=\overline{K}_\mu$ and, for $\nu\neq\mu$, $\alpha_\nu\not\leqq 0=\overline{K}_\nu$. Hence $\alpha_\nu\not\leqq\overline{K}_\nu$ $(\nu<k)$.

Case 2. $\alpha_\nu\leqq\lambda$ $(\nu<k)$. We choose a fixed set $A_0\subset S$ such that $\overline{A}_0=\alpha_0$. Generally, the letter A denotes sets such that $A\subset S$; $\overline{A}=\alpha_0$. Corresponding to every A, there is a real function $f_A(x)$, defined and strictly increasing for $x\in A_0$, such that $x\to f_A(x)$ is a mapping of A_0 on A. We extend the definition of f_A by putting $f_A(x)=0$ for $x\in L(A_0)$ and

$$f_A(x) = \sup [y \leqq x;\, y\in A_0] f_A(y) \qquad \text{for } x \notin L(A_0).$$

Then $f_A(x)$ is nondecreasing in S. The set A is uniquely determined by the function f_A and the set A_0. Let $D(A)$ be the set of those x_0 for which $f_A(x)$ is discontinuous at $x=x_0$. Then $|D(A)|\leqq\aleph_0$. The function f_A is uniquely determined by (i) the set $D(A)$ and (ii) the values of $f_A(x)$ for $x\in D(A)$ and (iii) the values of $f_A(x)$ for all rational x. Therefore

$$\left|\sum\{A\}\right| = \left|\sum\{f_A\}\right| \leqq |\lambda|^{3\aleph_0} = |\lambda| \leqq \left|\sum\{A\}\right|,$$

and $\left|\sum\{A\}\right|=|\lambda|=\aleph_n$, say. Now we can write $\sum\{A\}=\{A_{0\rho}:\rho<\omega_n\}$. By symmetry, we have, for every $\nu<k$, a set $\{A_{\nu\rho}:\rho<\omega_n\}$ whose elements are all subsets of S of type α_ν.

The set $N=\{(\nu,\rho):\nu<k;\ \rho<\omega_n\}$ satisfies $|N|=|k|\aleph_n=\aleph_n$. We order N in such a way that $\overline{N}=\omega_n$. Then we can find, inductively, $x_{\nu\rho}$ such that, for $(\nu,\rho)\in N$,

$$x_{\nu\rho}\in A_{\nu\rho} - \{x_{\mu\sigma}:(\mu,\sigma)<(\nu,\rho)\}.$$

For, $\left|\{(\mu,\sigma):(\mu,\sigma)<(\nu,\rho)\}\right|<|N|\leqq|A_{\nu\rho}|$. We have $x_{\nu\rho}\neq x_{\mu\sigma}$ for $(\nu,\rho)\neq(\mu,\sigma)$. Now, $S=\sum[\nu<k]K_\nu$, where $K_\nu=S-\{x_{\nu\rho}:\rho<\omega_n\}$. For, if $x\in S$, then, since $k\geqq 2$, there is $\nu<k$ such that $x\in K_\nu$. If, now, $\alpha_\nu\leqq\overline{K}_\nu$ for some $\nu<k$, then there is ρ such that $x_{\nu\rho}\in A_{\nu\rho}\subset K_\nu$, which is the required contradiction. This proves Theorem 29.

THEOREM 30. $|\lambda|\not\to(\aleph_1,\aleph_1)^r$ *for* $r\geqq 2$.

PROOF. The substance of this theorem is due to Sierpiński [15]. By Theorem 15 we need only consider the case $r=2$. Let $x<y$ be, as throughout this section, the order of S by magnitude, and let $x\ll y$ be an order of S such that $\overline{S}_{<<}=\omega_n$, where $|\lambda|=\aleph_n$. Then $[S]^2=K_0+'K_1$, where $K_0=\{\{x,y\}_<:x\ll y\}$. Now let $A\subset S$; $[A]^2\subset K_\nu$.

If $\nu=0$, then $\overline{A}_<\leqq\overline{S}_<=\lambda$; $\overline{A}_<=\overline{A}_{<<}\leqq\overline{S}_{<<}=\omega_n$, and hence $\overline{A}_<<\omega_1$; $|A|<\aleph_1$. If $\nu=1$, then $\overline{A}_>\leqq\overline{S}_>=\lambda^*$; $\overline{A}_>=\overline{A}_{<<}\leqq\overline{S}_{<<}=\omega_n$; $\overline{A}_><\omega_1$; $|A|<\aleph_1$.

This proves Theorem 30. We note that this theorem is, in fact, an easy corollary of [5, Example 4A].

7. **The general case.** We shall consider relations involving certain types of cardinal $\aleph_1$ as well as relations between types of any cardinal. We begin by proving a lemma. We establish this lemma in a form which is more general than will later be required, but in this form it seems to possess some interest of its own. We recall that a' denotes the cofinality cardinal belonging to a which was defined in §2.

LEMMA 1. *Let S be an ordered set, and $|S|'=\aleph_n$; ω_n, $\omega_n^* \not\leqq \bar{S}$. Then, corresponding to every rational number t, there is $S_t \subset S$ such that $|S_t|=|S|$; $S_t \subset L(S_u)$ for $t<u$.*

Sierpiński, in a letter to one of us, had already noted the weaker result that, *if* $|S|=\aleph_1$; ω_1, $\omega_1^* \not\leqq \bar{S}$, *then* $\eta \leqq \bar{S}$.

PROOF. *Case* 1. There is $A \subset S$ such that

$$|AL(x)| < |A| = |S| \qquad (x \in A).$$

Then we define x_ν for $\nu<\omega_n$ inductively as follows. Let $\nu_0<\omega_n$; $x_\nu \in A\ (\nu<\nu_0)$. Then, by definition of n,

$$\left|\sum[\nu < \nu_0](AL(x_\nu) + \{x_\nu\})\right| < |A|,$$

and hence there is $x_{\nu_0} \in A - \sum[\nu<\nu_0](L(x_0)+\{x_\nu\})$. Then $x_\mu<x_\nu$ $(\mu<\nu<\omega_n)$ and so $\omega_n \leqq \bar{S}$, which is false.

Case 2. There is $A \subset S$ such that

$$|AR(x)| < |A| = |S| \qquad (x \in A).$$

Then, by symmetry, the contradiction $\omega_n^* \leqq \bar{S}$ follows.

Case 3. There is $A \subset S$ such that

$$\min(|AL(x)|, |AR(x)|) < |A| = |S| \qquad (x \in A).$$

Then we put

$$A_0 = \{x : x \in A;\ |AL(x)| < |A|\},$$
$$A_1 = \{x : x \in A;\ |AR(x)| < |A|\}.$$

Then $A=A_0+A_1$.

Case 3.1. $|A_0|=|S|$. Then $|A_0L(x)| \leqq |AL(x)| < |S| = |A_0|$ $(x \in A_0)$, and hence, by Case 1, we find a contradiction.

Case 3.2. $|A_0| \neq |S|$. Then $|A_1|=|S|$ and, by symmetry, a contradiction follows.

We have so far proved that, if $A \subset S$; $|A|=|S|$, there is $z \in A$ such that $|AL(z)|=|AR(z)|=|S|$. Then $A=A'+A''$, where $A'=AL(z)$;

$|A'| = |A''| = |S|$; $A' \subset L(A'')$. By applying this result to A' we find a partition $A = A(0)+A(1)+A(2)$ such that

$$|A(\nu)| = |S| \ (\nu<3); \quad A(\nu) \subset L(A(\nu+1)) \qquad (\nu < 2).$$

Repeated application leads to sets

$$A(\lambda_0, \lambda_1, \cdots, \lambda_{k-1}) \qquad (k < \omega_0; \lambda_\nu < 3)$$

such that

$$|A(\lambda_0, \cdots, \lambda_{k-1})| = |S|;$$
$$A(\lambda_0, \cdots, \lambda_{k-1}) = \sum[\nu < 3]A(\lambda_0, \cdots, \lambda_{k-1}, \nu);$$
$$A(\lambda_0, \cdots, \lambda_{k-1}, \nu) \subset L(A(\lambda_0, \cdots, \lambda_{k-1}, \nu+1)) \qquad (\nu < 2).$$

Let N be the set of all systems $(\lambda_0, \cdots, \lambda_k)$ such that $k<\omega_0$; $\lambda_\nu \in \{0, 2\}$ $(\nu<k)$; $\lambda_k = 1$, ordered alphabetically. More accurately, if $p=(\lambda_0, \cdots, \lambda_k)$ and $q=(\mu_0, \cdots, \mu_l)$ are elements of N, then we put $p<q$ if $\sum[\nu<k]\lambda_\nu 3^{-\nu} < \sum[\nu<l]\mu_\nu 3^{-\nu}$. Then we have $A(\lambda_0, \cdots, \lambda_k) \subset L(A(\mu_0, \cdots, \mu_l))$, if $(\lambda_0, \cdots) < (\mu_0, \cdots)$. It now suffices to show that N is dense in itself. In fact, if $\{(\lambda_0, \cdots, \lambda_k), (\mu_0, \cdots, \mu_l)\}_< \subset N$, then

$$(\lambda_0, \cdots, \lambda_k) < (\mu_0, \cdots, \mu_{l-1}, 0, 2, 2, \cdots, 2, 1) < (\mu_0, \cdots, \mu_l),$$

provided only that the inner bracket contains a sufficiently large number of two's. Lemma 1 is proved.

THEOREM 31. *Suppose that ϕ is a type such that*

$$|\phi| > \aleph_0; \qquad \omega_1, \omega_1^* \not\leq \phi.$$

Let $\alpha<\omega_0 2$; $\beta<\omega_0^2$; $\gamma<\omega_1$. Then

(27) $$\phi \to (\alpha, \alpha, \alpha)^2,$$

(28) $$\phi \to (\alpha, \beta)^2,$$

(29) $$\phi \to (\omega_0, \gamma)^2,$$

(30) $$\phi \to (4, \alpha)^3.$$

THEOREM 32. *Let ϕ, α, γ be as in Theorem 31. Let S be an ordered set, $\overline{S}=\phi$, and $[S]^2 = K_0+K_1$. Then*

(a) *there is $V \subset S$ such that either*

$$\text{(i) } \overline{V} = \alpha; \quad [V]^2 \subset K_0,$$

or

$$\text{(ii) } \overline{V} = \gamma; \quad [V]^2 \subset K_1,$$

or

$$\text{(iii)} \ \overline{V} = \omega_0\gamma^*; \quad [V]^2 \subset K_1,$$

and

(b) *there is* $W \subset S$ *such that either*

$$\text{(i)} \ \overline{W} = \omega_0 + \overset{*}{\omega_0}; \quad [W]^2 \subset K_0,$$

or

$$\text{(ii)} \ \overline{W} = \gamma; \quad [W]^2 \subset K_1,$$

or

$$\text{(iii)} \ \overline{W} = \gamma^*; \quad [W]^2 \subset K_1.$$

In proving Theorems 31 and 32 we may assume that $|\phi| = \aleph_1$. There is m such that

$$4 \leqq m < \omega_0; \qquad \alpha \leqq \omega_0 + m; \qquad \beta \leqq \omega_0 m.$$

Let $\overline{S} = \phi$. The letters A, B, P, Q denote subsets of S, and we shall always suppose, in the proofs of the last two theorems, that

$$|A| = |B| = \aleph_1; \qquad \overline{P} = \overline{Q} = \omega_0.$$

Proof of Theorem 31, (29). Let $[S]^2 = K_0 + K_1$, and

(31) $$\omega_0 \notin [K_0].$$

We want to deduce that

(32) $$\gamma \in [K_1].$$

There is B such that

(33) $$|BR_0(x)| \leqq \aleph_0 (x \in B).$$

For otherwise there would be elements x_ν such that

$$x_0 \in S; \qquad |R_0(x_0)| = \aleph_1,$$
$$x_1 \in R_0(x_0); \qquad |R_0(x_0, x_1)| = \aleph_1,$$

generally, $x_\nu \in R_0(x_0, \cdots, x_{\nu-1})$,

$$|R_0(x_0, \cdots, x_\nu)| = \aleph_1 \qquad (\nu < \omega_0).$$

Then $[\{x_0, x_1, \cdots\}_<]^2 \subset K_0$, and hence $\omega_0 \in [K_0]$ which contradicts (31).

By hypothesis, $\omega_1, \omega_1^* \nleqq \overline{B}$. Hence, by Lemma 1,

$$\sum [t \text{ rational}] B(t) \subset B$$

for some sets $B(t)$ such that $B(t) \subset L(B(u))$ $(t<u)$.

There is a set T of rational numbers which, if ordered by magnitude, is of type γ. Let $T=\{t_\mu : \mu<\gamma\}$ where $t_\mu < t_\nu$ for $\mu<\nu<\gamma$. We define inductively elements $x_\nu(\nu<\gamma)$ as follows. Let $\nu_0<\gamma$, and suppose that $x_\nu \in B(\nu<\nu_0)$. Then, by (33),

$$\left| \sum [\nu < \nu_0] B R_0(x_\nu) \right| \leqq \aleph_0 < \left| B(t_{\nu_0}) \right|,$$

and therefore we can choose $x_{\nu_0} \in B(t_{\nu_0}) - \sum [\nu<\nu_0] R_0(x_\nu)$. Put $X=\{x_\nu : \nu<\gamma\}$. Then $\overline{X}=\gamma$; $[X]^2 \subset K_1$. Hence (32) holds, and (29) follows.

PROOF OF THEOREM 32 (a). Let the hypotheses be satisfied but suppose that (a) is false, i.e. that

$$\alpha \notin [K_0]; \qquad \gamma \notin [K_1]; \qquad \omega_0\gamma^* \notin [K_1]. \tag{34}$$

ASSUMPTION. If $A \subset S$, then there is $x_0 \in A$ such that

$$\left| A L_0(x_0) \right| > \aleph_0.$$

Then there are x_ν, $A_\nu(\nu \leqq m)$ such that

$$x_0 \in A_0 = S; \qquad A_0 L_0(x_0) = A_1; \qquad x_1 \in A_1; \qquad A_1 L_0(x_1) = A_2$$

and so on, up to

$$A_m = A_{m-1} L_0(x_{m-1}) = A_0 L_0(x_0, x_1, \cdots, x_{m-1}).$$

Then, by (29), $\overline{A}_m \to (\omega_0, \gamma)^2$; $\gamma \notin F_1(A_m)$, and hence $\omega_0 \in F_0(A_m)$. There is $P \subset A_m$ such that $[P]^2 \subset K_0$. Then $[P+\{x_0, \cdots, x_{m-1}\}]^2 \subset K_0$ which contradicts (34). Hence our assumption is false, and there is A such that

$$\left| A L_0(x) \right| \leqq \aleph_0 \qquad (x \in A). \tag{35}$$

By Lemma 1, there is $B(t) \subset A$, for rational t, such that $B(t) \subset L(B(u))$ $(t<\mu)$. There are rational numbers t_ν $(\nu<\gamma)$ such that $t_\mu > t_\nu$ $(\mu<\nu<\gamma)$. We define sets P_ν $(\nu<\gamma)$ as follows. Let $\nu_0<\gamma$, and suppose that $P_\nu \subset B(t_\nu)$ $(\nu<\nu_0)$. Then we may put, by (35),

$$B' = B(t_{\nu_0}) L_1\left(\sum [\nu < \nu_0] P_\nu\right).$$

Then, by (29), $\overline{B'} \to (\alpha, \omega_0)^2$; $\alpha \notin F_0(B')$; $\omega_0 \in F_1(B')$, and there is $P_{\nu_0} \subset B'$ such that $[P_{\nu_0}]^2 \subset K_1$. This defines P_ν for $\nu<\gamma$. Put $\sum [\nu<\gamma] P_\nu = X$. Then $\overline{X}=\omega_0\gamma^*$; $[X]^2 \subset K_1$. But this contradicts (34), and so (a) is proved.

PROOF OF THEOREM 32 (b). Let the hypotheses be satisfied but (b) be false. Then

(36) $\omega_0 + \omega_0^* \notin [K_0]; \quad \gamma \notin [K_1]; \quad \gamma^* \notin [K_1].$

Choose any A.

ASSUMPTION. $|AR_0(x)| \leq \aleph_0$ $(x \in A)$.

Then, by Lemma 1, there are sets $B(t) \subset A$, for rational t, such that $B(t) \subset L(B(u))$ $(t < u)$. There are rational numbers t_ν $(\nu < \gamma)$ such that $t_\mu < t_\nu$ $(\mu < \nu < \gamma)$. We define x_ν $(\nu < \gamma)$ as follows. Let $\nu_0 < \gamma$, and suppose that $x_\nu \in B(t_\nu)$ $(\nu < \nu_0)$. Then, by our assumption, there is

$$x_{\nu_0} \in B(t_{\nu_0}) - \sum [\nu < \nu_0] R_0(x_\nu).$$

Then the set $X = \{x_\nu : \nu < \gamma\}$ satisfies $\overline{X} = \gamma$; $[X]^2 \subset K_1$ which is a contradiction against (36). Hence our assumption is false, i.e., given any A, there is $x \in A$ such that $|AR_0(x)| = \aleph_1$. By symmetry, it follows that there also is $y \in A$ such that $|AL_0(y)| = \aleph_1$. By alternate applications of these two results we obtain elements x_ν, y_ν and sets A_ν, $B_\nu (\nu < \omega_0)$ such that the following conditions are satisfied.

$$x_0 \in S; \quad y_0 \in R_0(x_0) = B_0; \quad x_1 \in B_0 L_0(y_0) = A_1; \quad y_1 \in A_1 R_0(x_1) = B_1;$$

generally, for $\nu < \omega_0$,

$$x_{\nu+1} \in B_\nu L_0(y_\nu) = A_{\nu+1}; \qquad y_{\nu+1} \in A_{\nu+1} R_0(x_{\nu+1}) = B_{\nu+1}.$$

Then the set $\sum [\nu < \omega_0] \{x_\nu, y_\nu\} = D$ satisfies $\overline{D} = \omega_0 + \omega_0^*$; $[D]^2 \subset K_0$. This contradiction against (36) completes the proof of Theorem 32.

PROOF OF THEOREM 31, (27). Let $[S]^2 = K_0 + K_1 + K_2$,

(37) $$\alpha \notin [K_\nu] \qquad (\nu < 3).$$

Our aim is to deduce a contradiction. We shall reduce the general case to more and more special cases. For the sake of convenience of notation we shall use the same notation for the sets in question at each stage.

We put $K_{12} = K_1 + K_2$. The functions F_{12}, L_{12}, R_{12} refer to K_{12} in the same way as the functions F_ν, L_ν, R_ν refer to K_ν.

Let $A \subset S$. By Lemma 1, there are sets A_0, $A_1 \subset A$ such that $A_0 \subset L(A_1)$. Let $x_0 \in A_1$. Then $|AL(x_0)| = \aleph_1$, and there is $\nu_0 < 3$ such that $|AL_{\nu_0}(x_0)| = \aleph_1$. By repeating this argument we find numbers $\nu_\rho < 3$ and elements x_ρ $(\rho < \omega_0)$ such that

$$x_\rho \in SL_{\nu_0}(x_0) L_{\nu_1}(x_1) \cdots L_{\nu_{\rho-1}}(x_{\rho-1}),$$

$$|SL_{\nu_0}(x_0) \cdots L_{\nu_\rho}(x_\rho)| = \aleph_1 \qquad (\rho < \omega_0).$$

There are $\rho_0 < \rho_1 < \cdots \rho_{m-1} < \omega_0$ such that $\nu_{\rho_0} = \cdots = \nu_{\rho_{m-1}}$. We may assume $\nu_{\rho_0} = 0$. Put $L_0(x_{\rho_0}, \cdots, x_{\rho_{m-1}}) = A_0$.

ASSUMPTION 1. $\omega_0 \in F_0(A_0)$.

Then there is $P \subset A_0$ such that $[P]^2 \subset K_0$. Then $\alpha \leq \overline{C}$; $[C]^2 \subset K_0$, where $C = P + \{x_{\rho_\nu} : \nu < m\}$, which contradicts (37). Hence the Assumption 1 is false, and we have $\omega_0 \notin F_0(A_0)$. We may assume that

$$\omega_0 \notin [K_0]. \tag{38}$$

For a later application we remark that in what follows we may replace S by any nondenumerable subset of S without any of the conclusions becoming invalid.

Now let $A \subset S$. Then, by (29), $\overline{A} \to (\omega_0, \alpha)^2$. Also, $\omega_0 \to (\omega_0, \omega_0)^2$. Therefore, by Theorem 16, $\overline{A} \to (\omega_0, \omega_0, \alpha)^2$. Hence at least one of the following three relations holds.

$$\text{(i) } \omega_0 \in F_0(A), \qquad \text{(ii) } \omega_0 \in F_1(A), \qquad \text{(iii) } \alpha \in F_2(A).$$

Since (i) and (iii) are false, it follows that

$$\omega_0 \in F_1(A) \qquad (A \subset S). \tag{39}$$

By symmetry,

$$\omega_0 \in F_2(A) \qquad (A \subset S). \tag{40}$$

ASSUMPTION 2. There are x_ν, A_ν $(\nu < \omega_0)$ such that $x_0 \in A_0$; $A_0 R_0(x_0) = A_1$; $x_1 \in A_1$; $A_1 R_0(x_1) = A_2$; $x_2 \in A_2$, etc.

Then $[\{x_0, x_1, \cdots\}_<]^2 \subset K_0$ which contradicts (38). Hence the Assumption 2 is false, and there are $\nu_0 < \omega_0$; $x_\nu \in S$ $(\nu < \nu_0)$ such that we may put $A = R_0(x_0, \cdots, x_{\nu_0 - 1})$ and we then have $|AR_0(x)| \leq \aleph_0$ $(x \in A)$. We may assume that

$$|R_0(x)| \leq \aleph_0 \qquad (x \in S). \tag{41}$$

By Lemma 1, there are sets A, B such that $A \subset L(B)$. By (39), there is $P \subset A$ such that $[P]^2 \subset K_1$. For a later application we remark that at this stage we might have applied (40) in place of (39) and in this way could have interchanged the roles of K_1 and K_2. By (41), $|\sum [x \in P] R_0(x)| \leq \aleph_0$, and hence $|BR_{12}(P)| = \aleph_1$. Therefore we may assume

$$[P]^2 \subset K_1; \qquad P \subset L_{12}(S - P). \tag{42}$$

ASSUMPTION 3. If $Q \subset P$; $A \subset S$, then there is $x \in A$ such that

$$|QL_1(x)| = \aleph_0.$$

Now we argue as follows. By Lemma 1, there are sets $A_\nu \subset S - P$ such that $A_\mu \subset L(A_\nu)$ $(\mu < \nu < \omega_0 m)$. We define inductively x_ν, P_ν $(\nu < \omega_0 m)$ as follows. There is $x_0 \in A_0$ such that, if $P_0 = P$, we have $|P_0 L_1(x_0)| = \aleph_0$. Let $0 < \nu_0 < \omega_0 m$, and suppose that

$$x_\nu \in A_\nu; \quad P_\nu \subset P \qquad (\nu < \nu_0),$$
$$|P_\nu - P_\mu| < \aleph_0 \qquad (\mu < \nu < \nu_0).$$

Then we can write $[0, \nu_0] = \{\rho_\lambda : \lambda < \omega_0\}$. We can choose y_λ such that

$$y_\lambda \in P_{\rho_0} P_{\rho_1} \cdots P_{\rho_\lambda} - \{y_0, \cdots, y_{\lambda-1}\} \qquad (\lambda < \omega_0).$$

By (41) and Assumption 3, there is $x_{\nu_0} \in A_{\nu_0} - \sum [\nu < \nu_0] R_0(x_\nu)$ such that $|\{y_0, y_1, \cdots\} L_1(x_{\nu_0})| = \aleph_0$. We put $P_{\nu_0} = \{y_0, y_1, \cdots\} L_1(x_{\nu_0})$. Then, if $\nu < \nu_0$, there is $\lambda < \omega_0$ such that $\nu = \rho_\lambda$. Then $|P_{\nu_0} - P_\nu| \leqq |\{y_0, y_1, \cdots\} - P_{\rho_\lambda}| < \aleph_0$. This completes the definition of x_ν, P_ν $(\nu < \omega_0 m)$.

We have $|P_\nu - P_\mu| < \aleph_0$ $(\mu < \nu < \omega_0 m)$; $P_\nu \subset P L_1(x_\nu)$ $(\nu < \omega_0 m)$. Put $X = \{x_\nu : \nu < \omega_0 m\}$. Then, by (11), $[X]^2 \subset K_1 + K_2$; $X = \omega_0 m \rightarrow (m, \alpha)^2$.

Case 1. There is $D = \{x_{\mu_0}, \cdots, x_{\mu_{m-1}}\}_< \subset X$ such that $[D]^2 \subset K_1$. Then we put $P' = P_{\mu_{m-1}}$ and have, for $\tau < m$,

$$|P' - L_1(x_{\mu_\tau})| \leqq |P' - P_{\mu_\tau}| + |P_{\mu_\tau} - L_1(x_{\mu_\tau})| < \aleph_0 + 0.$$

By summing over τ we obtain $|P' - L_1(D)| < \aleph_0$. Hence we may put $P' L_1(D) = Q$, and we then have $\overline{Q+D} \geqq \alpha$; $[Q+D]^2 \subset K_1$ which contradicts (37).

Case 2. There is $D \subset X$ such that $\overline{D} = \alpha$; $[D]^2 \subset K_2$. This, again, contradicts (37). Hence the Assumption 3 is false, i.e., there are $P' \subset P$; $A' \subset S$ such that

$$|P' L_1(x)| < \aleph_0 \qquad (x \in A').$$

Then there is $A'' \subset A'$ such that the set $P' L_1(x)$ is constant for $x \in A''$. Then there is P'' such that $P' L_2(x) = P''$ $(x \in A'')$. We have therefore proved that there is P'', A'' such that

$$(43) \qquad [P'']^2 \subset K_1; \quad P'' \subset L_2(A'').$$

The whole argument from (38) onwards remains valid if S is replaced by any set A. Hence it follows from (43) that if $A \subset S$, then there are P, $A' \subset A$ such that

$$(44) \qquad [P]^2 \subset K_1; \quad P \subset L_2(A').$$

By Lemma 1, there are A_0, B_0 such that $A_0 \subset L(B_0)$. By repeated application of (44) we obtain sets P_ν, A'_ν $(\nu < \omega_0)$ such that

$$P_0 + A'_0 \subset A_0; \quad [P_0]^2 \subset K_1; \quad P_0 \subset L_2(A'_0),$$
$$P_1 + A'_1 \subset A'_0; \quad [P_1]^2 \subset K_1; \quad P_1 \subset L_2(A'_1),$$

generally, $P_\nu + A'_\nu \subset A_{\nu-1}$; $[P_\nu]^2 \subset K_1$; $P_\nu \subset L_2(A'_\nu)$ $(0 < \nu < \omega_0)$. Then $P_\nu \subset A_0 \subset L(B_0)$ $(\nu < \omega_0)$,

$$P_\mu \subset L_2(A'_\mu) \subset L_2(A'_{\nu-1}) \subset L_2(P_\nu) \qquad (\mu < \nu < \omega_0).$$

We put $B_1 = B_0 R_{12}(P_0 + P_1 + \cdots)$. Then we have the result that there are sets P_ν, B_1 ($\nu < \omega_0$) such that

$$(45) \qquad \begin{cases} [P_\nu]^2 \subset K_1; \quad P_\nu \subset L_{12}(B_1) & (\nu < \omega_0), \\ P_\mu \subset L_2(P_\nu) & (\mu < \nu < \omega_0). \end{cases}$$

Now let $\nu_0 < \omega_0$; $B_2 \subset B_1$; $P' \subset P_{\nu_0}$.

Assumption 4. $|P'L_2(x)| < \aleph_0$ ($x \in B_2$).

Then there is $B_3 \subset B_2$ such that the set $D = P'L_2(x)$ is constant for $x \in B_3$. By (39), there is $Q \subset B_3$ such that $[Q]^2 \subset K_1$. Then $[(P'-D)+Q]^2 \subset K_1$; $\omega_0 2 \in [K_1]$ which contradicts (37). Hence the Assumption 4 is false, i.e.

$$(46) \qquad \text{if } \nu_0 < \omega_0; \quad P' \subset P_{\nu_0}, \text{ then} \\ |\{x : x \in B_1;\ |P'L_2(x)| < \aleph_0\}| \leqq \aleph_0.$$

To B_1 the same argument applies as to S, from (38) onwards. The only change we make is that, after (41), we apply (40) instead of (39), so that now the roles of K_1 and K_2 are interchanged. We find sets Q_ν, $B_2 \subset B_1$ such that, in analogy to (45), (46), the following statements are true.

$$(47) \qquad \begin{matrix} [Q_\nu]^2 \subset K_2; \quad Q_\nu \subset L_{12}(B_2) & (\nu < \omega_0), \\ Q_\mu \subset L_1(Q_\nu) & (\mu < \nu < \omega_0). \end{matrix}$$

$$(48) \qquad \text{If } \nu_0 < \omega_0;\ Q' \subset Q_{\nu_0}, \text{ then} \\ |\{x : x \in B_2;\ |Q'L_1(x)| < \aleph_0\}| \leqq \aleph_0.$$

By Lemma 1, there is $B'_\nu \subset B_2$ ($\nu < \omega_0$) such that $B'_\mu \subset L(B'_\nu)$ ($\mu < \nu < \omega_0$). Let $P'_\nu \subset P_\nu$; $Q'_\nu \subset Q_\nu$ ($\nu < \omega_0$). Then, by (46), (48), there are at most $\aleph_0$ elements $x \in B_2$ such that at least one of the relations

$$|P'_\nu L_2(x)| < \aleph_0; \qquad |Q'_\nu L_1(x)| < \aleph_0$$

holds. By using this result repeatedly we find elements x_λ ($\lambda < \omega_0$) such that, for all $\nu < \omega_0$,

$$x_0 \in B'_0; \qquad |P_\nu L_2(x_0)| = |Q_\nu L_1(x_0)| = \aleph_0,$$
$$x_1 \in B'_1; \qquad |P_\nu L_2(x_0, x_1)| = |Q_\nu L_1(x_0, x_1)| = \aleph_0,$$

generally, $x_\lambda \in B'_\lambda$;

$$|P_\nu L_2(x_0, \cdots, x_\lambda)| = |Q_\nu L_1(x_0, \cdots, x_\lambda)| = \aleph_0 \qquad (\nu, \lambda < \omega_0).$$

Since $\omega_0 \to (\omega_0)^2_3$, there is a number $\nu < 3$ and a sequence $\lambda_0 < \lambda_1 < \cdots$;

$\lambda_\rho < \omega_0$, such that $[\{x_{\lambda_0}, x_{\lambda_1}, \cdots\}_<]^2 \subset K_\nu$. By (38), $\nu \neq 0$. We can choose y_μ, z_μ such that, for $\mu < \omega_0$,

$$y_\mu \in P_\mu L_2(x_0, x_1, \cdots, x_{\lambda_{m-1}}); \qquad z_\mu \in Q_\mu L_1(x_0, \cdots, x_{\lambda_{m-1}}).$$

Put $X = \{x_{\lambda_\rho} : \rho < m\}$; $Y = \{y_\mu : \mu < \omega_0\}$; $Z = \{z_\mu : \mu < \omega_0\}$.

Case 1. $\nu = 1$. Then $[Z+X]^2 \subset K_1$; $\alpha \in [K_1]$.

Case 2. $\nu = 2$. Then $[Y+X]^2 \subset K_2$; $\alpha \in [K_2]$. In either case, a contradiction against (37) follows. This proves (27).

PROOF OF THEOREM 31, (28). If $[S]^2 = K_0 + K_1$ and if we put $\gamma = \omega_0 m$ then we have, by Theorem 32 (a), either (i) $\alpha \in [K_0]$ or (ii) $\beta \leqq \gamma \in [K_1]$ or (iii) $\beta \leqq \omega_0 m \leqq \omega_0 \gamma^* \notin [K_1]$. This proves (28).

PROOF OF THEOREM 31, (30). Let $[S]^3 = K_0 +' K_1$,

$$4 \notin [K_0]; \qquad \alpha \notin [K_1]. \tag{49}$$

We shall deduce a contradiction.

By Theorem 2, there is $n < \omega_0$ such that $n \to (m, m)^3$, and p such that

$$(n-1)(1 + m + m(m-1)/2) < p < \omega_0. \tag{50}$$

By Lemma 1, there is $z_0 \in S$ such that

$$|L(z_0)|, \; |R(z_0)| > \aleph_0$$

and then there is $C \subset R(z_0)$ such that $\overline{C} = \eta$.

The following diagram shows the relative position in S and the inclusion relations between the various sets to be considered in the argument that follows. It might be of help to the reader.

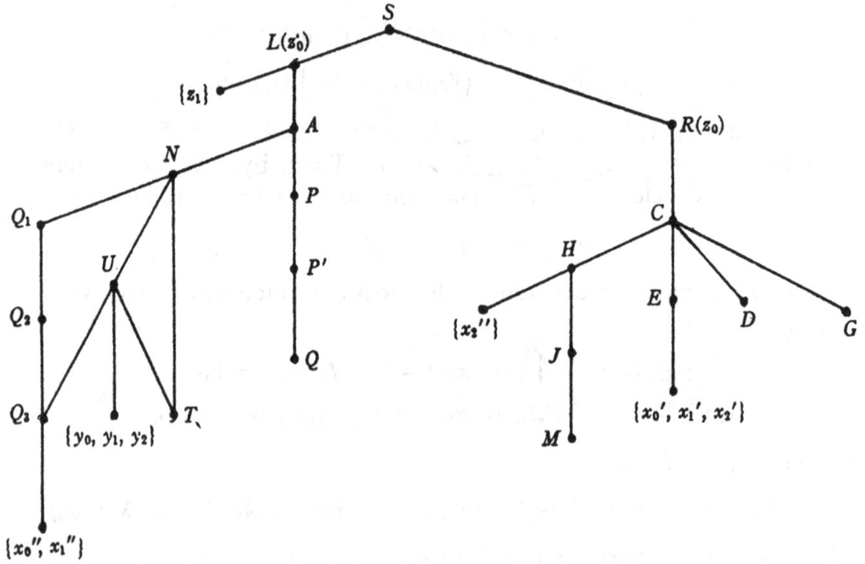

ASSUMPTION. If $D\in[C]^p$, then $|\prod[x_1, x_2\in D]\{x_0:x_0<z_0; \{x_0, x_1, x_2\}\notin K_0\}|\leqq\aleph_0$. Then there is $z_1<z_0$ such that

(51) if $D\in[C]^p$, then $\{z_1, x_1, x_2\}\in K_0$ for some $x_1, x_2\in D$.

Then $[C]^2=K_0'+K_1'$, where

$$K_\nu' = \{\{x_1, x_2\}:x_1, x_2\in C; \{z_1, x_1, x_2\}\in K_\nu\} \qquad (\nu<2).$$

By (11), $\overline{C}=\eta\geqq\omega_0 p\rightarrow(\omega_0+m, p)^2$. Hence there are two cases.

Case 1. There is $E\subset C$ such that $\overline{E}=\omega_0+m$; $[E]^2\subset K_0'$. Then, since, by (49), $\overline{E}=\omega_0+m\notin[K_1]$, there are $x_0', x_1', x_2'\in E$ such that $\{x_0', x_1', x_2'\}\in K_0$. Then $[\{z_1, x_0', x_1', x_2'\}_{\neq}]^3\subset K_0$ which contradicts (49).

Case 2. There is $G\in[C]^p$ such that $[G]^2\subset K_1'$. Then $\{z_1, x_1, x_2\}\notin K_0$ for all $x_1, x_2\in G$, which is a contradiction against (51).

It follows that our assumption is false, and that there are $H\in[C]^p$ and $A\subset L(z_0)$ such that

$$\{x_0, x_1, x_2\}\notin K_0 \qquad \text{for } x_0\in A;\ x_1, x_2\in H.$$

Put

$$V(x_0, x_1) = \{x_2:x_2\in H; \{x_0, x_1, x_2\}\in K_1\} \qquad \text{for } x_0, x_1\in A.$$

Then $[A]^2=K_0''+'K_1''$, where K_0'' is the set of all $\{x_0, x_1\}_<\subset A$ such that $|V(x_0, x_1)|\geqq n$. By (28), $\overline{A}\rightarrow(\omega_0, \omega_0+m)^2$. Hence there are two cases.

Case 1. There is $P\subset A$ such that $[P]^2\subset K_0''$. Then

$$|V(x_0, x_1)|\geqq n \qquad \text{for } \{x_0, x_1\}_<\subset P,$$
$$[P]^2 = \sum[W\subset H]K_W^{(3)},$$

where $K_W^{(3)}=\{\{x_0, x_1\}_<:x_0, x_1\in P;\ V(x_0, x_1)=W\}$. The number k of sets W is finite, and $\omega_0\rightarrow(\omega_0)_k^2$, by Theorem 1. Hence there are $P'\subset P$; $J\subset H$ such that $[P']^2\subset K_J^{(3)}$,

$$V(x_0, x_1) = J \qquad \text{for } \{x_0, x_1\}_<\subset P'.$$

Then

$$|J|\geqq n,$$
$$\{x_0, x_1, x_2\}\in K_1 \qquad \text{for } \{x_0, x_1\}_<\subset P';\ x_2\in J.$$

Since $[P']^3\subset K_0+K_1$ and, by Theorem 1, $\omega_0\rightarrow(\omega_0, \omega_0)^3$, there are $Q\subset P'$; $\nu<2$ such that $[Q]^3\subset K_\nu$. By (49), $\omega_0\notin[K_0]$. Hence $\nu=1$; $[Q]^3\subset K_1$.

Furthermore, $[J]^3\subset K_0+K_1$; $\overline{J}=n\rightarrow(m, m)^3$. Hence there are

$M \in [J]^m$; $\rho < 2$ such that $[M]^3 \subset K_\rho$. Since $m \geqq 4 \notin [K_0]$, we have $\rho = 1$. Then, in view of $Q \subset P' \subset P \subset A$; $M \subset J \subset H$,

$$[Q+M]^3 \subset K_1; \qquad \omega_0 + m = \overline{Q+M} \in [K_1]$$

which contradicts (49).

Case 2. There is $N \subset A$ such that $\overline{N} = \omega_0 + m$; $[N]^2 \subset K_1''$. Then

$$|V(x_0, x_1)| \leqq n - 1 \qquad \text{for } x_0, x_1 \in N.$$

Then

$$N = Q_1 + T; \qquad Q_1 \subset L(T); \qquad |T| = m.$$

We have $[Q_1]^2 = \sum' [\kappa < k_1] K_\kappa^{(1)}$, where $K_\kappa^{(1)} \neq 0$ $(\kappa < k_1)$, and two elements Z_0, Z_1 of $[Q_1]^2$ belong to the same $K_\kappa^{(1)}$ if, and only if, for every $x_2 \in H$, the sets $Z_0 + \{x_2\}$ and $Z_1 + \{x_2\}$ belong to the same class K_ν. Then $k_1 < \omega_0$ and, by $\omega_0 \rightarrow (\omega_0)_{k_1}^2$, there are $Q_2 \subset Q_1$; $\kappa_2 < k_1$ such that $[Q_2]^2 \subset K_{\kappa_2}^{(1)}$. This means that, for some $\rho(x_2) < 2$,

$$\{x_0, x_1, x_2\} \in K_{\rho(x_2)} \qquad \text{for } \{x_0, x_1\}_< \subset Q_2;\ x_2 \in H.$$

Similarly, we have

$$Q_2 = \sum{}' [\kappa < k_2] K_\kappa^{(2)}, \quad \text{where} \quad K_\kappa^{(2)} \neq 0 \qquad (\kappa < k_2),$$

and two elements x_{00} and x_{01} of Q_2 belong to the same $K_\kappa^{(2)}$ if, and only if, for every $x_1 \in T$; $x_2 \in H$, the two sets $\{x_{00}, x_1, x_2\}$ and $\{x_{01}, x_1, x_2\}$ belong to the same class K_ν. Then $k_2 < \omega_0$ and, by $\omega_0 \rightarrow (\omega_0)_{k_2}^1$, there are $Q_3 \subset Q_2$; $\kappa_3 < k_2$ such that $Q_3 \subset K_{\kappa_3}^{(2)}$. This means that, for some $\sigma(x_1, x_2) < 2$,

$$\{x_0, x_1, x_2\} \in K_{\sigma(x_1, x_2)} \qquad \text{for } x_0 \in Q_3;\ x_1 \in T;\ x_2 \in H.$$

Put $U = Q_3 + T$, and choose $\{x_0'', x_1''\}_< \subset Q_3$.

Consider any $x_2 \in H$ such that there are $x_0, x_1 \in U$ satisfying $\{x_0, x_1, x_2\}_< \in K_1$. If $x_0, x_1 \in Q_3$, then $\rho(x_2) = 1$; $\{x_0'', x_1'', x_2\} \in K_1$; $x_2 \in V(x_0'', x_1'')$. If $x_0 \in Q_3$; $x_1 \in T$, then $\sigma(x_1, x_2) = 1$; $\{x_0'', x_1, x_2\} \in K_1$; $x_2 \in V(x_0'', x_1)$. If $x_0, x_1 \in T$, then $\{x_0, x_1, x_2\} \in K_1$; $x_2 \in V(x_0, x_1)$. Hence, in any case,

$$x_2 \in V(x_0'', x_1'') + \sum [x \in T) V(x_0'', x) + \sum [\{x, y\}_< \subset T] V(x, y)$$

and therefore, in view of the definition of x_2 and the relations $|T| = m$ and (50),

$$\left| \sum [x_0, x_1 \in U] \{x_2 : x_2 \in H;\ \{x_0, x_1, x_2\}_< \in K_1\} \right|$$
$$\leqq (n-1) + (n-1)\binom{m}{1} + (n-1)\binom{m}{2} < p = |H|.$$

We deduce the existence of $x_2'' \in H$ such that

$$\{x_0, x_1, x_2''\} \notin K_1 \qquad \text{for all } x_0, x_1 \in U.$$

Since $\overline{U} = \omega_0 + m \notin [K_1]$, there are $y_0, y_1, y_2 \in U$ such that $\{y_0, y_1, y_2\} \in K_0$. But then $[\{y_0, y_1, y_2, x_2''\}_{\neq}]^3 \subset K_0$ which contradicts (49). This proves (30) and thus completes the proof of Theorems 31 and 32.

THEOREM 33. *Let* $\alpha < \omega_0 2$. *Then* $\omega_1 \rightarrow (\alpha, \alpha)^2$.

PROOF. Let $\overline{S} = \omega_1$; $[S]^2 = K_0 +' K_1$; $2 \leqq m < \omega_0$; $\alpha \leqq \omega_0 + m$,

$$\alpha \notin [K_\nu] \qquad (\nu < 2). \tag{52}$$

We have to deduce a contradiction. Let the conventions concerning the use of the letters A, B, P, Q be the same as in the proofs of Theorems 31 and 32. Choose any P.

ASSUMPTION. Let $[P]^2 \subset K_0$. Suppose that, if $P' \subset P$, then there is A such that

$$|P'L_0(x)| = \aleph_0 \qquad (x \in A).$$

Then we define x_ν, P_ν $(\nu < \omega_1)$ as follows. There is x_0 such that $|PL_0(x_0)| = \aleph_0$. Put $P_0 = PL_0(x_0)$. Now let $0 < \nu_0 < \omega_1$, and suppose that $x_\nu \in S$; $P_\nu \subset PL_0(x_\nu)$ $(\nu < \nu_0)$;

$$|P_\nu - P_\mu| < \aleph_0 \qquad (\mu < \nu < \nu_0).$$

Then we can write $[0, \nu_0] = \{\mu_\lambda : \lambda < \omega_0\}$. We can choose elements $y_\lambda (\lambda < \omega_0)$ such that $y_\lambda \in P_{\mu_0} P_{\mu_1} \cdots P_{\mu_\lambda} - \{y_\rho : \rho < \lambda\}$ $(\lambda < \omega_0)$. Put $P' = \{y_\lambda : \lambda < \omega_0\}$. Then, by our assumption, there is A such that $|P'L_0(x)| = \aleph_0$ $(x \in A)$. We can choose

$$x_{\nu_0} \in A - \sum [\nu < \nu_0](\{x_\nu\} + L(x_\nu)).$$

We put $P_{\nu_0} = P'L_0(x_{\nu_0})$. If, now, $\nu_1 < \nu_0$, then there is $\lambda < \omega_0$ such that $\nu_1 = \mu_\lambda$. Then

$$|P_{\nu_0} - P_{\nu_1}| \leqq |\{y_0, y_1, \cdots\} - P_{\mu_\lambda}| < \aleph_0.$$

Also, $P_{\nu_0} \subset PL_0(x_{\nu_0})$. This completes the inductive definition of x_ν, $P_\nu (\nu < \omega_1)$ such that

$$P_\mu \subset PL_0(x_\mu); \qquad |P_\nu - P_\mu| < \aleph_0 \qquad (\mu < \nu < \omega_1).$$

Put $X = \{x_\nu : \nu < \omega_1\}$. Then, by Theorem 23, $\overline{X} = \omega_1 > \omega_0 m \rightarrow (m, \omega_0 + m)^2$. Since, by (52), $\omega_0 + m \notin F_1(X)$, we have $m \in F_0(X)$, and there is $D \in [X]^m$ such that $[D]^2 \subset K_0$. Let $x_\rho = \max [x \in D] x$. Then, for any $x_\nu \in D$, $|P_\rho - L_0(x_\nu)| \leqq |P_\rho - P_\nu| + |P_\nu - L_0(x_\nu)| < \aleph_0$. Hence we may put $Q = P_\rho L_0(D)$, and then we have $\overline{Q + D} = \omega_0 + m \geqq \alpha$; $[Q + D]^2$

$\subset K_0$. This is a contradiction against (52). Therefore our assumption is false.

Now let $A\subset S$. Then, by Theorem 3, $|A| = \aleph_1 \rightarrow (\aleph_0, \aleph_1)^2$ and hence, by Theorem 14, $\overline{A} = \omega_1 \rightarrow (\omega_0, \omega_1)^2$. Since $\omega_1 \notin F_1(A)$, we conclude that $\omega_0 \in F_0(A)$, so that there is $P\subset A$ such that $[P]^2 \subset K_0$. As the assumption made above is false, there is $P'\subset P$ such that there are at most $\aleph_0$ elements x such that $|P'L_0(x)| = \aleph_0$. Then there is $A'\subset A$ such that $|P'L_0(x)| < \aleph_0$ $(x\in A')$. Hence there are at most $\aleph_0$ distinct sets $P'L_0(x)$ for varying values $x\in A'$, and there is $A''\subset A'$ and E such that

$$P'L_0(x) = E \qquad (x \in A'').$$

Since $|E| < \aleph_0$, we may put $P'' = P' - E$, and since $\overline{A''} = \omega_1$; $\overline{P''} = \omega_0$, we may put $A''' = A''R(P'')$. Now let $x\in A'''$; $y\in P''$. Then

$$x \in A''' \subset A''; \qquad y \notin E = P'L_0(x); \qquad y \notin L_0(x).$$

Also,

$$x \in A''' \subset R(P'') \subset R(y); x > y.$$

Hence

$$y \in L_1(x); \qquad P'' \subset L_1(A''').$$

So far we have proved that, given any A, there are sets P''', $A''\subset A$ such that $P''\subset L_1(A''')$; $[P'']^2 \subset K_0$, and, moreover, there are at most $\aleph_0$ elements x such that $|P''L_0(x)| = \aleph_0$.

By applying the last result repeatedly, starting with $A = S$, we obtain sets $P_\nu (\nu < \omega_0)$ such that

$$[P_\mu]^2 \subset K_0; \qquad P_\mu \subset L_1(P_\nu) \qquad (\mu < \nu < \omega_0).$$

There is Q_ν such that

$$|P_\nu L_0(x)| < \aleph_0 \qquad (\nu < \omega_0; x \in S - Q_\nu).$$

We can choose $B \subset S - \sum [\nu < \omega_0] Q_\nu$ such that $P_\nu \subset L(B)$ $(\nu < \omega_0)$. Then, by Theorem 23, $\overline{B} = \omega_1 > \omega_0 m \rightarrow (\omega_0 + m, m)^2$;

$$\omega_0 + m \notin F_0(B); \qquad m \in F_1(B),$$

and there is $D\in [B]^m$ such that $[D]^2 \subset K_1$. Then, for every $\nu < \omega_0$, $|\sum [x\in D] P_\nu L_0(x)| < \aleph_0$. Therefore we can choose $y_\nu \in P_\nu L_1(D)$ $(\nu < \omega_0)$. If we put $Q = \{y_\nu : \nu < \omega_0\}$ then $\overline{Q+D} = \omega_0 + m \geqq \alpha$; $[Q+D]^2 \subset K_1$. This contradiction against (52) completes the proof of Theorem 33.

The next theorem, while perhaps appearing to be of a rather special and complicated nature, is of interest in that it implies Theorem 7 (i) and Theorem 8. It may well be capable of further worthwhile applications.

THEOREM 34. *Let* α, β, γ *be ordinals, and* $\alpha \nrightarrow (\beta, \gamma)^2$. *Then there are ordinals* α_λ $(\lambda < \beta^-)$ *such that, if*

$$|k_\mu| = \prod [\lambda < \mu] |\alpha_\lambda| \qquad (\mu < \beta^-),$$

then

$$\alpha \nrightarrow (\alpha_0 + 1, \alpha_1 + 1, \cdots)^1_{\beta^-}; \qquad \alpha_\lambda \nrightarrow (\gamma)^1_{k_\lambda} \qquad (\lambda < \beta^-).$$

We begin by deducing (i) of Theorem 7 or, rather, a slightly stronger proposition, from Theorem 34.

COROLLARY 1. *Let* m *and* n *be such that* $\aleph_n^d \leqq \aleph_n$ $(d < \aleph_m)$. *Then* $\omega_{n+1} \rightarrow (\omega_m + 1, \omega_{n+1})^2$.

This implies, a fortiori, $\omega_{n+1} \rightarrow (\omega_m, \omega_{n+1})^2$ which, in its turn, by Theorem 14, is equivalent to Theorem 7 (i).

Deduction of Corollary 1 *from Theorem* 34. Let us suppose that $\omega_{n+1} \nrightarrow (\omega_m + 1, \omega_{n+1})^2$. Then, by Theorem 34, there are ordinals α_λ, k_λ such that $|k_\mu| = \prod [\lambda < \mu] |\alpha_\lambda|$ $(\mu < \omega_m)$;

(53) $$\omega_{n+1} \nrightarrow (\alpha_0 + 1, \alpha_1 + 1, \cdots)^1_{\omega_m},$$

(54) $$\alpha_\mu \nrightarrow (\omega_{n+1})^1_{k_\mu} \qquad (\mu < \omega_m).$$

Then, for $\lambda < \omega_m$,

(55) $$\alpha_\lambda < \omega_{n+1}.$$

For, let $\mu < \omega_m$, and suppose that (55) holds for $\lambda < \mu$. Then, using $|\mu| < \aleph_m$ and the hypothesis, we find that $|k_\mu| \leqq \aleph_n^{|\mu|} \leqq \aleph_n$. Now, by (54), we can write $|\alpha_\mu| = \sum [\nu < k_\mu] |\rho_\nu|$, where $\rho_\nu < \omega_{n+1}$ $(\nu < k_\mu)$. Hence $|\alpha_\mu| \leqq \aleph_n |k_\mu| \leqq \aleph_n$. This proves (55) for all $\lambda < \omega_m$. Now, by (53) and the obvious relation $m \leqq n$,

$$|\omega_{n+1}| \leqq \sum [\lambda < \omega_m] |\alpha_\lambda| \leqq \aleph_n \aleph_m = \aleph_n$$

which is the required contradiction.

COROLLARY 2. *Let* $\aleph_n' = \aleph_n$; $2^{\aleph_\nu} < \aleph_n$ *for all* $\nu < n$. *Then*

$$\omega_n \rightarrow (\beta, \omega_n)^2 \qquad (\beta < \omega_n).$$

By Theorem 14, this proposition implies Theorem 8.

Deduction of Corollary 2 *from Theorem* 34. Let $\beta < \omega_n$. Suppose that $\omega_n \nrightarrow (\beta, \omega_n)^2$. Then, by Theorem 34, there are ordinals α_λ, k_λ such that $|k_\mu| = \prod [\lambda < \mu] |\alpha_\lambda|$ $(\mu < \beta^-)$;

$$\omega_n \nrightarrow (\alpha_0 + 1, \cdots)^1_{\beta^-}; \qquad \alpha_\mu \nrightarrow (\omega_n)^1_{k_\mu} \qquad (\mu < \beta^-).$$

Let us assume that, for some $\mu<\beta^-$, we have $\alpha_\lambda<\omega_n$ $(\lambda<\mu)$. Then, putting $d=\sum[\lambda<\mu]|\alpha_\lambda|$, $d<\aleph_n$; $|k_\mu|\leqq d^{|\mu|}\leqq 2^{d|\mu|}<\aleph_n$; $|\alpha_\mu| = \sum[\nu<k_\mu]|\rho_\nu|$, where $\rho_\nu<\omega_n$. Then $|\alpha_\mu|<\aleph_n$. This proves, by induction, that $\alpha_\lambda<\omega_n$ $(\lambda<\beta^-)$. Now, $|\omega_n|=\sum[\lambda<\beta^-]|\sigma_\lambda|$, where $|\sigma_\lambda|\leqq|\alpha_\lambda|$. Therefore $|\omega_n|<\aleph_n$ which is the required contradiction.

The proof of Theorem 34 depends on a lemma.

LEMMA 2. *Let T be a well ordered set, and $[T]^2=K_0+K_1$. Then there is*[8] *a set $B=B(T)\subset T$ which has the following properties. We have $[B]^2\subset K_1$. If $x\in T-B$, then there is $y\in B$ such that $\{y, x\}_<\in K_0$.*

PROOF. We may assume $T\neq 0$. Choose l such that $|l|>|T|$. We define, inductively, y_λ $(\lambda<l)$ as follows. Let $\mu<l$; $y_\lambda\in T$ $(\lambda<\mu)$.

Case 1. There is $y\in T$ such that $\{y_\lambda, y\}\in K_1$ $(\lambda<\mu)$. Then we take as y_μ the first element y of this kind.

Case 2. If $y\in T$, then there is $\lambda<\mu$ such that $\{y_\lambda, y\}\notin K_1$. Then we have $\mu>0$. We put $y_\mu=y_0$.

Let $B=\{y_\lambda:\lambda<l\}$. Then there is $m<l$ such that

$$B = \{y_\lambda:\lambda < m\}; \qquad \{y_\lambda, y_\mu\}_< \in K_1 \quad (\lambda < \mu < m).$$

For, m is the least μ such that $0<\mu<l$; $y_\mu=y_0$. We have $[B]^2\subset K_1$. Now let $x\in T-B$. Then, by definition of y_m, there is a least $\mu<m$ such that $\{y_\mu, x\}\in K_0$. Then $\{y_\lambda, x\}\in K_1$ $(\lambda<\mu)$ and hence, by definition of y_μ, $x>y_\mu$. This proves Lemma 2.

PROOF OF THEOREM 34. There is an ordered set S such that

$$\bar{S} = \alpha; \qquad [S]^2 = K_0 + K_1; \qquad \beta \notin [K_0]; \qquad \gamma \notin [K_1].$$

We choose an ordinal ρ such that $|\rho|>|\alpha|$. Let $x\in S$. We define $f_\mu(x)$ $(\mu<\rho)$ as follows. Let $\nu<\rho$, and suppose that

$$f_\mu(x) \in S \ (\mu < \nu),$$
$$\{ f_\mu(x), x\}_< \in K_0 \quad \text{if} \quad \mu < \nu; \ f_\mu(x) \neq x.$$

Then we define $f_\nu(x)$ by the following rule. If $f_\mu(x)=x$ for some $\mu<\nu$, then put $f_\nu(x)=x$. Now let $f_\mu(x)\neq x$ $(\mu<\nu)$. Let T be the set of all $y\in S$ such that $\{f_\mu(x), y\}_<\in K_0$ $(\mu<\nu)$. Then $x\in T$. Let $B=B(T)$ be the set given in Lemma 2. Then $B\subset T$; $[B]^2\subset K_1$; $\bar{B}<\gamma$. If $x\in B$, then put $f_\nu(x)=x$. Now let $x\notin B$. Then, by Lemma 2, there is a first element $z\in B$ such that $\{z, x\}_<\in K_0$. We put $f_\nu(x)=z$. We have now defined $f_\nu(x)$ for $\nu<\rho$; $x\in S$, and we have

[8] In fact, there is exactly one such set.

$$\{f_\mu(x), f_\nu(x)\}_< \in K_0 \quad (\mu < \nu < \rho; f_\mu(x) \neq x);$$
$$f_\nu(x) \leqq x \qquad (\nu < \rho; x \in S).$$

If, for some x, $f_\nu(x) < x$ $(\nu < \rho)$, then the contradiction

$$|\rho| = |\{f_\nu(x) : \nu < \rho\}| \leqq |S| = |\alpha|$$

follows. Hence, given $x \in S$, there is $\sigma(x) < \rho$ such that

$$f_\nu(x) < x \quad (\nu < \sigma(x)); \qquad f_{\sigma(x)}(x) = x \quad (x \in S).$$

Then, for fixed x, $[\{f_\nu(x) : \nu \leqq \sigma(x)\}]^2 \subset K_0$,

$$\sigma(x) + 1 < \beta; \qquad \sigma(x) < \beta^-.$$

Put $M_\nu = \{f_\nu(x) : \sigma(x) \geqq \nu\}$ $(\nu < \rho)$. Then $M_0 \subset B(S)$; $\overline{M}_0 < \gamma$;

$$S = \sum [\nu < \beta^-] M_\nu; \qquad \alpha \not\rightarrow (\overline{M}_0 + 1, \overline{M}_1 + 1, \cdots)^1_{\beta^-}.$$

Let $0 < \nu < \beta^-$. Then

$$M_\nu = \sum [y_\mu \in M_\mu \text{ for } \mu < \nu] \{f_\nu(x) : \sigma(x) \geqq \nu; f_\lambda(x) = y_\lambda \text{ for } \lambda < \nu\}.$$

Now, for every choice of $y_\mu \in M_\mu$ $(\mu < \nu)$, the corresponding term in the last sum is a set contained in some set of the form $B(T)$. In order to see this, consider an element x such that $\sigma(x) \geqq \nu$; $f_\mu(x) = y_\mu$ $(\mu < \nu)$. We shall have $f_\mu(x) < x$ $(\mu < \sigma(x))$ and hence $f_\mu(x) < x$ $(\mu < \nu)$. Let T be the set used above in the inductive definition of $f_\nu(x)$, i.e. the set of all $y \in S$ such that $\{f_\mu(x), y\}_< \in K_0$ $(\mu < \nu)$. Then, by definition of $f_\nu(x)$, $x \in T$ and either

$$x \in B(T); \qquad f_\nu(x) = x \in B(T)$$

or

$$x \notin B(T); \qquad f_\nu(x) = z \in B(T).$$

In either case, $f_\nu(x) \in B(T)$. In fact, the set T does not depend on x since T is the set of all $y \in S$ such that

$$\{y_\mu, y\}_< \in K_0 \qquad (\mu < \nu).$$

All this proves that, given $y_\mu \in M_\mu$ $(\mu < \nu)$, there is T such that, whenever

$$x \in S; \qquad \sigma(x) \geqq \nu; \qquad f_\mu(x) = y_\mu \qquad (\mu < \nu),$$

then

$$f_\nu(x) \in B(T).$$

By definition of $B(T)$, we have $[B(T)]^2 \subset K_1$ and therefore $\overline{B(T)} < \gamma$.

Hence M_ν is a sum of $\prod[\mu<\nu]|M_\mu|$ sets each of type less than γ. This shows that $\overline{M}_\nu \nrightarrow (\gamma)^1_{k_\nu}$, when k_ν is any ordinal such that $|k_\nu| = \prod[\mu<\nu]|M_\mu|$. It now follows that the conclusion of Theorem 34 holds for $\alpha_\lambda = \overline{M}_\lambda$ $(\lambda<\beta^-)$.

THEOREM 35. *Suppose that* $\beta \geqq r \geqq 3$; $\beta, \beta^* \nleqq \alpha$; $s>(r-1)^2$. *Then, for any type* ϕ *such that* $|\phi| = |\alpha|$,

$$\phi \nrightarrow (s, \beta)^r. \tag{56}$$

COROLLARY. *If* $r \geqq 3$; $s>(r-1)^2$, *then*

$$\eta \nrightarrow (s, \omega_0 + 1)^r, \tag{57}$$

$$\phi \nrightarrow (s, \omega_1)^r, \tag{58}$$

where ϕ *is any type such that* $|\phi| = |\lambda|$.

The negative results (57) and (58) are not too far from the ultimate truth as is seen by comparing them with the following positive results. By Theorem 1,

$$\omega_0 \to (\omega_0, \omega_0, \cdots, \omega_0)^r_k \qquad (k < \omega_0). \tag{59}$$

By Theorem 31,

$$\phi \to (\omega_0, \gamma)^2 \qquad (\gamma < \omega_1), \tag{60}$$

where ϕ is any type such that $|\phi| > \aleph_0$; $\omega_1, \omega_1^* \nleqq \phi$.

PROOF OF THEOREM 35. The corollary follows by applying the theorem to the following two cases.

(i) $\beta = \omega_0 + 1$; $\alpha = \omega_0$; $\phi = \eta$,

(ii) $\beta = \omega_1$; $\alpha = \lambda$.

The proof of the theorem depends on the following lemma due to Erdös and Szekeres [7]. Throughout, we put

$$s = (r-1)^2 + 1.$$

LEMMA 3. *If* S *is an ordered set*, $r>0$, *and if* $z(\sigma) \in S$ $(\sigma<s)$, *then there is* $\{\sigma_0, \sigma_1, \cdots, \sigma_{r-1}\}_< \subset [0, s]$ *such that either*

$$z(\sigma_\rho) \leqq z(\sigma_{\rho+1}) \qquad (\rho + 1 < r)$$

or

$$z(\sigma_\rho) \geqq z(\sigma_{\rho+1}) \qquad (\rho + 1 < r).$$

We now prove the theorem. Let $\overline{S}_< = \phi$; $\overline{S}_{<<} = \alpha$. Then

$$[S]^r = K_0 +' K_1; \qquad K_1 = K_{10} + K_{11},$$

where

$$K_{10} = \{\{x_0, \cdots, x_{r-1}\}_<: \{x_0, \cdots, x_{r-1}\}_{<<} \subset S\},$$
$$K_{11} = \{\{x_0, \cdots, x_{r-1}\}_<: \{x_0, \cdots, x_{r-1}\}_{>>} \subset S\}.$$

Case 1. There is $A \in [S]^s$ such that $[A]^r \subset K_0$. Then, if $A = \{z(\sigma): \sigma < s\}$, an application of Lemma 3 shows the existence of $B \in [A]^r$ such that $B \in K_1$, which is a contradiction.

Case 2. There is $A \subset S$ such that $\overline{A}_< = \beta$; $[A]^r \subset K_1$. We shall prove that one of the two relations

$$[A]^r \subset K_{10}, \tag{61}$$
$$[A]^r \subset K_{11} \tag{62}$$

holds. If both (61) and (62) are false, then there are sets X, $Y \subset A$ such that

$$X = \{x_0, \cdots, x_{r-1}\}_< = \{x_0, \cdots, x_{r-1}\}_{<<}, \tag{63}$$
$$Y = \{y_0, \cdots, y_{r-1}\}_< = \{y_0, \cdots, y_{r-1}\}_{>>}. \tag{64}$$

Then there is $\sigma < r$ such that $x_\rho = y_\rho$ $(\rho < \sigma)$; $x_\sigma \neq y_\sigma$. We choose X and Y such that σ is as large as possible. We may assume that $x_\sigma < y_\sigma$. Then, if we suppose that $\sigma + 1 < r$, we find that $\{y_0, \cdots, y_{\sigma-1}, x_\sigma, y_\sigma, \cdots, y_{r-2}\}_<$ and therefore, by the maximum property of σ, $\{y_0, \cdots, y_{\sigma-1}, x_\sigma, y_\sigma, \cdots, y_{r-2}\}_{<<}$. Hence $\{y_0, \cdots, y_{\sigma-1}, y_\sigma, \cdots, y_{r-2}\}_{<<}$ and therefore, since $r-2>0$, $y_0 \ll y_{r-2}$ which contradicts (64). Therefore $\sigma+1=r$, and $x_0 = y_0 \gg y_1 = x_1$. But this contradicts (63). This shows that at least one of the relations (61), (62) holds. Now (61) implies $\beta = \overline{A}_< = \overline{A}_{<<} \leqq \overline{S}_{<<} = \alpha$, and (62) implies $\beta^* = \overline{A}_> = \overline{A}_{<<} \leqq \overline{S}_{<<} = \alpha$. Both conclusions contradict the hypothesis. We have proved that neither Case 1 nor Case 2 is possible, so that (56) is established. This proves Theorem 35.

THEOREM 36. (i) *If* $a_\nu < d$ $(\nu < n)$, *then* $\sum[\nu < n]a_\nu \nrightarrow (|n|^+, d)^2$.
(ii) $a \nrightarrow (a'^+, a)^2$ *for* $a > 1$.
(iii) $ab \nrightarrow (a^+, b^+)^2$ *for any* a, b.
(iv) $\aleph_n \nrightarrow (|n|^+, \aleph_n)^2$, *if* $n = n^- > 0$.

PROOF OF (i). Let $|A_\nu| = a_\nu (\nu < n)$; $S = \sum' [\nu < n] A_\nu$;

$$[S]^2 = K_0 + {}'K_1; \qquad K_1 = \sum[\nu < n][A_\nu]^2.$$

Then $[X]^2 \subset K_0$ implies $|XA_\nu| \leqq 1$ $(\nu < n)$; $|X| \leqq |n|$, and $[X]^2 \subset K_1$ implies the existence of $\nu < n$ such that $X \subset A_\nu$; $|X| \leqq |A_\nu| < d$. This proves (i).

PROOF OF (ii). By definition of a', the hypothesis of (i) holds for some a_ν, n, with $|n| = a'$; $d = a = \sum a_\nu$. Hence (ii) follows from (i).

PROOF OF (iii). Let $|n| = a$; $a_\nu = b(\nu < n)$. Then, by (i),

$$ab = \sum a_\nu \nrightarrow (a^+, b^+)^2.$$

PROOF OF (iv). Since $\aleph_n = \sum [\nu < n] \aleph_\nu$, (iv) follows from (i). This proves Theorem 36.

THEOREM 37. (i) *Let* $a \geqq \aleph_0$, *and let* b *be minimal such that* $a^b > a$. *Let* $a < \aleph_k' \leqq \aleph_k \leqq a^b$. *Then*

(65) $$\aleph_{\omega_k} \nrightarrow (b^+, \aleph_{\omega_k})^2.$$

A possible value for $\aleph_k$ *is* a^+.

(ii) $\aleph_{\omega_{m+1}} \nrightarrow (\aleph_{m+1}, \aleph_{\omega_{m+1}})^2$ *for all* m.

(iii) *If* $\aleph_0 < \aleph_k' \leqq \aleph_k \leqq 2^{\aleph_0}$, *then* $\aleph_{\omega_k} \nrightarrow (\aleph_1, \aleph_{\omega_k})^2$. *A possible value is* $k = 1$.

Deduction of (ii) *from* (i). Let $a = \aleph_m$, and let b be minimal such that $a^b > a$. Then, if $k = m+1$, we have $a < \aleph_k' = \aleph_k \leqq a^b$ and therefore, by (i), $\aleph_{\omega_{m+1}} \nrightarrow (b^+, \aleph_{\omega_{m+1}})^2$. This implies (ii).

Deduction of (iii) *from* (i). Put, in (i), $a = b = \aleph_0$. Then (iii) follows.

Before proving (i) we establish a lemma. For the sake of further applications later on the lemma is more general than is needed for the present purpose.

LEMMA 4. *If* $s \geqq 2$; $\beta_0, \beta_1^* \nleqq \alpha_0$; $|\alpha_0| = |\alpha_1|$, *then*

$$\alpha_1 \nrightarrow (\beta_0, \beta_1, s+1, s+1, \cdots, s+1)^s_{s!}.$$

PROOF. Let $\bar{S}_< = \alpha_1$; $\bar{S}_{<<} = \alpha_0$. Then to every set $X \in [S]^s$ there belongs a permutation $\pi(X): \lambda \to \sigma(\lambda)$ defined by

$$X = \{x_0, x_1, \cdots, x_{s-1}\}_< = \{x_{\sigma(0)}, x_{\sigma(1)}, \cdots, x_{\sigma(s-1)}\}_{<<}.$$

Let π_λ $(\lambda < s!)$ be all permutations of $[0, s]$ and, in particular,

$$\pi_0: \lambda \to \lambda; \qquad \pi_1: \lambda \to s - 1 - \lambda \qquad (\lambda < s).$$

Then $[S]^s = \sum [\nu < s!] K_\nu$, where $K_\nu = \{X : X \in [S]^s; \pi(X) = \pi_\nu\}$. Now suppose that $A \subset S$; $\nu < s!$; $[A]^s \subset K_\nu$. We shall deduce a contradiction in each of the three cases that follow and so establish the lemma.

Case 1. $\nu = 0$; $\bar{A}_< = \beta_0$. Then the contradiction $\beta_0 = \bar{A}_{<<} \leqq \bar{S}_{<<} = \alpha_0$ follows.

Case 2. $\nu = 1$; $\bar{A}_< = \beta_1$. Then the contradiction $\beta_1^* = \bar{A}_{<<} \leqq \bar{S}_{<<} = \alpha_0$ follows.

Case 3. $2 \leqq \nu < s!$; $\bar{A}_< = s+1$. Let $\pi_\nu: \lambda \to \sigma(\lambda)$. Then $A = \{x_0, x_1, \cdots, x_s\}_<$ and therefore, putting $y_\lambda = x_{1+\lambda}$ $(\lambda < s)$, we have

$$(66)\qquad \{x_0, x_1, \cdots, x_{s-1}\}_< = \{x_{\sigma(0)}, x_{\sigma(1)}, \cdots, x_{\sigma(s-1)}\}_{<<},$$

$$\begin{aligned}\{x_1, x_2, \cdots, x_s\}_< &= \{y_0, y_1, \cdots, y_{s-1}\}_< \\ (67)\qquad &= \{y_{\sigma(0)}, y_{\sigma(1)}, \cdots, y_{\sigma(s-1)}\}_{<<} \\ &= \{x_{1+\sigma(0)}, x_{1+\sigma(1)}, \cdots, x_{1+\sigma(s-1)}\}_{<<}.\end{aligned}$$

If $x_0 \ll x_1$, then alternate applications of (66) and (67) lead to $x_0 \ll x_1 \ll x_2 \ll \cdots \ll x_s$ and so to the contradiction $\pi_\nu = \pi_0$, while, similarly, the assumption $x_0 \gg x_1$ leads to the contradiction $\pi_\nu = \pi_1$. This proves the lemma.

Proof of Theorem 37, (i). Let $a = \aleph_m$; $b = \aleph_l$, and let F be the set of all mappings $\lambda \to h(\lambda)$ of $[0, \omega_l]$ into $[0, \omega_m]$. We order F by putting, for $h_0, h_1 \in F$, $h_0 \ll h_1$ if, and only if, there is $\lambda_0 < \omega_l$ such that

$$h_0(\lambda) = h_1(\lambda) \quad \text{for} \quad \lambda < \lambda_0; \qquad h_0(\lambda_0) < h_1(\lambda_0).$$

Then $|F| = a^b$, and we have, by[9] Lemma 2 of [6], if $a = \aleph_m$; $b = \aleph_l$,

$$(68)\qquad \omega_{m+1}, \overset{*}{\omega_{l+1}} \nleq \overline{F}_{<<} = \overline{F}, \quad \text{say}.$$

We can choose a set $X \in [F]^{\aleph_k}$. Let $x \to f(x)$ be a one-one mapping of X on $[0, \omega_k]$, and $S = \{(x, \nu) : x \in X; \nu < \omega_{f(x)}\}$. We order S alphabetically, by means of a relation $u < v$, and put $\overline{S}_< = \overline{S} = \phi$. Then

$$|S| = \sum [x \in X] \aleph_{f(x)} \leq \sum [\lambda < \omega_k] \aleph_{\omega_k} = \aleph_{\omega_k}.$$

On the other hand, if $d < \aleph_{\omega_k}$, then $d < \aleph_n$, for some $n < \omega_k$, and there is $x_0 \in X$ such that $f(x_0) > n$. Then $|S| \geq \aleph_{f(x_0)} > d$. Hence

$$(69)\qquad |\phi| = |S| = \aleph_{\omega_k}.$$

1. Let $S_1 \subset S$, and suppose that $\overline{S}_1$ is an ordinal. Put $X_1 = \sum [\nu < \omega_k] \cdot \{x : (x, \nu) \in S_1\}$. Then $\overline{X}_1$ is an ordinal, and $\overline{X}_1 \leq \overline{F}$. Hence, by (68), $\overline{X}_1 < \omega_{m+1}$; $|X_1| \leq \aleph_m = a < \aleph_k'$.

$$\sum [x \in X_1] |f(x)| < \aleph_k; \qquad \nu_1 = \sum [x \in X_1] f(x) < \omega_k;$$

$$\begin{aligned}|S_1| &= \sum [x \in X_1] |\{\nu : (x, \nu) \in S_1\}| \leq \sum [x \in X_1] \aleph_{f(x)} \\ &\leq \aleph_{\nu_1} |X_1| \leq \aleph_{\nu_1} \aleph_m < \aleph_{\omega_k};\end{aligned}$$

$$(70)\qquad \omega_{\omega_k} \nleq \phi.$$

2. Let $S_2 \subset S$, and suppose that $(\overline{S}_2)^*$ is an ordinal. Put $X_2 = \sum [\nu < \omega_k] \{x : (x, \nu) \in S_2\}$. Then $(\overline{X}_2)^*$ is an ordinal, and $(\overline{X}_2)^* \leq \overline{F}$. Hence, by (68), $(\overline{X}_2)^* < \omega_{l+1}$; $|X_2| \leq \aleph_l$. Put, for $x \in X_2$, $N(x) = \{\nu : (x, \nu) \in S_2\}$. Then $\overline{N(x)}$ is an ordinal. On the other hand,

[9] The authors are indebted to G. Kurepa for pointing out that the result of this lemma had already been obtained by F. Hausdorff, [9, Satz 14].

$(\overline{N(x)})^*$ is an ordinal, since $(\overline{N(x)})^* \leqq (\bar{S}_2)^*$. Hence $\overline{N(x)} < \omega_0$;

$$|S_2| = \sum[x \in X_2]|N(x)| \leqq \aleph_0 |X_2| \leqq \aleph_l; \qquad (\bar{S}_2)^* < \omega_{l+1};$$

(71) $$\overset{*}{\omega}_{l+1} \nleqq \phi.$$

3. We now apply Lemma 4 to the case

$$s = 2; \qquad \alpha_0 = \phi; \qquad \alpha_1 = \omega_{\omega_k}; \qquad \beta_0 = \omega_{\omega_k}; \qquad \beta_1 = \omega_{l+1}.$$

Its hypotheses are satisfied, by (70), (71), (69). We obtain $\omega_{\omega_k} \nrightarrow (\omega_{\omega_k}, \omega_{l+1})^2$. This implies (65), by Theorem 14, and completes the proof of Theorem 37.

REMARK. If $a \geqq \aleph_0$ and $\aleph_k = a^+$, then (i) of Theorem 37 yields a stronger result then (ii) of Theorem 36. For, first of all, we note that the hypothesis of (i) of Theorem 37 holds, since $a < a^+ = a^{+\prime} = \aleph_k' = \aleph_k \leqq a^b$. Hence the latter theorem gives

(72) $$\aleph_{\omega_k} \nrightarrow (b^+, \aleph_{\omega_k})^2.$$

On the other hand, (ii) of Theorem 36 gives

(73) $$\aleph_{\omega_k} \nrightarrow (\aleph_{\omega_k}^{\prime +}, \aleph_{\omega_k})^2.$$

It is known that, for any m,

(74) $$\aleph_{\omega_m}' = \aleph_m'.$$

Hence $\aleph_{\omega_k}^{\prime +} > \aleph_{\omega_k}' = \aleph_k' = a^{+\prime} = a^+ \geqq b^+$, and (72) is stronger than (73).

Since we were not able to find a reference for (74) we give, for the sake of completeness, a proof now.

Case 1. Let $\aleph_{\omega_m}' = \aleph_n < \aleph_m'$. Then $\aleph_{\omega_m} = \sum[\nu < \omega_n]\aleph_{\lambda_\nu}$, for some $\lambda_\nu < \omega_m$. Put $\lambda = \sum[\nu < \omega_n]\lambda_\nu$. Then, since $|\omega_n| < \aleph_m'$; $|\lambda_\nu| < \aleph_m$ we conclude that

$$|\lambda| = \sum[\nu < \omega_n]|\lambda_\nu| < \aleph_m; \qquad \aleph_n < \aleph_m' \leqq \aleph_m \leqq \aleph_{\omega_m};$$

$$\aleph_{\omega_m} \leqq \sum[\nu < \omega_n]\aleph_\lambda = \aleph_\lambda \aleph_n < \aleph_{\omega_m}$$

which is the desired contradiction.

Case 2. Let $\aleph_{\omega_m}' > \aleph_n = \aleph_m'$. Then $m > 0$, and $\aleph_m = \sum[\nu < \omega_n]\aleph_{\lambda_\nu}$ for some $\lambda_\nu < m$. Then $\sum[\nu < \omega_n]\aleph_{\omega_{\lambda_\nu}} = \aleph_l$ for some $l < \omega_m$; $\aleph_{\lambda_\nu} = |\omega_{\lambda_\nu}| \leqq |l|$; $\aleph_m = \sum[\nu < \omega_n]\aleph_{\lambda_\nu} \leqq |l|\aleph_n$; $m \leqq n$; $\aleph_{\omega_m} = \sum[\mu < \omega_m]\aleph_\mu$; $\aleph_{\omega_m}' \leqq |\omega_m| \leqq \aleph_n$ which, again, is a contradiction. This proves (74).

THEOREM 38. *If* $\aleph_m \nrightarrow (|\beta_0|, |\beta_1|, \cdots)_k^r$, *then*

(75) $$\omega_{m+1} \nrightarrow (\beta_0 + 1, \beta_1 + 1, \cdots)_k^{r+1}.$$

We give some applications of this theorem.

(i) If $|\beta| = |\gamma| = \aleph_{m+1}$, then

(76) $$\omega_{m+2} \nrightarrow (\beta + 1, \gamma + 1)^3.$$

For, let $a = \aleph_m$, and let b be minimal such that $a^b > a$. Then, by Theorem 7, $a^b \nrightarrow (a^+, b^+)^2$ and therefore $\aleph_{m+1} \nrightarrow (|\beta|, |\gamma|)^2$. Now (76) follows from Theorem 38.

(ii) If $\aleph_n' = \aleph_m$; $|\beta| = \aleph_{m+1}$; $|\gamma| = \aleph_{\omega_n}$, then

(77) $$\omega_{\omega_n+1} \nrightarrow (\beta + 1, \gamma + 1)^3.$$

In order to prove (77), we apply Theorem 36 (ii) to $a = \aleph_{\omega_n}$. We note that, by (74), $a' = \aleph_n' = \aleph_m$; $a'^+ = \aleph_{m+1}$. Hence, by Theorem 36, $\aleph_{\omega_n} \nrightarrow (|\beta|, |\gamma|)^2$, and (77) follows from Theorem 38.

(iii) If $|\beta| = \aleph_{k+1}$; $|\gamma| = \aleph_{\omega_{k+1}}$, then

(78) $$\omega_{\omega_{k+1}+1} \nrightarrow (\beta + 1, \gamma + 1)^3.$$

This follows immediately from Theorem 37 (ii) and an application of Theorem 38.

We note that on putting $n = k+1$ in (ii) above one obtains a result which is weaker than (78). For, (ii) becomes: if $\aleph_{k+1} = \aleph_m$; $|\beta| = \aleph_{k+2}$; $|\gamma| = \aleph_{\omega_{k+1}}$, then (78) holds.

The proof of Theorem 38 depends on a lemma.

LEMMA 5. *Let α be an ordinal. Suppose that β_ν $(\nu < k)$ and r are such that, whenever $\beta < \alpha$, then $\beta \nrightarrow (\beta_0, \beta_1, \cdots)_k^r$. Then*

(79) $$\alpha \nrightarrow (\beta_0 + 1, \beta_1 + 1, \cdots)_k^{r+1}.$$

PROOF. Let $\overline{S} = \alpha$; $x \in S$. Then $\overline{L(x)} < \alpha$, and hence, by hypothesis, there is a partition $[L(x)]^r = \sum [\nu < k] K_\nu(x)$ such that, whenever $X \subset L(x)$; $[X]^r \subset K_\nu(x)$, then $\overline{X} < \beta_\nu$. Put $K_\nu = \{A + \{x\} : x \in S; A \in K_\nu(x)\}$ $(\nu < k)$. Then $[S]^{r+1} = \sum [\nu < k] K_\nu$. If we now assume that

(80) $$S' \subset S; \quad [S']^{r+1} \subset K_\nu; \quad \overline{S'} = \beta_\nu + 1,$$

then $S' = S'' + \{x'\}$; $S'' \subset L(x')$; $\overline{S''} = \beta_\nu$; $[S'']^r \subset K_\nu(x')$ which contradicts the definition of $K_\nu(x')$. Hence (80) is impossible, and (79) follows.

PROOF OF THEOREM 38. If $\beta < \omega_{m+1}$, then $|\beta| \leq \aleph_m$, $|\beta| \nrightarrow (|\beta_0|, |\beta_1|, \cdots)_k^r$. By Theorem 13, this implies $\beta \nrightarrow (\beta_0, \beta_1, \cdots)_k^r$. Now (75) follows from Lemma 5.

THEOREM 39. (i) *If*

(81) $$l \rightarrow (\alpha_0, \alpha_1, \cdots)_k^r,$$

(82) $$|m| > \sum[\lambda < l]\,|k|^{|\lambda|^r},$$

then

(83) $$m \to (\alpha_0 + 1, \alpha_1 + 1, \cdots)_k^{r+1}.$$

(ii) *If* $r>0$; $\omega_n \to (\alpha_0, \alpha_1, \cdots)_k^r$, *and*

(84) $$2^{\aleph_\nu} \leqq \aleph_n \qquad \textit{for } \nu < n,$$

then

$$\omega_{n+1} \to (\alpha_0 + 1, \alpha_1 + 1, \cdots)_k^{r+1}.$$

(iii) *If* $|k| < \aleph_m'$; $r \geqq 0$, *and if* $2^{\aleph_\nu} \leqq \aleph_n$ *for* $m \leqq n < m+r$; $\nu < n$, *then*

(85) $$\omega_{m+r} \to (\omega_m + r)_k^{r+1},$$

(86) $$\omega_{m+r} \to (\omega_m + r)_k^r.$$

Deduction of (ii) *from* (i). Let the hypothesis of (ii) hold. If $|\alpha_\nu| < r$ for some $\nu < k$, then the assertion is trivial, by Theorem 22. Hence we may assume that $|\alpha_\nu| \geqq r$ $(\nu < k)$. Next, suppose that $|\alpha_\nu| = r$ for some $\nu < k$. Then, by Theorem 17, we may apply a suitable permutation to the system $\alpha_0, \alpha_1, \cdots$ so that for the new system, again denoted by $\alpha_0, \alpha_1, \cdots, \alpha_\nu, \cdots$ $(\nu < k)$, we have

$$\alpha_\nu = r \quad (\nu < k_0); \qquad |\alpha_\nu| > r \qquad (k_0 \leqq \nu < k).$$

Here k_0 is some ordinal, $0 < k_0 \leqq k$, and we can write $k = k_0 + k_1$. Then, by Theorem 20, (ii), the hypothesis implies $\omega_n \to (\alpha_{k_0}, \alpha_{k_0+1}, \cdots)_{k_1}^r$, and the assertion is implied by

$$\omega_{n+1} \to (\alpha_{k_0} + 1, \alpha_{k_0+1} + 1, \cdots)_{k_1}^{r+1}.$$

This shows that we may assume, without loss of generality, that $|\alpha_\nu| > r$ for $\nu < k$. Let us, now, suppose that $|k| \geqq \aleph_n$. Then, if $A = [0, \omega_n]$, we can write $[A]^r = \sum[\nu < k]K_\nu$, where $|K_\nu| \leqq 1$ for $\nu < k$. Since $\overline{A} \to (\alpha_0, \cdots)_k^r$, there are $X \subset A$, $\nu < k$ such that $\overline{X} = \alpha_\nu$; $[X]^r \subset K_\nu$. Then $|X| = |\alpha_\nu| > r$; $1 < |[X]^r| \leqq |K_\nu|$ which is a contradiction. This proves that $|k| < \aleph_n$.

Put

$$\sum[\lambda < \omega_n]\,|k|^{|\lambda|^r} = a.$$

It suffices to show that $|\omega_{n+1}| > a$.

If $n = 0$, then $a < \aleph_0$. If $n > 0$, then

$$a \leqq \sum[\lambda < \omega_n]2^{|k||\lambda|^r} \leqq \sum[\lambda < \omega_n]2^{\aleph_{\nu_\lambda}},$$

for some $\nu_\lambda < n$. Hence, by (84), $a \leqq \sum[\lambda < \omega_n]\aleph_n = \aleph_n$.

Deduction of (iii) *from* (ii). By definition of $\aleph'_m$, $\omega_m \to (\omega_m)^1_k$. Hence, by r applications of (ii), (85) follows. Now (86) follows from Theorem 15.

PROOF OF (i). Let $\overline{S}=m$; $[S]^{r+1}=\sum' [\nu<k]K_\nu$. We choose n such that $|n|>|m|$. Throughout this proof the letters κ, λ, ρ, σ denote ordinals less than n, and x, y and z elements of S. The relation

$$\{x_0, \cdots, x_r\} \equiv \{y_0, \cdots, y_r\}$$

expresses, by definition, the fact that, for some $\nu<k$,

$$\{x_0, \cdots, x_r\}, \{y_0, \cdots, y_r\} \in K_\nu.$$

We define $f_\kappa(x)\in S$ as follows. Let x be fixed, and suppose that, for some fixed λ, the elements $f_\kappa=f_\kappa(x)$ have already been defined for all $\kappa<\lambda$. Then we put $f_\lambda(x)=x$, if $f_\kappa=x$ for some $\kappa<\lambda$. If, on the other hand, $f_\kappa\neq x$ for $\kappa<\lambda$, then we define f_λ to be the first element y of $S-\{f_\kappa:\kappa<\lambda\}$ such that

$$(87)\quad \{f_{\kappa_0}, \cdots, f_{\kappa_{r-1}}, y\} \equiv \{f_{\kappa_0}, \cdots, f_{\kappa_{r-1}}, x\} \text{ for } \kappa_0 < \cdots < \kappa_{r-1}<\lambda.$$

This defines f_κ for all κ. We now prove that

$$(88)\qquad f_\lambda < f_\mu$$

if

$$(89)\qquad \lambda < \mu; \qquad f_\lambda \neq x.$$

First of all, (87) holds for $y=x$. Hence, by (89) and the definition of f_λ, we have $f_\lambda<x$. This proves (88) in the case when $f_\mu=x$. Now suppose that $f_\mu\neq x$. Then (87) holds for $y=f_\mu$ and, again, (88) follows. By (88) and $|n|>|m|$, there is $\rho(x)$ such that

$$f_\kappa(x) < f_\lambda(x) = x, \qquad \text{if } \kappa < \rho(x) \leqq \lambda.$$

Let, for $\kappa_0< \cdots <\kappa_{r-1}<\rho(x)$,

$$\{f_{\kappa_0}(x), \cdots, f_{\kappa_{r-1}}(x), x\} \in K_{\varrho(\kappa_0,\cdots,\kappa_{r-1},x)} = K'(\kappa_0, \cdots, x).$$

We now show that if x and z are such that

$$(90)\qquad \rho(x) = \rho(z),$$

$$(91)\qquad \begin{aligned} &K'(\kappa_0, \cdots, \kappa_{r-1}, x) \\ &\quad = K'(\kappa_0, \cdots, \kappa_{r-1}, z) \quad \text{for } \kappa_0 < \cdots <\kappa_{r-1} < \rho(x), \end{aligned}$$

then $x=z$. Let $\lambda\leqq\rho(x)$, and suppose that

$$(92)\qquad f_\kappa(x) = f_\kappa(z) \qquad \text{for } \kappa < \lambda.$$

Then $f_\lambda(x)$ is the first element y of $S-\{f_\kappa(x):\kappa<\lambda\}$ such that (87) holds, i.e.

$$\{f_{\kappa_0}(x), \cdots, f_{\kappa_{r-1}}(x), y\} \in K'(\kappa_0, \cdots, \kappa_{r-1}, x) \quad \text{for } \kappa_0 < \cdots < \kappa_{r-1} < \lambda.$$

Now, $f_\lambda(z)$ is defined by the same property, with z in place of x, and (90) and (91) show that $f_\lambda(x)=f_\lambda(z)$. We have thus proved, by induction, that $f_\kappa(x)=f_\kappa(z)$ for all $\kappa \leqq \rho(x)$. In particular, by (90),

$$x = f_{\rho(x)}(x) = f_{\rho(z)}(z) = z.$$

We next prove that $\rho(x_0) \geqq l$ for at least one x_0. Let us suppose, on the contrary, that $\rho(x)<l$ for all x. Let $\sigma<l$. Then the cardinal $a(\sigma)$ of the set $\{x:\rho(x)=\sigma\}$ is at most equal to the cardinal of the set of all functions $h(\kappa_0, \cdots, \kappa_{r-1})<k$, defined for $\kappa_0 < \cdots < \kappa_{r-1} < \sigma$. Hence

$$a(\sigma) \leqq |k|^{|\sigma|^r}; \qquad |m| = |S| = |\sum[\sigma<l]\{x:\rho(x)=\sigma\}|$$
$$\leqq \sum[\sigma<l]|k|^{|\sigma|^r},$$

which contradicts (82). This proves that $\rho(x_0) \geqq l$ for some suitable x_0. Put $S_0=\{f_\kappa(x_0):\kappa<l\}$. Then $\overline{S}_0=l$; $[S_0]^r=\sum[\nu<k]K_\nu''$, where $K_\nu''=\{A:A\in[S_0]^r;\ A+\{x_0\}\in K_\nu\}$ $(\nu<k)$. By (81), there are $S_1\subset S_0$; $\nu<k$ such that $\overline{S}_1=\alpha_\nu$; $[S_1]^r\subset K_\nu''$. Then the set $S_2=S_1+\{x_0\}$ satisfies $\overline{S}_2=\alpha_\nu+1$; $[S_2]^{r+1}\subset K_\nu$, and (83) follows. This completes the proof of Theorem 39.

COROLLARY OF THEOREM 39. *Given any r and any ordinals k, β_ν, there always exists an ordinal α such that*

$$\alpha \to (\beta_0, \beta_1, \cdots)_k^r.$$

For, if $r\leqq 1$, then any α can be taken such that $|\alpha| > \sum[\nu<k]|\beta_\nu|$, and the result for $r=2, 3, \cdots$ is obtained by applications of Theorem 39. The relation (5) shows that the last proposition becomes false if α and β_ν, instead of being ordinals, are allowed to be any order types. Later on (Theorem 45) the corollary will be extended, for $r=1$, to the case of arbitrary types β_ν.

We mention, without proof, the following further applications of Theorem 39 (i).

(a) $(2^a)^+ \to (a)_k^2$ if $|k|<a'$.

(b) $a^+ \to (a)_k^2$; $\omega_{n+1} \to (\omega_n+1)_k^2$,

if $a=\aleph_n$; $|k|<a'$, and $2^b \leqq a$ for all $b<a$.

(c) $(k^{\alpha_0+\cdots+\alpha_{k-1}-2k+1}+k-2)(k-1)^{-1} \to (\alpha_0, \cdots, \alpha_{k-1})_k^2$, if $2\leqq k$, $\alpha_0, \cdots, \alpha_{k-1}<\omega_0$.

If in Theorem 39 we take as k, l, α_ν finite numbers we obtain a result which implies Theorem 1 of [5]. Denote, in this special case, by

$$\rho_k(r;\alpha_0,\alpha_1,\cdots,\alpha_{k-1})$$

the least number n such that

$$n\rightarrow(\alpha_0,\cdots,\alpha_{k-1})_k^r.$$

Without loss of generality, we restrict ourselves to the case $k\geqq 2$; $0<r\leqq\alpha_\nu$. In [5, Theorem 1], an explicit upper estimate was given for the number $\rho_k(r;\alpha,\alpha,\cdots,\alpha)$, which, in that paper, was denoted by $R(k,r,\alpha)$.

Clearly, $\rho_k(1;\alpha_0,\cdots,\alpha_{k-1})=1+\alpha_0+\cdots+\alpha_{k+1}-k$. By Theorem 39,

$$\rho_k(r+1;\alpha_0+1,\alpha_1+1,\cdots,\alpha_{k-1}+1) \leqq 1+\sum[\lambda<\rho_k(r;\alpha_0,\cdots,\alpha_{k-1})]k^{\lambda^r}. \tag{93}$$

It is easily proved that, for $l<\omega_0$,

$$1+\sum[\lambda<l]k^{\lambda^r}\leqq k^{l^r}. \tag{94}$$

For, (94) holds for $l=0$, and if $0<m<\omega_0$, and (94) holds for $l=m-1$, then

$$1+\sum[\lambda<m]k^{\lambda^r}\leqq k^{(m-1)^r}+k^{(m-1)^r}\leqq k^{1+(m-1)^r}\leqq k^{m^r},$$

so that (94) holds for $l=m$. We have thus proved the following recurrence relation.

THEOREM 40. *If* $2\leqq k<\omega_0$; $0<r\leqq\alpha_\nu<\omega_0$ $(\nu<k)$, *and if* ρ_k *is defined as above, then*

$$\rho_k(r+1;\quad\alpha_0+1,\cdots,\alpha_{k-1}+1)\leqq k^{\rho_k(r;\alpha_0,\cdots,\alpha_{k-1})^r}.$$

In particular, we have, using the notation of [5],

$$R(k,r+1,\alpha+1)\leqq k^{R(k,r,\alpha)^r}\quad(k\geqq 2;0<r\leqq\alpha).$$

This is precisely the recurrence relation established in [5], from which the explicit estimate is deduced at once. This is no coincidence, as the method of proof of the present Theorem 39 is related to that used for proving Theorem 1 of [5].

Theorem 39 implies Theorem 4 (i), i.e.

$$(2^{\aleph_n})^+\rightarrow(\aleph_{n+1})^2_{\omega_n}. \tag{95}$$

For, clearly, $\aleph_{n+1}\rightarrow(\aleph_{n+1})^1_{\omega_n}$, and therefore $\omega_{n+1}\rightarrow(\omega_{n+1})^1_{\omega_n}$. Also,

$$\sum[\lambda<\omega_{n+1}]\,|\,\omega_n\,|^{|\lambda|}\leqq\aleph_n^{\aleph_n}\aleph_{n+1}=2^{\aleph_n}=\aleph_{m_0},$$

say. Hence, by Theorem 39 (i), $\omega_{m_0+1}\rightarrow(\omega_{n+1}+1)^2_{\omega_n}$, and (95) follows.

THEOREM 41. *If $r \geqq 3$, then, for all n,*

(96) $$\omega_{n+1} \nrightarrow (\omega_n + 2, \omega_0 + 1, r+1, r+1, \cdots, r+1)^{r}_{(r-1)!}.$$

As an application, consider the case $r=3$; $n=0$:

(97) $$\omega_1 \nrightarrow (\omega_0 + 2, \omega_0 + 1)^3.$$

This should be compared with:

$$\omega_1 \rightarrow (\omega_0 + 1)^{r}_{k} \qquad (k < \omega_0;\ r \geqq 0)$$

which follows from Theorem 39 (ii) and Theorem 1.

PROOF OF THEOREM 41. Let $\omega_n \leqq \beta < \omega_{n+1}$. We apply Lemma 4 to

$$s = r - 1; \quad \alpha_0 = \omega_n; \quad \alpha_1 = \beta; \quad \beta_0 = \omega_n + 1; \quad \beta_1 = \omega_0$$

and obtain

$$\beta \nrightarrow (\omega_n + 1, \omega_0, r, \cdots, r)^{r-1}_{(r-1)!}.$$

This holds, a fortiori, if $\beta < \omega_n$. Now Lemma 5 proves (96).

A type β is called *indecomposable* if the equation $\beta = \gamma + \delta$ implies that either $\gamma \geqq \beta$ or $\delta \geqq \beta$. It is known[10] that the indecomposable ordinals are those of the form ω_0^{λ}. The types η and λ are indecomposable. The next theorem asserts that in Lemma 4 the $s!-2$ classes corresponding to the entries $s+1$ in the partition relation may be suppressed in the special case when both β_0 and β_1 are indecomposable, at the cost, however, of raising the remaining entries slightly.

THEOREM 42. *Let $s \geqq 3$; $|\alpha_0| = |\alpha_1|$; $\beta_0, \beta_1^* \nleqq \alpha_0$, and suppose that β_0 and β_1 are indecomposable. Then*

(98) $$(s-3) + \alpha_1 \nrightarrow ((s-3) + \beta_0, (s-3) + \beta_1)^s.$$

PROOF. *Case* 1. $s=3$. Consider a set S with two orders such that $\bar{S}_< = \alpha_1$; $\bar{S}_{<<} = \alpha_0$. Then $[S]^3 = K_0 +' K_1$, where K_0 is the set of all sets $\{x_0, x_1, x_2\}_< = \{y_0, y_1, y_2\}_{<<} \subset S$ for which $x_\lambda \rightarrow y_\lambda$ is an even permutation of $[0, 3]$, i.e. one of the permutations 012, 120, 201. Now let us assume that

(99) $$\alpha_1 \rightarrow (\beta_0, \beta_1)^3.$$

It suffices to deduce a contradiction in each of the two cases that follow.

Case 1.1. There is $A \subset S$ such that $\bar{A}_< = \beta_0$; $[A]^3 \subset K_0$. Let x, y, z denote elements of A. Then $\{x, y, z\}_< = \{x_1, y_1, z_1\}_{<<}$ implies that x_1, y_1, z_1 is a cyclic permutation of x, y, z. Put $B = \{x : y \ll x$, whenever

[10] [13, §§75–78].

$y<x\}$; $C=A-B$. We shall prove three propositions about the two orders of A showing their effects on the partition $A=B+C$.

1. Let $x<y\in B$; $x\in C$. Then there is z such that $z<x\ll z$; $\{z, x, y\}_< = \{x, y, z\}_{<<}$; $z<y\ll z$; $y\in C$ which is false. Hence $x<y\in B$ implies $x\in B$, as well as $x\ll y$. Therefore $x>y\in C$ implies $x\in C$.

2. Let $x\in B$; $y\in C$; $x\ll y$. Then $x<y$. There is z such that $z<y\ll z$. Then $\{x, y, z\}_{<<}=\{z, x, y\}_<$; $z<x\ll z$; $x\in C$ which is false. Hence $x\in B$; $y\in C$ implies $y\ll x$.

3. Let $x, y\in C$; $x<y\ll x$. Then there is z such that $z<x\ll z$; $\{z, x, y\}_<=\{y, x, z\}_{<<}$ which contradicts the definition of K_0. Hence $x, y\in C$; $x<y$ implies $x\ll y$.

The results of 1, 2 and 3 show that, if we put $\overline{B}_<=\gamma_0$; $\overline{C}_<=\gamma_1$, then $\beta_0=\overline{A}_<=\gamma_0+\gamma_1$; $\overline{A}_{<<}=\gamma_1+\gamma_0$. Since β_0 is indecomposable, it follows that $\gamma_\nu\geqq\beta_0$ for some $\nu<2$. Then

$$\beta_0 \leqq \gamma_\nu \leqq \overline{A}_{<<} \leqq \overline{S}_{<<} = \alpha_0 \tag{100}$$

which is a contradiction.

Case 1.2. There is $A\subset S$ such that $\overline{A}_<=\beta_1$; $[A]^3\subset K_1$. Then $\{x, y, z\}_<=\{x_1, y_1, z_1\}_{<<}\subset A$ implies that x_1, y_1, z_1 is an odd permutation of x, y, z. This is equivalent to saying that $\{x, y, z\}_<=\{x_2, y_2, z_2\}_{>>}\subset A$ implies that x_2, y_2, z_2 is an even permutation of x, y, z. Hence the result of Case 1.1 holds if β_0 is replaced by β_1, and "$\ll$" by "$\gg$". We note that β_1^* is indecomposable. Hence, in place of (100) we have

$$\beta_1 \leqq \overline{A}_{>>} \leqq \overline{S}_{>>} = \overset{*}{\alpha_0}$$

which is a contradiction. This shows that the assumption (99) was false, i.e. that (98) holds.

Case 2. $s>3$. Then, by the result of Case 1, we have $\alpha_1\nrightarrow(\beta_0, \beta_1)^3$. By Theorem 15, this implies (98). This completes the proof of Theorem 42.

REMARK. If, in particular, β_0 and β_1 are ordinals, not zero, then $(s-3)+\beta_\nu=\beta_\nu$, so that (98) can be replaced by

$$(s - 3) + \alpha_1 \nrightarrow (\beta_0, \beta_1)^s.$$

We may also mention here the following corollary of two of our lemmas, in which λ is the type of the continuum.

(101) *If* $2^{\aleph_0} = \aleph_n$, *then* $\omega_{n+1} \nrightarrow (\omega_1 + 1, \omega_1 + 1)^3$.

PROOF. Let $\omega_n\leqq\alpha_1<\omega_{n+1}$. Then, by Lemma 4, with $\alpha_0=\lambda$, we have $\alpha_1\nrightarrow(\omega_1, \omega_1)^2$. By Lemma 5 this leads to (101).

THEOREM 43. *If* $r<s\leqq\beta_0$; $\alpha\nrightarrow(\beta_0)^r_k$; $\beta_1\rightarrow(s)^r_k$, *then*

$$\alpha \rightarrow (\beta_0, \beta_1)^s.$$

This proposition remains valid if the types α, β_0, β_1 *are replaced by cardinals.*

PROOF. Let $r<s\leqq\beta_0$; $\alpha\rightarrow(\beta_0,\beta_1)^s$; $\beta_1\rightarrow(s)^r_k$. We have to deduce that

(102) $$\alpha \rightarrow (\beta_0)^r_k.$$

Let $\bar{S}=\alpha$; $[S]^r=\sum'[\nu<k]K_\nu$. Then $[S]^s=K_0'+'K_1'$, where

$$K_0' = \sum[\nu < k]\{A : A \in [S]^s; \quad [A]^r \subset K_\nu\}.$$

Then there are $B\subset S$; $\lambda<2$ such that $[B]^s\subset K_\lambda'$; $\bar{B}=\beta_\lambda$. If $\lambda=1$, then $\bar{B}\rightarrow(s)^r_k$, and therefore there are $A\in[B]^s$; $\nu<k$ such that $[A]^r\subset K_\nu$. Then $A\in K_0'$; $A\notin K_1'$, which is false. Hence $\lambda=0$. Let $\{X, Y\}_{\neq}\subset[B]^r$. Then we can write $X=\{x_0, \cdots, x_{r-1}\}$; $Y=\{x_m, \cdots, x_{m+r-1}\}$, where $1\leqq m\leqq r$; $\{x_0, \cdots, x_{m+r-1}\}_{\neq}$. Put

$$X_\mu = \{x_\mu, \cdots, x_{\mu+r-1}\} \qquad (\mu \leqq m).$$

Now let $\mu<m$. Then, since $|B|=|\beta_0|\geqq s>r$, there is $Y_\mu\in[B]^s$ such that $X_\mu+X_{\mu+1}\subset Y_\mu$. But $Y_\mu\in K_0'$, so that X_μ, $X_{\mu+1}\in[Y_\mu]^r\subset K_{\nu_\mu}$, for some $\nu_\mu<k$. Then $\nu_0=\nu_1=\cdots=\nu_{m-1}$; $X=X_0\in K_{\nu_0}$; $Y=X_m\in K_{\nu_{m-1}}=K_{\nu_0}$. Since X and Y are arbitrary, it follows that ν_0 is independent of X and Y, and that $[B]^r\subset K_{\nu_0}$. This proves (102). The analogous theorem, with cardinals in place of types, is proved by means of the obvious modifications of the above argument.

Applications of Theorem 43. (a) Let λ be the type of the continuum and $|\lambda|=\aleph_n$. Then, by Theorem 30, $\aleph_n\nrightarrow(\aleph_1)^2_2$. Also, as is easily verified,

(103) $$6 \rightarrow (3)^2_2.$$

Hence, by Theorem 43, $\aleph_n\nrightarrow(\aleph_1, 6)^3$, and therefore $\omega_n\nrightarrow(\omega_1, 6)^3$. Now, by Theorem 15, $\omega_n\nrightarrow(\omega_1, r+3)^r$ $(r\geqq3)$ follows and therefore, finally,

(104) $$2^{\aleph_0} \nrightarrow (\aleph_1, r + 3)^r \qquad (r \geqq 3).$$

(b) By (97),

(105) $$\omega_1 \nrightarrow (\omega_0 + 2)^3_2.$$

By (103) and Theorem 39, we have

(106) $$m \rightarrow (4)^3_2,$$

where $m=1+\sum[\mu<6]2^{\mu^2}<2^{26}$. It now follows from (105) and (106), by Theorem 43, that $\omega_1 \not\rightarrow (\omega_0+2, 2^{26})^4$ and therefore, by Theorem 15, that

$$(107)\qquad \omega_1 \not\rightarrow (\omega_0 + 2, 2^{26} + r - 4)^r \qquad (r \geqq 4).$$

We now give a new proof of the theorem of Dushnik and Miller [2].[11] Our proof bears some resemblance to the original proof but can, we think, be followed more easily.

THEOREM 44. *If* $a \geqq \aleph_0$, *then* $a \rightarrow (\aleph_0, a)^2$.

PROOF. We use induction with respect to a. By Theorem 1, the assertion is true for $a = \aleph_0$. We assume that $n > 0$ and that the assertion is true for $a < \aleph_n$, and we let

$$|S| = b = \aleph_n > \aleph_0; \qquad [S]^2 = K_0 + K_1.$$

We suppose that

$$\text{if} \quad X \in [S]^{\aleph_0}, \quad \text{then} \quad [X]^2 \not\subset K_0,$$

and we want to find $Y \in [S]^b$ such that $[Y]^2 \subset K_1$.

There is a maximal set $A = \{x_\nu : \nu < l\} \subset S$ such that $l < \omega_0$,[12]

$$(108)\qquad \begin{aligned} &x_\nu \in U_0(x_0, \cdots, x_{\nu-1}),\\ &|U_0(x_0, \cdots, x_\nu)| = b \end{aligned} \qquad (\nu < l).$$

For, the relations (108) imply that $[A]^2 \subset K_0$. Put $B = U_0(A)$. Then

$$(109)\qquad |BU_0(x)| < |B| = b \qquad (x \in B).$$

Case 1. $b' = b$. Then we define x_ν $(\nu < \omega_n)$ as follows. Let $\nu < \omega_n$, and suppose that $x_\mu \in B$ $(\mu < \nu)$. Then, by (109),

$$\left| \sum [\mu < \nu](\{x_\mu\} + BU_0(x_\mu)) \right| < |B|,$$

and therefore there is $x_\nu \in B - \sum[\mu < \nu](\{x_\mu\} + U_0(x_\mu))$. We may put $Y = \{x_\nu : \nu < \omega_n\}$.

Case 2. $b' = \aleph_m < b$. Then $b = \sum[\mu < \omega_m] b_\mu$, where $b_\mu < b$. Let $x \in B$. Then there is a first ordinal $\rho(x) < \omega_m$ such that $|BU_0(x) < b_{\rho(x)}$, since otherwise we would obtain the contradiction $b = \sum[\mu < \omega_m] b_\mu \leqq |BU_0(x)| \aleph_m < b$. Now put

$$B(\tau) = \{x : x \in B; \rho(x) = \tau\} \qquad (\tau < \omega_m).$$

We define, by induction, τ_μ, X_μ $(\mu < \omega_m)$ as follows. Let, for some $\nu < \omega_m$,

---

[11] Theorem 3, (i).

[12] The symbol U_0 was defined in §2.

$$\tau_\mu < \omega_m; \qquad X_\mu \in [B(\tau_\mu)]^{b_\mu} \qquad\qquad (\mu < \nu).$$

Then, by definition of m,

$$|\sum [\mu < \nu; x \in X_\mu](\{x\} + BU_0(x))| \leqq \sum [\mu < \nu; x \in X_\mu](1 + b_{\tau_\mu})$$
$$= \sum [\mu < \nu](1 + b_{\tau_\mu})b_\mu < b.$$

Hence $|D| = b$, where $D = B - \sum[\mu<\nu;\, x\in X_\mu](\{x\} + U_0(x))$. There is a first ordinal $\tau_\nu < \omega_m$ such that $|DB(\tau_\nu)| < b_\nu$, since otherwise we would have the contradiction

$$b = |D| = \sum [\mu < \omega_m] |DB(\mu)| \leqq b_\nu \aleph_m < b.$$

Now we can choose $X_\nu \in [DB(\tau_\nu)]^{b_\nu}$. Then $X_\nu \subset U_1(X_\mu)$ $(\mu < \nu < \omega_m)$. By the induction hypothesis there is $Y_\mu \in [X_\mu]^{b_\mu}$ such that $[Y_\mu]^2 \subset K_1$ $(\mu < \omega_m)$. Then we may put $Y = \sum[\mu < \omega_m] Y_\mu$. This proves Theorem 44.

Our next theorem may be considered as providing, in the case $r=1$, an extension of the Corollary of Theorem 39 to the case of arbitrary types β_ν, not necessarily ordinals. In view of (5), the extension to values $r \geqq 2$ is false.

Let $k>0$, and consider any types β_ν $(\nu<k)$. Let $\overline{B}_\nu = \beta_\nu$, and denote by P the cartesian product of the sets B_ν, i.e. the set of all mappings $\nu \rightarrow x_\nu \in B_\nu$ defined for $\nu < k$. We order P alphabetically and call the order type π of P the *alphabetical product* of the types β_ν and write

$$\pi = \prod \times [\nu < k] \beta_\nu.$$

This multiplication has been considered by Hausdorff [10].

THEOREM 45. *If* $k>0$, *then* $\prod \times [\nu<k]\beta_\nu \rightarrow (\beta_0, \beta_1, \cdots)_k^1$.

PROOF. In spite of its somewhat complicated appearance the proof is, in fact, very simple, as can be seen by following it in the case $k=2$ or $k=3$. Let $P = \sum[\nu<k]K_\nu$. We want to find $X \subset P$ and $\nu_0 < k$ such that $\overline{X} = \beta_{\nu_0}$; $X \subset K_{\nu_0}$. We use the notation $(z_0, z_1, \cdots, \hat{z}_n)$ for the system of all z_ν such that $\nu < n$, ordered according to increasing ν. Thus

$$P = \{\{x_0, x_1, \cdots, \hat{x}_k\} : x_\nu \in B_\nu \text{ for } \nu < k\}.$$

Case 1. There is $\nu < k$ such that the following condition is satisfied. There is a system of elements $x_\mu \in B_\mu$ $(\mu < \nu)$ such that, given any $x_\nu \in B_\nu$, there is some system $z(x_\nu) = (x_0, x_1, \cdots, \hat{x}_k)$ which belongs to K_ν. Then we may put $X = \{z(x) : x \in B_\nu\}$.

Case 2. There is no ν such that the condition of Case 1 holds. Then, given any $\nu < k$ and any elements $x_\mu \in B_\mu$ $(\mu < \nu)$, there is a function $f_\nu(x_0, x_1, \cdots, \hat{x}_\nu) \in B_\nu$ such that, for any choice of $x_\lambda \in B_\lambda$ $(\nu < \lambda < k)$,

$$(x_0, x_1, \cdots, \hat{x}_\nu, f_\nu(x_0, \cdots, \hat{x}_\nu), x_{\nu+1}, \cdots, \hat{x}_k) \notin K_\nu.$$

In particular, the function $f_0(\hat{x}_0)$ is constant. Then we define, inductively, elements $y_\nu (\nu < k)$ as follows. We put, for $\nu < k$, $y_\nu = f_\nu(y_0, \cdots, \hat{y}_\nu)$. Then $y_\mu \in B_\mu (\mu < k)$, and hence there is $\nu < k$ such that $(y_0, \cdots, \hat{y}_k) \in K_\nu$. But then

$$(y_0, y_1, \cdots, \hat{y}_\nu, f_\nu(y_0, \cdots, \hat{y}_\nu), y_{\nu+1}, \cdots, \hat{y}_k) \in K_\nu$$

which contradicts the definition of f_ν. The theorem is proved.

8. **Canonical partition relations.** Let S be an ordered set, and consider a partition

$$[S]^r = \sum' [\nu < k] K_\nu. \qquad (110)$$

To every such *disjoint* partition there belongs an equivalence relation Δ on $[S]^r$ defined by the rule that elements X, Y of $[S]^r$ are equivalent for Δ, in symbols:

$$X \equiv Y(\cdot \Delta)$$

if, and only if, there is $\nu < k$ such that $X, Y \in K_\nu$. This equivalence relation Δ is unaltered if the classes K_ν are renumbered in any way. The partition (110) and the corresponding equivalence relation Δ is called *canonical*[13] if there is a system $(\epsilon_0, \epsilon_1, \cdots, \epsilon_{r-1})$ of numbers $\epsilon_\rho < 2$ such that, for $X = \{x_0, \cdots, x_{r-1}\}_< \subset S$; $Y = \{y_0, \cdots, y_{r-1}\}_< \subset S$, we have $X \equiv Y(\cdot \Delta)$ if, and only if, $x_\rho = y_\rho$ for every $\rho < r$, such that $\epsilon_\rho = 1$.

The canonical equivalence relation defined by means of the numbers ϵ_ρ is denoted by $\Delta^r_{\epsilon_0 \epsilon_1 \cdots \epsilon_{r-1}}$. The *canonical partition relation*

$$\alpha \rightarrow *(\beta)^r \qquad (111)$$

has, by definition, the following meaning. Whenever $\bar{S} = \alpha$, and (110) is any disjoint partition, with any arbitrary k, then there is $B \subset S$ such that $\bar{B} = \beta$, and such that the equivalence relation Δ belonging to (110), if restricted to $[B]^r$, coincides with some canonical equivalence relation $\Delta^r_{\epsilon_0 \cdots \epsilon_{r-1}}$. The main result of [4] is expressible in the form $\omega_0 \rightarrow *(\omega_0)^r$. The problem arises of finding canonical partition relations between types other than ω_0. The main difference between canonical and noncanonical relations derives from the fact that if the

[13] [4; 5]. The notation used in the present note differs slightly from that used in the earlier papers.

canonical relation (111) holds then a certain choice of a subset of S can be made irrespective of the number $|k|$ of classes of (110). The relation $\omega_0 \to * (\omega_0)^1$ is equivalent to the statement that, *if a denumerable set S is arbitrarily split into nonoverlapping subsets S_ν, then there are either infinitely many nonempty subsets S_ν, or else at least one of the subsets S_ν is infinite.*

The following theorem establishes a connection between canonical and noncanonical partition relations.

THEOREM 46. (i) *Let q_s denote the number of distinct equivalence relations which can be defined on the set $[0, s]$. Let*

$$(112) \qquad s = \binom{2r}{r}; \qquad |\beta| > 2r; \qquad \alpha \to (\beta)^{2r}_{q_s}.$$

Then $\alpha \to * (\beta)^r$. If $|\beta| > 4$; $\alpha \to (\beta)^4_{203}$, then $\alpha \to * (\beta)^2$.

(ii) *If m, $r \geq 0$, and $2^{\aleph_\nu} \leq \aleph_n$ for $m \leq n < m+2r+1$; $\nu < n$, then* $\omega_{m+2r+1} \to * (\omega_m + 2r + 1)^{r+1}$.

REMARK. The first few values of q_s are:

$$q_0 = 1; \quad q_1 = 1; \quad q_2 = 2; \quad q_3 = 5; \quad q_4 = 15; \quad q_5 = 52; \quad q_6 = 203.$$

A rough estimate for all s is

$$q_s \leq 2^{\binom{s}{2}},$$

obtained by observing that an equivalence relation is fixed if for $\mu < \nu < s$ it is decided whether $\mu \equiv \nu$ or $\mu \not\equiv \nu$. It is easy to prove that, for $s > 0$,

$$q_{s+1} = \binom{s}{0} q_s + \binom{s}{1} q_{s-1} + \cdots + \binom{s}{s} q_0$$

and hence $q_s \leq s!$.

Deduction of (ii) *from* (i). By Theorem 39 (iii), we have

$$\omega_{m+2r+1} \to (\omega_m + 2r + 1)^{2r+2}_k \qquad (k < \omega_0),$$

and the conclusion follows from Theorem 46 (i).

PROOF OF (i). Suppose that (112) holds. Let $\overline{S} = \alpha$, and consider any disjoint partition (110). Let Δ be the equivalence relation on $[S]^r$ which belongs to (110). Our first aim is to define a certain equivalence relation Δ^* on $[S]^{2r}$.

Let $[[0, 2r]]^r = \{P_0, P_1, \cdots, P_{s-1}\}_{\neq}$. Then

$$s = \binom{2r}{r}.$$

Let $X = \{x_0, \cdots, x_{2r-1}\}_< \subset S$. Define the equivalence relation $\Delta(X)$ on $[0, s]$ by putting, for $\mu < \nu < s$,

$$\mu \equiv \nu(\cdot \Delta(X))$$

if, and only if, $\{x_\lambda : \lambda \in P_\mu\} \equiv \{x_\lambda : \lambda \in P_\nu\}(\cdot \Delta)$. Put, for $X, Y \in [S]^{2r}$,

$$X \equiv Y(\cdot \Delta^*)$$

if, and only if, $\Delta(X) = \Delta(Y)$.

Now, by definition of q_s, Δ^* has at most q_s nonempty classes. By (112), there is $B \subset S$ such that $\overline{B} = \beta$, and any two elements of $[B]^{2r}$ are equivalent for Δ^*. This means that, in the terminology of [4], Δ is invariant in $[B]^r$ (cf. [4, p. 253]). Choose any $A \subset B$ such that $2r < |A| < \aleph_0$, which is possible since $|\beta| > 2r$. Then Δ is invariant in $[A]^r$ and hence, by [4, Theorem 2], canonical in $[A]^r$. Thus there is a canonical equivalence relation $\Delta(A)$ on $[B]^r$ such that $\Delta = \Delta(A)$ on $[A]^r$. It only remains to show that $\Delta(A)$ is independent of A.

Let $A_0, A_1 \subset B$; $2r < |A_0|, |A_1| < \aleph_0$, and let $\Delta = \Delta(A_0) = \Delta^r_{\epsilon_0 \cdots \epsilon_{r-1}}$ on $[A_0]^r$, $\Delta = \Delta(A_1) = \Delta^r_{\eta_0 \cdots \eta_{r-1}}$ on $[A_1]^r$. Then $2r < |A_0 + A_1| < \aleph_0$, and hence, for some $\kappa_\rho < 2$, $\Delta = \Delta(A_0 + A_1) = \Delta^r_{\kappa_0 \cdots \kappa_{r-1}}$ on $[A_0 + A_1]^r$. Let $A_0 = \{y_0, y_1, \cdots, y_{2r-1}, \cdots\}_<$. Consider the sets

$$P = \{y_{2\lambda} : \lambda < r\}; \qquad Q = \{y_{2\lambda+1-\epsilon_\lambda} : \lambda < r\}.$$

By definition of $\Delta^r_{\epsilon_0 \cdots \epsilon_{r-1}}$, we have $P \equiv Q(\cdot \Delta^r_{\epsilon_0 \cdots \epsilon_{r-1}})$. Hence

$$P \equiv Q(\cdot \Delta); \qquad P \equiv Q(\cdot \Delta^r_{\kappa_0 \cdots \kappa_{r-1}}); \qquad \kappa_\rho \leqq \epsilon_\rho \ (\rho < r).$$

Similarly, by considering the sets P and $Q' = \{y_{2\lambda+1-\kappa_\lambda} : \lambda < r\}$, we find that $P \equiv Q'(\cdot \Delta^r_{\kappa_0 \cdots \kappa_{r-1}})$; $P \equiv Q'(\cdot \Delta)$;

$$P \equiv Q'(\cdot \Delta^r_{\epsilon_0 \cdots \epsilon_{r-1}}); \qquad \epsilon_\rho \leqq \kappa_\rho (\rho < r).$$

Hence $\epsilon_\rho = \kappa_\rho$ for all ρ. For reasons of symmetry, $\eta_\rho = \kappa_\rho$, and so, finally $\epsilon_\rho = \eta_\rho$ $(\rho < r)$. Therefore $\Delta(A)$ is independent of A, which means that Δ is canonical in the whole set $[B]^r$. This proves Theorem 46.

9. **Polarised partition relations.** The relation $a \rightarrow (b_0, b_1)^2$ refers to a partition of the set of all pairs of elements of a set S of cardinal a. Instead of pairs of elements of one and the same set we shall now consider pairs of elements, one from each of two sets S_0, S_1. The relation $a \rightarrow (b_0, b_1)^2$ can be thought of as referring, in the terminology of combinatorial graph theory (linear combinatorial topology), to decompositions of the complete graph of cardinal a into two subgraphs. The new kind of relation to be defined now refers to decom-

positions of the "complete" even graph of cardinal-pair a_0, a_1, i.e. the graph obtained by joining every "point" of a set of cardinal a_0 to every point of a disjoint set of cardinal a_1.

More generally, we introduce the notation

$$[S_0, S_1, \cdots, S_{t-1}]^{r_0, r_1, \cdots, r_{t-1}}$$

for the cartesian product of the t sets $[S_\lambda|^{r_\lambda}$, i.e. we put

$$[S_0, \cdots, S_{t-1}]^{r_0, \cdots, r_{t-1}} = \{(X_0, \cdots, X_{t-1}) : X_\lambda \in [S_\lambda]^{r_\lambda} \text{ for } \lambda < t\}.$$

We shall always have $0 < t < \omega_0$. The introduction of this set leads naturally to the following definition of a corresponding partition relation. The relation

$$\begin{pmatrix} a_0 \\ a_1 \\ \cdot \\ \cdot \\ \cdot \\ a_{t-1} \end{pmatrix} \to \begin{pmatrix} b_{00} & b_{01} & \cdots \\ b_{10} & b_{11} & \cdots \\ \cdot & \cdot & \cdots \\ \cdot & \cdot & \\ \cdot & \cdot & \\ b_{t-1,0} & b_{t-1,1} & \cdots \end{pmatrix}_k^{r_0, r_1, \cdots, r_{t-1}} \tag{113}$$

has, by definition, the following meaning. Whenever $|S_\lambda| = a_\lambda$ for $\lambda < t$, and

$$[S_0, \cdots, S_{t-1}]^{r_0, \cdots, r_{t-1}} = \sum [\nu < k] K_\nu,$$

then there are sets $B_\lambda \subset S_\lambda$ and an ordinal $\nu < k$ such that $|B_\lambda| = b_{\lambda\nu}$ for $\lambda < t$, and $[B_0, \cdots, B_{t-1}]^{r_0, \cdots, r_{t-1}} \subset K_\nu$. There is a similar kind of relation involving order types which will, however, not be considered here. If the number of classes is finite we write in the rows on the right hand side of (113) a last element. The negation of (113) is obtained by replacing $\to$ by $\nrightarrow$. If, for $\lambda < t$, $b_{\lambda 0} = b_{\lambda 1} = \cdots$, we use the obvious abbreviation for (113). Cf. Theorem 49.

The passage from our former type of partition relation i.e. the case $t = 1$, to the more general kind (113) bears a certain resemblance to the process of polarisation used in the theory of algebraic forms, which accounts for the name *polarised partition relation* suggested for relations of the form (113).

We shall deduce some results for polarised relations but will not develop the theory to the same extent as was done in the case of the nonpolarised relation.

In [11] one of us has considered polarised canonical parition relations between finite numbers, and also involving ω_0.

THEOREM 47. *If $a' = b'$, then*

(114) $$\binom{a}{b} \nrightarrow \begin{pmatrix} a & 1 \\ 1 & b \end{pmatrix}^{1,1}.$$

In particular, (114) *holds if* $1 < a, b < \aleph_0$.

THEOREM 48. *The relation*

(115) $$\binom{\aleph_0}{b} \rightarrow \begin{pmatrix} \aleph_0 & \aleph_0 \\ \aleph_0 & b \end{pmatrix}^{1,1}$$

holds if, and only if, either $b = 0$ *or* $b' > \aleph_0$. *In particular,*

(116) $$\binom{\aleph_0}{2^{\aleph_0}} \rightarrow \begin{pmatrix} \aleph_0 & \aleph_0 \\ \aleph_0 & 2^{\aleph_0} \end{pmatrix}^{1,1}.$$

PROOF OF THEOREM 47. If $1 < a < \aleph_0$, then $b' = a' = 2$; $1 < b < \aleph_0$. Let, in this case, $A = [0, a]$; $B = [0, b]$. Then $[A, B]^{1,1} = K_0 +' K_1$, where

$$K_0 = \{(\kappa, \lambda) : \kappa < a; \lambda < b; \kappa + \lambda \text{ even}\}.$$

Then, for any $\kappa < a$, the pairs $(\kappa, 0)$ and $(\kappa, 1)$ lie in different classes K_ν, and the same holds for $(0, \lambda)$ and $(1, \lambda)$. This proves (114). Now let $a \geqq \aleph_0$. Then $a' = b' = \aleph_n$, say, and we can write

$$a = \sum [\nu < \omega_n] a_\nu; \qquad b = \sum [\nu < \omega_n] b_\nu,$$

where $a_\nu < a$; $b_\nu < b$. Let $A = \sum' [\nu < \omega_n] A_\nu$; $B = \sum' [\nu < \omega_n] B_\nu$; $|A_\nu| = a_\nu$; $|B_\nu| = b_\nu$. Then $[A, B]^{1,1} = K_0 + K_1$, where

$$K_0 = \{(x, y) : x \in A_\mu; \quad y \in B_\nu; \quad \mu < \nu < \omega_n\},$$
$$K_1 = \{(x, y) : x \in A_\mu; \quad y \in B_\nu; \quad \nu \leqq \mu < \omega_n\}.$$

Let $X \in [A]^a$; $y_0 \in B$. Then $y_0 \in B_\nu$ for some $\nu < \omega_n$. We have $|A_\mu X| \leqq |A_\mu| < a$ and hence, by definition of n, $\sum [\mu \leqq \nu] |A_\mu X| < a = |X|$. Therefore we can choose $x_0 \in X - \sum [\mu \leqq \nu] A_\mu$. Then $(x_0, y_0) \notin K_0$, and and hence $[X, \{y_0\}]^{1,1} \not\subset; K_0$. By symmetry, it follows that, if $Y \in [B]^b$; $x_1 \in A$, then $[\{x_1\}, Y]^{1,1} \not\subset K_1$. This proves (114) and completes the proof of Theorem 47.

PROOF OF THEOREM 48. If $b = 0$, then (115) obviously holds. If $0 < b < \aleph_0$, then (115) is false, as is seen by considering the partition of the set in question in which $K_1 = 0$. Next, if $b' = \aleph_0'$, then, by Theorem 47,

$$\binom{\aleph_0}{b} \nrightarrow \begin{pmatrix} \aleph_0 & 1 \\ 1 & b \end{pmatrix}^{1,1}.$$

Hence, a fortiori, (115) is false. It remains to prove (115) under the assumption $b' > \aleph_0$.

Let $|A| = \aleph_0$; $|B| = b$; $[A, B]^{1,1} = K_0 + K_1$. We may suppose that

(117) $\quad (X, Y) \in [A, B]^{\aleph_0, b}$ implies $[X, Y]^{1,1} \not\subset K_1$.

Let $(X, Y) \in [A, B]^{\aleph_0, b}$.

1. Put $Y_0 = \sum [x \in X] \{y : (x, y) \in K_0\}$. Then $[X, Y = Y_0]^{1,1} \subset K_1$, and hence, by (117), $|Y - Y_0| < b$,

$$\sum [x \in X] |\{y : y \in Y; (x, y) \in K_0\}| \geq |YY_0| = b.$$

Since $b' > \aleph_0 = |X|$, this implies the existence of $x_0 \in X$ such that

$$|\{y : y \in Y; (x_0, y) \in K_0\}| = b.$$

Put

$$\phi(X, Y) = x_0; \qquad \psi(X, Y) = \{y : y \in Y; (x_0, y) \in K_0\}.$$

Then

$$\phi(X, Y) \in X; \qquad \psi(X, Y) \in [Y]^b,$$
$$[\{\phi(X, Y)\}, \psi(X, Y)]^{1,1} \subset K_0.$$

2. Put $f(y) = \{x : x \in X; (x, y) \in K_0\}$ $(y \in Y)$. If

(118) $\quad y \in Y$ implies $|f(y)| < \aleph_0$,

then $b = |Y| = \sum [P \subset X; |P| < \aleph_0] |\{y : y \in Y; f(y) = P\}|$. But, since there are only $\aleph_0$ distinct sets P, and $b' > \aleph_0$, there is $P_1 \subset X$ such that $|P_1| < \aleph_0$; $|Y_1| = b$, where $Y_1 = \{y : y \in Y; f(y) = P_1\}$. Then $[X - P_1, Y_1]^{1,1} \subset K_1$, which contradicts (117). Therefore (118) is false, and there is $y_1 \in Y$ such that $|f(y_1)| = \aleph_0$. Put $\phi_1(X, Y) = y_1$; $\psi_1(X, Y) = f(y_1)$. Then $\phi_1(X, Y) \in Y$; $\psi_1(X, Y) \in [X]^{\aleph_0}$;

$$[\psi_1(X, Y), \{\phi_1(X, Y)\}]^{1,1} \subset K_0.$$

3. We define sequences $x_\nu, y_\nu, X_\nu, Y_\nu$ $(\nu < \omega_0)$ as follows.

$$x_0 = \phi(A, B);\ Y_0 = \psi(A, B);\ y_0 = \phi_1(A - \{x_0\}, Y_0);\ X_0 = \psi_1(A - \{x_0\}, Y_0).$$

For $0 < \nu < \omega_0$, we put

$$x_\nu = \phi(X_{\nu-1}, Y_{\nu-1} - \{y_{\nu-1}\}); \qquad Y_\nu = \psi(X_{\nu-1}, Y_{\nu-1} - \{y_{\nu-1}\});$$
$$y_\nu = \phi_1(X_{\nu-1} - \{x_\nu\}, Y_\nu); \qquad X_\nu = \psi_1(X_{\nu-1} - \{x_\nu\}, Y_\nu).$$

Then

$$x_\nu \in X_{\nu-1} \subset X_{\nu-2} - \{x_{\nu-1}\} \subset X_{\nu-3} - \{x_{\nu-2}, x_{\nu-1}\} \subset \cdots$$
$$\subset X_0 - \{x_1, \cdots, x_{\nu-1}\} \subset A - \{x_0, \cdots, x_{\nu-1}\};$$

$$y_\nu \in Y_\nu \subset Y_{\nu-1} - \{y_{\nu-1}\} \subset \cdots \subset Y_0 - \{y_0, \cdots, y_{\nu-1}\}$$
$$\subset B - \{y_0, \cdots, y_{\nu-1}\};$$
$$[\{x_\nu\}, Y_\nu]^{1,1} \subset K_0; \; [\{x_\nu\}, \{y_\nu, y_{\nu+1}, \cdots\}]^{1,1} \subset K_0;$$
$$[X_\nu, \{y_\nu\}]^{1,1} \subset K_0; \; [\{x_{\nu+1}, x_{\nu+2}, \cdots\}, \{y_\nu\}]^{1,1} \subset K_0;$$
$$(x_\mu, y_\nu) \in K_0 \; (\mu, \nu < \omega_0).$$

This proves (115). Finally, as is well known [13, p. 135], $(2^{\aleph_0})' > \aleph_0$, so that (116) is a special case of (115). This proves Theorem 48.

We introduce the notation

$$\left\{ \begin{matrix} a \\ b \end{matrix} \right\} = |[A]^b|,$$

where A is a set such that $|A| = a$. If $a, b < \aleph_0$, then

$$\left\{ \begin{matrix} a \\ b \end{matrix} \right\}$$

is the ordinary binomial coefficient

$$\binom{a}{b}.$$

The following lemma is probably well known.

LEMMA 6. *If* $a \geqq \aleph_0$, *then*

$$\left\{ \begin{matrix} a \\ b \end{matrix} \right\} = a^b \text{ for } b \leqq a \quad \text{and} \quad \left\{ \begin{matrix} a \\ b \end{matrix} \right\} = 0 \text{ for } b > a.$$

PROOF. The result is obvious for $b = 0$ and for $b > a$. Now let $0 < b < a$. Choose n such that $|n| = b$, and A_ν, A such that $|A_\nu| = a$ for $\nu < n$, and $A = \sum' [\nu < n] A_\nu$. Then $|A| = ab = a$. If $X \in [A]^b$, then $X = \{x_\nu : \nu < n\}$. Every x_ν has a possible values. Hence

$$\left\{ \begin{matrix} a \\ b \end{matrix} \right\} \leqq a^b.$$

On the other hand, if $y_\nu \in A_\nu$ for $\nu < n$, then $Y = \{y_\nu : \nu < n\} \in [A]^b$, and the set of all such Y has a cardinal a^b. Hence

$$\left\{ \begin{matrix} a \\ b \end{matrix} \right\} \geqq a^b,$$

and the lemma follows.

THEOREM 49. *Suppose that* $0<s<t<\aleph_0$; $k>0$;

$$|l| = |k|^{\binom{a_0}{r_0}\binom{a_1}{r_1}\cdots\binom{a_{s-1}}{r_{s-1}}},$$

$$(119)\qquad \begin{pmatrix} a_0 \\ \vdots \\ a_{s-1} \end{pmatrix} \to \begin{pmatrix} b_0 \\ \vdots \\ b_{s-1} \end{pmatrix}_k^{r_0,\cdots,r_{s-1}},$$

$$(120)\qquad \begin{pmatrix} a_s \\ \vdots \\ a_{t-1} \end{pmatrix} \to \begin{pmatrix} b_s \\ \vdots \\ b_{t-1} \end{pmatrix}_l^{r_s,\cdots,r_{t-1}}.$$

Then

$$(121)\qquad \begin{pmatrix} a_0 \\ \vdots \\ a_{t-1} \end{pmatrix} \to \begin{pmatrix} b_0 \\ \vdots \\ b_{t-1} \end{pmatrix}_k^{r_0,\cdots,r_{t-1}}.$$

PROOF. We use the notation of the partition calculus explained in the proof of Theorem 25. In addition, if Δ is a partition of M, and $M'\subset M$, then the relation

$$|\Delta| \leqq a \quad \text{in} \quad M'$$

expresses the fact that the number of classes of Δ containing at least one element of M' is at most a. (119) and (120) imply that $b_\lambda \leqq a_\lambda$ for $\lambda<t$. If $b_\lambda<r_\lambda$ for some $\lambda<t$, then (121) holds trivially. Hence we may assume that

$$(122)\qquad r_\lambda \leqq b_\lambda \leqq a_\lambda \qquad \text{for } \lambda < t.$$

Let $|A_\lambda| = a_\lambda$ for $\lambda<t$, and consider any equivalence relation Δ on $[A_0, \cdots, A_{t-1}]^{r_0,\cdots,r_{t-1}}$ such that $|\Delta| \leqq |k|$. Our aim is to find $B_\lambda \in [A_\lambda]^{b_\lambda} (\lambda<t)$ such that

$$(123)\qquad |\Delta| \leqq 1 \quad \text{in} \quad [B_0, \cdots, B_{t-1}]^{r_0,\cdots,r_{t-1}}.$$

Put, for $X_\lambda \in [A_\lambda]^{r_\lambda}$ $(s\leqq\lambda<t)$,

$$\Delta_1(X_s, \cdots, X_{t-1}) = \prod \Delta(X_0, \cdots, X_{t-1}),$$

where the last product is extended over all systems $(X_0, \cdots, X_{s-1}) \in [A_0, \cdots, A_{s-1}]^{r_0,\cdots,r_{s-1}}$. By (122), this product has at least one factor. It follows that $|\Delta_1| \leqq |l|$. Hence, by (120), there is $B_\lambda \in [A_\lambda]^{b_\lambda}$ $(s\leqq\lambda<t)$ such that

$$|\Delta_1| \leqq 1 \quad \text{in} \quad [B_s, \cdots, B_{t-1}]^{r_s,\cdots,r_{t-1}}.$$

By (122), we can choose

$$Y_\lambda \in [B_\lambda]^{r_\lambda} \qquad (s \leqq \lambda < t).$$

Put, for $X_\lambda \in [A_\lambda]^{r_\lambda}$ $(\lambda < s)$,

$$\Delta_2(X_0, \cdots, X_{s-1}) = \Delta(X_0, \cdots, X_{s-1}, Y_s, \cdots, Y_{t-1}).$$

Then $|\Delta_2| \leqq |k|$, and therefore, by (119), there is $B_\lambda \in [A_\lambda]^{b_\lambda}$ $(\lambda < s)$ such that $|\Delta_2| \leqq 1$ in $[B_0, \cdots, B_{s-1}]^{r_0, \cdots, r_{s-1}}$. By (122), we can choose $Y_\lambda \in [B_\lambda]^{r_\lambda}$ $(\lambda < s)$. Then, for any $X_\lambda \in [B_\lambda]^{r_\lambda}$ $(\lambda < t)$,

$$\begin{aligned}(X_0, \cdots, X_{t-1}) &\equiv (X_0, \cdots, X_{s-1}, Y_s, \cdots, Y_{t-1}) \\ &\equiv (Y_0, \cdots, Y_{s-1}, Y_s, \cdots, Y_{t-1})(\cdot\Delta).\end{aligned}$$

This proves (123) and so establishes Theorem 49.

We note the following special case of Theorem 49.

Corollary. *If*

$$a_0 \to (b_0)^1_k; \quad a_1 \to (b_1)^1_{|k|^{a_0}}$$

then

$$\binom{a_0}{a_1} \to \binom{b_0}{b_1}^{1,1}_k.$$

We give some applications of this last result.

(a) If $0 < d < \aleph_0 \leqq a_1$, then

$$\binom{2d-1}{a_1} \to \binom{d}{a_1}^{1,1}_2.$$

This is best possible in the sense that, if $2d-1$ is replaced by $2d-2$, the last relation becomes false. We even have, as is easily seen,

$$\binom{2d-2}{a_1} \nrightarrow \binom{d}{1}^{1,1}_2 \qquad (0 < d < \aleph_0 \leqq a_1).$$

(b) If $a_0' > |k| > 0$; $a_1' > |k|^{a_0}$, then

$$\binom{a_0}{a_1} \to \binom{a_0}{a_1}^{1,1}_k.$$

In particular, if we assume that $2^{\aleph_0} = \aleph_1$, then

$$\binom{\aleph_0}{\aleph_2} \to \binom{\aleph_0}{\aleph_2}^{1,1}_2.$$

More generally, if $2^{\aleph_n}=\aleph_{n+1}$, then

$$\binom{\aleph_n}{\aleph_{n+2}} \to \binom{\aleph_n}{\aleph_{n+2}}_2^{1,1}.$$

This is best possible in the following strong sense.

(c) If $a' \leqq |k|$; $b>0$, then

$$\binom{a}{b} \nrightarrow \binom{a}{1}_k^{1,1}.$$

To prove (c), choose $n \leqq k$ such that $|n|=a'$. Then $a=\sum[\nu<n]a_\nu$, where $a_\nu<a$. Choose sets A_ν, A, B such that $|A_\nu|=a_\nu$ $(\nu<n)$; $A=\sum'[\nu<n]A_\nu$; $|B|=b$, and put $[A, B]^{1,1}=\sum'[\nu<k]K_\nu$, where $K_\nu=[A_\nu, B]^{1,1}$ $(\nu<n)$. If, now, $0\neq X\subset A$; $y\in B$; $[X, \{y\}]^{1,1}\subset K_\nu$ for some $\nu<k$, then $\nu<n$; $X\subset A_\nu$; $|X|<a$. This proves (c).

The following theorem is a corollary of a result due to Sierpiński.

THEOREM 50. *If* $2^{\aleph_0}=\aleph_1$, *then*

$$\binom{\aleph_0}{\aleph_1} \nrightarrow \binom{\aleph_0}{\aleph_1}_2^{1,1}.$$

PROOF. Let $A=[0, \omega_0]$; $B=[0, \omega_1]$. According to Sierpiński[14] the assumption $2^{\aleph_0}=\aleph_1$ implies, and is, in fact, equivalent to, the existence of a sequence of functions $f_\lambda(y)\in B$ $(\lambda\in A)$, defined for $y\in B$, such that, given any $Y\in[B]^{\aleph_1}$, there is $\lambda_0\in A$ such that $\{f_\lambda(y):y\in Y\}=B$ $(\lambda_0\leqq\lambda<\omega_0)$. Then $[A, B]^{1,1}=K_0+'K_1$, where $K_0=\{(\lambda, y):\lambda\in A; y\in B; f_\lambda(y)=0\}$. If, now, $(X, Y)\in[A, B]^{\aleph_0,\aleph_1}$, then, by the property of the functions f_λ, there is $\lambda\in X$; $y_0, y_1\in Y$ such that $f_\lambda(y_\nu)=\nu$ $(\nu<2)$. Then $(\lambda, y_\nu)\in[X, Y]^{1,1}K_\nu$ $(\nu<2)$. This proves the assertion.

THEOREM 51. *If* $a, b>1$;

$$\binom{a}{b} \to \binom{a}{b}_k^{1,1}.$$

Then

$$\binom{a'}{b'} \to \binom{a'}{b'}_k^{1,1}.$$

PROOF. Let l and m be such that $|l|=a'$; $|m|=b'$. Then $a=\sum[\lambda<l]a_\lambda$; $b=\sum[\mu<m]b_\mu$, where $a_\lambda<a$; $b_\mu<b$. Put $A'=[0, l]$; $B'=[0, m]$, and suppose that $[A', B']^{1,1}=\sum[\nu<k]K'_\nu$. Then we choose sets A_λ, A, B_μ, B such that $|A_\lambda|=a_\lambda$ $(\lambda<l)$; $|B_\mu|=b_\mu$ $(\mu<m)$;

[14] [14], French translation in [16]. See also [1].

$A=\sum'[\lambda<l]A_\lambda$; $B=\sum'[\mu<m]B_\mu$. Then $|A|=a$; $|B|=b$; $[A,B]^{1,1}=\sum[\nu<k]K_\nu$, where $K_\nu=\{(x,y):x\in A_\lambda;\ y\in B_\mu;\ (\lambda,\mu)\in K'_\nu\}$ $(\nu<k)$. By hypothesis, there is $(X,Y)\in[A,B]^{a,b}$ and $\nu<k$ such that $[X,Y]^{1,1}\subset K_\nu$. Put

$$A''=\{\lambda:\lambda<l;\ A_\lambda X\neq 0\};\qquad B''=\{\mu:\mu<m;\ B_\mu Y\neq 0\}.$$

Then $a=|X|=\sum[\lambda\in A'']|A_\lambda X|$; $|A_\lambda X|\leq|A_\lambda|<a$. Hence, by definition of a', $|A''|\geq a'$. But, $|A''|\leq|l|=a'$. Hence $|A''|=a'$ and, by symmetry, $|B''|=b'$.

Let $\lambda\in A''$; $\mu\in B''$. Then we can choose $x\in A_\lambda X$; $y\in B_\mu Y$, and we then have $(x,y)\in[X,Y]^{1,1}\subset K_\nu$; $(\lambda,\mu)\in K'_\nu$. Hence $[A'',B'']^{1,1}\subset K'_\nu$; $(A'',B'')\in[A',B']^{a',b'}$, and Theorem 51 follows.

COROLLARY. *If* $a>1$, *then*

$$\binom{a'}{a}\nrightarrow\binom{a'}{a}^{1,1}_2.$$

For, if

$$\binom{a'}{a}\rightarrow\binom{a'}{a}^{1,1}_2,$$

then, by Theorem 51 and the known equation $a''=a'$, we conclude that

$$\binom{a'}{a'}\rightarrow\binom{a'}{a'}^{1,1}_2$$

which contradicts Theorem 47.

We may mention that there is an obvious extension of Theorem 51 to relations

$$\begin{pmatrix}a_0\\ \vdots\\ a_{t-1}\end{pmatrix}\rightarrow\begin{pmatrix}a_0\\ \vdots\\ a_{t-1}\end{pmatrix}\qquad\text{for any } t.$$

In conclusion, we collect some polarised partition relations involving the first three infinite cardinals. They follow from Theorems 47–50. We put $\aleph_0=a$; $\aleph_1=b$; $\aleph_2=d$.

$$\binom{a}{a}\nrightarrow\begin{pmatrix}a&1\\1&a\end{pmatrix}^{1,1};\ \binom{b}{b}\nrightarrow\begin{pmatrix}b&1\\1&b\end{pmatrix}^{1,1};$$

$$\binom{d}{d}\nrightarrow\begin{pmatrix}d&1\\1&d\end{pmatrix}^{1,1}\qquad\text{(Theorem 47);}$$

$$\binom{a}{b}\rightarrow\begin{pmatrix}a&a\\a&b\end{pmatrix}^{1,1};\ \binom{a}{d}\rightarrow\begin{pmatrix}a&a\\a&d\end{pmatrix}^{1,1}\qquad\text{(Theorem 48).}$$

If $2^a = b$, then

$$\binom{a}{d} \to \begin{pmatrix} a & a \\ d & d \end{pmatrix}^{1,1} \qquad \text{(Theorem 49)}$$

and

$$\binom{a}{b} \nrightarrow \begin{pmatrix} a & a \\ b & b \end{pmatrix}^{1,1} \qquad \text{(Theorem 50).}$$

It seems curious that the continuum hypothesis should enable us both to strengthen

$$\binom{a}{d} \to \begin{pmatrix} a & a \\ a & d \end{pmatrix}^{1,1}$$

to

$$\binom{a}{d} \to \begin{pmatrix} a & a \\ d & d \end{pmatrix}^{1,1}$$

and to show that

$$\binom{a}{b} \to \begin{pmatrix} a & a \\ a & b \end{pmatrix}^{1,1}$$

cannot be strengthened to

$$\binom{a}{b} \to \begin{pmatrix} a & a \\ b & b \end{pmatrix}^{1,1}.$$

References

1. F. Bagemihl and H. D. Sprinkle, *On a proposition of Sierpiński*, Proc. Amer. Math. Soc. vol. 5 (1954) pp. 726–728.

2. B. Dushnik and E. W. Miller, *Partially ordered sets*, Amer. J. Math. vol. 63 (1941) p. 605.

3. P. Erdös, *Some set-theoretical properties of graphs*, Revista Universidad Nacional de Tucuman, Serie A vol. 3 (1942) pp. 363–367.

4. P. Erdös and R. Rado, *A combinatorial theorem*, J. London Math. Soc. vol. 25 (1950) pp. 249–255.

5. ———, *Combinatorial theorems on classifications of subsets of a given set*, Proc. London Math. Soc. (3) vol. 2 (1952) pp. 417–439.

6. ———, *A problem on ordered sets*, J. London Math. Soc. vol 28 (1953) pp. 426–438.

7. P. Erdös and G. Szekeres, *A combinatorial problem in geometry*, Compositio Math. vol. 2 (1935) pp. 463–470.

8. P. Hall, *On representations of sub-sets*, J. London Math. Soc. vol. 10 (1934) pp. 26–30.

9. F. Hausdorff, *Grundzüge einer theorie der geordneten Mengen*, Math. Ann. vol. 65 (1908) pp. 435–506.

10. ———, *Mengenlehre*, 3d ed., 1944, §16.

11. R. Rado, *Direct decompositions of partitions*, J. London Math. Soc. vol. 29 (1954), pp. 71–83.

12. F. P. Ramsey, *On a problem of formal logic*, Proc. London Math. Soc. (2) vol. 30 (1930) pp. 264–286.

13. W. Sierpiński, *Leçons sur les nombres transfinis*, Paris, 1928.

14. ———, *O jednom problemu G Ruzjevića koji se odnosi na hipotezu kontinuuma*, Glas Srpske Kraljevske Akademije vol. 152 (1932) pp. 163–169.

15. ———, *Sur un problème de la théorie des relations*, Annali R. Scuola Normale Superiore de Pisa Ser. 2 vol. 2 (1933) pp. 285–287.

16. ———, *Concernant l'hypothèse du continu*, Académie Royale Serbe. Bulletin de l'Académie des Sciences Mathématiques et Naturelles. A. Sciences Mathématiques et Physiques vol. 1 (1933) pp. 67–73.

17. A. Tarski, *Quelques théorèmes sur les alephs*, Fund. Math. vol. 7 (1925) p. 2.

HEBREW UNIVERSITY OF JERUSALEM AND
UNIVERSITY OF READING

Reprinted from
Bull. Amer. Math. Soc. **62** (1956), 427–489

MAXIMAL FLOW THROUGH A NETWORK

L. R. FORD, JR. AND D. R. FULKERSON

Introduction. The problem discussed in this paper was formulated by T. Harris as follows:

"Consider a rail network connecting two cities by way of a number of intermediate cities, where each link of the network has a number assigned to it representing its capacity. Assuming a steady state condition, find a maximal flow from one given city to the other."

While this can be set up as a linear programming problem with as many equations as there are cities in the network, and hence can be solved by the simplex method **(1)**, it turns out that in the cases of most practical interest, where the network is planar in a certain restricted sense, a much simpler and more efficient hand computing procedure can be described.

In §1 we prove the minimal cut theorem, which establishes that an obvious upper bound for flows over an arbitrary network can always be achieved. The proof is non-constructive. However, by specializing the network (§2), we obtain as a consequence of the minimal cut theorem an effective computational scheme. Finally, we observe in §3 the duality between the capacity problem and that of finding the shortest path, via a network, between two given points.

1. The minimal cut theorem. A *graph* G is a finite, 1-dimensional complex, composed of *vertices* $a, b, c, \ldots, e$, and *arcs* $\alpha(ab), \beta(ac), \ldots, \delta(ce)$. An arc $\alpha(ab)$ *joins* its end vertices a, b; it passes through no other vertices of G and intersects other arcs only in vertices. A *chain* is a set of distinct arcs of G which can be arranged as $\alpha(ab)$, $\beta(bc)$, $\gamma(cd), \ldots, \delta(gh)$, where the vertices $a, b, c, \ldots, h$ are distinct, i.e., a chain does not intersect itself; a chain *joins* its end vertices a and h.

We distinguish two vertices of G: a, the *source*, and b, the *sink*.[1] A *chain flow* from a to b is a couple $(C; k)$ composed of a chain C joining a and b, and a non-negative number k representing the flow along C from source to sink.

Each arc in G has associated with it a positive number called its *capacity*. We call the graph G, together with the capacities of its individual arcs, a *network*. A *flow* in a network is a collection of chain flows which has the property that the sum of the numbers of all chain flows that contain any arc is no greater than the capacity of that arc. If equality holds, we say the arc is *saturated* by the flow. A chain is *saturated* with respect to a flow if it contains

Received September 20, 1955.

[1]The case in which there are many sources and sinks with shipment permitted from any source to any sink is obviously reducible to this.

a saturated arc. The value of a flow is the sum of the numbers of all the chain flows which compose it.

It is clear that the above definition of flow is not broad enough to include everything that one intuitively wishes to think of as a flow, for example, sending trains out a dead end and back or around a circuit, but as far as effective transportation is concerned, the definition given suffices.

A *disconnecting set* is a collection of arcs which has the property that every chain joining a and b meets the collection. A disconnecting set, no proper subset of which is disconnecting, is a *cut*. The *value* of a disconnecting set D (written $v(D)$) is the sum of the capacities of its individual members. Thus a disconnecting set of minimal value is automatically a cut.

THEOREM 1. (Minimal cut theorem). *The maximal flow value obtainable in a network N is the minimum of $v(D)$ taken over all disconnecting sets D.*

Proof. There are only finitely many chains joining a and b, say n of them. If we associate with each one a coordinate in n-space, then a flow can be represented by a point whose jth coordinate is the number attached to the chain flow along the jth chain. With this representation, the class of all flows is a closed, convex polytope in n-space, and the value of a flow is a linear functional on this polytope. Hence, there is a maximal flow, and the set of all maximal flows is convex.

Now let S be the class of all arcs which are saturated in every maximal flow.

LEMMA 1. *S is a disconnecting set.*

Suppose not. Then there exists a chain $\alpha_1, \alpha_2, \ldots, \alpha_m$ joining a and b with $\alpha_i \notin S$ for each i. Hence, corresponding to each α_i, there is a maximal flow f_i in which α_i is unsaturated. But the average of these flows,

$$f = \frac{1}{m}\sum f_i,$$

is maximal and α_i is unsaturated by f for each i. Thus the value of f may be increased by imposing a larger chain flow on $\alpha_1, \alpha_2, \ldots, \alpha_m$, contradicting maximality.

Notice that the orientation assigned to an arc of S by a positive chain flow of a maximal flow is the same for all such chain flows. For suppose first that (C_1, k_1), (C_2, k_2) are two chain flows occurring in a maximal flow f, $k_1 \geqslant k_2 > 0$, where

$$C_1 = \alpha_1(a\, a_1),\ \alpha_2(a_1 a_2), \ldots, \alpha_j(a_{j-1}, a_j), \ldots, \alpha_r(a_{r-1}, b)$$

$$C_2 = \beta_1(a\, b_1),\ \beta_2(b_1 b_2), \ldots, \beta_k(b_{k-1}, b_k), \ldots, \beta_s(b_{s-1}, b),$$

and $\alpha_j(a_{j-1}, a_j) = \beta_k(b_{k-1}, b_k) \in S$, $a_{j-1} = b_k$, $a_j = b_{k-1}$. Then

$$C_1' = \alpha_1, \alpha_2, \ldots, \alpha_{j-1}, \beta_{k+1}, \ldots, \beta_s$$

$$C_2' = \beta_1, \beta_2, \ldots, \beta_{k-1}, \alpha_{j+1}, \ldots, \alpha_r$$

contain chains C_1'', C_2'' joining a and b, and another maximal flow can be obtained from f as follows. Reduce the C_1 and C_2 components of f each by k_2, and increase each of the C_1'' and C_2'' components by k_2. This unsaturates the arc α_j, contradicting its definition as an element of S. On the other hand, if (C_1, k_1), (C_2, k_2) were members of distinct maximal flows f_1, f_2, consideration of $f = \frac{1}{2}(f_1 + f_2)$ brings us back to the former case. Hence, the arcs of S have a definite orientation assigned to them by maximal flows. We refer to that vertex of an arc $\alpha \in S$ which occurs first in a positive chain flow of a maximal flow as the *left* vertex of α.

Now define a *left arc* of S as follows: an arc α of S is a left arc if and only if there is a maximal flow f and a chain $\alpha_1, \alpha_2, \ldots, \alpha_k$ (possibly null) joining a and the left vertex of α with no α_i saturated by f. Let L be the set of left arcs of S.

LEMMA 2. *L is a disconnecting set.*

Given an arbitrary chain $\alpha_1(a\,a_1), \alpha_2(a_1a_2), \ldots, \alpha_m(a_{m-1}\,b)$ joining a and b, it must intersect S by Lemma 1. Let $\alpha_t(a_{t-1}, a_t)$ be the first $\alpha_i \in S$. Then for each α_i, $i < t$, there is a maximal flow f_i in which α_i is unsaturated. The average of these flows provides a maximal flow f in which $\alpha_1, \alpha_2, \ldots, \alpha_{t-1}$ are unsaturated. It remains to show that this chain joins a to the left vertex of α_t, i.e., a_{t-1} is the left vertex of α_t. Suppose not. Then the maximal flow f contains a chain flow

$$[\beta_1(ab_1),\ \beta_2(b_1, b_2), \ldots, \beta_r(b_{r-1}, b); k],\ k > 0,\ \beta_s = \alpha_t,\ b_{s-1} = a_t,\ b_s = a_{t-1}.$$

Let the amount of unsaturation in f of α_i $(i = 1, \ldots, t-1)$, be $k_i > 0$. Now alter f as follows: decrease the flow along the chain $\beta_1, \beta_2, \ldots, \beta_r$ by $\min\,[k, k_i] > 0$ and increase the flow along the chain contained in

$$\alpha_1, \alpha_2, \ldots, \alpha_{t-1}, \beta_{s+1}, \ldots, \beta_r$$

by this amount. The result is a maximal flow in which α_t is unsaturated, a contradiction. Hence $\alpha_t \in L$.

LEMMA 3. *No positive chain flow of a maximal flow can contain more than one arc of L.*

Assume the contrary, that is, there is a maximal flow f_1 containing a chain flow

$$[\beta_1(ab_1),\ \beta_2(b_1b_2), \ldots, \beta_r(b_{r-1}, b); k],\ k > 0,$$

with arcs $\beta_i, \beta_j \in L$, β_i occurring before β_j, say, in the chain. Let f_2 be that maximal flow for which there is an unsaturated chain

$$\alpha_1(aa_1),\ \alpha_2(a_1, a_2), \ldots, \alpha_s(a_{s-1}, b_{j-1})$$

from a to the left vertex of β_j. Consider $f = \frac{1}{2}(f_1 + f_2)$. This maximal flow contains the chain flow $[\beta_1, \beta_2, \ldots, \beta_r; k']$ with $k' \geqslant \frac{1}{2}k$, and each $\alpha_i (i = 1, \ldots, s)$ is unsaturated by $k_i > 0$ in f. Again alter f: decrease the flow along

$\beta_1, \beta_2, \ldots, \beta_r$ by min $[k', k_i] > 0$ and increase the flow along the chain contained in $\alpha_1, \alpha_2, \ldots, \alpha_s, \beta_j, \ldots, \beta_r$ by the same amount, obtaining a maximal flow in which β_i is unsaturated, a contradiction.

Now to prove the theorem it suffices only to remark that the value of every flow is no greater than $v(D)$ where D is any disconnecting set; and on the other hand we see from Lemma 3 and the definition of S that in adding the capacities of arcs of L we have counted each chain flow of a maximal flow just once. Since by Lemma 2 L is a disconnecting set, we have the reverse inequality. Thus L is a minimal cut and the value of a maximal flow is $v(L)$.

We shall refer to the value of a maximal flow through a network N as the *capacity* of N (cap (N)). Then note the following corollary of the minimal cut theorem.

COROLLARY. *Let A be a collection of arcs of a network N which meets each cut of N in just one arc. If N' is a network obtained from N by adding k to the capacity of each arc of A, then* cap $(N') =$ cap $(N) + k$.

It is worth pointing out that the minimal cut theorem is not true for networks with several sources and corresponding sinks, where shipment is restricted to be from a source to its sink. For example, in the network (Fig. 1) with shipment from a_i to b_i and capacities as indicated, the value of a minimal disconnecting set (i.e., a set of arcs meeting all chains joining sources and corresponding sinks) is 4, but the value of a maximal flow is 3.

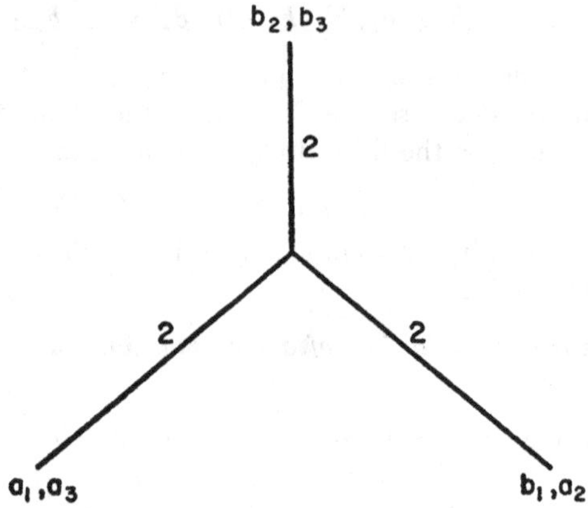

Fig. 1

2. A computing procedure for source-sink planar networks.[2] We say that a network N is *planar* with respect to its source and sink, or briefly, N is *ab*-planar, provided the graph G of N, together with arc ab, is a planar

[2]It was conjectured by G. Dantzig, before a proof of the minimal cut theorem was obtained, that the computing procedure described in this section would lead to a maximal flow for planar networks.

graph **(2; 3)**. (For convenience, we suppose there is no arc in G joining a and b.) The importance of ab-planar networks lies in the following theorem.

THEOREM 2. *If N is ab-planar, there exists a chain joining a and b which meets each cut of N precisely once.*

Proof. We may assume, without loss of generality, that the arc ab is part of the boundary of the outside region, and that G lies in a vertical strip with a located on the left bounding line of the strip, b on the right. Let T be the chain joining a and b which is top-most in N. T has the desired property, as we now show. Suppose not. Then there is a cut D, at least two arcs of which are in T. Let these be α_1 and α_2, with α_1 occurring before α_2 in following T from a to b. Since D is a cut, there is a chain C_1 joining a and b which meets D in α_1 only. Similarly there is a chain C_2 meeting D in α_2 only. Let C_2' be that part of C_2 joining a to an end point of α_2. It follows from the definition of T that C_1 and C_2' must intersect. But now, starting at a, follow C_2' to its last intersection with C_1, then C_1 to b. We thus have a chain from a to b not meeting D, contradicting the fact that D is a cut.

Symmetrically, of course, the bottom-most chain of N has the same property.

Notice that this theorem is not valid for networks which are not ab-planar. A simple example showing this is provided by the "gas, water, electricity" graph (Fig. 2), in which every chain joining a and b meets some cut in three arcs.

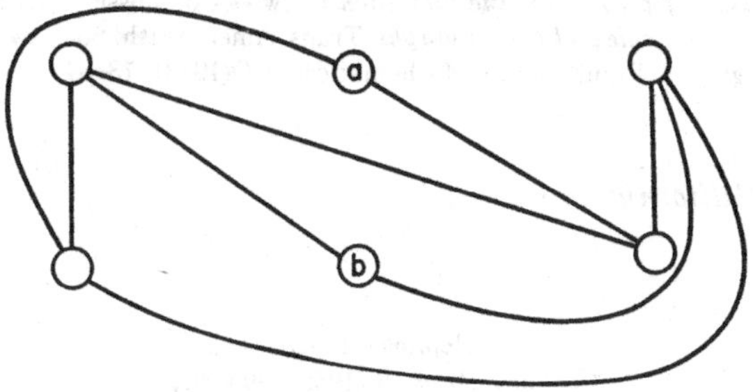

Fig. 2

Theorem 2 and the corollary to Theorem 1 provide an easy computational procedure for determining a maximal flow in a network of the kind here considered. Simply locate a chain having the property of Theorem 2; this can be done at a glance by finding the two regions separated by arc ab, and taking the rest of the boundary of either region (throwing out portions of the boundary where it has looped back and intersected itself, so as to get a chain). Impose as large a chain flow $(T; k)$ as possible on this chain, thereby saturating one or more of its arcs. By the corollary, subtracting k from each capacity in T reduces the capacity of N by k. Delete the saturated arcs, and proceed as before. Eventually, the graph disconnects, and a maximal flow has been constructed.

3. A minimal path problem. For source-sink planar networks, there is an interesting duality between the problem of finding a chain of minimal capacity-sum joining source and sink and the network capacity problem, which lies in the fact that chains of N joining source and sink correspond to cuts (relative to two particular vertices) of the dual[3] of N and vice versa. More precisely, suppose one has a network N, planar relative to two vertices a and b, and wishes to find a chain joining a and b such that the sum of the numbers assigned to the arcs of the chain is minimal. An easy way to solve this problem is as follows. Add the arc ab, and construct the dual of the resulting graph G. Let a' and b' be the vertices of the dual which lie in the regions of G separated by ab. Assign each number of the original network to the corresponding arc in the dual. Then solve the capacity problem relative to a' and b' for the dual network by the procedure of §2. A minimal cut thus constructed corresponds to a minimal chain in the original network.

[3]The *dual* of a planar graph G is formed by taking a vertex inside each region of G and connecting vertices which lie in adjacent regions by arcs. See **(2; 3)**.

REFERENCES

1. G. B. Dantzig, *Maximization of a linear function of variables subject to linear inequalities: Activity analysis of production and allocation* (Cowles Commission, 1951).
2. H. Whitney, *Non-separable and planar graphs*, Trans. Amer. Math. Soc., *34* (1932), 339–362.
3. ———, *Planar graphs*, Fundamenta Mathematicae, *21* (1933), 73–84.

Rand Corporation,
Santi Monica, California

Reprinted from
Canad. J. Math. **8** (1956), 399–404

ON PICTURE-WRITING*

G. PÓLYA, Stanford University

To write "sun", "moon" and "tree" in picture-writing, one draws simply a circle, a crescent and some simplified, conventionalized picture of a tree, respectively. Picture-writing was used by some tribes of red Indians and it may well be that more advanced systems of writing evolved everywhere from this primitive system. And so picture-writing may be the ultimate source of the Greek, Latin and Gothic alphabets, the letters of which we currently use as mathematical symbols. I wish to observe that also the primitive picture-writing may be of some use in mathematics. In what follows, I wish to show how the method of generating functions, important in Combinatory Analysis, can be quite intuitively evolved from "figurate series" the terms of which are pictures (or, more precisely, variables represented by pictures).

Picture-writing is easy to use on paper or blackboard, but it is clumsy and expensive to print. Although I have presented several times the contents of the following pages orally, I hesitated to print it.† I am indebted to the editor of the MONTHLY who encouraged me to publish this article.

I shall try to explain the general idea by discussing three particular examples the first of which, although the easiest, will be very broadly treated.

1.1. In how many ways can you change one dollar? Let us generalize the proposed question. Let P_n denote the number of ways of paying the amount of n cents with five kinds of coins: cents, nickels, dimes, quarters and half-dollars. The "way of paying" is determined if, and only if, it is known how many coins of each kind are used. Thus, $P_4=1$, $P_5=2$, $P_{10}=4$. It is appropriate to set $P_0=1$. The problem stated at the outset requires us to compute P_{100}. More generally, we wish to understand the nature of P_n and eventually devise a procedure for computing P_n.

It may help to visualize the various possibilities. We may use no cent, or just 1 cent, or 2 cents, or 3 cents, or $\cdots$. These alternatives are schematically pictured in the first line of Figure 1;** "no cent" is represented by a square which may remind us of an empty desk. The second line pictures the alternatives: using no nickel, 1 nickel, 2 nickels, $\cdots$. The following three lines represent in the same way the possibilities regarding dimes, quarters and half-dollars. We have to choose one picture from the first line, then one picture from the second line, and so on, choosing just one picture from each line; combining (juxtaposing) the five pictures so selected, we obtain a manner of paying. Thus, Figure 1 exhibits directly the alternatives regarding each kind of coin and, indirectly, all manners of paying we are concerned with.

* Address presented at the meeting of the Association in Athens, Ga., March 16, 1956.

† I used it, however, in research. See **2**, especially p. 156, where the "figurate series" are introduced in a closely related, but somewhat different, form. (Numerals in boldface indicate the references at the end of the paper.)

** A photo of actual coins would be more effective here but too clumsy in the following figures.

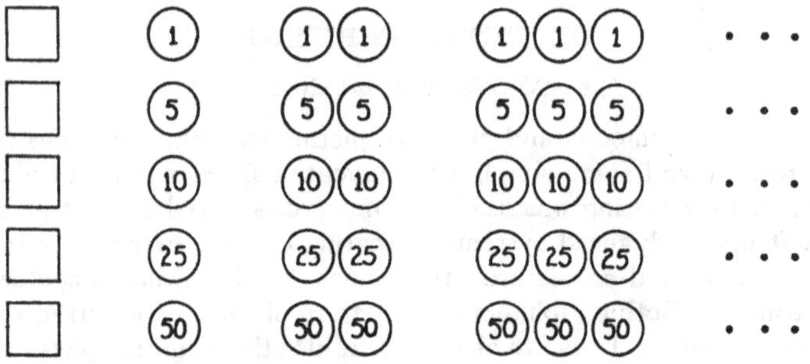

FIG. 1. A complete survey of alternatives.

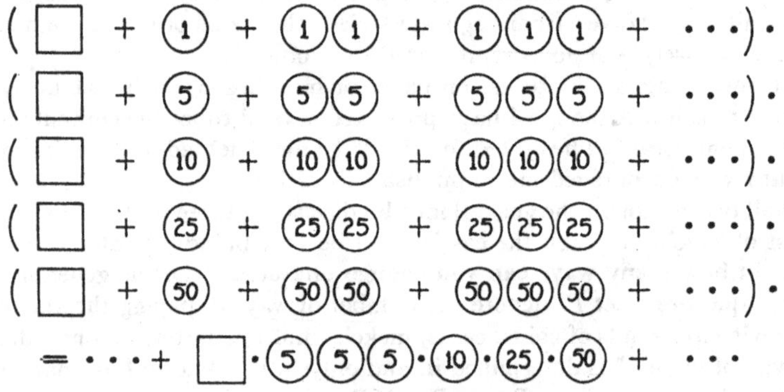

FIG. 2. Genesis of the figurate series.

The main discovery consists in observing that, in fact, we combine the pictures in Figure 1 according to certain rules of algebra: if we conceive each line of Figure 1 as the *sum* of the pictures contained in it and we consider the *product* of these five (infinite) sums, in short, if we pass from Figure 1 to Figure 2, and we develop the product, the terms of this development will represent the various manners of paying we are concerned with. The one term of the product exhibited in the last line of Figure 2 as an example represents one manner of paying one dollar (putting down no cents, three nickels, one dime, one quarter and one half-dollar). The sum of all such terms is an infinite series of pictures; each picture exhibits one manner of paying, different terms represent different manners of paying, and the whole series of pictures, appropriately called the *figurate series*, displays all manners of paying that we have to consider when we wish to compute the numbers P_n.

1.2. Yet this way of conceiving Figure 2 raises various difficulties. First, there is a theoretical difficulty: in which sense can we add and multiply pictures? Then, there is a practical difficulty: how can we pick out conveniently from the whole figurate series the terms counted by P_n, that is, those cases in which the sum paid amounts to just n cents?

We avoid the theoretical difficulty if we employ the pictures, these symbols of a primitive writing, as we are used to employing the letters of more civilized alphabets: we regard each picture as the symbol for a variable or *indeterminate*.†

To master the other difficulty, we need one more essential idea: we substitute for each "pictorial" variable (that is, variable represented by a picture) a power of a new variable x, the *exponent* of which is the *joint value of the coins* represented by the picture, as it is shown in detail by Figure 3. The third line of Figure 3 shows a lucky coincidence: we have conceived the three juxtaposed nickels as *one picture*, as the symbol of one variable (corresponding to the use of precisely three nickels). For this variable we have to substitute x^{15} according to our general rule; yet even if we substitute for each of the juxtaposed coins the correct power of x and consider the product of these juxtaposed powers, we arrive at the same final result x^{15}.

$$\text{①} = x, \quad \text{⑤} = x^5, \quad \text{⑩} = x^{10}, \quad \text{㉕} = x^{25}, \quad \text{㊿} = x^{50},$$

$$\square = x^0 = 1,$$

$$\text{⑤⑤⑤} = x^5 x^5 x^5 = x^{15},$$

$$\square \cdot \text{⑤} \cdot \text{⑤} \cdot \text{⑤} \cdot \text{⑩} \cdot \text{㉕} \cdot \text{㊿} = x^{100},$$

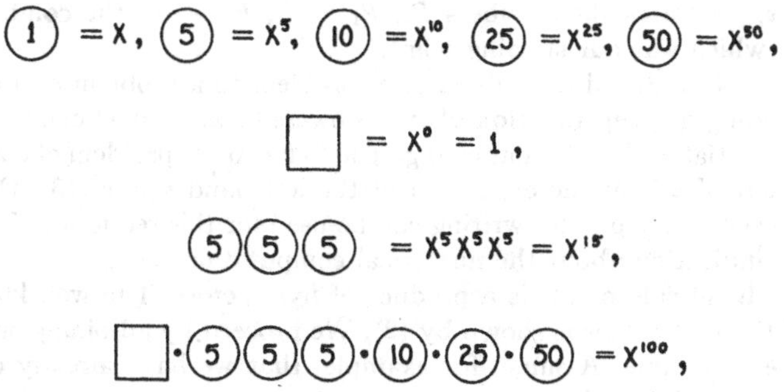

FIG. 3. Powers of one variable substituted for variables represented by pictures.

The last line of Figure 3 is very important. It shows by an example (see the last line of Fig. 2) how the described substitution affects the general term of the figurate series. Such a term is the product of 5 pictures (pictorial variables). For each factor a power of x is substituted whose exponent is the value in cents of that factor; the exponent of the product, obtained as a sum of 5 exponents, will be the joint value of the factors. And so the substitution indicated by Figure 3 changes each term of the figurate series into a power x^n. As the figurate series represents each manner of paying just once, the exponent n arises precisely P_n times so that (after suitable rearrangement of the terms) the whole figurate series goes over into

---

† In a formal presentation it may be advisable to restrict the term "picture" to denote a (visible, written or printed) symbol that stands for an indeterminate; in the present introductory, rather informal, address the word is now and then more loosely used.

Let us pass over two somewhat touchy points: the infinity of variables and the convergence of the series in which they arise. Both are considered in certain advanced theories and both are momentary. They will be eliminated by the next step.

(1) $$P_0 + P_1x + P_2x^2 + \cdots + P_nx^n + \cdots.$$

In this series the coefficient of x^n enumerates the different manners of paying the amount of n cents, and so (1) is suitably called the *enumerating series.*

The substitution indicated by Figure 3 changes the first line of Figure 2 into a geometric series:

(2) $$1 + x + x^2 + x^3 + \cdots = (1 - x)^{-1}.$$

In fact, this substitution changes each of the first five lines of Figure 2 into some geometric series and the equation indicated by Figure 2 goes over into

(3) $$(1 - x)^{-1}(1 - x^5)^{-1}(1 - x^{10})^{-1}(1 - x^{25})^{-1}(1 - x^{50})^{-1} = P_0 + P_1x + P_2x^2 + \cdots + P_nx^n + \cdots.$$

We have succeeded in expressing the sum of the enumerating series. This sum is usually termed the *generating function*; in fact, this function, expanded in powers of x, generates the numbers $P_0, P_1, \cdots, P_n, \cdots$, the combinatorial meaning of which was our starting point.

1.3. We have reduced a combinatorial problem to a problem of a different kind: expanding a given function of x in powers of x. In particular, we have reduced our initial problem about changing a dollar to the problem of computing the coefficient of x^{100} in the expansion of the left hand side of (3). Our main goal was to show how picture-writing can be used for this reduction. Yet let us add a brief indication about the numerical computation.

The left hand side of (3) is a product of five factors. The well known expansion of the first factor is shown by (2). We proceed by adjoining successive factors, one at a time. Assume, for example, that we have already obtained the expansion of the product of the first two factors:

$$(1 - x)^{-1}(1 - x^5)^{-1} = a_0 + a_1x + a_2x^2 + \cdots,$$

and we wish to go on hence to three factors:

$$(1 - x)^{-1}(1 - x^5)^{-1}(1 - x^{10})^{-1} = b_0 + b_1x + b_2x^2 + \cdots.$$

It follows that

$$(b_0 + b_1x + b_2x^2 + \cdots)(1 - x^{10}) = a_0 + a_1x + a_2x^2 + \cdots.$$

Comparing the coefficient of x^n on both sides, we find that

(4) $$b_n = b_{n-10} + a_n$$

(set $b_m = 0$ if $m < 0$). By (4), we can conveniently compute the coefficients b_n by recursion if the a_n are already known, and the series (3) can be obtained from (2) in four successive steps each of which is similar to the one we have just discussed.

We add a table that shows the computation of P_{50}. This table exhibits the coefficient of x^n for some values of n in five different expansions. The head of each column shows the value of n, the beginning of each row the last factor taken into account; the bottom row would show P_n for $n=0, 5, 10, \cdots, 50$ *if* we had computed it. Yet the table registers only the steps needed for computing the answer to our initial question and yields $P_{50}=50$; that is, one can pay 50 cents in exactly 50 different ways. We leave it to the reader to continue the computation and verify that $P_{100}=292$; he can also try to justify the procedure of computation directly without resorting to the enumerating series.*

Table to compute P_{50}

$n=0$		5	10	15	20	25	30	35	40	45	50
$(1-x)^{-1}$	1	1	1	1	1	1	1	1	1	1	1
$(1-x^5)^{-1}$	1	2	3	4	5	6	7	8	9	10	11
$(1-x^{10})^{-1}$	1	2	4	6	9	12	16		25		36
$(1-x^{25})^{-1}$	1					13					49
$(1-x^{50})^{-1}$	1										50

2.1. Dissect a convex polygon with n sides into $n-2$ triangles by $n-3$ diagonals and compute D_n, the number of different dissections of this kind. Examining first the simplest particular cases helps to understand the problem. We easily see that $D_4=2$, $D_5=5$; of course $D_3=1$.

The solution is indicated by the parts (I), (II), and (III) of Figure 4. After the broad discussion of the foregoing solution it should not be difficult to understand the indications of Figure 4.

Part (I) of Figure 4 hints the key idea: we build up the dissections of any polygon that is not a triangle from the dissections of other polygons which have fewer sides. For this purpose, we emphasize one of the sides of the polygon, place it horizontally at the bottom and call it the *base*. One of the triangles into which the polygon is dissected has the base as side; we call this triangle Δ. In the given polygon there are two smaller polygons, one to the left, the other to the right, of Δ. For example, the top line of Figure 4 (I) shows an octagon in which there is a quadrilateral to the left, and a pentagon to the right, of Δ, both suitably dissected. As the figure suggests, we can generate this dissection of the octagon by starting from Δ and placing on it, from both sides, the two other appropriately pre-dissected polygons. We may hope that building up the dissections in this manner will be useful.

In exploring the prospects of this idea, we may run into an objection: there are cases, such as the one displayed in the second line of Figure 4 (I), in which the partial polygon on a certain side of Δ does not exist. Yet we can parry this objection: yes, the partial polygon on that side of Δ (the left side in the case of the figure) *does* exist, but it is degenerate; it is reduced to a mere *segment*.

* For the usual method of deriving the generating function, *cf.* 1, Vol. 1, p. 1, Problem 1.

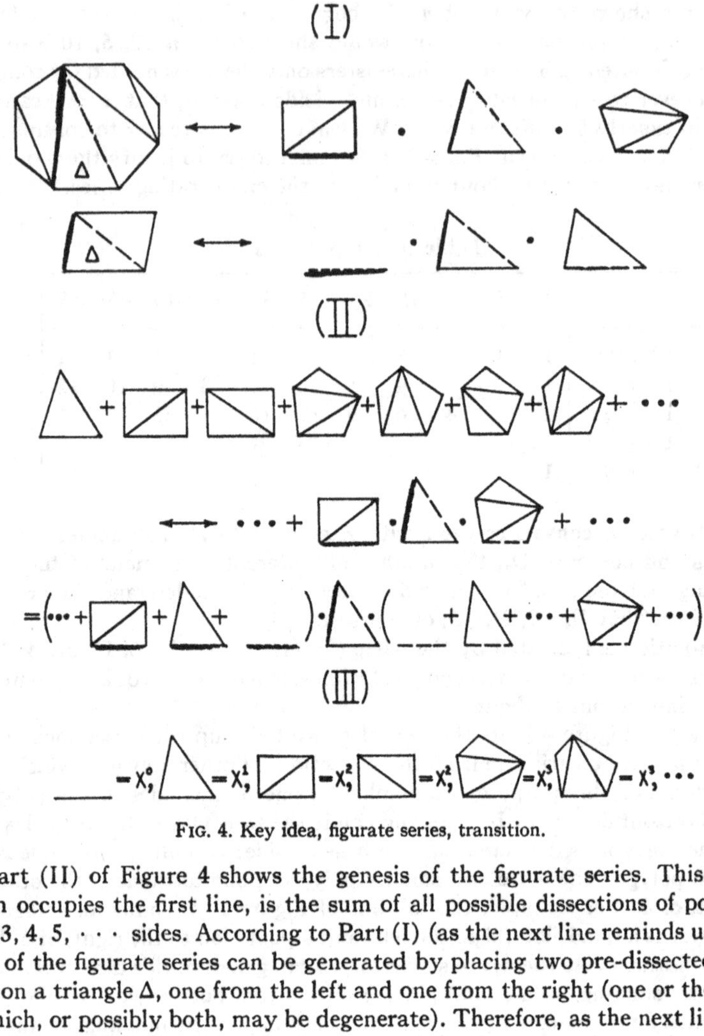

FIG. 4. Key idea, figurate series, transition.

Part (II) of Figure 4 shows the genesis of the figurate series. This series, which occupies the first line, is the sum of all possible dissections of polygons with 3, 4, 5, $\cdots$ sides. According to Part (I) (as the next line reminds us) each term of the figurate series can be generated by placing two pre-dissected polygons on a triangle Δ, one from the left and one from the right (one or the other of which, or possibly both, may be degenerate). Therefore, as the next line (the last of Figure 4 (II)) indicates, the terms of the figurate series are in one-one correspondence with the expansion of a product of three factors: the middle factor is just a triangle, the other two factors are equal to the figurate series augmented by the segment.

2.2. Part (III) of Figure 4 hints the transition from the figurate series to the enumerating series. Following the pattern set by Figure 3 and Section 1.2, we substitute for each dissection (more precisely, for the variable represented by that dissection) a power of x the exponent of which is the number of triangles in that dissection. This substitution, indicated by Figure 4 (III), changes the figurate series into

$$D_3x + D_4x^2 + D_5x^3 + \cdots + D_nx^{n-2} + \cdots = E(x), \tag{5}$$

where $E(x)$ stands for enumerating series. The relation displayed by Figure 4 (II) goes over into

$$E(x) = x[1 + E(x)]^2. \tag{6}$$

This is a quadratic equation for $E(x)$ the solution of which is

$$\begin{aligned} E(x) &= D_3x + D_4x^2 + D_5x^3 + \cdots + D_nx^{n-2} + \cdots \\ &= \frac{1 - 2x - [1 - 4x]^{1/2}}{2x} \\ &= x + 2x^2 + \cdots. \end{aligned} \tag{7}$$

In fact, to arrive at (7), we have to discard the other solution of the quadratic equation (6) which becomes ∞ for $x=0$.

2.3. We have reduced our original problem which was to compute D_n to a problem of a different kind: to find the coefficient of x^{n-2} in the expansion of the function (7) in powers of x.* This latter is a routine problem which we need not discuss broadly. We obtain from (7), using the binomial formula and straightforward transformations, that for $n \geqq 3$

$$D_n = -\frac{1}{2}\binom{1/2}{n-1}(-4)^{n-1} = \frac{2}{2}\,\frac{6}{3}\,\frac{10}{4}\cdots\frac{4n-10}{n-1}.$$

3.1. A (topological) *tree* is a connected system of two kinds of objects, *lines* and *points*, that contains no closed path. A certain point of the tree in which just one line ends is called the *root* of the tree, the line starting from the root the *trunk*, any point different from the root a *knot*. In Figure 5 the root is indicated by an arrow, and each knot by a small circle. Our problem is: *compute T_n, the number of different trees with n knots.*

It makes no difference whether the lines are long or short, straight or curved, drawn on the paper to the left or to the right: only the difference in (topological) connection is relevant. Examining the simplest cases may help the reader to understand the intended meaning of the problem; it is easily seen that $T_1=1$, $T_2=1$, $T_3=2$, $T_4=4$, $T_5=9$.†

* For a more usual method *cf.* **3**, Vol. 1, p. 102, Problems 7, 8, and 9.

† The trees here considered should be called more specifically *root-trees;* see **4**, Vol. 11, p. 365. Their definition which is merely hinted here is elaborated in **2**, pp. 181–191; see also the passages there quoted of **5**. It may be, however, sufficient and in some respects even advantageous if, at a first reading, the reader takes the definition "intuitively" and supplements it by examples. Observe that in Cayley's first paper on the subject, **4**, Vol. 3, pp. 242–246, the definition of a tree is not even attempted. Chemistry is one of the sources of the notion "tree": if the points stand for atoms and the connecting lines for valencies, the tree represents a chemical compound.

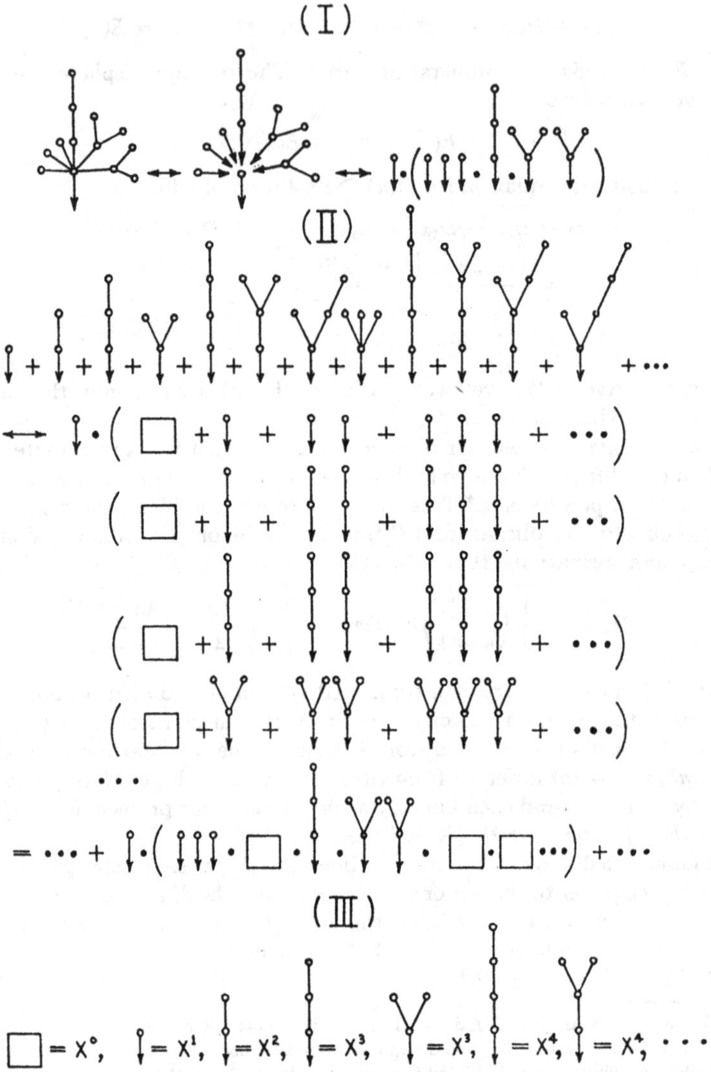

FIG. 5. Key idea, figurate series, transition.

The solution is indicated by the three parts of Figure 5 the general arrangement of which is closely similar to that of Figure 4. The reader should try to understand the solution by merely looking at Figure 5 and observing relevant analogies with all the foregoing figures. He may, however, fall back upon the following brief comments.

The simplest tree consists of root, trunk and just one knot. The key idea is to build up any tree different from the simplest tree from other trees which have fewer knots. For this purpose we conceive, as Figure 5 (I) shows, the "main branches" of any tree as trees (with fewer knots) inserted into the upper endpoint (the only knot) of the trunk. Therefore, as Figure 5 (I) further shows, we can conceive of any tree as the juxtaposition of the simplest tree and of several pictures, each of which consists of one, or two, or more *identical* trees; observe the analogy with the last line of Figure 2.

Part (II) of Figure 5 displays the figurate series: the infinite sum of all different trees. Its genesis is similar to, but more complex than, that of the figurate series of Figure 2. In Figure 2 we see a product of five "virtually geometric" series; in Figure 5 we see a product of an infinity of "virtually geometric" series, multiplied by an initial one term factor (the simplest tree, the common trunk of all trees).

3.2. Part (III) of Figure 5 displays the substitution that changes the figurate series into the enumerating series. By this substitution, each "virtually geometric" series arising in Figure 5 (II) goes over into a proper geometric series the sum of which is known, and the whole relation displayed by Figure 5 (II) goes over into the remarkable relation due to Cayley*

$$\begin{aligned} &T_1x + T_2x^2 + T_3x^3 + \cdots + T_nx^n + \cdots \\ &\quad = x(1-x)^{-T_1}(1-x^2)^{-T_2}(1-x^3)^{-T_3}\cdots(1-x^n)^{-T_n}\cdots. \end{aligned} \tag{8}$$

3.3. By expanding the right hand side of Equation (8) in powers of x and comparing the coefficient of x^n on both sides, we obtain a recursion formula, that is, an expression for T_n in terms of $T_1, T_2, \cdots, T_{n-1}$ for $n \geqq 2$. The reader should work out the first cases and verify by analytical computation the values T_n for $n \leqq 5$ which he found before by geometrical experimentation.

References

1. G. Pólya and G. Szegö, Aufgaben und Lehrsätze aus der Analysis, 2 volumes, Berlin, 1925.
2. G. Pólya, Acta Mathematica, vol. 68 (1937), pp. 145–254.
3. G. Pólya, Mathematics and Plausible Reasoning, 2 volumes, Princeton, 1954.
4. A. Cayley, Collected Mathematical Papers, 13 volumes, Cambridge, 1889–1898.
5. D. König, Theorie der endlichen und unendlichen Graphen, Leipzig, 1936.

* This form is slightly different from that given in 4, Vol. 3, pp. 242–246. For other forms see 2, p. 149.

Reprinted from
Amer. Math. Monthly **63** (1956), 689–697

A THEOREM ON FLOWS IN NETWORKS

DAVID GALE

1. Introduction. The theorem to be proved in this note is a generalization of a well-known combinatorial theorem of P. Hall, [4].

HALL'S THEOREM. *Let $S_1, S_2, \cdots, S_n$ be subsets of a set X. Then a necessary and sufficient condition that there exist distinct elements $x_1, \cdots, x_n$, such that $x_i \in S_i$ is that the union of every k sets from among the S_i contain at least k elements.*

The result has a simple interpretation in terms of transportation networks. A certain article is produced at a set X of origins, and is demanded at n destinations $y_1, \cdots, y_n$. Certain of the origins x are "connected" to certain of the destinations y making it possible to ship one article from x to y.

PROBLEM. *Under what conditions is it possible to ship articles to all the destinations y?*

An obvious reinterpretation of Hall's theorem shows that this is possible if and only if every k of the destinations are connected to at least k origins.

We shall now give a verbal statement of the generalization to be proved. A more formal statement will be given in the next section.

Let N be an arbitrary network or graph. To each node x of N corresponds a real number $d(x)$, where $|d(x)|$ is to be thought of as the demand for or the supply of some good at x according as $d(x)$ is positive or negative. To each edge (x, y) corresponds a nonnegative real number $c(x, y)$, the capacity of this edge, which assigns an upper bound to the possible flow from x to y.

The demands $d(x)$ are called *feasible* if there exists a flow in the network such that the flow along each edge is no greater than its capacity, and the net flow into (out of) each node is at least (at most) equal to the demand (supply) at that node.

An obviously necessary condition for the demands $d(x)$ to be feasible is the following.

For every collection S of nodes the sum of the demands at the nodes

Received September 24, 1956. The results of this paper were discovered while the author was working as a consultant for the RAND Corporation. A later revision was partially supported by an O. N. R. contract.

of S must not exceed the sum of the capacities of the edges leading into S.

If this condition were not satisfied it would clearly be impossible to satisfy the aggregate demand of the subset S. The principal theorem of this paper shows that conversely, if the above condition is satisfied, then the demands $d(x)$ are feasible.

Hall's theorem drops out as a special case of this result if one applies it to the particular network described in the paragraph above and makes use of the known fact (see [1]) that transportation problems of this type with integral constraints have integral solutions. However, the simple inductive argument which works in [4] does not seem to generalize to yield a proof of our theorem. Our approach is in fact quite different and is based on the "minimum cut" theorem of Ford and Fulkerson, [2], [1].

In the next section we give a formal statement of the problem and prove the principal theorem. The final section is devoted to the treatment of a special case for which the "feasibility criterion" yields a very simple method for computing solutions.

2. The principal theorem. We proceed to define in a more formal manner the objects to be discussed.

DEFINITIONS. A *network* $[N,c]$ consists of a finite set of *nodes* N and a *capacity function* c on $N \times N$ where $c(x, y)$ is a nonnegative real number or plus infinity.

A *flow* f on $[N, c]$ is a function f on $N \times N$ such that

$$f(x, y)+f(y, x)=0\,, \tag{1}$$

$$f(x, y)\leq c(x, y) \qquad \text{for all } x, y \in N\,. \tag{2}$$

A *demand* d on $[N, c]$ is simply a real valued function on N.

Note that we do not require the function c to be symmetric, thus the maximum allowable flow from x to y need not be the same as that from y to x. Condition (1) above corresponds to the usual convention that the net flow from x to y is the negative of the net flow from y to x.

We shall save writing many summation symbols in what follows by adopting the following convenient notation.

NOTATION. If S is a subset of N and d a function on N, we write

$$d(S)=\sum_{x\in S} d(x)\,.$$

If S and T are subsets of N and f a function on $N \times N$ we write

$$f(S, T) = \sum_{x \in S,\, y \in T} f(x, y).$$

From these definitions it follows at once that if U and V are disjoint subsets of N then

$$(3) \qquad d(U \cup V) = d(U) + d(V)$$

$$f(S, U \cup V) = f(S, U) + f(S, V).$$

In particular, denoting the complement of S by S' we have,

$$f(N, T) = f(S, T) + f(S', T) \qquad \text{for all } S \subset N.$$

In this notation (1) and (2) are clearly equivalent to

$$(1') \qquad f(A, A) = 0;$$

and

$$(2') \qquad f(A, B) \leq c(A, B) \qquad \text{for all } A, B \subset N.$$

The above notation is natural to our problem, for if d is a demand function then $d(S)$ is simply the aggregate demand of the set S, and if f is a flow then $f(S, T)$ represents the net flow from S into T.

DEFINITION. A demand d is called *feasible* if there exists a flow f such that

$$(4) \qquad f(N, x) \geq d(x) \qquad \text{for } x \in N.$$

This condition states that the flow into each node must be at least equal to the demand at that node. However (1) and (4) together imply

$$f(x, N) \leq -d(x)$$

so that we are also requiring the flow out of each node to be at most equal to the supply at that node (recalling that a negative demand represents a supply).

Finally we note that from (3) it follows that (4) is equivalent to

$$(4') \qquad f(N, S) \geq d(S) \qquad \text{for all } S \subset N.$$

We can now give a simple statement of our main result.

FEASIBILITY THEOREM. *The demand d is feasible if and only if for every subset $S \subset N$*

$$(5) \qquad d(S') \leq c(S, S').$$

Proof. The necessity of (5) is obvious, for if d is feasible then there is a flow f such that

$$d(S') \leq f(N, S') = f(S, S') + f(S', S') = f(S, S') \leq c(S, S').$$

The proof of sufficiency depends on the "minimum cut theorem" of Ford and Fulkerson, which we shall now state and prove in our own formulation. While our proof is little more than a translation of the above authors' second proof [3] into our notation, we record it here, nevertheless, both for the sake of completeness and because it is substantially shorter than any proof published heretofore.

DEFINITION. Let $[N, c]$ be a network and let s and s' be two distinguished nodes (s=source, s'=sink). A *flow from s to s'* is a flow such that

$$f(N, x) = 0 \qquad \text{for } x \neq s,\ x \neq s'. \tag{6}$$

Let F denote the set of all flows from s to s'.

A *cut* (S, S') of N with respect to s and s' is a partition of N into sets S and S' such that $s \in S$, $s' \in S'$.

Let Q denote the set of all such cuts.

MINIMUM CUT THEOREM. *For any network* $[N, c]$

$$\max_F f(s, N) = \min_Q c(S, S'),$$

Proof. First note that for any flow $f \in F$ and cut $(S, S') \in Q$ we have

$$\begin{aligned} f(s, N) &= f(s, N) + \sum_{x \in S-s} f(x, N) = f(s, N) + f(S-s, N) \\ &= f(S, N) = f(S, S) + f(S, S') = f(S, S') \leq c(S, S'). \end{aligned} \tag{7}$$

Hence, it remains only to show that equality is attained in (7) for some flow and cut.

Let $\bar{f} \in F$ be a flow such that $\bar{f}(s, N)$ is a maximum. Let S consist of s and all nodes x such that there exists a *chain* $\sigma = (x_0, x_1, \cdots, x_n)$ of distinct nodes with $x_0 = s$, $x_n = x$ and $c(x_{i-1}, x_i) - \bar{f}(x_{i-1}, x_i) > 0$, $i = 1, \cdots, n$. Now s' is not in S, for, if it were, there would be a chain σ as above with $x = s'$. But then letting

$$\mu = \min\,[c(x_{i-1}, x_i) - \bar{f}(x_{i-1}, x_i)],$$

one could superimpose a flow of μ along the chain σ on top of the flow $\bar{f}$, contradicting the maximality of $\bar{f}$.

The above argument shows that (S, S') is a cut, and we conclude the proof by observing that $\bar{f}(s, N)=c(S, S')$, for if not, then from (7), $\bar{f}(S, S')<c(S, S')$, hence for some $x \in S$ and $y \in S'$ we would have $c(x, y)-\bar{f}(x, y)>0$, but since $x \in S$ there is a chain $\sigma=(s, x_1, \cdots, x)$ which could be extended to a chain $\sigma'=(s, x_1, \cdots, x, y)$, contrary to the fact that $y \in S'$. This completes the proof.

Proof of feasibility theorem. Consider a new network $[\bar{N}, \bar{c}]$ where $\bar{N}$ consists of N plus two additional nodes s and s'. Let $U \subset N$ be all nodes x such that $d(x) \leq 0$. Then $\bar{c}$ is defined by the rules

$$\begin{aligned} &\bar{c}(x, y)=c(x, y) && \text{for } x, y \in N, \\ &\bar{c}(s, x)=-d(x) && \text{for } x \in U, \\ &\bar{c}(x, s')=d(x) && \text{for } x \in U', \\ &\bar{c}(x, y)=0 && \text{otherwise.} \end{aligned}$$

We now assert that the cut $(\bar{N}-s', s')$ is a minimal cut of $[\bar{N}, \bar{c}]$, for let $\bar{S}$, and $\bar{S}'$ be any cut of $[\bar{N}, \bar{c}]$ and let $S=\bar{S}-s$, $S'=\bar{S}'-s'$. From the definition above we have

$$\begin{aligned} \bar{c}(\bar{S}, \bar{S}') &= c(S, S')+\bar{c}(s, S')+\bar{c}(S, s') \\ &= c(S, S')-d(S' \cap U)+d(S \cap U'), \end{aligned}$$

$$\bar{c}(\bar{N}-s', s')=d(U')=d(S' \cap U')+d(S \cap U');$$

and subtracting we get

$$\begin{aligned} \bar{c}(\bar{N}-s', s')-\bar{c}(\bar{S}, \bar{S}') &= d(S' \cap U')+d(S' \cap U)-c(S, S') \\ &= d(S')-c(S, S') \leq 0, \end{aligned}$$

the last inequality being the hypothesis (5), and the assertion is proved.

Now, from the Minimum Cut Theorem, there is a flow $\bar{f}$ from s to s' on $[\bar{N}, \bar{c}]$ such that

$$\bar{f}(\bar{N}-s', s')=\bar{c}(\bar{N}-s', s')=d(U'),$$

hence

$$\bar{f}(x, s')=d(x) \qquad \text{for all } x \in U'. \tag{8}$$

Let f be $\bar{f}$ restricted to $N \times N$. Then f is clearly a flow and it remains to show that f satisfies (4). If $x \in U'$ then

$$0=\bar{f}(x, \bar{N})=f(x, N)+\bar{f}(x, s')=f(x, N)+d(x),$$

hence

$$f(N, x) = d(x) . \tag{9}$$

If $x \in U$ then

$$0 = \bar{f}(\bar{N}, x) = f(N, x) + \bar{f}(s, x) \leq f(N, x) + \bar{c}(s, x) = f(N, x) - d(x) ,$$

so

$$f(N, x) \geq d(x) , \tag{10}$$

and (9) and (10) together show that f satisfies (4), completing the proof.

REMARK. We wish to call attention to the following important fact. We have at no point in what has been said thus far made use of the assumption that the functions d, c and f were real valued. In fact, all definitions and proofs go through verbatim if the real numbers are replaced by any ordered Abelian group, in particular, the group of integers. One useful consequence of this remark is the fact that if a network with integer valued demand and capacity functions admits a feasible flow then this flow may also be chosen to be integer valued. We shall make use of this fact in the next section.

There is a second formulation of the Feasibility Theorem which is sometimes convenient. In the network $[N, c]$ let U be as above the set of nodes x such that $d(x) \leq 0$.

THEOREM. *The demand* d *is feasible if and only if for every set* $Y \subset U'$ *there exists a flow* f_Y *such that*

$$f_Y(N, x) \geq d(x) \qquad \text{for } x \in U \tag{11}$$

$$f_Y(N, Y) \geq d(Y) . \tag{12}$$

Proof. The necessity is obvious. To prove sufficiency we show that (11) and (12) imply (5).

Let (S, S') be a partition of N and let $X = U \cap S$, $X' = U \cap S'$, $Y = U' \cap S$, $Y' = U' \cap S'$. Then from (11) there exists $f_{Y'}$ such that

$$d(X') \leq f_{Y'}(N, X') = f_{Y'}(X \cup Y, X') + f_{Y'}(Y', X') ,$$

and from (12),

$$d(Y') \leq f_{Y'}(N, Y') = f_{Y'}(X \cup Y, Y') + f_{Y'}(X', Y') .$$

Adding these inequalities we get

$$d(S')=d(X')+d(Y')=f_{Y'}(X\cup Y,\ X')+f_{Y'}(X\cup Y,\ Y')$$
$$=f_{Y'}(X\cup Y,\ X'\cup Y')=f(S,\ S')\leq c(S,\ S')\ ,$$

which is exactly (5).

3. An example. As an illustration of the feasibility theorem, consider the following problem.

(I). *Let $a_1, \cdots, a_m$ and $b_1, \cdots, b_n$ be two sets of positive integers. Under what conditions can one find integers $\alpha_{ij}=0$ or 1, such that*

$$\sum_{i=1}^{m} \alpha_{ij} \geq b_j$$

and

$$\sum_{j=1}^{n} \alpha_{ij} \leq a_i\ ,$$

for all i and j?

As a concrete illustration, suppose n families are going on a picnic in m busses, where the jth family has b_j members and the ith bus has a_i seats. When is it possible to seat all passengers in such a way that no two members of the same family are in the same bus?

In the case $\sum a_i = \sum b_i$ the problem becomes that of filling an $m\times n$ matrix M with zeros and ones so that the rows and columns shall have prescribed sums.

The feasibility theorem gives a simple necessary and sufficient condition for the problem to have a solution. In order to state if we need the following.

DEFINITION. Let $\{a_i\}$ be a nonincreasing sequence of nonnegative integers $a_1, a_2, \cdots$, such that all but a finite number of the a_i are zero. Let

$$S_j = \{a_i | a_i \geq j\}$$

where j is a positive integer and let s_j be the number of elements in S_j. The sequence of numbers $\{s_j\}$ clearly satisfies the same conditions as the sequence $\{a_i\}$; it is called the *dual sequence* of the sequence $\{a_i\}$ and is denoted by $\{a_i\}^*$.

It is clear that $\{a_i\}^*$ determines $\{a_i\}$ since the integer a_i occurs exactly $s_{a_i} - s_{a_i+1}$ times in $\{a_i\}$. Actually the correspondence between $\{a_i\}$ and $\{a_i\}^*$ is completely dual in the following sense.

THEOREM. $\{a_i\}^{**} = \{a_i\}$.

This result will not be needed in the sequel and its proof is left as an exercise. However, its validity can be made quite obvious by means of a simple pictorial representation.

Let each number a_i be represented by a row of dots, and write these rows in a vertical array so that a_{i+1} lies under a_i, thus:

$$\begin{array}{l} a_1 \cdot\cdot\cdot\cdot\cdot\cdot \\ a_2 \cdot\cdot\cdot\cdot\cdot \\ a_3 \cdot\cdot\cdot\cdot\cdot \\ a_4 \cdot\cdot\cdot \\ a_5 \cdot \end{array}$$

It is then clear that the dual number s_j is simply the number of dots in the jth column of the array.

We can now give the criterion for the feasibility of Problem I. Henceforth for convenience we shall assume the numbers a_i and b_j are indexed in decreasing order, and shall define $a_i=0$ for $i>m$, $b_j=0$ for $j>n$.

THEOREM. *Let* $\{s_j\}=\{a_i\}^*$. *Then Problem* 1 *is feasible if and only if*

$$\sum_{j=1}^{k} b_j \leq \sum_{j=1}^{k} s_j\,, \qquad \textit{for all integers } k\,.$$

Proof. We may interpret (I) as a flow problem. Let N be a network consisting of $m+n$ nodes $x_1, \cdots, x_m$ and $y_1, \cdots, y_n$, and let $c(x_i, y_i)=1$ for all i and j, $c=0$ otherwise. Let $d(x_i)=-a_i$ and $d(y_j)=b_j$. One easily verifies that the feasibility of (I) is equivalent to the feasibility of the demand d.

We shall show that d is feasible by applying the second theorem of the previous section. Let Y be a subset of k nodes y_j, say $Y=\{y_{j_1}, \cdots, y_{j_k}\}$. We now compute the maximum possible flow into Y. Because all capacities are unity this maximal flow f_Y is achieved by shipping as much as possible from each node x_i into the set Y. Thus, the flow from x_i to Y is $\min[a_i, k]$ and the total flow into Y is

$$f_Y(N, Y)=\sum_{i=1}^{m} \min[a_i, k]\,.$$

We now assert

$$\sum_{i=1}^{m} \min[a_i, k]=\sum_{j=1}^{k} s_j\,, \tag{13}$$

which is proved by induction on k. It is clear from the definition that

$$\sum_{i=1}^{m} \min[a_i, 1] = m = s_1 .$$

Now

$$\min[a_i, k+1] = \begin{cases} \min[a_i, k] & \text{for } a_i \leq k \\ \min[a_i, k]+1 & \text{for } a_i \geq k+1, \text{ or } a_i \in S_{k+1}, \end{cases}$$

hence,

$$\sum_{i=1}^{m} \min[a_i, k+1] = \sum_{i=1}^{m} \min[a_i, k] + s_{k+1} ,$$

and (13) follows from the induction hypothesis.

The second feasibility theorem now states that the problem is feasible if and only if

$$\sum_{r=1}^{k} b_{j_r} \leq \sum_{j=1}^{k} s_j ,$$

and since the b_j are indexed in decreasing order, the conclusion of the theorem follows.

It is interesting that for this particular problem there is a simple "n-step" method for actually filling out the matrix of α_{ij}'s. Such procedures are sufficiently rare in programming theory so that it seems worth while to present it here.

The procedure is the following: If the problem is feasible then $b_1 \leq s_1$ and hence $a_1, \cdots, a_{b_1} \geq 1$ (recall that the a_i's are indexed in descending order). Let $\alpha_{i1}=1$ for $i \leq b_1$, $\alpha_{i1}=0$ for $i > b_1$. Now consider the new problem, (I)$'$, with the matrix M' having m rows and $n-1$ columns, $j=2, \cdots, n$, with $a'_i = a_i - \alpha_{i1}$ and $b'_j = b_j$. We assert that (I)$'$ is again feasible so that by repeating the process we will eventually fill out the whole matrix.

To show that (I)$'$ is feasible we must prove, for any k,

$$\sum_{j=2}^{k+1} b_j \leq \sum_{j=1}^{k} s'_j = \sum_{i=1}^{m} \min[s'_i, k] ,$$

where $\{s'_i\}$ is the dual sequence to $\{a'_i\}$. The expression on the right can be rewritten

$$\sum_{i=1}^{m} \min[a'_i, k] = \sum_{i=1}^{b_1} \min[a_i - 1, k] + \sum_{i=b_1+1}^{m} \min[a_i, k] .$$

We must now consider two cases.

Case 1. $s_{k+1} \geq b_1$. Then $a_i - 1 \geq k$ for $i \leq b_1$ and hence $\min[a_i - 1, k]$

$=k=\min[a_i, k]$, so that we get

$$\sum_{j=1}^{k} s'_j = \sum_{i=1}^{m} \min[a'_i, k] = \sum_{i=1}^{m} \min[a_i, k] = \sum_{j=1}^{k} s_j \geq \sum_{j=1}^{k} b_j \geq \sum_{j=2}^{k+1} b_j .$$

Case 2. $s_{k+1} < b_1$. Then for $i \leq s_{k+1}$, $a_i \geq k+1$ so $a_i - 1 \geq k$ and $\min[a_i - 1, k] = k = \min[a_1, k]$. For $s_{k+1} < i \leq b_1$, $a_i \leq k$, so $\min[a_i - 1, k] = \min[a_i, k] - 1$, hence,

$$\sum_{i=1}^{m} \min[a'_i, k] = \sum_{i=1}^{m} \min[a_i, k] - b_1 + s_{k+1} = \sum_{j=1}^{k+1} s_j - b_1 \geq \sum_{i=2}^{k+1} b_j ,$$

since

$$\sum_{j=1}^{k+1} s_j \geq \sum_{j=1}^{k+1} b_j$$

by the feasibility condition. The proof is now complete.

In terms of the picnic problem, the n families should be seated in n stages according to the following simple rule: at each stage distribute the largest unseated family among those busses having the greatest number of vacant seats.

REFERENCES

1. G. B. Dantzig and D. R. Fulkerson, *On the max-flow min-cut theorem of networks*, Ann. of Math. Study No. **38**, *Contributions to linear inequalities and related topics*, edited by H. W. Kuhn and A. W. Tucker, 215-221.
2. L. R. Ford, Jr., and D. R. Fulkerson, *Maximal flow through a network*, Canad. J. Math. **8** (1956), 399-404.
3. ———, *A simple algorithm for finding maximal network flows aud an application to the Hitchcock problem*, Canad. J. Math. **9** (1957), 210-218.
4. P. Hall. *On Representatives of Subsets*, J. London Math. Soc., **10** (1935), 26-30.

THE RAND CORPORATION AND
BROWN UNIVERSITY

Reprinted from
Pacific J. Math. 7 (1957), 1073–1082

COMBINATORIAL PROPERTIES OF MATRICES OF ZEROS AND ONES

H. J. RYSER

1. Introduction. This paper is concerned with a matrix A of m rows and n columns, all of whose entries are 0's and 1's. Let the sum of row i of A be denoted by r_i $(i = 1, \ldots, m)$ and let the sum of column i of A be denoted by s_i $(i = 1, \ldots, n)$. It is clear that if τ denotes the total number of 1's in A

$$\tau = \sum_{i=1}^{m} r_i = \sum_{i=1}^{n} s_i.$$

With the matrix A we associate the *row sum vector*

$$R = (r_1, \ldots, r_m),$$

where the ith component gives the sum of row i of A. Similarly, the *column sum vector* S is denoted by

$$S = (s_1, \ldots, s_n).$$

We begin by determining simple arithmetic conditions for the construction of a (0, 1)-matrix A having a given row sum vector R and a given column sum vector S. This requires the concept of majorization, introduced by Muirhead. Then we apply to the elements of A an elementary operation called an interchange, which preserves the row sum vector R and column sum vector S, and prove that any two (0, 1)-matrices with the same R and S are transformable into each other by a finite sequence of such interchanges. The results may be rephrased in the terminology of finite graphs or in the purely combinatorial terms of set and element. Applications to Latin rectangles and to systems of distinct representatives are studied.

2. Maximal matrices and majorization. Let

$$\delta_i = (1, \ldots, 1, 0, \ldots, 0)$$

be a vector of n components with 1's in the first r_i positions, and 0's elsewhere. A matrix of the form

$$\bar{A} = \begin{bmatrix} \delta_1 \\ \vdots \\ \delta_m \end{bmatrix}$$

is called *maximal*, and we refer to $\bar{A}$ as the *maximal form* of A. The maximal $\bar{A}$ may be obtained from A by a rearrangement of the 1's in the rows of A. Also by inverse row rearrangements one may construct the given A from $\bar{A}$.

Received July 1, 1956. This work was sponsored in part by the Office of Ordnance Research.

Let $\bar{R} = (\bar{r}_1, \ldots, \bar{r}_m)$ and $\bar{S} = (\bar{s}_1, \ldots, \bar{s}_n)$ be the row sum and column sum vectors of $\bar{A}$. Evidently

$$R = \bar{R}.$$

Moreover, it is clear that the row sum vector R uniquely determines $\bar{A}$, and hence $\bar{S}$. Indeed, $\tau = \sum r_i = \sum \bar{s}_i$ constitute conjugate partitions of τ.

Consider two vectors $S = (s_1, \ldots, s_n)$ and $S^* = (s_1^*, \ldots, s_n^*)$, where the s_i and s_i^* are nonnegative integers. The vector S is *majorized* by S^*,

$$S < S^*,$$

provided that with the subscripts renumbered (**5**; **3**):

(1) $$s_1 \geq \ldots \geq s_n, \; s_1^* \geq \ldots \geq s_n^*;$$

(2) $$s_1 + \ldots + s_i \leq s_1^* + \ldots + s_i^*, \qquad i = 1, \ldots, n-1;$$

(3) $$s_1 + \ldots + s_n = s_1^* + \ldots + s_n^*.$$

For the vectors S and $\bar{S}$ associated with the matrices A and $\bar{A}$, respectively, we prove that

$$S < \bar{S}.$$

We renumber the subscripts of the s_i of A so that

$$s_1 \geq s_2 \geq \ldots \geq s_n.$$

For $\bar{A}$, we already have

$$\bar{s}_1 \geq \bar{s}_2 \geq \ldots \geq \bar{s}_n.$$

Now A must be formed from $\bar{A}$ by a shifting of 1's in the rows of $\bar{A}$. But for each $i = 1, \ldots, n-1$, the total number of 1's in the first i columns of $\bar{A}$ cannot be increased by a shifting of 1's in the rows of $\bar{A}$. Hence

$$s_1 + \ldots + s_i \leq \bar{s}_1 + \ldots + \bar{s}_i,$$

$i = 1, \ldots, n-1$. Moreover,

$$s_1 + \ldots + s_n = \bar{s}_1 + \ldots + \bar{s}_n,$$

whence we conclude that $S < \bar{S}$.

THEOREM 2.1[1]. *Let the matrix $\bar{A}$ be maximal and have column sum vector $\bar{S}$. Let S be majorized by $\bar{S}$. Then by rearranging 1's in the rows of $\bar{A}$, one may construct a matrix A having column sum vector S.*

Without loss of generality, we may assume that the column sums of A satisfy $s_1 \geq s_2 \geq \ldots \geq s_n$. We construct the desired A inductively by columns by a rearrangement of the 1's in the rows of $\bar{A}$.

[1]*Added in proof.* The author has been informed recently that Theorem 2.1 was obtained independently by Professor David Gale. His investigations concerning this theorem and certain generalizations are to appear in the Pacific Journal of Mathematics.

By hypothesis, $S \prec \bar{S}$, whence $s_1 \leqslant \bar{s}_1$. If $s_1 = \bar{s}_1$, we leave the first column of $\bar{A}$ unchanged. Suppose that $s_1 < \bar{s}_1$. We may rearrange 1's in the rows of $\bar{A}$ to obtain s_1 1's in the first column, unless

$$\bar{s}_2 > s_1, \ldots, \bar{s}_n > s_1.$$

But if these inequalities hold, then

$$\bar{s}_1 + \ldots + \bar{s}_n > ns_1 \geqslant s_1 + \ldots + s_n = \bar{s}_1 + \ldots + \bar{s}_n,$$

which is a contradiction.

Let us suppose then that the first t columns of A have been constructed, and let us proceed to the construction of column $t + 1$. We have then given an m by n matrix

$$[\eta_1, \ldots, \eta_t, \quad \eta_{t+1}, \ldots, \eta_n],$$

where the number of 1's in column η_i is s_i $(i = 1, \ldots, t)$. Let the number of 1's in column η_j be s'_j $(j = t + 1, \ldots, n)$. We may suppose that

$$s'_{t+1} \geqslant s'_{t+2} \geqslant \ldots \geqslant s'_n.$$

Two cases arise.

Case 1. $$s_{t+1} < s'_{t+1}.$$

In this case, remove 1's from column η_{t+1} by row rearrangements, and place the 1's in columns $\eta_{t+2}, \ldots, \eta_n$. If sufficiently many 1's may be removed from η_{t+1} in this manner, then we are finished. Suppose then that there remain e 1's in column $t + 1$, with

$$s_{t+1} < e \leqslant s'_{t+1},$$

and that no further 1's may be removed by this procedure. Then there must exist an integer $w \geqslant 0$ such that

$$s_{t+1} + \ldots + s_n = (n - t)e + w.$$

But

$$\begin{aligned} &s_{t+1} < e, \\ &s_{t+2} \leqslant s_{t+1} < e, \\ &\cdot \\ &\cdot \\ &\cdot \\ &s_n < e. \end{aligned}$$

Therefore

$$(n - t)e + w = s_{t+1} + \ldots + s_n < (n - t)e,$$

which is a contradiction.

Case 2. $$s'_{t+1} < s_{t+1}.$$

By row rearrangements, insert 1's into column η_{t+1} from columns $\eta_{t+2}, \ldots, \eta_n$. If sufficiently many 1's may be inserted in this manner, then we are finished. Suppose then that there remain e 1's in column $t + 1$ with

$$s'_{t+1} \leqslant e < s_{t+1},$$

and that no further 1's may be inserted by our procedure.

Let the matrix at this stage of the construction process be denoted by

$$[e_{rs}].$$

If now

$$e_{r,t+1} = 0,$$

then

$$e_{rj} = 0, \qquad (j = t + 1, \ldots, n).$$

Suppose that some

$$e_{rj} = 1, \qquad j \geqslant t + 2.$$

Then either

$$e_{rk} = 1, \qquad (k = 1, \ldots, t + 1),$$

or else for some k, $1 \leqslant k \leqslant t$,

$$e_{rk} = 0.$$

Consider the case in which $e_{rk} = 0$. Since $s_k \geqslant s_{t+1} > e$, there must exist

$$e_{pk} = 1, \quad e_{p,t+1} = 0.$$

Interchanging $e_{rj} = 1$ and $e_{rk} = 0$, and interchanging $e_{pk} = 1$ and $e_{p,t+1} = 0$, we see that $s_1, \ldots, s_t$ are left unaltered, and that e is increased by 1.

Continue to increase e by transformations of this variety. Suppose that all such transformations have been applied and that e still satisfies

$$s'_{t+1} \leqslant e < s_{t+1}.$$

But now it is no longer possible to move a 1 from columns $t + 2, \ldots, n$ into columns $1, 2, \ldots, t + 1$. This means that

$$s_1 + \ldots + s_t + e = \bar{s}_1 + \ldots + \bar{s}_t + \bar{s}_{t+1}.$$

But then

$$s_1 + \ldots + s_{t+1} \leqslant \bar{s}_1 + \ldots + \bar{s}_{t+1} = s_1 + \ldots + s_t + e,$$

whence $s_{t+1} \leqslant e$, which is a contradiction. This completes the proof.

The preceding theorem has a variety of applications. For example, let the (0, 1)-matrix A of m rows and n columns contain exactly $\tau = km$ 1's, where k is a positive integer. Let the column sum vector of A be $S = (s_1, \ldots, s_n)$. Then there exists an m by n matrix A^* composed of 0's and 1's with exactly k 1's in each row, and column sum vector S. For let $\bar{A}$ be m by n, with all 1's in the first k columns and 0's elsewhere. If $\bar{S}$ denotes the column sum vector of $\bar{A}$, then $S \prec \bar{S}$, and the desired A^* may be constructed from $\bar{A}$.

In this connection we mention the following result arising in the study of the completion of Latin rectangles (**1**; **7**). Let A be a (0, 1)-matrix of r rows and n columns, $1 \leqslant r < n$. Let there be k 1's in each row of A, and let the column sums of A satisfy $k - (n - r) \leqslant s_i \leqslant k$. Then $n - r$ rows of 0's and 1's may be adjoined to A to obtain a square matrix with exactly k 1's in each row and column (**7**). To prove this it suffices to construct an $n - r$ by n matrix A^* of 0's and 1's with exactly k 1's in each row, and column sum vector $(k - s_1, \ldots, k - s_n)$. By the remarks of the preceding paragraph, such a construction is always possible.

3. Interchanges. We return now to the m by n matrix A composed of 0's and 1's, with row sum vector R and column sum vector S. We are concerned with the 2 by 2 submatrices of A of the types

$$A_1 = \begin{bmatrix} 1 & 0 \\ 0 & 1 \end{bmatrix} \text{ and } A_2 = \begin{bmatrix} 0 & 1 \\ 1 & 0 \end{bmatrix}.$$

An *interchange* is a transformation of the elements of A that changes a specified minor of type A_1 into type A_2, or else a minor of type A_2 into type A_1, and leaves all other elements of A unaltered. Suppose that we apply to A a finite number of interchanges. Then by the nature of the interchange operation, the resulting matrix A^* has row sum vector R and column sum vector S.

THEOREM 3.1. *Let A and A^* be two m by n matrices composed of* 0'*s and* 1'*s, possessing equal row sum vectors and equal column sum vectors. Then A is transformable into A^* by a finite number of interchanges.*

The proof is by induction on m. For $m = 1$ and 2, the theorem is trivial. The induction hypothesis asserts the validity of the theorem for two (0, 1)-matrices of size $m - 1$ by n.

We attempt to transform the first row of A into the first row of A^* by interchanges. If we are successful, the theorem follows at once from the induction hypothesis. Suppose that we are not successful and that we denote the transformed matrix by A'. For notational convenience, we simultaneously permute the columns of A' and A^* and designate the first row of A' by

$$(\delta_r, \eta_s, \delta_t, \eta_t)$$

and the first row of A^* by

$$(\delta_r, \eta_s, \eta_t, \delta_t).$$

Here δ_r and δ_t are vectors of all 1's with r and t components, respectively, and η_s and η_t are 0 vectors with s and t components, respectively. Thus we have been successful in obtaining agreement between the two rows in the positions labelled δ_r and η_s, but have been unable to obtain agreement in the positions labelled δ_t and η_t. We may suppose, moreover, that these 2t positions of disagreement are the minimal number of disagreements obtainable among all attempts to transform the first row of A into the first row of A^* by interchanges.

Let A'_{m-1} and A^*_{m-1} denote the matrices composed of the last $m - 1$ rows of A' and A^*, respectively. The row sum vectors of A'_{m-1} and A^*_{m-1} are equal. Also corresponding columns of A'_{m-1} and A^*_{m-1} below the positions labelled δ_r and η_s have equal sums. Let α_i denote the $(r + s + i)$th column of A'_{m-1}, and let β_i denote the $(r + s + t + i)$th column of A'_{m-1}, where $i = 1, \ldots, t$. Let $\alpha_1^*, \ldots, \alpha_t^*$ and $\beta_1^*, \ldots, \beta_t^*$ denote the corresponding columns of A^*_{m-1}. Let a_i, b_i, a_i^*, b_i^* denote the column sums of $\alpha_i, \beta_i, \alpha_i^*, \beta_i^*$, respectively.

Now in A'_{m-1} we cannot have simultaneously a 0 in the position determined by row j and column α_i and a 1 in the position determined by row j and column β_i. For if this were the case, we could perform an interchange and reduce the $2t$ disagreements in the first row of A'. Hence $a_i \geqslant b_i$. Moreover, $a_i^* = a_i + 1$ and $b_i^* = b_i - 1$, whence

$$a_i^* - b_i^* = a_i - b_i + 2 \geqslant 2.$$

In A^*_{m-1}, consider columns α_i^* and β_i^*. There exists a row of A^*_{m-1} that has a 1 in column α_i^* and a 0 in column β_i^*. Replace the 1 by 0 and the 0 by 1, and let such a replacement be made for each $i = 1, \ldots, t$. We obtain in this way a new matrix $\tilde{A}_{m-1}$ whose row and column sum vectors are equal to those of A'_{m-1}. By the induction hypothesis, we may transform A'_{m-1} into $\tilde{A}_{m-1}$ by interchanges. However, these interchanges applied to A' will allow us to perform further interchanges and make the first rows of the transformed A' and A^* coincide. Hence the theorem follows.

Let $\mathfrak{A}$ denote the class of all (0, 1)-matrices of m rows and n columns, with row sum vector R and column sum vector S. The *term rank* ρ of A in $\mathfrak{A}$ is the order of the greatest minor of A with a nonzero term in its determinant expansion **(6)**. This integer is also equal to the minimal number of rows and columns that contain collectively all of the nonzero elements of A **(4)**. A (0, 1)-matrix $A = [a_{rs}]$ may be considered an incidence matrix distributing n elements $x_1, \ldots, x_n$ into m sets $S_1, \ldots, S_m$. Here $a_{ij} = 1$ or 0 according as x_j is or is not in S_i. From this point of view the term rank of a matrix is a generalization of the concept of a system of distinct representatives for subsets $S_1, \ldots, S_m$ of a finite set N **(2)**. Indeed, the subsets $S_1, \ldots, S_m$ possess a system of distinct representatives if and only if $\rho = m$.

THEOREM 3.2. *Let $\tilde{\rho}$ be the minimal and $\bar{\rho}$ the maximal term rank for the matrices in $\mathfrak{A}$. Then there exists a matrix in $\mathfrak{A}$ possessing term rank ρ, where ρ is an arbitrary integer on the range*

$$\tilde{\rho} \leqslant \rho \leqslant \bar{\rho}.$$

For an interchange applied to a matrix in $\mathfrak{A}$ either changes the term rank by 1 or else leaves it unaltered. But by Theorem 3.1, we may transform the matrix of term rank $\bar{\rho}$ into the matrix of term rank $\tilde{\rho}$. This implies that there exists a matrix in $\mathfrak{A}$ of term rank ρ.

REFERENCES

1. Marshall Hall, *An existence theorem for Latin squares*, Bull. Amer. Math. Soc., *51* (1945), 387–388.
2. P. Hall, *On representatives of subsets*, J. Lond. Math. Soc., *10* (1935), 26–30.
3. G. H. Hardy, J. E. Littlewood, and G. Pólya, *Inequalities* (Cambridge, 1952).
4. Dénes König, *Theorie der endlichen und unendlichen Graphen* (New York, 1950).
5. R. F. Muirhead, *Some methods applicable to identities and inequalities of symmetric algebraic functions of n letters*, Proc. Edinburgh Math. Soc., *21* (1903), 144–157.
6. Oystein Ore, *Graphs and matching theorems*, Duke Math. J., *22* (1955), 625–639.
7. H. J. Ryser, *A combinatorial theorem with an application to Latin rectangles*, Proc. Amer. Math. Soc., *2* (1951), 550–552.

Ohio State University

Reprinted from
Canad. J. Math. **9** (1957), 371–377

GRAPH THEORY AND PROBABILITY

P. ERDÖS

A well-known theorem of Ramsay **(8; 9)** states that to every n there exists a smallest integer $g(n)$ so that every graph of $g(n)$ vertices contains either a set of n independent points or a complete graph of order n, but there exists a graph of $g(n) - 1$ vertices which does not contain a complete subgraph of n vertices and also does not contain a set of n independent points. (A graph is called complete if every two of its vertices are connected by an edge; a set of points is called independent if no two of its points are connected by an edge.) The determination of $g(n)$ seems a very difficult problem; the best inequalities for $g(n)$ are **(3)**

$$(1) \qquad 2^{\frac{1}{2}n} < g(n) \leqslant \binom{2n-2}{n-1}.$$

It is not even known that $g(n)^{1/n}$ tends to a limit. The lower bound in (1) has been obtained by combinatorial and probabilistic arguments without an explicit construction.

In our paper **(5)** with Szekeres $f(k, l)$ is defined as the least integer so that every graph having $f(k, l)$ vertices contains either a complete graph of order k or a set of l independent points ($f(k, k) = g(k)$). Szekeres proved

$$(2) \qquad f(k, l) \leqslant \binom{k+l-2}{k-1}.$$

Thus for

$$k = 3, f(3, l) \leqslant \binom{l+1}{2}.$$

I recently proved by an explicit construction that $f(3, l) > l^{1+c_1}$ **(4)**. By probabilistic arguments I can prove that for $k > 3$

$$(3) \qquad f(k, l) > l \binom{k+l-2}{k-1}^{c_2},$$

which shows that (2) is not very far from being best possible.

Define now $h(k, l)$ as the least integer so that every graph of $h(k, l)$ vertices contains either a closed circuit of k or fewer lines, or that the graph contains a set of l independent points. Clearly $h(3, l) = f(3, l)$.

By probabilistic arguments we are going to prove that for fixed k and sufficiently large l

$$(4) \qquad h(k, l) > l^{1+1/2k}.$$

Further we shall prove that

Received December 13, 1957.

(5) $$h(2k+1, l) < c_3\, l^{1+1/k},\ h(2k+2, l) < c_3\, l^{1+1/k}.$$

A graph is called r chromatic if its vertices can be coloured by r colours so that no two vertices of the same colour are connected; also its vertices cannot be coloured in this way by $r-1$ colours. Tutte (1, 2) first showed that for every r there exists an r chromatic graph which contains no triangle and Kelly (6) showed that for every r there exists an r chromatic graph which contains no k-gon for $k \leqslant 5$. (Tutte's result was rediscovered several times, for instance, by Mycielski (7). It was asked if such graphs exist for every k.) Now (4) clearly shows that this holds for every k and in fact that there exists a graph of n vertices of chromatic number $> n^{\epsilon}$ which contains no closed circuit of fewer than k edges.

Now we prove (4). Let n be a large number,

$$0 < \epsilon < \frac{1}{k}$$

is arbitrary. Put $m = [n^{1+\epsilon}]$ ($[x]$ denotes the integral part of x, that is, the greatest integer not exceeding x), $p = [n^{1-\eta}]$ where $0 < \eta < \epsilon/2$ is arbitrary. Let $\mathfrak{G}^{(n)}$ be the complete graph of n vertices $x_1, x_2, \ldots, x_n$ and $\mathfrak{G}^{(p)}$ any of its complete subgraphs having p vertices. Clearly we can choose $\mathfrak{G}^{(p)}$ in $\binom{n}{p}$ ways. Let

$$\mathfrak{G}_\alpha^{(n)},\ 1 \leqslant \alpha \leqslant \binom{\binom{n}{2}}{m}$$

be an arbitrary subgraph of $\mathfrak{G}^{(n)}$ having m edges (the number of possible choices of α is clearly as indicated).

First of all we show that for almost all α $\mathfrak{G}_\alpha^{(n)}$ has the property that it has more than n common edges with every $\mathfrak{G}^{(p)}$. Almost all here means: for all α's except for

$$o\binom{\binom{n}{2}}{m}.$$

Let the vertices of $\mathfrak{G}^{(p)}$ be $x_1, x_2, \ldots, x_p$. The number of graphs $\mathfrak{G}_\alpha^{(n)}$ containing not more than n of the edges (x_i, x_j), $1 \leqslant i < j \leqslant p$ equals by a simple combinatorial reasoning

$$\sum_{l=0}^{n} \binom{\binom{p}{2}}{l}\binom{\binom{n}{2}-\binom{p}{2}}{m-l} < (n+1)\binom{\binom{p}{2}}{n}\binom{\binom{n}{2}-\binom{p}{2}}{m}$$

$$< p^{2n}\binom{\binom{n}{2}-\binom{m}{2}}{m} < \binom{\binom{n}{2}}{m} p^{2n}\left(1-\frac{\binom{p}{2}}{\binom{n}{2}}\right)^m < \binom{\binom{n}{2}}{m} p^{2n}\left(1-\frac{p^2}{n^2}\right)^m$$

$$< \binom{\binom{n}{2}}{m} p^{2n} \exp\left(-\frac{mp^2}{n^2}\right).$$

Now the number of possible choices for $\mathfrak{G}^{(p)}$ is

$$\binom{n}{p} < n^p < p^n.$$

Thus the number of α's for which there exists a $\mathfrak{G}^{(p)}$ so that $\mathfrak{G}^{(p)} \cap \mathfrak{G}_\alpha^{(n)}$ has not more than n^ϵedges is less than ($\eta < \epsilon/2$)

$$\binom{\binom{n}{2}}{m} p^{3n} \exp(-n^{1+\epsilon-2\eta}) = o\binom{\binom{n}{2}}{m}$$

as stated.

Unfortunately almost all of these graphs $\mathfrak{G}_\alpha^{(n)}$ contain closed circuits of length not exceeding k (in fact almost all of them contain triangles). But we shall now prove that almost all $\mathfrak{G}_\alpha^{(n)}$ contain fewer than n/k closed circuits of length not exceeding k.

The number of graphs $\mathfrak{G}_\alpha^{(n)}$ which contain a given closed circuit (x_1, x_2), $(x_2, x_3), \ldots, (x_l, x_1)$ clearly equals

$$\binom{\binom{n}{2} - l}{m - l}.$$

The circuit is determined by its vertices and their order—thus there are $n(n-1)\ldots(n-l+1)$ such circuits. Therefore the expected number of closed circuits of length not exceeding k equals

$$\binom{\binom{n}{2}}{m}^{-1} \sum_{l=3}^{k} l! \binom{n}{l} \binom{\binom{n}{2} - l}{m - l} < (1 + o(1)) \sum_{l=3}^{k} n^l \left(\frac{m}{\binom{n}{2}}\right)^l$$

$$< (1 + o(1))\, n^k \frac{(2m)^k}{n^{2k}} = o(n)$$

since $\epsilon < 1/k$. Therefore, by a simple and well-known argument, the number of the α's for which $\mathfrak{G}_\alpha^{(n)}$ contains n/k or more closed paths of length not exceeding k is

$$o\binom{\binom{n}{2}}{m},$$

as stated.

Thus we see that for almost all α $\mathfrak{G}_\alpha^{(n)}$ has the following properties: in every $\mathfrak{G}^{(p)}$ it has more than n edges and the number of its closed circuits having k or fewer edges is less than n/k. Omit from $\mathfrak{G}_\alpha^{(n)}$ all the edges contained in a closed circuit of k or fewer edges. By what has just been said we omit fewer than n edges. Thus we obtain a new graph $\mathfrak{G}_\alpha'^{(n)}$ which by construction does not contain a closed circuit of k or fewer edges. Also clearly $\mathfrak{G}_\alpha'^{(n)} \cap \mathfrak{G}^{(p)}$

is not empty for every $\mathfrak{G}^{(p)}$. Thus the maximum number of independent points in $\mathfrak{G}_\alpha'^{(n)}$ is less than $p = [n^{1-\eta}]$, or

$$h(k, [n^{1-\eta}]) > n$$

which proves (4).

By more complicated arguments one can improve (4) considerably; thus for $k = 3$ I can show that for every $\epsilon > 0$ and sufficiently large l

$$f(3, l) = h(3, l) > l^{2-\epsilon},$$

which by (2) is very close to the right order of magnitude.

At the moment I am unable to replace the above "existence proof" by a direct construction.

By using a little more care I can prove by the above method the following result: there exists a (sufficiently small) constant c_4 so that for every k and l

$$(6) \qquad h(k, l) > c_4\, l^{1+\frac{1}{3k}}.$$

(If $k > c \log l$ (6) is trivial since $h(k, l) \geqslant l$.)

From (6) it is easy to deduce that to every r there exists a c_5 so that for $n > n_0(r, c_5)$ there exists an r chromatic graph of n vertices which does not contain a closed circuit of fewer than $[c_5 \log n]$ edges. I am not sure if this result is best possible.

We do not give the details of the proof of (3) since it is simpler than that of (4). For $k = 3$ (3) follows from (4). If $k > 3$, put

$$m = c_6[n^{2-\frac{2}{k-1}}]$$

and denote by $\mathfrak{G}_\alpha^{(n)}$ the "random" graph of m edges. By a simple computation it follows that for sufficiently small c_6, $\mathfrak{G}_\alpha^{(n)}$ does not contain a complete graph of order k for more than

$$0 \cdot 9 \binom{\binom{n}{2}}{m}$$

values of α, and that for more than this number of values of α $\mathfrak{G}_\alpha^{(n)}$ does not contain a set of $c_7 n^{2/k-1} \log n$ independent points ($c_7 = c_7(c_6)$ is sufficiently large). Thus

$$f(k, c_7 n^{2/k-1} \log n) > n,$$

which implies (3) by a simple computation.

Now we prove (5). It will clearly suffice to prove the first inequality of (5). We use induction on l. Let there be given a graph $\mathfrak{G}$ having $h(2k + 1, l) - 1$ vertices which does not contain a closed circuit of $2k + 1$ or fewer edges and for which the maximum number of independent points is less than l. If every point of $\mathfrak{G}$ has order at least $[l^{1/k}] + 2$ (the order of a vertex is the number of edges emanating from it) then, starting from an arbitrary point, we reach in k steps at least l points, which must be all distinct since otherwise $\mathfrak{G}$ would have to contain a closed circuit of at most $2k$ edges. The endpoints thus obtained must be independent, for if two were connected by an edge $\mathfrak{G}$ would contain a closed circuit of $2k + 1$ edges. Thus $\mathfrak{G}$ would have a set of at least l independent points, which is false.

Thus $\mathfrak{G}$ must have a vertex x_1 of order at most $[l^{1/k}] + 1$. Omit the vertex x_1 and all the vertices connected with it. Thus we obtain the graph $\mathfrak{G}'$ and x_1 is not connected with any point of $\mathfrak{G}'$, thus the maximum number of independent points of $\mathfrak{G}'$ is $l - 1$, or $\mathfrak{G}'$ has at most $h(2k + 1, l - 1) - 1$ vertices, hence

$$h(2k + 1, l) \leqslant h(2k + 1, l - 1) + [l^{1/k}] + 2$$

which proves (5).

References

1. Blanche Descartes, *A three colour problem*, Eureka (April, 1947). (Solution March, 1948.)
2. ——— *Solution to Advanced Problem no. 4526*, Amer. Math. Monthly, *61* (1954), 352.
3. P. Erdös, *Some remarks on the theory of graphs*, B.A.M.S. *53*, (1947), 292–4.
4. ——— *Remarks on a theorem of Ramsey*, Bull. Research Council of Israel, Section F, *7* (1957).
5. P. Erdös and G. Szekeres, *A combinatorial problem in geometry*, Compositio Math. *2* (1935) 463–70.
6. J. B. Kelly and L. M. Kelly, *Paths and circuits in critical graphs*, Amer. J. Math., *76* (1954), 786–92.
7. J. Mycielski, *Sur le colorage des graphs*, Colloquium Math. *3* (1955), 161–2.
8. F. P. Ramsay, *Collected papers*, 82–111.
9. T. Skolem, *Ein kombinatorischer Satz mit Anwendung auf ein logisches Entscheidungs problem*, Fund. Math. *20* (1933), 254–61.

University of Toronto
and
Technion, Haifa

Reprinted from
Canad. J. Math. **11** (1959), 34–38

Kasteleyn, P. W.
1961

Physica 27
1209–1225

THE STATISTICS OF DIMERS ON A LATTICE

I. THE NUMBER OF DIMER ARRANGEMENTS ON A QUADRATIC LATTICE

by P. W. KASTELEYN

Koninklijke/Shell-Laboratorium, Amsterdam, Nederland
(Shell Internationale Research Maatschappij N.V.)

Synopsis

The number of ways in which a finite quadratic lattice (with edges or with periodic boundary conditions) can be fully covered with given numbers of "horizontal" and "vertical" dimers is rigorously calculated by a combinatorial method involving Pfaffians. For lattices infinite in one or two dimensions asymptotic expressions for this number of dimer configurations are derived, and as an application the entropy of a mixture of dimers of two different lengths on an infinite rectangular lattice is calculated. The relation of this combinatorial problem to the Ising problem is briefly discussed.

§ 1. *Introduction.* Combinatorial problems relating to a regular space lattice arise in the theory of various physical phenomena. One of these problems is the "*arrangement problem*", which plays a role in the explanation of the non-ideal thermodynamic behaviour of liquids consisting of molecules of different size with zero energy of mixing (athermal mixtures). In the investigations devoted to this problem (most of which have been discussed critically by Guggenheim [1])) much attention has been paid to the so-called quasi-crystalline model. One considers a regular lattice consisting of points (sites, vertices) connected by bonds. This lattice is fully covered with *monomers* (molecules occupying one site) and rigid or flexible *polymers* (molecules occupying several sites connected by bonds); the latter may be dimers, trimers etc., but also "high polymers". If the energy of mixing is zero the thermodynamic properties of this system can be calculated from the *combinatorial factor*, i.e. the number of ways of arranging given numbers of monomers and polymers on the lattice.

The same combinatorial problem arises in the cell-cluster theory of the liquid state [2]). There one divides the volume of a liquid into a set of cells of which the centres form a regular lattice, and one considers situations in which, by the removal of certain interfaces between cells, a number of double cells, triple cells etc. have been formed. For the calculation of the free energy of the liquid one then has to determine the number of ways in which a given volume can be divided into given numbers of single, double, triple cells etc.; the equivalence of this combinatorial factor with the former one is obvious.

A two-dimensional form of the arrangement problem is encountered in the theory of adsorption of diatomic, triatomic etc. molecules on a regular surface. The empty sites of the surface then play the role of "monomers".

As in many problems of this sort, it is easy to find the most general solution for a one-dimensional lattice [2]), it appears very difficult to find a more or less general solution for two-dimensional lattices, whereas for three dimensions any exact solution seems extremely remote. Therefore one generally uses approximation methods; we refer to the work of Fowler and Rushbrooke [3]) (who also made some rigorous calculations on two- and three-dimensional infinite strips of finite width), Chang, Flory, Huggins, Miller, Guggenheim (for detailed references see ref. 1), Orr [4]), Rushbrooke, Scoins and Wakefield[5]), and Cohen, De Boer and Salsburg[2]). Recently, Green and Leipnik[6]) claimed to have found a rigorous solution for the case of monomer-dimer mixtures on a two-dimensional lattice, but Fisher and Temperley [7]), and Katsura and Inawashiro [7]) proved that their results were not correct.

In this paper we present a rigorous solution to the above-mentioned combinatorial problem for a very special case, viz. that of a *two-dimensional quadratic lattice, completely covered with dimers* (in terms of graph theory we ask for the number of "perfect matchings" of the lattice [8])). Both the absence of monomers and the dimension of the lattice form serious restrictions, but it is hoped that the present investigation may be useful as a first step. The situation has some resemblance to that of the combinatorial problem connected with the Ising model of cooperative phenomena [9]), for which an exact solution has been given for an equally special case [10–13]). It will be shown that the two problems are to a certain extent analogous. In § 2 and § 3 we shall develop the method of solution for the case of a finite lattice imbedded in a plane (i.e. a rectangle with edges), and in § 4 for that of a lattice imbedded in a torus (i.e. with periodic boundary conditions). In § 5 an alternative method will be sketched. As an application, the entropy of a certain mixture of dimers is calculated in § 6.

It is intended to treat in a subsequent paper the statistics of dimers on other two-dimensional lattices, to discuss boundary effects and to make some remarks on three-dimensional lattices.

§ 2. *The planar quadratic lattice.* Consider a planar quadratic $m \times n$ lattice Q_{mn} to which one can attach dimers (figures consisting of two linked vertices) in such a way that every dimer occupies two lattice points connected by a bond. We indicate the lattice points by (i, j) or p $(i = 1, \ldots, m;$ $j = 1, \ldots, n;\ p = 1, \ldots, mn)$, the number of "horizontal" dimers (occupying two points (i, j) and $(i+1, j)$) by N_2 and the number of "vertical" dimers (occupying two points (i, j) and $(i, j+1)$) by N_2'. If $g(N_2, N_2')$ is the *combinatorial factor*, i.e. the number of ways of covering the lattice with dimers so that every site is covered by one and only one dimer vertex, we ask for the *configuration generating function*

$$Z_{mn}(z, z') = \sum'_{N_2, N_2'} g(N_2, N_2')\, z^{N_2} z'^{N_2'}, \tag{1}$$

where the sum runs over all combinations N_2, N_2' satisfying $2(N_2 + N_2') = mn$; if desired, the counting variables z and z' may be viewed as activities and Z_{mn} as the configurational partition function. At least one of the two numbers m and n has to be even; let m be even. We shall refer to the arrangement of dimers occupying the pairs of sites p_1 and p_2, p_3 and p_4, p_5 and p_6, etc. as to the *configuration* $C = |p_1; p_2|\, p_3; p_4\, |p_5; p_6| \ldots |p_{mn-1}; p_{mn}|$. A simple but important configuration is

$$C_0 = |1,1; 2,1|3,1; 4,1| \ldots |m-1, 1; m, 1|1,2; 2,2| \ldots |m-1, n; m, n|,$$

which we shall call the *standard configuration* (fig. 1*a*). We could, however, represent this arrangement of dimers equally well by $|2,1; 1,1|3,1; 4,1| \ldots$ or by $|4,1; 3,1|1,1; 2,1| \ldots$ etc. To make the representation unique we order the points of the lattice row after row by choosing the p-numbering as follows:

$$(i, j) \leftrightarrow p = (j-1)\, m + i, \tag{2}$$

and we introduce the convention that the points of a configuration shall be indicated in the following ("canonical") order:

$$p_1 < p_2;\ p_3 < p_4;\ \ldots;\ p_{mn-1} < p_{mn}; \tag{3a}$$

$$p_1 < p_3 < \ldots < p_{mn-1}. \tag{3b}$$

By analogy to the determinantal approach to the Ising problem developed by Kac and Ward [11]) we shall try to construct a mathematical form consisting of a series of terms each of which corresponds uniquely to one configuration and has the "weight" $z^{N_2} z'^{N_2'}$ of this configuration. The conditions (3) strongly suggest that this form should be a *Pfaffian* rather than a determinant. A Pfaffian is a number attributed to a triangular array of coefficients $a(k; k')$ ($k = 1, \ldots, N$; $k' = 1, \ldots, N$; $k < k'$; N even) in the following way [14]):

$$\mathrm{Pf}\{a(k; k')\} = \sum'_{\mathrm{P}} \delta_{\mathrm{P}} a(k_1; k_2)\, a(k_3; k_4) \ldots a(k_{N-1}; k_N), \tag{4}$$

where the sum runs over those permutations $k_1, k_2, \ldots, k_N$ of the numbers $1, 2, \ldots, N$ which obey

$$k_1 < k_2;\ k_3 < k_4;\ \ldots;\ k_{N-1} < k_N;\ k_1 < k_3 < \ldots < k_{N-1}, \tag{3'}$$

and where δ_{P} is the parity of the permutation P, i.e. -1 or $+1$ according as P is an odd or an even permutation. Pfaffians have been introduced into physics by Caianello and Fubini [15]) and in lattice-combinatorial problems by Hurst and Green [13]).

We shall now show that it is possible indeed to define a triangular array of elements $D(p; p')$ so that

$$Z_{mn}(z, z') = \mathrm{Pf}\{D(p; p')\}. \tag{5}$$

We begin by noting that if we define $D(p; p') = 0$ for all pairs of sites $(p; p')$ that are not connected by a bond, all terms in the Pfaffian that would not correspond to a dimer configuration will vanish. Next we put the coefficients $D(p; p')$ corresponding to pairs of sites that are connected by a horizontal or a vertical bond equal, in absolute magnitude, to z and z', respectively. In this way we get all configurations represented by a term of the proper weight; the conditions (3) and (3′) ensure that the correspondence is one-to-one.

Finally, in order that all configurations are counted positively we have to choose the signs of the non-zero elements such that the product $D(p_1; p_2)\, D(p_3; p_4) \ldots$ has the same sign as the parity δ_P. It is evident that the product corresponding to the standard configuration C_0 has to be positive. Now, from C_0 one can obtain any arbitrary configuration C in the following way. We draw a picture of the lattice in which every dimer of the configuration C_0 is represented by a dotted line and every dimer of the configu-

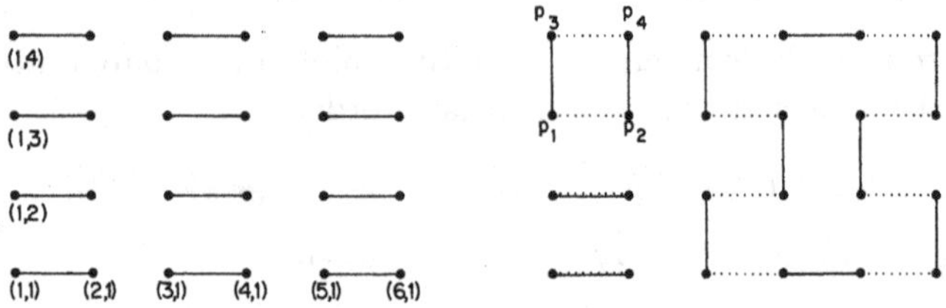

(a) Fig. 1. (b)

(a) The standard configuration C_0 of the planar lattice Q_{64}

(b) The construction of a new configuration C (full lines) from C_0 (dotted lines)

guration C by a full line (fig. 1b). Since any lattice point is the endpoint of just one line of each type the resulting figure consists of: a) pairs of sites connected both by a dotted line and by a full line; b) closed polygons consisting of alternating dotted and full lines (to be called C_0-bonds and non-C_0-bonds). If we then take the configuration C_0 and we shift in each of the "alternating" polygons all dimers clockwise or counter-clockwise by one step, C_0 goes over into C.

By analogy to the Ising problem [9]) one might expect that each of the polygons (representing a cyclic permutation of an even number of lattice sites) would contribute a factor -1 to the parity δ_P of the permutation P corresponding to C. In fact, this is true , but owing to the restrictions on the permutations occurring in a Pfaffian, the proof is not so simple as in the Ising problem. Consider e.g. the small square in fig. 1*b*. According to (3′) its vertices occur in the term representing C_0 as ... $D(p_1; p_2)$ ... $D(p_3; p_4)$ ... and in the term representing C as ... $D(p_1; p_3)\,D(p_2; p_4)$ .... Obviously the change from $p_1p_2p_3p_4$ to $p_1p_3p_2p_4$ is not a cyclic permutation of the four points along the square. It can, however, be considered as the product of the following permutations: $p_1p_2p_3p_4 \to p_1p_2p_4p_3$ (putting the points into a cyclic order corresponding to the square) $\to p_2p_4p_3p_1$ (permuting the four points cyclically) $\to p_2p_4p_1p_3$ (ordering the points p_1 and p_3 according to (3*a*)) $\to p_1p_3p_2p_4$ (ordering the pairs $(p_1; p_3)$ and $(p_2; p_4)$ according to (3*b*)). The resulting permutation is odd; apparently the parities of the re-ordering permutations just compensate each other.

To show that this is true in general we first take a configuration which differs from C_0 only in the position of the dimers on one polygon. Consider a column of "C_0-bonds" which is crossed by this polygon. If we describe a closed path along the polygon, we cross this column as many times in the "forward direction" (i.e., in the direction of increasing p) as in the "backward direction". This is true for any column of C_0-bonds, and therefore, if the path contains r forward steps along C_0-bonds it must also contain r backward steps along C_0-bonds. By a similar argument combined with the alternation of the two types of bonds, we can show that it also contains r forward and r backward steps along non-C_0-bonds. In the terms of the Pfaffian which correspond to C_0 and C, on the other hand, all points of the polygon occur in the order of increasing p. Consequently the permutation which changes the "C_0-term" into the "C-term" can be considered as the product of: 1) the reversal of the r pairs of sites (C_0-bonds) for which the canonical order is opposite to the required cyclic order; 2) the rearrangement of the $2r$ C_0-bonds which is needed to get all polygon vertices in the cyclic order; 3) the cyclic permutation of these $4r$ vertices; 4) the reversal of the r pairs of sites which now violate (3*a*); 5) the rearrangement of the $2r$ pairs which is needed to satisfy (3*b*). Each reversal within a pair contributes a factor -1 to the parity of the resulting permutation, a reshuffling of the pairs a factor $+1$, and the cyclic permutation a factor $(-1)^{4r-1}$. We thus find that the total parity of the permutation is $(-1)^r(+1)(-1)^{4r-1}(+1)(-1)^r = -1$. If there is more than one polygon, we can perform the corresponding permutations consecutively; each polygon then contributes a factor -1.

It shall now be indicated how these factors can be compensated for. First we remark that since no two C_0-bonds have a point in common, an alternating polygon can neither intersect itself nor cross or touch other polygons. Therefore we shall not encounter such difficulties as arose in the corresponding step of Kac and Ward's method [11) 12)].

Any alternating polygon can be considered as built up from horizontal strips of connected unit squares. From the requirement that during the cyclic shift the opposite sides of a connected figure are shifted in opposite directions, combined with the alternation of C_0-bonds and non-C_0-bonds we conclude that both the numbers of unit squares in a strip and the number of strips are odd (cf. fig. 1*b*). Consequently, each alternating polygon encircles an odd number of unit squares, and it will be sufficient to choose the signs of the $D(p; p')$ such that among the four bonds bounding a unit square there is an odd number having a negative $D(p; p')$; we have further seen that the standard configuration has to appear with a positive sign. This can be realized e.g. by attributing minus signs to the coefficients of the vertical bonds between lattice sites of odd i. We thus get the following set of coefficients

$$\begin{aligned} &D(i, j; i+1, j) = +z && \text{for } 1 \le i \le m-1,\ 1 \le j \le n, \\ &D(i, j; i, j+1) = (-1)^i z' && \text{for } 1 \le i \le m, \qquad 1 \le j \le n-1, \\ &D(i, j; i', j') \quad = 0 && \text{otherwise} \end{aligned} \tag{6}$$

The equations (5) and (6) are sufficient to derive the configuration generating function $Z_{mn}(z, z')$. It should be remarked that throughout this paper we assume that the dimers are symmetric. If they were asymmetric all elements $D(p; p')$ would have to be multiplied by 2, since any pair of sites might then be occupied in two distinguishable ways.

§ 3. *The evaluation of the Pfaffian.* For the evaluation of $\mathrm{Pf}\{D(p; p')\}$ we make use of the property of a Pfaffian that its square is equal to the determinant of the skew-symmetric matrix to which the given triangular array of coefficients can be extended [14]). That is, in our case,

$$Z^2_{mn}(z, z') = [\mathrm{Pf}\ \mathsf{D}]^2 = \det \mathsf{D}, \tag{7}$$

where D is the matrix given by (6) together with the requirement of skew symmetry:

$$D(i, j; i', j') = -D(i', j'; i, j), \tag{8}$$

and Pf D stands for $\mathrm{Pf}\{D(p; p')\}$.

If D were a completely periodic matrix, it could easily be brought into a diagonal form, viz. by a Fourier-type similarity transformation [9]), and the calculation of the determinant would be straightforward. However, the truncated edges of the lattice Q_{mn} disturb the periodicity of the matrix. Fortunately, it is still possible to bring it into a "nearly diagonal" form. We write D as the sum of two direct products of a $m \times m$ matrix and $n \times n$ matrix:

$$\mathsf{D} = z(\mathsf{Q}_m \times \mathsf{E}_n) + z'(\mathsf{F}_m \times \mathsf{Q}_n), \tag{9}$$

where E is the unit matrix,

$$Q=\begin{bmatrix} 0 & 1\,0\,0 & . & . & . & 0\,0 \\ -1 & 0\,1\,0 & . & . & . & 0\,0 \\ 0 & -1\,0\,1 & . & . & . & 0\,0 \\ . & . \; . \; . & . & . & . & . \; . \\ 0 & 0\,0\,0 & . & . & . & 0\,1 \\ 0 & 0\,0\,0 & . & . & . & -1\,0 \end{bmatrix}, F=\begin{bmatrix} -1\,0 & 0 & . & . & . & 0\,0 \\ 0\,1 & 0 & . & . & . & 0\,0 \\ 0\,0 & -1 & . & . & . & 0\,0 \\ . \; . & . & . & . & . & . \; . \\ 0\,0 & 0 & . & . & . & -1\,0 \\ 0\,0 & 0 & . & . & . & 0\,1 \end{bmatrix}, \tag{10}$$

and the indices indicate the order of the matrices. It can be verified that Q_n can be diagonalized by a similarity transformation $\tilde{Q}_n = U_n^{-1}Q_nU_n$ with the matrix U_n given by

$$\begin{aligned} U_n(l;l') &= \{2/(n+1)\}^{\frac{1}{2}}\, \mathrm{i}^l \sin\{ll'\,\pi/(n+1)\}, \\ U_n^{-1}(l;l') &= \{2/(n+1)\}^{\frac{1}{2}}\, (-\mathrm{i})^{l'} \sin\{ll'\pi/(n+1)\}; \end{aligned} \tag{11}$$

the diagonal elements of $\tilde{Q}_n$ are the eigenvalues $2\mathrm{i}\cos\{l\pi/(n+1)\}$ of $Q_n(l = 1, \ldots, n)$. On the other hand, this transformation obviously leaves E_n invariant. Q_m can be diagonalized analogously by a transformation with U_m, but this transformation disturbs the diagonal form of F_m, although not seriously. Transforming D with the direct product $U = U_m \times U_n$ we find that $\tilde{D} = U^{-1}DU$ has the following elements:

$$\tilde{D}(k,l;k',l') = 2\mathrm{i}z\, \delta_{k,k'}\, \delta_{l,l'} \cos\{k\pi/(m+1)\} - 2\mathrm{i}z'\, \delta_{k+k',m+1}\, \delta_{l,l'} \times \cos\{l\pi/(n+1)\};$$

i.e. the only non-zero elements are grouped in 2×2 blocks along the diagonal. Thus the determinant is readily found:

$$\det D = \det \tilde{D} = \prod_{k=1}^{\frac{1}{2}m} \prod_{l=1}^{n} \begin{vmatrix} 2\mathrm{i}z \cos \dfrac{k\pi}{m+1} & -2\mathrm{i}z' \cos \dfrac{l\pi}{n+1} \\ -2\mathrm{i}z' \cos \dfrac{l\pi}{n+1} & -2\mathrm{i}z \cos \dfrac{k\pi}{m+1} \end{vmatrix}, \tag{12}$$

and, from (7) and (12), we find the following expression for the configuration generating function of the lattice Q_{mn}:

$$\boxed{\begin{aligned} Z_{mn}(z,z') &= \prod_{k=1}^{\frac{1}{2}m} \prod_{l=1}^{n} 2\left[z^2 \cos^2 \frac{k\pi}{m+1} + z'^2 \cos^2 \frac{l\pi}{n+1}\right]^{\frac{1}{2}} = \\ &= \begin{cases} 2^{\frac{1}{2}mn} \displaystyle\prod_{k=1}^{\frac{1}{2}m} \prod_{l=1}^{\frac{1}{2}n} \left[z^2 \cos^2 \frac{k\pi}{m+1} + z'^2 \cos^2 \frac{l\pi}{n+1}\right], & (n \text{ even}) \\ 2^{\frac{1}{2}m(n-1)}\, z^{\frac{1}{2}m} \displaystyle\prod_{k=1}^{\frac{1}{2}m} \prod_{l=1}^{\frac{1}{2}(n-1)} \left[z^2 \cos^2 \frac{k\pi}{m+1} + z'^2 \cos^2 \frac{l\pi}{n+1}\right]. & (n \text{ odd}) \end{cases} \end{aligned}} \tag{13}$$

In writing down the expression valid for odd n use has been made of the relation

$$\prod_{k=1}^{\frac{1}{2}m} 2\cos\{k\pi/(m+1)\} = 1,$$

which is a particular case of the identity

$$\prod_{k=1}^{\frac{1}{2}m} 4\left[u^2 + \cos^2 \frac{k\pi}{m+1}\right] \equiv \frac{[u + (1+u^2)^{\frac{1}{2}}]^{m+1} - [u - (1+u^2)^{\frac{1}{2}}]^{m+1}}{2(1+u^2)^{\frac{1}{2}}} \, ; \quad (14)$$

this identity holds (for even m) because the two members represent two polynomials in u of the same degree m, with the same zeros, and with equal coefficients of the leading term. For numerical calculations it is sometimes useful to perform, with the aid of (14), the product over k in (13). In this way we find

$$Z_{mn}(z, z') = z^{\frac{1}{2}mn} \prod_{l=1}^{[\frac{1}{2}n]} \frac{\left\{ \left[\zeta \cos \frac{l\pi}{n+1} + \left(1+\zeta^2 \cos^2 \frac{l\pi}{n+1}\right)^{\frac{1}{2}}\right]^{m+1} - \left[\zeta \cos \frac{l\pi}{n+1} - \left(1+\zeta^2 \cos^2 \frac{l\pi}{n+1}\right)^{\frac{1}{2}}\right]^{m+1} \right\}}{2\left(1+\zeta^2 \cos^2 \frac{l\pi}{n+1}\right)^{\frac{1}{2}}}, \quad (15)$$

where $\zeta = z'/z$ and $[\frac{1}{2}n] = \frac{1}{2}n$ or $\frac{1}{2}(n-1)$ according as n is even or odd.

In the limit $m \to \infty$, i.e. for *infinitely long strips* of finite width n, we get

$$Z_n(z, z') = \lim_{m\to\infty} \{Z_{mn}(z, z')\}^{1/m} = z^{\frac{1}{2}n} \prod_{l=1}^{\frac{1}{2}n} \left[\zeta \cos \frac{l\pi}{n+1} + \left(1+\zeta^2 \cos^2 \frac{l\pi}{n+1}\right)^{\frac{1}{2}}\right]. \quad (16)$$

Finally we have, in the limit of an *infinitely large lattice*:

$$\begin{aligned} Z(z, z') &= \lim_{m,n\to\infty} \{Z_{mn}(z, z')\}^{1/mn} = \\ &= \exp\left\{\pi^{-2} \int_0^{\pi/2} d\omega \int_0^{\pi/2} d\omega' \ln 4[z^2 \cos^2\omega + z'^2 \cos^2\omega']\right\} = \qquad (17) \\ &= z^{\frac{1}{2}} \exp\left\{\pi^{-1} \int_0^{\pi/2} d\omega \ln [\zeta \cos \omega + (1 + \zeta^2 \cos^2\omega)^{\frac{1}{2}}]\right\}. \end{aligned}$$

For $|z'| \leq |z|$ we may expand the latter integrand in terms of ζ, integrate term by term, and sum the resulting series, which gives:

$$\begin{aligned} \ln Z(z, z') &= \tfrac{1}{2} \ln z + \pi^{-1} \sum_{j=0}^{\infty} (-1)^j (2j+1)^{-2} \zeta^{2j+1} = \\ &= \tfrac{1}{2} \ln z + \pi^{-1} \int_0^{\zeta} dx\, x^{-1} \arctan x = \qquad (18) \\ &= \tfrac{1}{2} \ln z' + \pi^{-1} \int_0^{1/\zeta} dx\, x^{-1} \arctan x. \end{aligned}$$

From the equivalence of the last two expressions (which is easily proved) and the analogous derivation in the case $|z'| \geq |z|$ it follows that either of them may be used for all values of z and z'. Using the relation $\arctan x = (2i)^{-1}[\ln(1+ix) - \ln(1-ix)]$ and introducing the function

$\Lambda_2(x) = (2i)^{-1}[L_2(ix) - L_2(-ix)]$, where $L_2(u) = -\int_0^u dx\, x^{-1} \ln(1-x)$ is Euler's dilogarithm [16]), we finally arrive at the following expressions for the limit of the configurational partition function per site:

$$\boxed{\begin{aligned}\ln Z(z,z') &= \tfrac{1}{2}\ln z + \pi^{-1}\Lambda_2(z'/z)\\ &= \tfrac{1}{2}\ln z' + \pi^{-1}\Lambda_2(z/z').\end{aligned}} \qquad (19)$$

By substituting $z = z' = 1$ in equations (13) or (15), (16) and (19) we immediately find the *total number of dimer arrangements*, $g(\frac{1}{2}mn)$, and its asymptotic behaviour. One is sometimes interested in the "*molecular freedom*" φ_2 of the dimers defined [3]) as the number of arrangements per dimer:

$$\varphi_2 = \{g(\tfrac{1}{2}mn)\}^{2/mn} = \{\sum_{N_2, N_2'}{}' g(N_2, N_2')\}^{2/mn} = \{Z_{mn}(1,1)\}^{2/mn} \qquad (20)$$

In particular, for the infinite lattice we find

$$\boxed{\varphi_2^{(\infty)} = Z^2(1,1) = \exp\{2\pi^{-1}\Lambda_2(1)\} = \exp\{2G/\pi\} = 1.791\,622\,812\ldots} \qquad (21)$$

where

$$G = 1^{-2} - 3^{-2} + 5^{-2} - 7^{-2} + \ldots = 0.915\,965\,594\ldots \text{ (Catalan's constant)}.$$

Several approximate values for $\varphi_2^{(\infty)}$ have, in more or less explicit form, been given in the literature. From Flory's theory of polymers [17]) one can derive a value which corresponds to a "Bragg-Williams" or "random mixing" approximation, Chang [18]) and Cohen *e.a.* [2]) used a "1st Bethe-Kikuchi" or "quasi-chemical" approximation, Orr [4]) worked out a "2nd Bethe approximation", Miller [19]) calculated a lower bound, and Fowler and Rushbrooke [3]) obtained a very close estimate by extrapolating their exact results for infinite quadratic strips of widths up to 8*) (which are, of course, included as special cases in our expression (16)). The various results are summarized in table II (p. 1220).

§ 4. *The toroidal quadratic lattice.* In this section we shall investigate the changes brought forward by introducing *periodic boundary conditions,* i.e. by winding the lattice on a torus. We shall call the toroidal $m \times n$ lattice $Q_{mn}^{(t)}$, and the corresponding generating functions $Z_{mn}^{(t)}(z,z')$, $Z_n^{(t)}(z,z')$ and $Z^{(t)}(z,z')$.

One difference with the case of a planar lattice is that $D(m,j;1,j)$ and $D(i,n;i,1)$ should no longer be taken to be zero but equal, in absolute

*) Miller's criticism [19]) of the calculations of Fowler and Rushbrooke rests on a wrong interpretation of the method and is therefore not valid.

magnitude, to z and z', respectively; we can still choose the signs of these elements. Further we have now to distinguish four classes of configurations. The first class comprises those configurations that can be derived from the standard configuration (which we take identical to that of § 2) by cyclic shifts along polygons not looping the torus either in horizontal or in vertical direction, or, more generally, looping the torus an even number of times in both directions; let us call them (e, e) configurations. In an analogous

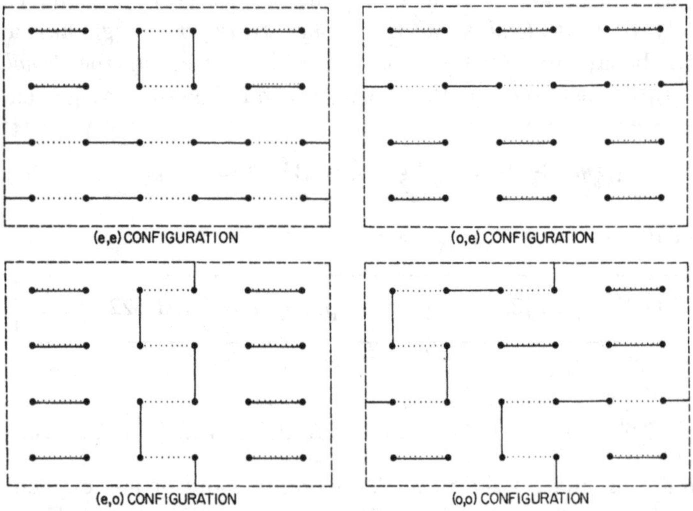

Fig. 2. Configurations from the four configuration classes of the toroidal lattice $Q_{64}^{(t)}$

way we define (o, e), (e, o) and (o, o) configurations (cf. fig. 2; e = even, o = odd, first symbol refers to horizontal loops). If we define, for $1 \leq i \leq m$, $1 \leq j \leq n$,

$$D(m, j; 1, j) = +z; D(i, n; i, 1) = (-1)^i z', \tag{22.1}$$

i.e. if we make the matrix, now to be called $\mathbf{D}_1$, periodical in both directions, it can be shown that Pf $\mathbf{D}_1$ counts only the (e, e) configurations correctly; the others are counted with the wrong sign. In order to get all configurations correctly counted we use a device introduced by Potts and Ward [20]) in their treatment of the corresponding Ising problem. We define three other matrices:

$$\mathbf{D}_2 \text{ with } D(m, j; 1, j) = +z; D(i, n; i, 1) = (-1)^{i+1} z', \tag{22.2}$$

$$\mathbf{D}_3 \text{ with } D(m, j; 1, j) = -z; D(i, n; i, 1) = (-1)^i z', \tag{22.3}$$

$$\mathbf{D}_4 \text{ with } D(m, j; 1, j) = -z; D(i, n; i, 1) = (-1)^{i+1} z', \tag{22.4}$$

and we remark that Pf $\mathbf{D}_2$ counts all configurations correctly except those from the class (o, e), Pf $\mathbf{D}_3$ all configurations except the class (e, o) and

Pf D_4 all configurations except the class (o, o). These counting rules are summarized in table I:

TABLE I

Counting of configurations on a toroidal lattice				
Class of configurations	Sign of corresponding terms in Pf D_μ			
	D_1	D_2	D_3	D_4
(e, e)	+	+	+	+
(o, e)	−	−	+	+
(e, o)	−	+	−	+
(o, o)	−	+	+	−

It is evident from this table that if we put

$$Z^{(t)}_{mn}(z, z') = \tfrac{1}{2}(-\text{Pf } D_1 + \text{Pf } D_2 + \text{Pf } D_3 + \text{Pf } D_4), \tag{23}$$

all configurations are counted with the right sign so that we have obtained the analogue of eq. (5) for a toroidal quadratic lattice.

The evaluation of eq. (23) runs parallel to that of eq. (5), the only difference being the occurrence of the matrices

$$\begin{bmatrix} 0 & 1\,0\,0 & \ldots & 0 & -1 \\ -1 & 0\,1\,0 & \ldots & 0 & 0 \\ 0 & -1\,0\,1 & \ldots & 0 & 0 \\ \cdot & \cdot\,\cdot\,\cdot & \ldots & \cdot & \cdot \\ 0 & 0\,0\,0 & \ldots & 0 & 1 \\ 1 & 0\,0\,0 & \ldots & -1 & 0 \end{bmatrix} \quad \text{and} \quad \begin{bmatrix} 0 & 1\,0\,0 & \ldots & 0 & 1 \\ -1 & 0\,1\,0 & \ldots & 0 & 0 \\ 0 & -1\,0\,1 & \ldots & 0 & 0 \\ \cdot & \cdot\,\cdot\,\cdot & \ldots & \cdot & \cdot \\ 0 & 0\,0\,0 & \ldots & 0 & 1 \\ -1 & 0\,0\,0 & \ldots & -1 & 0 \end{bmatrix}$$

instead of Q (cf. eq. (9) and (10)). They can be diagonalized successively by a transformation with the matrices

$$\begin{aligned} V(l, l') &= (1/n)^{\frac{1}{2}} \exp\{2ll'\,\pi i/n\}, \\ V^-(l, l') &= (1/n)^{\frac{1}{2}} \exp\{l(2l' - 1)\pi i/n\}. \end{aligned} \tag{24}$$

Proceeding as in § 3 we find

$$\begin{aligned} Z^{(t)}_{mn}(z, z') = &-\tfrac{1}{2}\prod_{k=1}^{\frac{1}{2}m} \prod_{l=1}^{n} 2[z^2 \sin^2\{2k\pi/m\} + z'^2 \sin^2\{2l\pi/n\}]^{\frac{1}{2}} + \\ &+\tfrac{1}{2}\prod_{k=1}^{\frac{1}{2}m} \prod_{l=1}^{n} 2[z^2 \sin^2\{2k\pi/m\} + z'^2 \sin^2\{(2l-1)\,\pi/n\}]^{\frac{1}{2}} + \\ &+\tfrac{1}{2}\prod_{k=1}^{\frac{1}{2}m} \prod_{l=1}^{n} 2[z^2 \sin^2\{(2k-1)\,\pi/m\} + z'^2 \sin^2\{2l\pi/n\}]^{\frac{1}{2}} + \\ &+\tfrac{1}{2}\prod_{k=1}^{\frac{1}{2}m} \prod_{l=1}^{n} 2[z^2 \sin^2\{(2k-1)\,\pi/m\} + z'^2 \sin^2\{(2l-1)\,\pi/n\}]^{\frac{1}{2}}. \end{aligned} \tag{25}$$

The first term of the right-hand member is easily seen to be equal to zero.

If desired, this equation can be put into a form analogous to (15) with the aid of the following identities, valid for even m and non-negative values of u:

$$\prod_{k=1}^{\frac{1}{2}m} 2[u^2+\sin^2\{2k\pi/m\}]^{\frac{1}{2}} \equiv [u+(1+u^2)^{\frac{1}{2}}]^{\frac{1}{2}m}-[-u+(1+u^2)^{\frac{1}{2}}]^{\frac{1}{2}m},$$
$$\prod_{k=1}^{\frac{1}{2}m} 2[u^2+\sin^2\{(2k-1)\pi/m\}]^{\frac{1}{2}} \equiv [u+(1+u^2)^{\frac{1}{2}}]^{\frac{1}{2}m}+[-u+(1+u^2)^{\frac{1}{2}}]^{\frac{1}{2}m}. \tag{26}$$

For $m \to \infty$, i.e. for *infinite cylindrical strips*, the second and fourth term of eq. (25) can be shown to be dominant and equal, and we find

$$Z_n^{(t)}(z,z') = z^{\frac{1}{2}n} \prod_{l=1}^{[\frac{1}{2}n]} \left[\zeta \sin\frac{(2l-1)\pi}{n} + \left(1+\zeta^2 \sin^2\frac{(2l-1)\pi}{n}\right)^{\frac{1}{2}}\right], \tag{27}$$

which is to be compared with eq. (16), valid for planar strips. In the limit $n \to \infty$ we finally obtain eq. (17) again. The values for the molecular freedom φ_2 in cylindrical strips of widths up to 8 calculated by Fowler and Rushbrooke [3]) can be found as special cases from eq. (27). In table II we list the various values of φ_2 (exact and approximate) calculated for strips and for the infinite lattice.

TABLE II

The molecular freedom φ_2 for planar and toroidal quadratic $\infty \times n$ lattices

n	planar lattice	toroidal lattice	method
1	1.000	—	
2	1.618	2.414	Fowler and Rushbrooke [3]) and this paper (in those cases where ref. 3 gave no or less accurate results, those of the present method have been recorded)
3	1.551	1.686	
4	1.685	1.932	
5	1.658	1.754	
6	1.716	1.849	
7	1.701	1.772	
8	1.732	1.823	
∞	1.471		Flory [17]) (approximate)
∞	1.687		Chang [18]) (approximate)
∞	1.736		Orr [4]) (approximate)
∞	1.63		Miller [19]) (approximate)
∞	1.8		Fowler and Rushbrooke [3]) (approximate)
∞	1.791 623 ...		this paper

§ 5. *Alternative approaches.* We saw in § 2 that the number of dimer configurations on a lattice is equal to the number of alternating polygons (defined with respect to a standard configuration) on that lattice. The strong analogy with the Ising problem suggests that a solution of the present problem is possible which follows more closely the method of Kac and Ward referred to above. This can indeed be developed, and again one obtains the configuration generating function Z_{mn} as the square root of a determinant.

We shall not go into details but instead mention another closely related method for calculating Z_{mn}.

It follows from the considerations of § 2 that there is a one-to-one correspondence between alternating polygons on Q_{mn} and closed paths on a corresponding *oriented lattice* Q^{or}_{mn}, sketched in fig. 3. In this lattice any bond may be traversed in only one direction: the C_0-bonds in the direction of increasing (decreasing) i for odd (even) values of j, the non-C_0-bonds in the direction from the "head" of a C_0-bond to the "tail" of another C_0-bond.

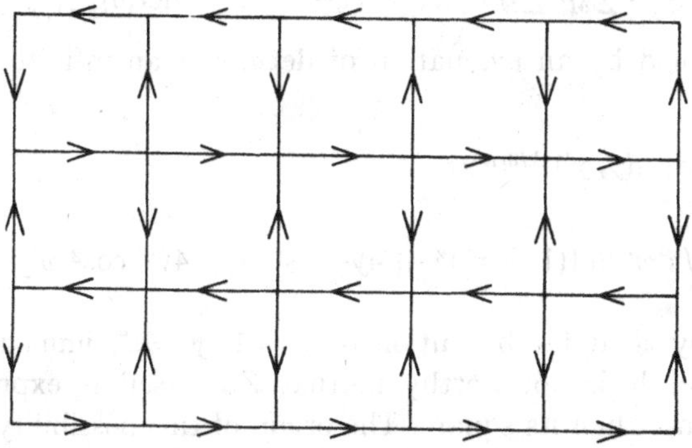

Fig. 3. The oriented lattice Q^{or}_{64}.

We now form a $mn \times mn$ matrix d whose rows and columns correspond to the sites of Q^{or}_{mn}. We define, for all combinations of indices which represent lattice points,

$$\begin{aligned}
&d(i,j;i,j) &&= 1 \\
&d(i,j;i+1,j) &&= y \qquad \text{for } j \text{ odd}, \\
&d(i,j;i-1,j) &&= y \qquad \text{for } j \text{ even}, \\
&\left.\begin{matrix} d(i,j;i,j+1) \\ d(i,j;i,j-1) \end{matrix}\right\} &&= \begin{cases} +y' & \text{for } i \text{ even}, j \text{ odd}, \\ -y' & \text{for } i \text{ odd}, j \text{ even}, \end{cases} \\
&d(i,j;i',j') &&= 0 \qquad \text{otherwise},
\end{aligned} \tag{28}$$

i.e. we attach weight factors y and $\pm y'$ to horizontal and vertical oriented bonds, respectively, and a factor 1 to each lattice site on its own. Thus in $\det d$ each term will correspond to a configuration of closed paths on Q^{or}_{mn} (cf. ref. 9). A term representing a permutation consisting of ν cycles (each permuting an even number of points) occurs with a factor $(-1)^\nu$; since in a determinant, as contrasted with a Pfaffian, there is no restriction on the order of the indices of the elements, the difficulty mentioned in § 2 does not arise here. The argument of § 2 shows again that the difference in sign between permutations consisting of odd and even numbers of cycles is compensated for by the negative signs attributed to the vertical bonds between sites with odd values of i. So $\det d$ is just the path generating function $H_{mn}(y, y')$ for the lattice Q^{or}_{mn}:

$$\det d = \Sigma_M \Sigma_{M'} h(M, M')\, y^M y'^{M'} = H_{mn}(y, y'),$$

$h(M, M')$ being the number of ways of combining M horizontal and M' vertical steps to closed paths. According to § 2 such a combination may be considered as representing a configuration of M' vertical dimers, and hence $\frac{1}{2}mn - M'$ horizontal dimers. It follows that

$$Z_{mn}(z, z') = \Sigma_{M'}\, g(\tfrac{1}{2}mn - M', M')\, z^{\frac{1}{2}mn-M'} z'^{M'} = \\ = z^{\frac{1}{2}mn} \Sigma_{M'} \Sigma_M h(M, M')\, \zeta^{M'} = z^{\frac{1}{2}mn} \det(d)_{y=1,\, y'=\zeta}. \tag{29}$$

This is confirmed by an evaluation of $\det d$. For an infinite lattice e.g. one finds

$$H(y, y') = \lim_{m,n\to\infty} \{H_{mn}(y, y')\}^{1/mn} = \\ = \exp\{\pi^{-2} \int_0^{\pi/2} d\omega \int_0^{\pi/2} d\omega' \ln [(1 - y^2)^2 + 4y^2 \cos^2 \omega + 4y'^2 \cos^2 \omega']\}; \tag{30}$$

multiplication by $z^{\frac{1}{2}}$ and substitution of $y = 1$, $y' = \zeta$, immediately lead to (17). This result is noteworthy in that Z_{mn} itself is expressed as a determinant rather than its square. The origin of this possibility lies in the fact that the quadratic lattice can, in the well-known way, be divided into two sublattices, that of the "odd" and that of the "even" sites (or, in terms of graph theory, that it is "*dichromatic*" [8]); this ensures the possibility of working with the oriented lattice Q^{or}_{mn}. The algebraic root lies in the possibility of writing certain Pfaffians as a determinant (cf. Muir [14]), vol. IV p. 263). For the triangular lattice, on the other hand, the method of this section cannot be used, whereas that of § 2 still works, as we hope to show in the envisaged sequel to this paper.

§ 6. *The entropy of a system of dimers on a rectangular lattice.* Consider a planar rectangular $m \times n$ lattice, i.e. a lattice whose horizontal and vertical bonds differ in length. Let the lattice be covered entirely by two sorts of dimers: N_2 dimers which fit only into horizontal positions (i.e. which can occupy two sites (i, j) and $(i + 1, j)$), and N_2' dimers fitting only into vertical positions. If the energy of mixing of these dimers is zero, the configurational free energy of the mixture is completely determined by the *entropy of mixing*, i.e. by the combinatorial factor $g(N_2, N_2')$. This quantity can be calculated from the configuration generating function with the aid of Cauchy's formula:

$$g(N_2, N_2') = (2\pi i)^{-1} \oint d\zeta\, \zeta^{-(N_2'+1)} Z_{mn}(1, \zeta), \tag{31}$$

where the path of integration encircles the origin but excludes the singularities of $Z_{mn}(1, \zeta)$. We shall introduce $x = N_2/\frac{1}{2}mn$ and $x' = N_2'/\frac{1}{2}mn = 1 - x$.

For large m and n we can evaluate (31) by the saddle-point method. We find

$$\lim_{m,n\to\infty} \{g(\tfrac{1}{2}mnx, \tfrac{1}{2}mnx')\}^{1/mn} = \gamma(x), \tag{32}$$

where $\gamma(x)$ is given by the following two equivalent expressions:

$$\begin{aligned}\ln \gamma(x) &= \pi^{-1}\Lambda_2 (\tan \tfrac{1}{2}\pi x) - \tfrac{1}{2}x \ln(\tan \tfrac{1}{2}\pi x) = \\ &= \pi^{-1}\Lambda_2(\tan \tfrac{1}{2}\pi x') - \tfrac{1}{2}x' \ln (\tan \tfrac{1}{2}\pi x').\end{aligned} \tag{33}$$

In fig. 4 the reduced entropy per dimer of this "interlocking mixture", $\sigma = S/\tfrac{1}{2}mnk = 2 \ln \gamma(x)$, which corresponds to Flory's "entropy of disorientation" [17]), is plotted against x. For comparison we have also plotted the entropy of mixing for an "ideal" or "random mixture", i.e. of a system where to each single lattice site a horizontal dimer (available fraction: x) or a vertical dimer (fraction: $x' = 1 - x$) is attached in a random manner,

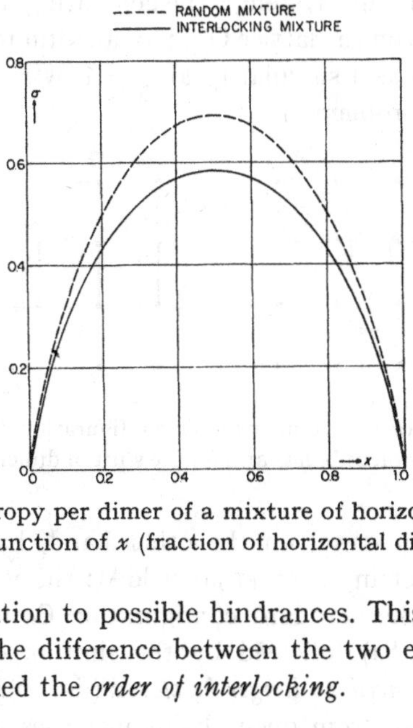

Fig. 4. The reduced entropy per dimer of a mixture of horizontal and vertical dimers as a function of x (fraction of horizontal dimers).

without paying attention to possible hindrances. This quantity is equal to $-x \ln x - x' \ln x'$. The difference between the two entropies is a measure of what might be called the *order of interlocking*.

§ 7. *Concluding remarks.* We have seen that the generating function $Z_{mn}(z, z')$ for dimer configurations on a planar quadratic lattice can be written in the form of a Pfaffian. The corresponding skew-symmetric matrix could by a similarity transformation be brought into a nearly diagonal form, and its determinant, which is the square of the Pfaffian, evaluated. The asymptotic behaviour of $Z_{mn}(z, z')$ for large lattices was found to be described by equation (17), which can also be written as

$$Z(z, z') = \exp\{(2\pi)^{-2} \int_0^\pi \mathrm{d}\omega \int_0^\pi \mathrm{d}\omega' \ln 2[z^2 + z'^2 + z^2 \cos \omega + z'^2 \cos \omega']\}. \tag{34}$$

The same result was found when periodic boundary conditions were introduced. The effect of boundary conditions is, however, not entirely trivial and will be discussed in more detail in a subsequent paper.

The right-hand member of (34) has a remarkable resemblance to Onsager's expression for the partition function per spin of a rectangular Ising system [9][10] *). A more detailed examination reveals that $Z(z, z')$ as a function of $\zeta = z'/z$ has no singular points on the real positive axis; it corresponds to Onsager's partition function at the critical point (or critical line, if the strengths of the horizontal and vertical interactions vary with respect to each other). This fact might tempt one to conjecture that the more general problem of monomer-dimer mixtures would be the analogue of the Ising problem at arbitrary temperatures, and hence rigorously solvable. However, this is not the case. It is easy to see that the true analogue of the partition function of an Ising system is the generating function $H_{mn}(y, y') = \det \mathsf{d}$ for closed paths on the lattice $\mathcal{Q}_{mn}^{\mathrm{or}}$; for an infinite lattice, the function $H(y, y')$ given by (30) has a singularity at $y = 1$, which is just the value of interest for the dimer problem.

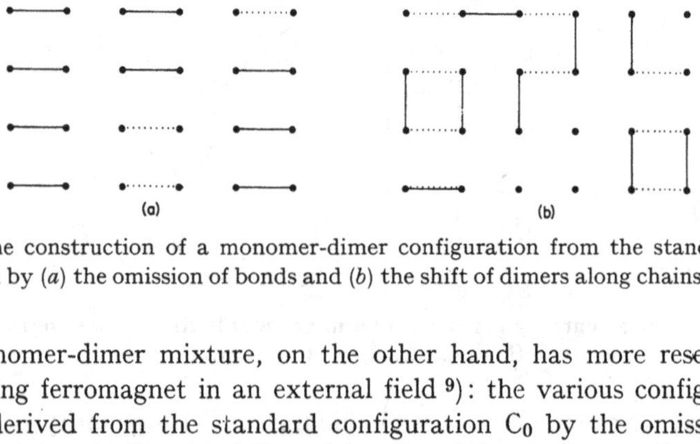

Fig. 5. The construction of a monomer-dimer configuration from the standard configuration by (*a*) the omission of bonds and (*b*) the shift of dimers along chains of bonds.

A monomer-dimer mixture, on the other hand, has more resemblance to an Ising ferromagnet in an external field [9]): the various configurations can be derived from the standard configuration C_0 by the omission of a number of bonds (fig. 5*a*) followed by the shift of dimers along certain chains of bonds, *which need no longer be closed* (fig. 5*b*). The contribution to the combinatorial factor from open chains increases with the ratio of the activity of two monomers to that of a dimer, whereas in the Ising case it increases with the ratio of the activity of +spins to that of −spins, i.e. with the magnetic field. Since this general Ising problem has as yet resisted all attempts at a rigorous solution one suspects that the monomer-dimer problem will also be very hard to solve. On the other hand, a better insight into the latter problem might throw some new light on the former.

---

*) In this connection it may be remarked that the algebraic introduction which Hurst and Green [13]) need for re-deriving Onsager's results by the Pfaffian method can be avoided by deriving them along the lines of the present paper.

Note added in proof.

After the submission of this paper the author received preprints of a short communication by H. N. V. Temperley and M. E. Fisher (to be published in Phil. Mag.) and of an article by M. E. Fisher (to be published in Phys. Rev.), both on the statistics of dimers on a quadratic lattice. Following the lines used by Hurst and Green [13]) in the discussion of the Ising problem, the authors obtain results identical to those of the present paper. They discuss in more detail the asymptotic behaviour of these results for large lattices, and in addition make some remarks on monomer-dimer mixtures. On the other hand, they restrict themselves to planar quadratic lattices, and their method – although formally equivalent to that developed above – seems less suited to generalization to other two-dimensional lattices. We hope to comment upon these papers in more detail later on.

Received 28-6-61

REFERENCES

1) Guggenheim, E. A., Mixtures, Clarendon Press, Oxford (1952) Chapter X.
2) Cohen, E. G. D., De Boer, J. and Salsburg, Z. W., Physica **21** (1955) 137.
3) Fowler, R. H. and Rushbrooke, G. S., Trans. Faraday Soc. **33** (1937) 1272.
4) Orr, W. J. C., Trans. Faraday Soc. **40** (1944) 306.
5) Rushbrooke, G. S., Scoins, H. I. and Wakefield, A. J., Discussions Faraday Soc. **15** (1953) 57.
6) Green, H. S. and Leipnik, R., Rev. mod. Phys. **32** (1960) 129.
7) Fisher, M. E. and Temperley, H. N. V., Rev. mod. Phys. **32** (1960) 1029.
Katsura, S. and Inawashiro, S., Rev. mod. Phys. **32** (1960) 1031.
8) Berge, C., Théorie des graphes et ses applications, Dunod, Paris (1958) 175, 30.
9) Newell, G. F. and Montroll, E. W., Rev. mod. Phys. **25** (1953) 352.
Domb, C., Adv. in Phys. **9** (1960) 149, in particular § 3.
10) Onsager, L., Phys. Rev. **65** (1944) 117.
11) Kac, M. and Ward, J. C., Phys. Rev. **88** (1952) 1332.
12) Sherman, S., J. math. Phys. **1** (1960) 202.
13) Hurst, C. A. and Green, H. S., J. chem. Phys. **33** (1960) 1059.
14) Muir, T., Contributions to the History of Determinants, London (1930).
Scott, R. F. and Mathews, G. B., Theory of Determinants, Cambridge University Press, New York (1904) 93.
15) Caianello, E. R. and Fubini, S., Nuovo Cimento **9** (1952) 1218.
16) Gröbner, W. and Hofreiter, N., Integraltafel II, Springer Verlag, Wien & Innsbruck (1950) 72.
17) Flory, P. J., J. chem. Phys. **10** (1942) 51.
18) Chang, T. S., Proc. roy. Soc., London, A **169** (1939) 512.
19) Miller, A. R., Proc. Camb. phil. Soc. **38** (1942) 109.
20) Potts, R. B. and Ward, J. C., Progr. theor. Phys. **13** (1955) 38.

Errata to "The statistics of dimers on a lattice. I. The number of dimer arrangements on a quadratic lattice" by P.W. Kasteleyn, Physica 27 (1961) 1209-1225.

P. 1220, line 6: after "dominant and equal" a footnote sign should be added, referring to the following footnote, to be placed at the bottom of the page:

> provided n is even. For odd n the second term vanishes while the third and fourth terms are equal, and a factor 2 should be inserted into the right-hand side of eq. (27).

P. 1222, eq. (30): the last term in the right-hand side should read $4y^2y'^2\cos^2\omega'$ and not $4y'^2\cos^2\omega'$.

LONGEST INCREASING AND DECREASING SUBSEQUENCES

C. SCHENSTED

This paper deals with finite sequences of integers. Typical of the problems we shall treat is the determination of the number of sequences of length n, consisting of the integers $1, 2, \ldots, m$, which have a longest increasing subsequence of length α. Throughout the first part of the paper we will deal only with sequences in which no numbers are repeated. In the second part we will extend the results to include the possibility of repetition. Our results will be stated in terms of standard Young tableaux.

PART I

Definition. A standard Young tableau of order n is an arrangement of n distinct natural numbers in rows and columns so that the numbers in each row and in each column form increasing sequences, and so that there is an element of each row (column) in the first column (row) and there are no gaps between numbers.

Example.

$$\begin{matrix} 2 & 4 & 7 \\ 3 & 8 & \\ 5 & 9 & \end{matrix} \qquad \text{(order} = 7\text{)}$$

Definition. The *shape* of a standard tableau is an arrangement of squares with one square replacing each number in the standard tableau.

Example. The shape of $\begin{matrix} 2 & 4 & 7 \\ 3 & 8 & \\ 5 & 9 & \end{matrix}$ is as shown in Figure 1.

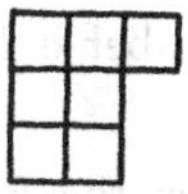

FIG. 1.

Received June 23, 1959; in revised form August 29, 1960. This work was conducted by Project MICHIGAN under Department of the Army Contract (DA-36-069-SC-78801), administered by the U.S. Army Signal Crops.

The author would like to thank W. Richardson, G. Rabson, T. Curtz, I. Schensted, R. Thrall, and J. Riordan for illuminating discussions concerning this problem, and E. Graves for calculations which contributed to the solution. The problem originated as one aspect of a paper on sorting theory by R. Bear and P. Brock, *Natural sorting*, The University of Michigan, Willow Run Laboratories, Project MICHIGAN Report 2144-278-T, submitted for publication in Soc. Ind. App. Math.

One reason that standard tableaux are so useful to us is that it is easy to compute the number of standard tableaux of a given shape either by means of a simple recurrence relation, or by means of the following elegant result; Frame, Robinson, and Thrall **(1)**.

THEOREM. *The number of standard tableaux of a given shape containing the integers* 1, 2, . . . , *n is*

$$\frac{n!}{\prod_{j=1}^{n} h_j} \tag{1}$$

Here the h_j are the hook lengths, that is, the number of elements counting from the bottom of a column to a given element and then to the right end of the row.

Example. To compute the number of standard tableaux of the shape shown in Figure $2(a)$, we first find the hook lengths, which are shown in Figure

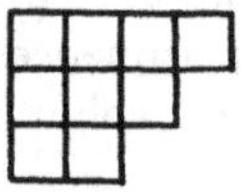

FIG. 2(a).

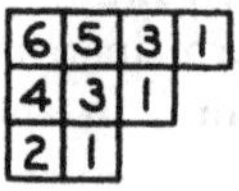

FIG. 2(b).

$2(b)$. Then we find that the number of standard tableaux of this shape is

$$\frac{9!}{6\cdot 5\cdot 3\cdot 1\cdot 4\cdot 3\cdot 1\cdot 2\cdot 1} = 168.$$

Definition. $S \leftarrow x$ is defined as the array obtained from the standard tableau, S, by means of the following steps:

(i) Insert x in the first row of S either by displacing the smallest number which is larger than x, or if no number is larger than x, by adding x at the end of the first row.

(ii) If x displaced a number from the first row, then insert this number in the second row either by displacing the smallest number which is larger than it or by adding it at the end of the second row.

(iii) Repeat this process row by row until some number is added at the end of a row.

In the above steps "adding at the end of the row" is interpreted as putting in the first column in the given row if the row does not yet have any entries in it. We define $x \rightarrow S$ similarly except that we replace the word "row" by the word "column" throughout.

Example. If

$$S = \begin{matrix} 2\,4\,7 \\ 3\,8 \\ 5\,9 \end{matrix}$$

then

$$S \leftarrow 6 = \begin{matrix} 2\,4\,6 \\ 3\,7 \\ 5\,8 \\ 9 \end{matrix} \quad \text{and} \quad 6 \rightarrow S = \begin{matrix} 2\,4\,7 \\ 3\,8 \\ 5\,9 \\ 6 \end{matrix}$$

LEMMA 1. *$S \leftarrow x$ and $x \rightarrow S$ are standard tableaux.*

Proof. Since the proofs for $S \leftarrow x$ and $x \rightarrow S$ are similar we consider only $S \leftarrow x$.

First we note that if two consecutive rows of S have the same length, and if a number is displaced from the first of these two rows, then it will either displace the number which was standing under it or else some number to its left, and thus will not be added at the end of the row. Thus a row cannot be made longer than the row above it and $S \leftarrow x$ cannot fail to be a standard tableau on account of its shape. Thus we have only to prove that the numbers in each row and column still form increasing sequences.

A number is inserted into a row in such a place that the number to its left (if any) is smaller, and the number to its right (if any) is larger. Thus the numbers in each row form increasing sequences.

The number (if any) which ends up below a number which is inserted at a new position is either the number which it displaced, which is therefore larger, or else the number which previously stood below the number which it displaced, which is larger still.

When a number is displaced from one row to the next it ends up either in the position directly beneath the one in which it originally stood, or else further to the left (since it is smaller than the number which previously stood underneath it). Thus it is either under the number which displaced it, which is therefore smaller, or else a number to the left of it, which is smaller still.

The last two paragraphs show that two consecutive numbers in a column form an increasing sequence if either of them has just been inserted into its present position. If neither of them has just been inserted, then they are the numbers which were previously there in S and which therefore are in increasing order. Hence the columns also form increasing sequences and the proof of the lemma is completed.

Definition. The *P-symbol* corresponding to a sequence of distinct integers $x_1x_2 \ldots x_n$ is the standard tableau $(\ldots((x_1 \leftarrow x_2) \leftarrow x_3) \ldots \leftarrow x_n)$. The *Q-symbol* corresponding to the same sequence is the array which is obtained by putting k in the square which is added to the shape of the P-symbol when x_k is inserted in the P-symbol.

Examples.

Sequence	3	3 5	3 5 4	3 5 4 9	3 5 4 9 8	3 5 4 9 8 2	3 5 4 9 8 2 7
P-symbol	3	3 5	3 4	3 4 9	3 4 8	2 4 8	2 4 7
			5	5	5 9	3 9	3 8
						5	5 9
Q-symbol	1	1 2	1 2	1 2 4	1 2 4	1 2 4	1 2 4
			3	3	3 5	3 5	3 5
						6	6 7

LEMMA 2. *The Q-symbol corresponding to an arbitrary sequence is a standard tableau.*

Proof. Since the Q-symbol has the same shape as the P-symbol, and since the P-symbol is a standard tableau, the shape of the Q-symbol is legitimate. Each digit added to the Q-symbol is larger than all of the previous digits, and in particular is larger than the digits above it and to its left. Hence the numbers in each row and column form increasing sequences, and the lemma is established.

LEMMA 3. *There is a one-to-one correspondence between sequences made with the n distinct integers $x_1, x_2, \ldots, x_n$ and ordered pairs of standard tableaux of the same shape—the first containing $x_1, x_2, \ldots, x_n$ and the second containing* $1, 2, \ldots, n$.

Proof. Given a sequence, the P-symbol and Q-symbol are uniquely determined standard tableaux of the type mentioned in the lemma. Given a pair of standard tableaux of the appropriate types we can find the unique sequence which could have them for a P-symbol and Q-symbol as follows: The position of the largest number in the second tells us which number was added on to a row of the first without displacing another number when the last digit was inserted. This must have been displaced from the previous row by the largest number which is smaller than it (there always will be at least one number smaller than it in the preceding row since the one directly above it is smaller). This in turn must have been displaced from the next row up. Finally we get to the first row and discover what number was inserted into it. This is the last digit of the sequence. We now also know what the P-symbol and Q-symbol were before the last digit was inserted. Thus we can repeat the procedure to find the next to the last digit of the sequence. This proves the lemma.

Note. Since there are $n!$ possible sequences of $x_1, x_2, \ldots, x_n$, Lemma 3 shows that there are $n!$ ordered pairs of standard tableaux of order n such that the shapes of tableaux in each pair are the same, but the shapes of tableaux in different pairs are not necessarily the same. This fact is already known **(2)**. Of course, the number of ordered pairs of standard tableaux of a given shape is equal to the square of the number of standard tableaux of that shape, which is given in turn by Expression (1).

Definition. The jth *basic subsequence* of a given sequence consists of the digits which are inserted into the jth place in the first row of the P-symbol.

LEMMA 4. *Each basic subsequence is a decreasing subsequence.*

Proof. Each number in the jth basic subsequence, on insertion in the first row displaces the previous member of the jth basic subsequence, which must therefore be larger than the present member.

LEMMA 5. *Given any member of the* jth *basic subsequence, we can find a member of the* $(j-1)$st *basic subsequence which is smaller and which occurs further to the left in the given sequence.*

Proof. The number in the $(j-1)$st place in the first row, when the given member of the jth basic subsequence is inserted, *is* such a member of the $(j-1)$st basic subsequence.

THEOREM 1. *The number of columns in the P-symbol (or the Q-symbol) is equal to the length of the longest increasing subsequence of the corresponding sequence.*

Proof. The number of columns is the same as the number of basic subsequences. By Lemma 4 there can be at most one member of each basic subsequence in any increasing subsequence. By Lemma 5 we can construct an increasing subsequence with one element from each basic subsequence, Q.E.D.

Note. The proof shows us how to actually obtain in increasing subsequence of maximal length.

LEMMA 6. $(x \to S) \leftarrow y = x \to (S \leftarrow y)$.

Proof. Suppose first, that of all the digits in x, y, and S, the largest is y. We represent S schematically by Figure 3. There are two cases of interest.

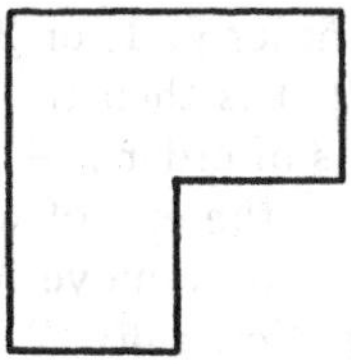

FIG. 3.

The square added to the shape of S in $x \to S$ is in the first row, or it is not. We represent $x \to S$ schematically in these two cases by Figure 4(a) and 4(b) respectively, where x' is the number added to the end of some column without displacing another number when we form $x \to S$. It is easily verified

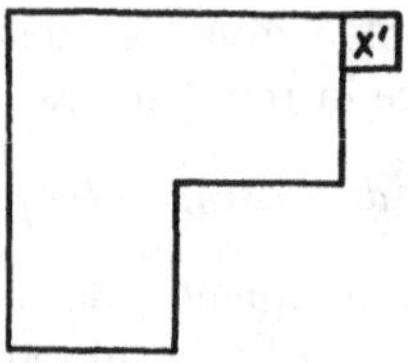

FIG. 4(a).

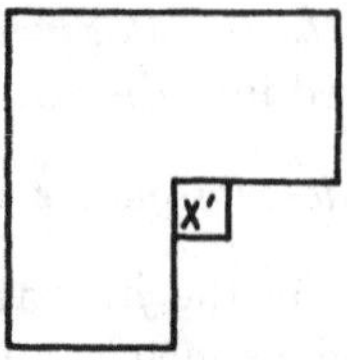

FIG. 4(b).

that in the first case the final result is as shown in Figure 5(a) and in the second case the result is that of Figure 5(b).

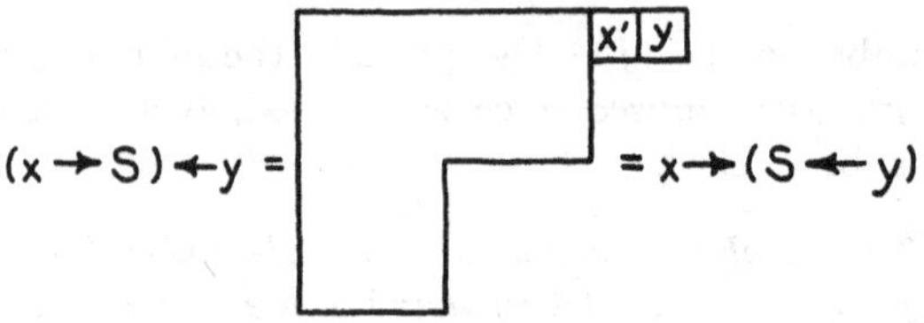

FIG. 5(a).

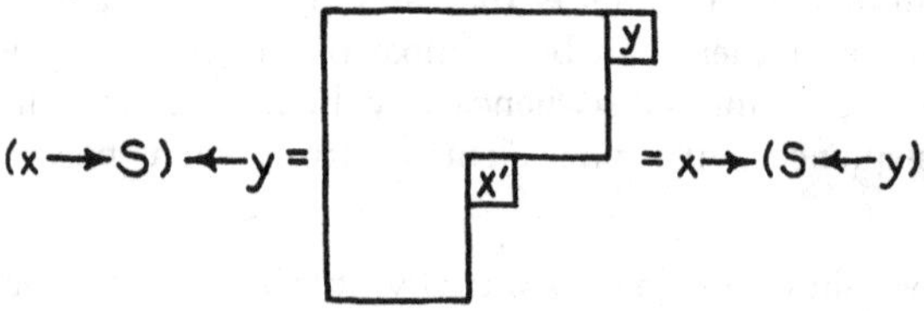

FIG. 5(b).

This proves the lemma if y is the largest number involved, and the proof is similar if x is the largest number involved.

Suppose now that, of all the digits in x, y, and S, the largest is N, and that N is in S. In this case we use induction. The lemma can be easily verified by direct calculation if S is of order 0, 1, or 2. We assume the lemma true for S of order n, and prove that it is then true for S of order $n + 1$.

Let us suppose, then, that S is of order $n + 1$. Now, since N is the largest number in S, we see that N is at the end of whatever row it is in, and also at the end of its column. Thus, if we remove N from S we will obtain a new standard tableau, S', of order n. Now since N is larger than any of the other numbers, it can never displace any of them, and hence the presence or absence of N cannot have any influence on the position of the other numbers. Thus $(x \to S) \leftarrow y$ will be the same as $(x \to S') \leftarrow y$ except that N is added somewhere, and $x \to (S \leftarrow y)$ will be the same as $x \to (S' \leftarrow y)$ except for the addition of N. However, since S' is of order n, we have by assumption

$$(x \to S') \leftarrow y = x \to (S' \leftarrow y).$$

Thus we have only to prove that N occupies the same position in $(x \rightarrow S) \leftarrow y$ and $x \rightarrow (S \leftarrow y)$ to prove the lemma. The truth of this can be easily verified for each of the possible cases which can arise as to the relative locations of N, x', and y'. Here $x'(y')$ is the number which is added to some column (row) without displacing another number when we form $x \rightarrow S'(S' \leftarrow y)$. In making these verifications it is necessary to keep the following facts in mind.

If x' and y' do not fall into the same square, then we represent S', $x \rightarrow S'$, and $S' \leftarrow y$ schematically by Figure 6(*a*), 6(*b*), and 6(*c*) respectively. The shape of $(x \rightarrow S') \leftarrow y$ must have a square added to the shape of

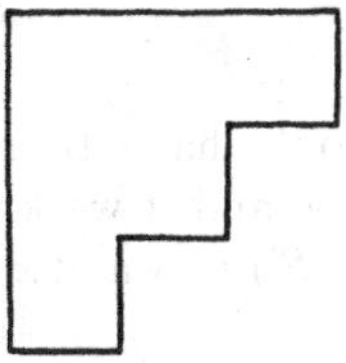

FIG. 6(*a*).

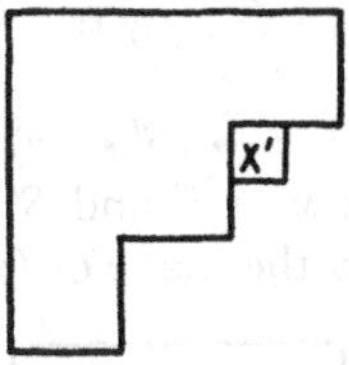

FIG. 6(*b*).

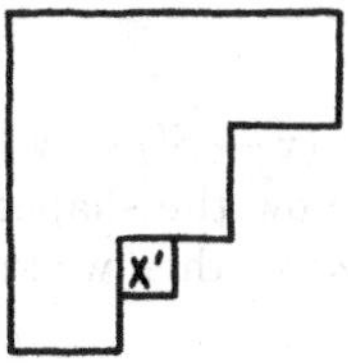

FIG. 6(*c*).

$x \rightarrow S'$, and the shape of $x \rightarrow (S' \leftarrow y)$ must have a square added to the shape of $S' \leftarrow y$. By assumption $(x \rightarrow S') \leftarrow y = x \rightarrow (S' \leftarrow y)$ so that the shape of $(x \rightarrow S') \leftarrow y$ and $x \rightarrow (S' \leftarrow y)$ must be Figure 7.

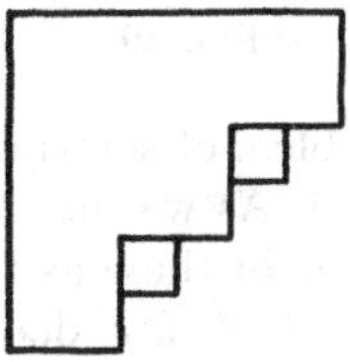

FIG. 7.

If x' (in $x \rightarrow S'$) and y' (in $S' \leftarrow y$) occupy the same position then we schematically represent S', $x \rightarrow S'$, and $S' \leftarrow y$ by Figure 8(*a*), 8(*b*), and 8(*c*) respectively. Here the shaded parts of $x \rightarrow S'$ and $S' \leftarrow y$ are the

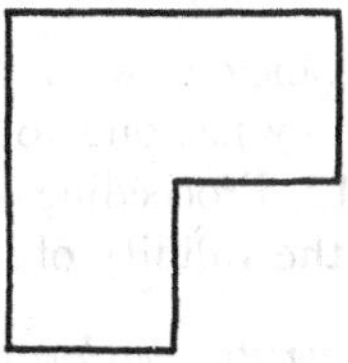

FIG. 8(*a*).

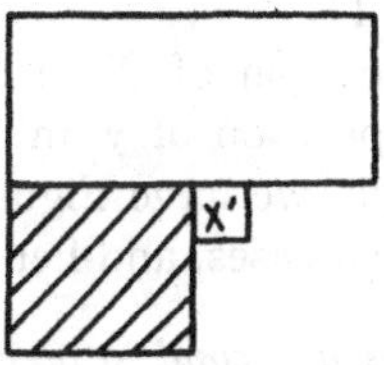

FIG. 8(*b*).

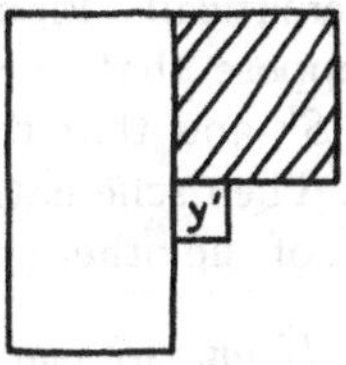

FIG. 8(*c*).

regions where numbers could have been displaced. Now let us suppose that $y' > x'$. Then when we insert y into $x \rightarrow S'$ the same numbers will be displaced in each row as were displaced when we inserted y into S, until we displace y'.

In $S' \leftarrow y$ we would have put y' where x' is, but $y' > x'$, thus y' will be added at the end of the row containing x', and the shape of $(x \rightarrow S') \leftarrow y$ (and hence of $x \rightarrow (S' \leftarrow y)$) will be Figure 9. If we had had $x' > y'$, then

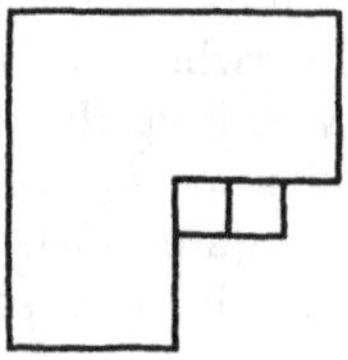

FIG. 9.

the shape of $(x \rightarrow S') \leftarrow y$ and $x \rightarrow (S' \leftarrow y)$ would have been Figure 10. Thus, if we know the shapes of $x \rightarrow S'$ and $S' \leftarrow y$, and if we know whether $x' > y'$ or $x' < y'$, then we know the shape of $(x \rightarrow S') \leftarrow y$ and $x \rightarrow (S' \leftarrow y)$.

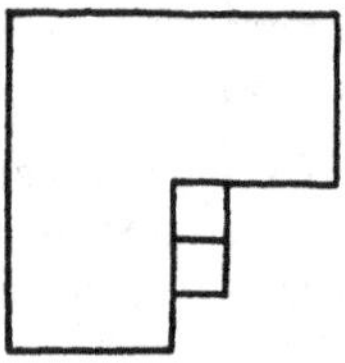

FIG. 10.

Now we can return to the problem of showing that N has the same position in $(x \rightarrow S) \leftarrow y$ and $x \rightarrow (S \leftarrow y)$. As we mentioned there are several special cases. We will consider only three of these as the others go in the same way. First suppose that the position of N in S does not coincide with either the position of x' in $x \rightarrow S'$ or the position of y' in $S' \leftarrow y$. Then N will never be displaced and it will have the same position in $(x \rightarrow S) \leftarrow y$ and $x \rightarrow (S \leftarrow y)$ as it does in S.

Next suppose that the position of N in S coincides with the position of x' in $x \rightarrow S'$, and that the position of y' in $S' \leftarrow y$ lies to the left of this. Then we have schematically Figure 11.

Finally suppose that the position of N in S coincides with the position of x' in $x \rightarrow S'$, and that the position of y' in $S' \leftarrow y$ lies one column to the right of this. Then schematically we have Figure 12. Proceeding similarly we can verify all of the other special cases, and hence the validity of Lemma 6.

LEMMA 7. *If one sequence is a second sequence written backwards, then P-symbol of the first is obtained from the P-symbol of the second by interchanging rows and columns.*

Proof. First we note that $x \rightarrow y = x \leftarrow y$ since if $x < y$ they are both xy and if $x > y$ they are both $\begin{smallmatrix} y \\ x \end{smallmatrix}$. Now we define $P(x_1, x_2, \ldots, x_n) \equiv (\ldots((x_1$

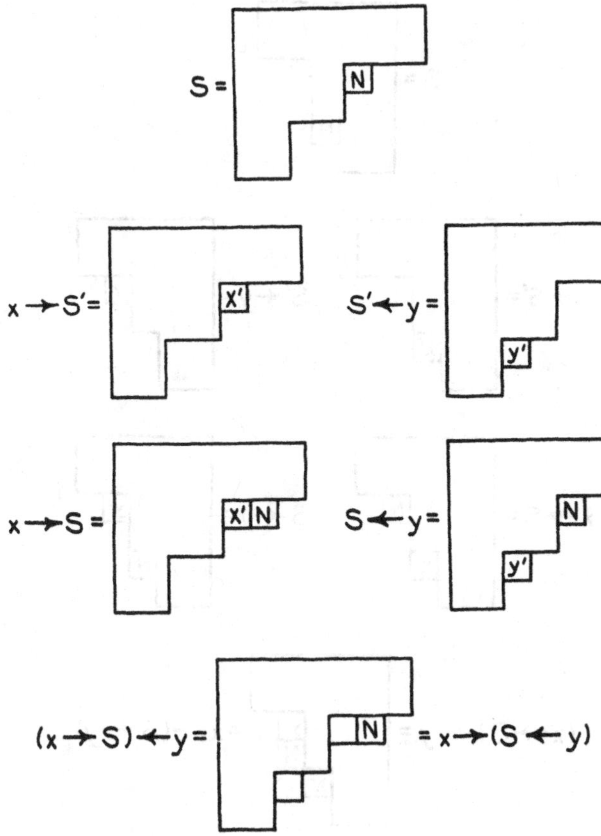

FIG. 11.

$\leftarrow x_2) \leftarrow x_3) \ldots \leftarrow x_n)$ and $\tilde{P}(x_1, x_2, \ldots, x_n) \equiv (x_1 \to \ldots (x_{n-2} \to (x_{n-1} \to x_n)) \ldots)$. Next we assume that $P(x_1, x_2, \ldots, x_{n-1}) = \tilde{P}(x_1, x_2, \ldots, x_{n-1})$ and that $P(x_1, x_2, \ldots, x_n) = \tilde{P}(x_1, x_2, \ldots, x_n)$ and prove that $P(x_1, x_2, \ldots, x_n, x_{n+1}) = \tilde{P}(x_1, x_2, \ldots, x_n, x_{n+1})$. (We have just shown that $P(x_1, x_2) = x_1 \leftarrow x_2 = x_1 \to x_2 = \tilde{P}(x_1, x_2)$, furthermore $P(x_1) = x_1 = \tilde{P}(x_1)$.) We have

$$\begin{aligned} P(x_1, x_2, \ldots, x_n, x_{n+1}) &= P(x_1, x_2, \ldots, x_n) \leftarrow x_{n+1} \\ &= \tilde{P}(x_1, x_2, \ldots, x_n) \leftarrow x_{n+1} \\ &= [x_1 \to \tilde{P}(x_2, \ldots, x_n)] \leftarrow x_{n+1} \\ &= x_1 \to [\tilde{P}(x_2, \ldots, x_n) \leftarrow x_{n+1}] \\ &= x_1 \to [P(x_2, \ldots, x_n) \leftarrow x_{n+1}] \\ &= x_1 \to P(x_2, \ldots, x_n, x_{n+1}) \\ &= x_1 \to \tilde{P}(x_2, \ldots, x_n, x_{n+1}) \\ &= \tilde{P}(x_1, x_2, \ldots, x_n, x_{n+1}). \end{aligned}$$

Of these lines, the second, fifth, and seventh follow by assumption, and the

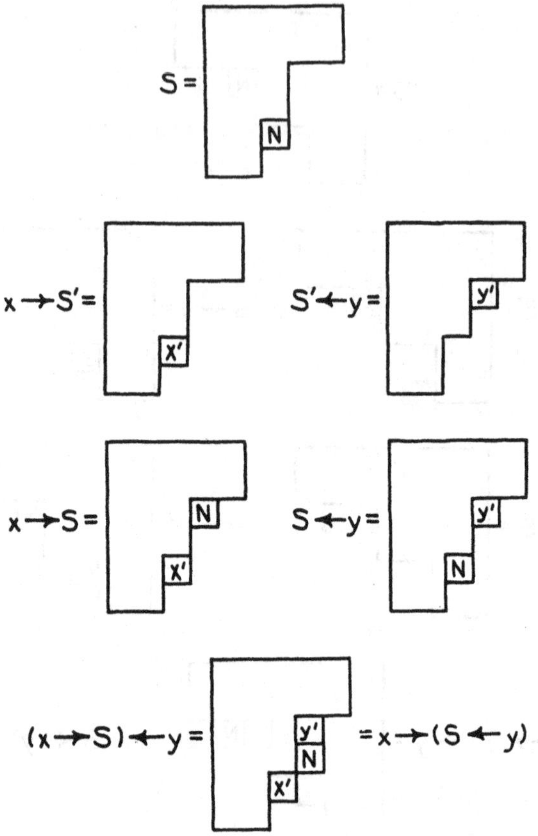

FIG. 12.

fourth from Lemma 6. Now $P(x_1, \ldots, x_n)$ is the P-symbol for the sequence $x_1, x_2, \ldots, x_n$, while $\tilde{P}(x_1, x_2, \ldots, x_n)$ is the P-symbol for the sequence $x_n, \ldots, x_2, x_1$ with rows and columns interchanged. Hence the lemma follows.

Note. It must not be assumed that Lemma 7 holds for Q-symbols.

THEOREM 2. *The number of rows in the P-symbol (or the Q-symbol) is equal to the length of the longest decreasing subsequence of the corresponding sequence.*

Proof. This follows immediately from Theorem 1 and Lemma 7, since writing a sequence backwards changes increasing subsequences into decreasing subsequences.

THEOREM 3. *The number of sequences consisting of the distinct numbers $x_1, x_2, \ldots, x_n$, and having a longest increasing subsequence of length α and a longest decreasing subsequence of length β, is the sum of the squares of the numbers of standard tableaux with shapes having α columns and β rows.*

Proof. Follows immediately from Lemma 3 and Theorems 1 and 2 (see also the note to Lemma 3).

Example. To find the number of permutations of $1, 2, 3, \ldots, 25$ having a longest decreasing subsequence of length three and a longest increasing subsequence of length 21 we note that the only allowed shapes with 25 squares, 21 columns, and 3 rows are those of Figure 13.

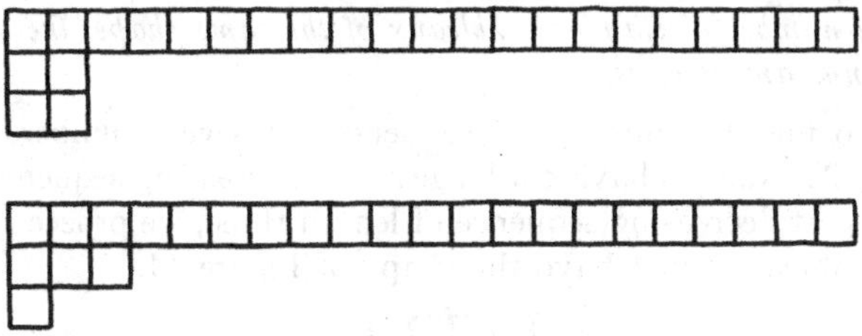

FIG. 13.

By the Frame–Robinson–Thrall theorem, the corresponding numbers of standard tableaux are 21,000 and 31,350 respectively. Thus the desired number of permutations is

$$21{,}000^2 + 31{,}350^2 = 1{,}423{,}822{,}500.$$

PART II

We now want to consider sequences in which some of the numbers are repeated. We can obtain the properties of such sequences in terms of sequences without repetitions by a simple artifice. Suppose the smallest number appears p times in the sequence, the next smallest q times, etc. We replace the p occurrences of the smallest number by the numbers $1, 2, \ldots, p$ (in this order), the q occurrences of the next number by $p + 1, p + 2, \ldots, p + q$, etc. Then the decreasing subsequences of the two sequences will be in one-to-one correspondence, while the increasing subsequences of the new sequence will be in one-to-one correspondence with the non-decreasing subsequences of the original sequence.

Example. Given the sequence 3 3 2 3 4 1, we replace 1 by 1, 2 by 2, the three 3's by 4, 5, 6, and 4 by 7. The result is 4 5 2 6 7 1. The latter sequence has a decreasing subsequence 5 2 1 which corresponds to a decreasing subsequence 3 2 1 in the original and an increasing subsequence 4 5 6 7 which corresponds to a non-decreasing subsequence 3 3 3 4 in the original.

If we construct the P-symbol for the derived sequence, and map the numbers in it back to the numbers in the original sequence, then we get a modified standard tableau in which repeated numbers are allowed, the numbers in each column form an increasing sequence, and the numbers in each row form a non-decreasing sequence. Since the numbers in the Q-symbol refer to the order of addition of spaces to the P-symbol, the Q-symbols of the two sequences will be identical.

We can get modified forms of each of the results in Part I. The main result, Theorem 3, now takes the form:

THEOREM 4. *The number of sequences of $x_1, x_2, \ldots, x_n$ having a longest non-decreasing sequence of length α and a longest decreasing sequence of length β is the sum of the products of the number of modified standard tableaux of a given shape with the number of standard tableaux of the same shape, the shapes each having α columns and β rows.*

Example. To find the number of sequences of seven numbers consisting entirely of 1's, 2's, and 3's having a longest non-decreasing sequence of length four and a longest decreasing sequence of length three, we proceed as follows. The possible tableaux must have the shape of Figure 14.

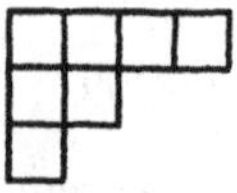

FIG. 14.

The possible modified standard tableaux are

			1 1 1 1 2 2 3 ,	1 1 1 1 2 3 3 ,
1 1 1 2 2 2 3 ,	1 1 1 2 2 3 3 ,	1 1 1 3 2 2 3 ,	1 1 1 3 2 3 3 ,	1 1 2 2 2 2 3 ,
1 1 2 2 2 3 3 ,	1 1 2 3 2 2 3 ,	1 1 2 3 2 3 3 ,	1 1 3 3 2 2 3 ,	1 1 3 3 2 3 3 ,
1 2 2 2 2 3 3 ,	1 2 2 3 2 3 3 ,	1 2 3 3 2 3 3 .	They are 15 in number.	

By the Frame–Robinson–Thrall theorem the number of standard tableaux of this shape is 35. Hence the number of sequences of the desired type is $15 \times 35 = 525$.

As a further example we will work out explicit formulae for binary sequences (sequences consisting of 0's and 1's). In this case the modified standard tableaux have the general form of Figure 15, where the bracketed region can have any division of 0's and 1's (the 0's preceding the 1's, of course).

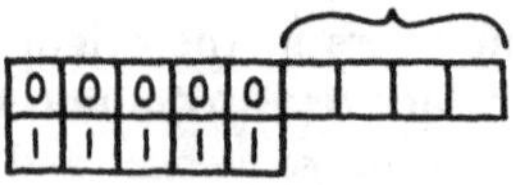

FIG. 15.

Let n be the number of digits in the sequence. Let m be the length of the longest non-decreasing subsequence. Then there are no sequences for which $m < n/2$. If $m = n$ the longest decreasing subsequence is of length 1. If $n/2 \leqslant m < n$, the longest decreasing subsequence is of length 2.

The number of possible modified tableaux is $2m - n + 1$. The number of standard tableaux is

$$(2m - n + 1)\frac{n!}{(m+1)!(n-m)!}.$$

Thus the number of binary sequences of n digits with a longest non-decreasing subsequence of length m is

$$\frac{n!(2m-n+1)^2}{(m+1)!(n-m)!}.$$

Note. Since the total number of binary sequences is 2^n we have

$$2^n = \sum_{m > n/2}^{n} \frac{n!(2m-n+1)^2}{(m+1)!(n-m)!}.$$

In the above derivation we allowed all possible binary sequences. Theorem 4 also readily solves the problem if the number of 0's and 1's in the sequence is fixed. In this case there is at most one modified tableau and thus the number of sequences of n digits with a longest non-decreasing subsequence of length m is

$$\frac{n!(2m-n+1)}{(m+1)!(n-m)!}$$

with the additional restriction that the number, p, of 0's in the sequence must satisfy $n - m \leqslant p \leqslant m$.

Note. This shows that

$$\binom{n}{p} = \sum_{m=\max(p,n-p)}^{n} \frac{n!(2m-n+1)}{(m+1)!(n-m)!}.$$

Throughout Part II we could have dealt equally well with increasing and non-increasing subsequences rather than decreasing and non-decreasing subsequences.

References

1. J. S. Frame, G. de B. Robinson, and R. M. Thrall, *The hook graphs of the symmetric group*, Can. J. Math., *6* (1954), 316.
2. D. E. Rutherford, *Substitutional analysis* (Edinburgh University Press, 1948), p. 26.

Institute for Defence Analysis
Princeton

Reprinted from
Canad. J. Math. **13** (1961), 179–191

ON A THEOREM OF R. JUNGEN

M. P. SCHÜTZENBERGER

Let us recall the following elementary result in the theory of analytic functions in one variable.

THEOREM (R. JUNGEN [7]). *If a is rational and b algebraic their Hadamard product c is algebraic; if, further, b is rational, c also is rational.*

For several variables, Jungen's proof shows that the theorem is still true for the Bochner-Martin [2] Hadamard product. It does not hold for the Cameron-Martin [3] and for the Haslam-Jones [6] Hadamard products. In this note we give a version of Jungen's theorem which is valid for a restricted interpretation of the notions involved when a and b are formal power series in a finite number of noncommuting variables.

1. **Notations.** Let R be a fixed not necessarily commutative ring with unit 1. For any finite set Z, $F(Z)$ is the free monoid generated by Z and $R_{pol}(Z)$ is the free module on $F(Z)$ over R. An element a of $R_{pol}(Z)$ will usually be written in the form $a=\sum\{(a,f)\cdot f: f\in F(Z)\}$ where the coefficients (a,f) are in R; $R_{pol}(Z)$ is graded in the usual manner and $\pi_n a=\sum\{(a,f)\cdot f: f\in F(Z), \deg f\leq n\}$. We identify R with $\pi_0 R_{pol}(Z)$. $R_{pol}(Z)$ is also a ring with product $aa' =\sum\{(a,f')(a',f'')\cdot f: f,f',f''\in F(Z), f=f'f''\}$.

It is well known (cf., e.g., [4; 3]) that these notions extend to the ring $R(Z)$ of the formal power series (with coefficients in R) in the noncommuting variables $z\in Z$; $R(Z)$ is topologized in the same manner as a ring of commutative formal power-series and $aa' =\lim_{n,n'\to\infty}(\pi_n a)(\pi_{n'}a')$. Any $b\in R^*(Z)=\{a\in R(Z): \pi_0 a=0\}$ has a quasi-inverse $(-b)^*=\lim_{n\to\infty}\sum_{n'<n}(-b)^{n'}$. If a is invertible, $a^{-1}=(1+b^*)(\pi_0 a^{-1})$ where $b=-(\pi_0 a^{-1})(a-\pi_0 a)\in R^*(Z)$. We shall say that $S^*\subset R^*(Z)$ is *rationally closed* if $r, r'\in R$, $b, b'\in S^*$ imply $rb+b'r'$, bb', $b^*\in S^*$. If this is so, the set of those elements a of $R(Z)$ such that $a-\pi_0 a\in S^*$ is a ring containing the inverses of its invertible elements.

DEFINITION 1. $R^*_{rat}(X)$ is the least rationally closed subset (of $R(X)$) containing X.

Now let $Y=\{y_j\}$ be a set of a finite number M of new variables and $R^M(X\cup Y)$ (resp. $R^M_{pol}(X\cup Y)$) the cartesian product of M copies

Received by the editors December 6, 1961.

of the R-module $R(X \cup Y)$ (resp. $R^M_{\text{pol}}(X \cup Y)$). For each $q = (q_1, \cdots, q_m) \in R^M(X \cup Y)$, $\pi_n q = (\pi_n q_1, \cdots, \pi_n q_m)$. If $q \in R^{*M}(X \cup Y)$ (i.e., if $\pi_0 q = 0$) let λ_q be the homomorphism of the monoid $F(X \cup Y)$ into the multiplicative monoid structure of $R(X \cup Y)$ that is induced by $\lambda_q x = x$ if $x \in X$ and $\lambda_q y_j = q_j$ if $y_j \in Y$. Since $\pi_0 q = 0$, λ_q can be extended to an endomorphism of the R-module $R(X \cup Y)$ by $\lambda_q a = \sum\{(a, f)\lambda_q f : f \in F(X \cup Y)\}$; also, $\lambda_q p = (\lambda_q p_1, \cdots, \lambda_q p_M)$ for any $p \in R^M(X \cup Y)$.

We shall say that $p \in R^{*M}(X \cup Y)$ is a *proper system* if $(p_j, y_{j'}) = 0$ for all $j, j' \leq M$. Then, if $q \in R^{*M}(X)$, $\lambda_q p \in R^{*M}(X)$ and $\pi_{n+1}\lambda_q p = \pi_{n+1}\lambda_{\pi_n q} p$ for all n. Consider now the infinite sequence $p(0) = 0$, $p(1) = \lambda_{p(0)} p, \cdots, p(m+1) = \lambda_{p(m)} p, \cdots$. Trivially, $\pi_{m'} p(m') = \pi_{m'} p(m' + m'') \in R^{*M}(X)$ for $m' = 0$ and all m''. If these relations hold for $m' \leq m$, they still hold for $m+1$ because

$$\pi_{m+1} p(m+1) = \pi_{m+1}\lambda_{p(m)} p = \pi_{m+1}\lambda_{\pi_m p(m)} p = \pi_{m+1}\lambda_{\pi_m p(m+m'')} p$$
$$= \pi_{m+1}\lambda_{p(m+m'')} p = \pi_{m+1} p(m+1+m'').$$

Hence, $p(\infty) = \lim_{m \to \infty} p(m)$ exists and it satisfies $p(\infty) \in R^{*M}(X)$, $\pi_0 p(\infty) = 0$, $p(\infty) = \lambda_{p(\infty)} p$. In fact, $p(\infty)$ is the only element to satisfy these equations because if $\pi_0 p' = 0$ and $p' = \lambda_{p'} p$, any relation $\pi_m p(\infty) = \pi_m p'$ implies $\pi_{m+1} p' = \pi_{m+1}\lambda_{\pi_m p'} p = \pi_{m+1}\lambda_{\pi_m p(\infty)} p = \pi_{m+1} p(\infty)$. For this reason we call $p(\infty)$ *the solution* of p.

DEFINITION 2. $R^*_{\text{alg}}(X)$ is the least subset (of $R^*(X)$) that contains every coordinate of the solution of any proper system having its coordinates in $R^*_{\text{pol}}(X \cup Y)$.

(REMARK. It can easily be shown that $R^*_{\text{alg}}(X)$ is rationally closed and that it contains every coordinate of the solution of any proper system having its coordinates in $R^*_{\text{alg}}(X \cup Y)$.)

DEFINITION 3. For any

$$a, b \in R(X), \quad a \odot b = \sum\{(a, f)(b, f) \cdot f : f \in F(X)\}.$$

2. Main result.

Property 2.1. The element a of $R^*(X)$ belongs to $R^*_{\text{rat}}(X)$ if and only if there exists a finite integer $N \geq 2$ and a homomorphism μ of $F(X)$ into the multiplicative monoid of $R^{N \times N}$ (the ring of the $N \times N$ matrices with entries in R) such that $a = \sum\{\mu f_{1,N} \cdot f : f \in F(X)\}$ (abbreviated as $\sum \mu f_{1,N} \cdot f$).

PROOF. (1) *The condition is necessary.* This is trivial if $a = \pi_1 a$. Hence it suffices to show that for any $r, r' \in R$, $a = \sum \mu f_{1,N} \cdot f$ and $a' = \sum \mu' f_{1,N'} \cdot f$ one can construct suitable homomorphisms giving $ra + a'r'$, aa' and a^*. This is done below, defining the homomorphisms by their restriction to X.

Addition. Let $N''=N+N'+2$ and $\mu''x\in R^{N''\times N''}$ defined for each $x\in X$ by

$$\mu''x_{i,1} = \mu''x_{N'',i} = 0 \qquad \text{for } 1 \leqq i \leqq N'';$$

$$\mu''x_{1,i+1} = r\mu x_{1,i} \text{ and } \mu''x_{i+1,N''} = \mu x_{i,N} \qquad \text{for } 1 \leqq i \leqq N;$$

$$\mu''x_{1,i+N+1} = \mu'x_{1,i} \text{ and } \mu''x_{i+N+1,N''} = \mu'x_{i,N'}\cdot r' \qquad \text{for } 1 \leqq i \leqq N';$$

$$\mu''x_{i,i'} = \text{the direct sum of } \mu x \text{ and } \mu'x \qquad \text{for } 2 \leqq i, i' \leqq N'' - 1;$$

$$\mu''x_{1,N''} = r\mu x_{1,N} + \mu'x_{1,N'}r'.$$

The verification is trivial.

Product. Let $N''=N+N'$ and define $\nu f\in R^{N''\times N''}$ for each $f\in F(X)$ by $\nu f_{i,i'}=\mu f_{i,N}$ if $f\neq 1$, $1\leqq i\leqq N$, $i'=N+1$; $\nu f_{i,i'}=0$, otherwise. Then, if $\mu''x=\bar\mu x+\nu x$ where $\bar\mu x$ is the direct sum of μx and $\mu'x$, one has for each $f = x^{(1)}x^{(2)} \cdots x^{(n)}$, $\mu''f = \bar\mu f + \sum\{\bar\mu f'\nu x^{(i)}\bar\mu f'' : f'x^{(i)}f'' = f\}$. Since $\nu f x^{(i)}=\bar\mu f \nu x^{(i)}$ and $(\nu f'''\bar\mu f'')_{1,N''}=0$ when $f''=1$, one has $\mu''f_{1,N''}=\sum\{(\mu f'_{1,N})(\mu' f''_{1,N'}):f'f''=f\}$. Hence, $\sum \mu''f_{1,N''}\cdot f=aa'$.

Quasi-inverse. Let $N''=N$ and define $\nu f\in R^{N\times N}$ for each $f\in F(X)$ by $\nu f_{i,i'}=\mu f_{i,N}$ if $f\neq 1$, $1\leqq i\leqq N$, $i'=1$; $\nu f_{i,i'}=0$, otherwise. Then $\mu''x = \mu x + \nu x$ and since $\mu f \nu x = \nu f x$ identically one has $\mu''f = \sum \nu f^{(1)}\nu f^{(2)} \cdots \nu f^{(k)}\mu f^{(k+1)}$ where the summation is over all the factorisations $f=f^{(1)}f^{(2)} \cdots f^{(k+1)}$ of f in an arbitrary number of factors. The $(1, N)$ entry of any of these products is zero unless all its factors are different from 1 and under this condition, it is equal to $\mu f^{(1)}_{1,N}\, \mu f^{(2)}_{1,N} \cdots \mu f^{(k+1)}_{1,N}$. Hence, $\sum \mu''f_{1,N}\cdot f=\sum_{n>0} a^n = a^*$ and the first part of the proof is completed.

(2) *The condition is sufficient.* We say that the proper system p is *linear* if for each $j\leqq M$, $p_j=q_{j,0}+\sum_{j'} q_{j,j'}y_{j'}$ where all the q's belong to $R^*_{\text{rat}}(X)$ and we verify that all coordinates of the solution of such a system belong to $R^*_{\text{rat}}(X)$.

This is trivial if $M=1$ because $p(\infty)=(1-q_{1,1})^{-1}q_{1,0}(=(1+q^*_{1,1})q_{1,0})$. If it is true for $M'<M$ it is still true for M. Indeed, because $p(\infty)_M =(1-q_{M,M})^{-1}(q_{M,0}+\sum_{j<M} q_{M,j'}p(\infty)_{j'})$, the proper linear system p' defined by $p'_j = p_j - q_{j,M}y_M + q_{j,M}p_M$ for $j < M$ and $p'_M =(1-q_{M,M})^{-1}(p_M - q_{M,M}y_M)$ is such that $p(\infty)=p'(\infty)$. Since its first $M-1$ coordinates do not involve y_M the result follows from the induction hypothesis.

Now, given a homomorphism μ of $F(X)$ into $R^{M\times M}$, the M elements $a_j=\sum\{\mu f_{j,M}\cdot f: f\in F(X),\ f\neq 1\}$ are such that $(a_j, xf) = \sum_{j'} \mu x_{j,j'}(a_{j'}, f)$. Hence $(a_1, \cdots, a_M)$ is the solution of the linear proper system such that $q_{j,0} = \sum\{\mu x_{j,M}\cdot x: x \in X\}$, $q_{j,j'} = \sum\{\mu x_{j',j}\cdot x: x\in X\}$ for each j, j' and 2.1 is proved.

We now consider two subrings R' and R'' of R that commute element-wise.

Property 2.2. If $a=\sum\mu' f_{1,N}\cdot f\in R'^*_{\text{rat}}(X)$ where μ' is a homomorphism into $R'^{N\times N}$ and if $b=p(\infty)_1\in R''^*_{\text{alg}}(X)$ where the proper system p has its coordinates in $R''^*_{\text{pol}}(X\cup Y)$, then $a\odot b\in R^*_{\text{alg}}(X)$. If, further, $b\in R''^*_{\text{rat}}(X)$ then $a\odot b\in R^*_{\text{rat}}(X)$.

PROOF. We verify first the case of $b\in R''^*_{\text{rat}}(X)$, i.e., of $b=\sum\mu'' f_{1,N''}\cdot f$ for some N'' and μ''. Then $a\odot b=\sum(\mu'\otimes\mu'')f_{1,NN''}\cdot f$ where the kroneckerian product $\mu'\otimes\mu''$ is a homomorphism of $F(X)$ into $R^{NN''\times NN''}$ because R' and R'' commute and the result is proved.

For the general case we denote by $K(Z)$ for any set Z the ring of the $N\times N$ matrices with entries in $R(Z)$. We shall have to consider several homomorphisms of module $\sigma: R^M(Z')\to K^M(Z'')$ where Z' and Z'' are two finite sets. In each case σ is defined by a mapping $Z'\to K(Z'')$ which is extended in a natural fashion to a homomorphism of the monoid $F(Z')$ into the multiplicative structure of $K(Z'')$. Then for each

$$a=(a_1,\cdots,a_M)\in R^M(Z'),\quad \sigma a_j=\sum\{(a_j,g)\cdot\sigma g: g\in F(Z')\}$$

and $\sigma a=(\sigma a_1,\cdots,\sigma a_M)$.

More specifically, $\mu: R^M(X)\to K^M(X)$ is induced by a mapping $\mu: X\to K(X)$ such that the entries of each μx belong to $R'^*(X)$.

For each $q\in R''^{*M}(X)$, $\lambda_{\mu q}: R(X\cup Y)\to K^M(X)$ is induced by $\lambda_{\mu q}f=\mu f$ if $f\in F(X)$ and $\lambda_{\mu q}y_j=\mu q_j$ if $y_j\in Y$. Hence, since R' and R'' commute element-wise, $\mu\lambda_q g=\lambda_{\mu q}g$ for each $g\in F(X\cup Y)$ (with λ_q as previously defined). Consequently, $\mu\lambda_q p=\lambda_{\mu q}p$ for any $p\in R''^M(X\cup Y)$.

Let now $Z=\{z_{j,i,i'}\}(1\le j\le M;\ 1\le i,\ i'\le N)$, a set of $M\times N\times N$ new variables and $\nu: R^M(X\cup Y)\to K^M(X\cup Z)$ induced by $\nu f=\mu f$ if $f\in F(X)$, $\nu y_j=$ the $N\times N$ matrix with entries $z_{j,i,i'}$ if $y_j\in Y$. Also $\lambda_{\nu q}: R(X\cup Z)\to R(X)$ is induced by $\lambda_{\nu q}f=f$ if $f\in F(X)$ and $\lambda_{\nu q}z_{j,i,i'}=(\nu q_j)_{i,i'}$ if $z_{j,i,i'}\in Z$. We extend $\lambda_{\nu q}$ to a homomorphism $K^M(X\cup Z)\to K^M(X)$ by defining $\lambda_{\nu q}\boldsymbol{m}$ for any $\boldsymbol{m}\in K(X\cup Z)$ as the $N\times N$ matrix with entries $\lambda_{\nu q}(\boldsymbol{m}_{i,i'})$.

Because R' and R'' commute, $\lambda_{\mu q}g=\lambda_{\nu q}\nu g$ for each $g\in F(X\cup Y)$ and, consequently, $\lambda_{\mu q}p=\lambda_{\nu q}\nu p$ for each $p\in R''^{*M}(X\cup Y)$. Hence, if p is a proper M-dimensional system with coordinates in $R''^*(X\cup Y)$ we have $\mu p(\infty)=\mu\lambda_{p(\infty)}p=\lambda_{\mu p(\infty)}p$. Since μ and ν coincide on $R''^{*M}(X)$, we have also $\mu p(\infty)=\nu p(\infty)=\lambda_{\mu p(\infty)}p=\lambda_{\nu p(\infty)}\nu p$.

However, the $M\times N\times N$ elements $p'_{j,i,i'}=(\nu p_j)_{i,i'}$ all belong to $R^*(X\cup Z)$ and they constitute a proper system p' of dimension MN^2. Thus, by construction, $(\mu p(\infty)_j)_{i,i'}=p'(\infty)_{j,i,i'}$ identically. If, further, $p \in R''^{*M}_{\text{pol}}(X \cup Y)$ all the entries appearing in νp belong to $R^*_{\text{pol}}(X \cup Z)$ and then finally $(\mu p(\infty)_j)_{i,i'} \in R^*_{\text{alg}}(X)$.

This completes the proof because

$$a \odot b = \sum \{(b, f)\mu' f_{1,N} \cdot f : f \in F(X)\}$$
$$= \sum \{(b, f)\mu f_{1,N} : f \in F(X)\} = \mu b_{1,N}$$

where for each $x \in X$, μ is defined by $\mu x_{i,i'} = \mu' x_{i,i'} \cdot x$.

REMARK 1. Definitions 1, 2, and 3 and the computations of this section used only the structure of monoid of the additive groups considered. Hence, the results are still valid when an arbitrary *semiring* S is taken in place of R. For S consisting of two Boolean elements, Jungen's theorem and its special case for b rational have been obtained in a different form by Y. Bar-Hillel, M. Perles and E. Shamir [1] (also by S. Ginsburg and G. F. Rose [5]) and by S. Kleene [8] respectively as by-products of more sophisticated theories.

REMARK 2. Let $R = C$, the field of complex numbers; and p a proper system of dimension M. Introducing $4M$ new symbols z_j and replacing each y_j by $z_{4j} + iz_{4j+1} - z_{4j+2} - iz_{4j+3}$ in the p_js we can deduce from p a new system of dimension $4M$ in which all the coefficients are non-negative real numbers and whose solution is simply related to $p(\infty)$.

Assume now that $p \in C^{*M}_{\text{pol}}(X \cup Y)$ has only real non-negative coefficients and denote by α a homomorphism of $C_{\text{pol}}(X \cup Y)$ into C. Because of the assumption that $(p_j, y_{j'}) = (p_j, 1) = 0$, identically, we can find an $\epsilon > 0$ such that $|\alpha p_j| < \epsilon$ for all j when $|\alpha x| \leqq \epsilon$ and $|\alpha y| \leqq 2\epsilon$ for all $x \in X$ and $y \in Y$. Since the sequence $\alpha p(0)$, $\alpha p(1)$, $\cdots$, $\alpha p(n)$, $\cdots$ is monotonically increasing it converges to a finite solution (cf., e.g., [10]).

Hence, the canonical epimorphism of $C_{\text{pol}}(X \cup Y)$ onto the ring of the ordinary (commutative) polynomials can be extended to an epimorphism of $C_{\text{alg}}(X)$ onto the ring of the Taylor series of the algebraic functions.

Acknowledgment. Acknowledgment is made to the Commonwealth Fund for the grant in support of the visiting professorship of biomathematics in the Department of Preventive Medicine at Harvard Medical School.

REFERENCES

1. Y. Bar-Hillel, M. Perles and E. Shamir, *On formal properties of simple phrase structure grammars*, Technical Report No. 4. Information System Branch, Office of Naval Research, 1960.

2. S. Bochner and W. T. Martin, *Singularities of composite functions in several variables*, Ann. of Math. **38** (1938), 293–302.

3. R. H. Cameron and W. T. Martin, *Analytic continuation of diagonals*, Trans. Amer. Math. Soc. **44** (1938), 1–7.

4. K. T. Chen, R. H. Fox and R. C. Lyndon, *Free differential calculus*. IV, Ann. of Math. (2) **68** (1958), 81–95.

5. S. Ginsburg and G. F. Rose, *Operations which preserve definability*, System Development Corporation, Santa Monica, Calif., SP-511, October, 1961.

6. U. S. Haslam-Jones, *An extension of Hadamard multiplication theorem*, Proc. London Math. Soc. II. Ser. **27** (1928), 223–232.

7. R. Jungen, *Sur les series de Taylor n'ayant que des singularités algébrico-logarithmiques sur leur cercle de convergence*, Comment. Math. Helv. **3** (1931), 226–306.

8. S. Kleene, *Representation of events in nerve nets and finite automata*, Automata Studies, Princeton Univ. Press, Princeton, N. J., 1956.

9. M. Lazard, *Lois de groupes et analyseurs*, Ann. Sci. Ecole Norm. Sup. (4) **72** (1955), 299–400.

10. A. M. Ostrowski, *Solutions of equations and systems of equations*, Academic Press, New York, 1960.

Harvard Medical School

Reprinted from
Proc. Amer. Math. Soc. **13** (1962), 885–890

REGULARITY AND POSITIONAL GAMES

BY
A. W. HALES AND R. I. JEWETT

Reprinted from the
TRANSACTIONS OF THE AMERICAN MATHEMATICAL SOCIETY
Vol. 106, No. 2, pp. 222–229
February, 1963

REGULARITY AND POSITIONAL GAMES

BY
A. W. HALES AND R. I. JEWETT

1. **Introduction.** Suppose X is a set, $\mathscr{S}$ a collection of sets (usually subsets of X), and N is a cardinal number. Following the terminology of Rado [1], we say $\mathscr{S}$ is N-regular in X if, for any partition of X into N parts, some part has as a subset a member of $\mathscr{S}$. If $\mathscr{S}$ is n-regular in X for each integer n, we say $\mathscr{S}$ is regular in X.

For example, let $X = \{1,2,\cdots,mn-n+1\}$ and $\mathscr{S}$ be all m element subsets of X (hereafter designated $X^{(m)}$). Then $\mathscr{S}$ is n-regular in X, but not $(n+1)$-regular. Another example is the famous theorem of Ramsey which states that given integers k, m, n, there exists an integer p such that, if $A = \{1,2,\cdots,p\}$, then $\{B^{(k)} : B \in A^{(m)}\}$ is n-regular in $A^{(k)}$. The concept of regularity is useful in analyzing certain types of games, as we shall see in §3. In §2, we shall give some general results and discuss related problems.

2. **Regularity.** One of the first problems in this area was proposed at Göttingen in 1927. The problem was as follows: If the positive integers are split into two parts, does one part contain arithmetic progressions of arbitrary length? B. L. van der Waerden solved this and a more general problem. He proved that, given integers m and n, there exists an integer p such that the set of all arithmetic progressions of length m is n-regular in $\{1,2,\cdots,p\}$ [2]. This will be a consequence of Theorem 1. First we shall give some preliminaries.

DEFINITION. If $\mathscr{S}$ and $\mathscr{T}$ are collections of sets, let $\mathscr{S} \otimes \mathscr{T}$ be the collection of all sets $A \times B$, where A is in $\mathscr{S}$ and B is in $\mathscr{T}$.

LEMMA 1. *Let M and N be cardinal numbers. Let $\mathscr{S}$ be N-regular in X, a set of cardinality M, and let $\mathscr{T}$ be N^M-regular in Y. Then $\mathscr{S} \otimes \mathscr{T}$ is N-regular in $X \times Y$.*

Proof. Let P be a set of cardinality N. Then a partition of $X \times Y$ into N parts can be represented by a function f from $X \times Y$ into P. For each $y \in Y$, f defines a function f_y from X into P given by

$$f_y(x) = f(x,y).$$

Since there are N^M such functions the mapping $y \to f_y$ induces a partition of Y into N^M parts. One of these parts contains as a subset a member T of $\mathscr{T}$. That is, for all y, $y' \in T$

Received by the editors July 5, 1961 and, in revised form, December 26, 1961.

$$f_y = f_{y'},$$

$$f(x,y) = f(x,y') \qquad (x \in X).$$

Choose $y_0 \in T$. Then f_{y_0} partitions X into N parts and hence $\exists S \in \mathcal{S}$, $p \in P$ such that

$$f_{y_0}(x) = p \qquad (x \in S).$$

But then

$$f_y(x) = p \qquad (x \in S,\ y \in T).$$

That is,

$$f(x,y) = p \qquad (x \in S,\ y \in T)$$

which was to be shown.

DEFINITION. Let X and Y be sets, $\mathcal{S}$ and $\mathcal{T}$ collections of sets. Then a mapping $f\colon X \to Y$ is called *provincial with respect to* $\mathcal{S}$ *and* $\mathcal{T}$ in case when $A \in \mathcal{S}$, $A \subseteq X$ there exists a set $B \in \mathcal{T}$, $B \subseteq Y$ such that $B \subseteq f(A)$.

LEMMA 2. *Let $f\colon X \to Y$ be provincial with respect to $\mathcal{S}$ and $\mathcal{T}$. Then if $\mathcal{S}$ is N-regular in X, $\mathcal{T}$ is N-regular in Y.*

Proof. Let P be a set of cardinality N and $g\colon Y \to P$. Then $g(f)\colon X \to P$ and there exist $p \in P$ and $A \in \mathcal{S}$ such that $A \subseteq X$ and $g(f(A)) = \{p\}$. But there is a $B \in \mathcal{T}$, $B \subseteq Y$ such that $B \subseteq f(A)$. So $g(B) = \{p\}$.

LEMMA 3. *Let X be a semigroup, $\mathcal{S} \subseteq 2^X$. Suppose for each positive integer k, $\mathcal{S}$ is k-regular in a finite subset of X. Then for each n,*

$$\mathcal{S}_n = \{A_1 A_2 \cdots A_n : A_i \in \mathcal{S}\}$$

is regular in X.

Proof. We induct on n. Suppose $\mathcal{S}_{n-1}$ is regular in X and $k \geq 1$. Then there is an integer m and $B \in X^{(m)}$ such that $\mathcal{S}$ is k-regular in B. Since $\mathcal{S}_{n-1}$ is k^m-regular in X, $\mathcal{S} \otimes \mathcal{S}_{n-1}$ is k-regular in $B \times X \subseteq X \times X$. But the mapping $(x,y) \to xy$ of $X \times X$ into X is clearly provincial with respect to $\mathcal{S} \otimes \mathcal{S}_{n-1}$ and $\mathcal{S}_n$, and thus $\mathcal{S}_n$ is k-regular in X.

Let W be a fixed set and $t \notin W$. Let X be the free semigroup on the set W. A functional f is a mapping of W into X which can be described as follows. For some positive integer n there is an n-tuple $\alpha = (\alpha_1, \alpha_2, \cdots, \alpha_n)$ of elements of $W \cup \{t\}$ in which t appears at least once, such that, for $w \in W$, $f(w)$ is the result of replacing the t by a w and multiplying (in X) the n components of the new n-tuple.

For example, if $W = \{1,2,3,4\}$ and $\alpha = (1, t, 3, t, t, 2, 1, t)$ then the corresponding functional f would satisfy

$$f(4) = 14344214.$$

Suppose that $f_1, f_2, \cdots, f_n$ are functionals. Let $\phi : W^n \to f_1(W)f_2(W)\cdots f_n(W)$ be defined by

$$\phi(w_1 w_2 \cdots w_n) = f_1(w_1)f_2(w_2)\cdots f_n(w_n).$$

We see that if g is a functional of "length" n then there is a functional h such that

$$\phi(g(w)) = h(w) \qquad (w \in W).$$

Loosely, we have $\phi(g(t)) = h(t)$.

If $A \subseteq W$, $R \subseteq W^n$ and R is a member of $\{f(A) : f \text{ is a functional}\}$ then $\phi(R)$ is also a member of this collection. Thus, relative to this collection, ϕ is a provincial map of A^n onto $f_1(A)f_2(A)\cdots f_n(A)$.

THEOREM 1. *If A is a finite subset of W the collection $\{f(A)$: f is a functional$\}$ is regular in X.*

Proof. Let $(I_{m,j})$ be the statement: If $B \in W^{(m)}$ there exists an integer p such that $\{f(B) : f \text{ is a functional}\}$ is j-regular in B^p.

We will prove these statements by induction and thus prove the theorem. It is clear that $(I_{m,1})$ and $(I_{1,j})$ are true. Assume further that for $n > 1$ and $k \geqq 1$, $(I_{n,k})$ and $(I_{n-1,j})$ are true for all j.

Let $A \in W^{(n)}$. Pick $a \in A$ and let $B = A - \{a\}$. By assumption, there is an integer r such that $\{f(A): f \text{ is a functional}\}$ is k-regular in A^r. By $(I_{n-1,j})$, the fact that B^s is finite for each s, and Lemma 3, we see that $\{f_0(B)f_1(B)\cdots f_r(B): f_i \text{ is a functional}\}$ is $(k+1)$-regular in the subsemigroup of X generated by B, namely, $B^1 \cup B^2 \cup \cdots \cup B^s \cup \cdots$. The B^s are disjoint, and if $f_0, f_1, \cdots, f_r$ are functionals, then the set $f_0(B)f_1(B)\cdots f_r(B)$, all of whose elements have the same "length," meets at most one of the B^s. In such a situation, $(k+1)$-regularity in the union implies $(k+1)$-regularity in one of the parts. Thus there is an integer q such that

$$\{f_0(B)f_1(B)\cdots f_r(B)\}$$

is $(k+1)$-regular in B^q. We will use the integer q to verify $(I_{n,k+1})$.

Let $A^q = P_0 \cup P_1 \cup \cdots \cup P_k$. This defines a partition of B^q and so there are functionals $g_0, g_1, \cdots, g_r$ such that $g_0(B)g_1(B)\cdots g_r(B)$ is contained in one of the parts, say P_0, and also in B^q. Thus each "entry" in a g_i is an element of $B \cup \{t\}$. Since $B \subseteq A$, we can conclude that $g_0(A)g_1(A)\cdots g_r(A) \subseteq A^q$. Now the mapping $\phi: A^r \to g_0(a)g_1(A)\cdots g_r(A)$ defined by $\phi(w_1 w_2 \cdots w_r) = g_0(a)g_1(w_1)\cdots g_r(w_r)$ is provincial with respect to $\{f(A) : f \text{ is a functional}\}$. Thus, by $(I_{n,k})$, $\{f(A)\}$ is k-regular in $g_0(a)g_1(A)\cdots g_r(A)$.

If $g_0(a)g_1(A)\cdots g_r(A)$ is disjoint from P_0 we are done. If not, there are elements $a_1, a_2, \cdots, a_r$ of A such that

$$x = g_0(a)g_1(a_1)\cdots g_r(a_r) \in P_0.$$

Suppose $x = v_1 v_2 \cdots v_q \in A^q$. Define $\alpha = (\alpha_1, \alpha_2, \cdots, \alpha_q)$ by

$$\alpha_i = \begin{cases} v_i & \text{if } v_i \in B, \\ t & \text{if } v_i = a. \end{cases}$$

Note that t appears in f_0, so a appears in $f_0(a)$, and hence t appears in α. Then α represents a functional g for which

$$g(a) = x$$

and $g(B) \subseteq g_0(B)g_1(B) \cdots g_r(B) \subseteq P_0$. Since $g(A) = g(B) \cup \{g(a)\}$, we have $g(A) \subseteq P_0$, and the theorem is proved.

COROLLARY. *Let S be a finite subset of a commutative semigroup H. Then the collection of all sets*

$$\{a + nx : x \in S\},$$

where $a \in H$ and n is a positive integer, is regular in H.

Proof. Let X be the free semigroup on S. Then the mapping

$$a_1 a_2 \cdots a_n \rightarrow a_1 + a_2 + \cdots + a_n$$

is provincial.

COROLLARY (VAN DER WAERDEN). *For any partition of the positive integers into a finite number of parts, one of the parts contains arithmetic progressions of arbitrary length.*

The stronger statement proved by van der Waerden and mentioned above is clear from the proof of Theorem 1. This suggests a general result which we have given as a corollary to Theorem 2. Theorem 2 is proved in Rado [3]. The proof is essentially an application of Tychonoff's Theorem, as shown by Gottschalk [4].

THEOREM 2. *Let X and Γ be sets, and for every finite subset A of X let f_A be a function from A into Γ. Suppose that for each $x \in X$, $\Gamma_x = \{f_A(x) : A$ is a finite subset of X containing $x\}$ is finite. Then there exists a function F from X into Γ with the property that, given any finite subset A of X, there exists a finite subset B of X such that F and f_B agree on A.*

COROLLARY. *Let X be a set and $\mathscr{S}$ a collection of finite sets. Then if $\mathscr{S}$ is n-regular in X for some positive integer n, $\mathscr{S}$ is n-regular in a finite subset of X.*

Proof. Let P be a set with n elements. Suppose $\mathscr{S}$ is not n-regular on any finite subset of X. Then for each finite subset A of X, there is a function f_A from A into P that is not constant on any member of $\mathscr{S}$. By Theorem 2 there is a function F from X into P that is not constant on any member of $\mathscr{S}$, a contradiction.

The above corollary suggests a general problem. Let M and N be cardinals. Does there exist a cardinal P having the following property? If X is a set and $\mathscr{S}$ is a collection of sets each of cardinality less than M, and $\mathscr{S}$ is N-regular in

X, then $\mathcal{S}$ is N-regular in a subset of X of cardinality less than P. A simple example shows that if $M > 1$ and $N > 1$ are integers, $P = \aleph_0$ is best possible. The corollary says that if $M = \aleph_0$, and $N > 1$ is finite, then $P = \aleph_0$ is sufficient. No further results in this area are known.

Another problem is the "rectangle" problem. Let M, N, and P be cardinals. For what pairs (R, S) of cardinals (if any) is the following true? If X has cardinality R and Y has cardinality S, then $X^{(M)} \otimes Y^{(n)}$ is P-regular in $X \times Y$. That is, if an $R \times S$ rectangle is partitioned into P parts, one part contains an $M \times N$ rectangle. From Lemma 1, such pairs always exist. For example, if $M=2$, $P=2$, $N=5$, $P=2$, then $R=3$ and $S=40$ is sufficient. The "minimal" pairs (R, S) for given M, N, and P are not known in general. A particularly interesting case occurs when $M = N = \aleph_0$ and $P = 2$. It is easily seen that $(\aleph_0, \aleph_0)$ does not work and $(\aleph_0, 2^{2^{\aleph_0}})$ does. The sufficiency of $(\aleph_0, 2^{\aleph_0})$ or, for that matter, $(2^{\aleph_0}, 2^{\aleph_0})$ is an open question.

3. **Positional games.** By a positional game we shall mean a game played by n players on a "board" (finite set) X with which is associated a collection $\mathcal{S}$ of subsets of X. The rules are that each player, in turn, claims as his own a previously unclaimed "square" (element) of X. The game proceeds either until one player has claimed every element of some $S \in \mathcal{S}$, in which case he wins, or until every element has been claimed, but no one has yet won, in which case the game is a tie. The most familiar example of such a game is "Tick-Tack-Toe." Another is the Oriental game "Go Moku."

It is known from game theory that, in a finite two-player perfect information game, either one player has a forced win or each player can force a tie [5].

LEMMA 4. *In a positional game involving 2 players, where $\mathcal{S}$ is 2-regular in X, the first player has a forced win.*

Proof. Since no tie can occur, one player has a forced win. Assume the second player has a forced win. But then the first player can force a win by (1) making his first move at random, and (2) thereafter following the optimum strategy for the second player, ignoring the last random move, and playing again at random if this is impossible. Since having made an extra move cannot possibly hurt, this will give the first player a win, a contradiction. Therefore, the first player has a forced win.

The following result in combinatorial analysis is due to Philip Hall [6].

LEMMA 5. *Let $S_1, S_2, \cdots, S_n$ be an indexed collection of finite sets. Then (A) and (B) are equivalent.*

(A) *There exist $s_1, s_2, \cdots, s_n$ such that each $s \in S_i$ and $s_i \neq s_j$ if $i \neq j$.*

(B) *For each $F \subseteq \{1, \cdots, n\}$, the set $\bigcup_{i \in F} S_i$ has at least as many elements as F.*

If condition (A) is satisfied, we say $S_1, \cdots, S_n$ have distinct representatives. We will use this lemma to exhibit a tying strategy for the second player in certain positional games.

LEMMA 6. *Let X be the board of a 2-player positional game with winning sets $\mathscr{S} = \{S_1, \cdots, S_n\}$. For $k = 1, 2, \cdots, n$ let $T_{2k-1} = T_{2k} = S_k$. Then if $T_1, \cdots, T_{2n}$ have distinct representatives, the the second player can force a tie.*

Proof. Let the representatives be $t_1, t_2, \cdots, t_{2n}$. Consider the sets $\{t_1, t_2\}$, $\{t_3, t_4\}, \cdots, \{t_{2n-1}, t_{2n}\}$. Observe that in order to win the first player must have both elements of at least one of these sets. Since the second player can easily prevent this, he can force a tie.

LEMMA 7. *Let $\mathscr{S} \subseteq 2^X$, where X is finite. Let n be the size of the smallest member of $\mathscr{S}$. Let m be the size of the largest set of the form $\{S \in \mathscr{S} : x \in S\}$ where $x \in X$. If $n \geqq 2m$, then in the corresponding 2-player positional game, the second player can force a tie.*

Proof. By a simple counting argument, Lemmas 5 and 6 can be applied to obtain the desired result.

The rest of this paper will be concerned with a particular class of 2-player positional games, namely generalizations of Tick-Tack-Toe. The traditional Tick-Tack-Toe game is played on a 3×3 array of points in the plane. For positive integers k and n, the "k^n-game" is played on a $k \times k \times \cdots \times k$ (n times) array of points in n-space. If we choose as a board the set

$$X = \{(a_1, a_2, \cdots, a_n) \ : \ 1 \leqq a_i \leqq k \text{ for all } i\},$$

then S is in $\mathscr{S}$, the collection of winning sets (paths), in case $\mathscr{S}$ consists of k points in a straight line. An equivalent characterization of $S \in \mathscr{S}$ would be that the elements of S, in some order, are $\alpha_1, \alpha_2, \cdots, \alpha_k$ where $\alpha_i = (a_{i1}, \cdots, a_{in})$ and, for each j, the sequence $(a_{1j}, a_{2j}, \cdots, a_{kj})$ is one of the following:

$$\begin{array}{lll}
(1, & 1, \cdots, & 1) \\
(2, & 2, \cdots, & 2) \\
\cdot & \cdot & \cdot \\
\cdot & \cdot & \cdot \\
\cdot & \cdot & \cdot \\
(k, & k, \cdots, & k) \\
(1, & 2, \cdots, & k) \\
(k, & k-1, \cdots, & 1).
\end{array}$$

In this case we say $\alpha_1, \alpha_2, \cdots, \alpha_k$ are in a natural order (there are two such orders).

In traditional Tick-Tack-Toe, the second player can achieve a tie. In the 3^3-game, however, the first player has a forced win (in fact, no tie position exists).

Thus, in the 3-dimensional games sold on the market, k is usually 4. Our previous results enable us to draw some conclusions about the existence of winning and tying strategies in the general case.

THEOREM 4. (a) *If* $k \geqq 3^n-1$ *(k odd) or if* $k \geqq 2^{n+1}-2$ *(k even), then the second player can force a tie in the k^n-game.*

(b) *For each k, there exists n_k such that the first player can force a win in the k^n-game if $n \geqq n_k$.*

Proof. (a) If k is odd, there are at most $(3^n-1)/2$ paths through any point and this bound is achieved only at the center point. If k is even, the bound is 2^n-1. The result follows readily from Lemma 7. This suggests that the center point is the optimum move for the first player if k is odd.

(b) In Theorem 1, let $W=\{1,2,\cdots,k\}$. Note that if f is a functional then $f(W)$ is a path, but the converse is not true. Now $(I_{k,2})$ and Lemma 4 yield the result.

We conjecture that the bounds in Theorem 4(a) can be improved by a direct application of Lemmas 5 and 6. It seems possible that $k \geqq 2(2^{1/n}-1)^{-1}$, i.e., that the total number of points be greater than the total number of paths, can be shown to be sufficient in this way.

Even though, in some k^n-games, the second player cannot force a tie, a tie position may still exist, i.e., $\mathscr{S}$ (the collection of paths) may not be 2-regular in X (the board). The bounds of Theorem 4(a) apply, but much more can be said.

THEOREM 5. *If $k \geqq n+1$, then in the k^n-game the collection of paths is not 2-regular in the board.*

Let k be fixed. For each n let the k^n-game board X_n be the set of n-tuples on $\{1,2,\cdots,k\}$. Designate the elements of $GF(2)$ by $\{0,1\}$. Any partition of X_n into two parts can be represented (in two ways) as a function from X_n into $GF(2)$. Let $f: X_m \to GF(2)$ and $g: X_n \to GF(2)$ represent partitions. Then we define $f \oplus g$: $X_{m+n} \to GF(2)$ by

$$(f \oplus g)\,(a_1,\cdots,a_{m+n}) = f(a_1,\cdots,a_m) + g(a_{m+1},\cdots,a_{m+n})$$

where addition on the right takes place in $GF(2)$. Thus $f \oplus g$ represents a partition of X_{m+n} into two parts. Note that "$\oplus$" is an associative operation on functions from the X_i into $GF(2)$.

Proof of Theorem 5. Let $V_1, V_2, \cdots, V_m$ be k-dimensional vectors over $GF(2)$, that is, functions from X_1 into $GF(2)$. Define

$$(a_1,a_2,\cdots,a_k)' = (a_k,a_{k-1},\cdots,a_1).$$

Suppose that for each choice of

$$(i=1,\cdots,k)$$

the vector

$$r_1 w_1 + r_2 w_2 + \cdots + r_m w_m$$

is neither all zeros nor all ones. Then from the above discussion it can be seen that

$$V_1 \oplus V_2 \oplus \cdots \oplus V_m$$

represents a partition of X_m no part of which contains a path.

The theorem will be proved if for each k, $k-1$ such vectors can be found. The desired constructions are obvious extensions of the following two examples for odd and even k.

For $k=5$:

$$\begin{array}{l}(1,\ 0,\ 0,\ 0,\ 1)\\(0,\ 1,\ 0,\ 1,\ 0)\\(1,\ 0,\ 0,\ 0,\ 0)\\(0,\ 1,\ 0,\ 0,\ 0)\end{array}$$

For $k=6$:

$$\begin{array}{l}(1,\ 0,\ 0,\ 0,\ 0,\ 1)\\(0,\ 1,\ 0,\ 0,\ 1,\ 0)\\(1,\ 0,\ 0,\ 0,\ 0,\ 0)\\(0,\ 1,\ 0,\ 0,\ 0,\ 0)\\(0,\ 0,\ 1,\ 0,\ 0,\ 0).\end{array}$$

REFERENCES

1. R. Rado, *Note on combinatorial analysis*, Proc. London Math. Soc. (2) **48** (1943–45), 122–160.

2. A. Y. Khinchin, *Three pearls of number theory*, Graylock Press, Rochester, 1952, pp. 11–12

3. R. Rado, *Axiomatic treatment of rank in infinite sets*, Canad. J. Math. **1** (1949), 338.

4. W. H. Gottschalk, *Choice functions and Tychonoff's Theorem*, Proc. Amer. Math. Soc. **2** (1951), 172.

5. D. Blackwell and M. A. Girshick, *Theory of games and statistical decisions*, Wiley, New York, 1954, p. 21.

6. P. Hall, *On representatives of subsets*, J. London Math. Soc, **10** (1935), 26–30.

California Institute of Technology,
Pasadena, California
University of Oregon,
Eugene, Oregon

On well-quasi-ordering finite trees

By C. ST. J. A. NASH-WILLIAMS
King's College, Aberdeen

(*Received* 9 *March* 1963)

Abstract. A new and simple proof is given of the known theorem that, if $T_1, T_2, \dots$ is an infinite sequence of finite trees, then there exist i and j such that $i < j$ and T_i is homeomorphic to a subtree of T_j.

A *quasi-ordered set* is a set Q on which a reflexive and transitive relation $\leqslant$ is defined. Q and Q' will denote quasi-ordered sets. An infinite sequence $q_1, q_2, \dots$ of elements of Q will be called *good* if there exist positive integers i, j such that $i < j$ and $q_i \leqslant q_j$; if not, the sequence will be called *bad*. A quasi-ordered set Q is *well-quasi-ordered* (*wqo*) if every infinite sequence of elements of Q is good.

A *graph* G consists (for our purposes) of a **finite** set $V(G)$ of elements called *vertices* of G and a subset $E(G)$ of the Cartesian product $V(G) \times V(G)$. The elements of $E(G)$ are called *edges* of G. If $(\xi, \eta) \in E(G)$, we call η a *successor* of ξ. If $\xi, \eta \in V(G)$, a $\xi\eta$*-path* is a sequence $\xi_0, \dots, \xi_n$ of vertices of G such that $\xi_0 = \xi$, $\xi_n = \eta$ and $(\xi_{i-1}, \xi_i) \in E(G)$ for $i = 1, \dots, n$. The sequence with sole term ξ is accepted as a $\xi\xi$-path. If there exists a $\xi\eta$-path, we say that η *follows* ξ. For the purposes of this paper, a *tree* is a graph T possessing a vertex $\rho(T)$ (called its *root*) such that, for every $\xi \in V(T)$, there exists a unique $\rho(T)\xi$-path in T. The letter T (with or without dashes or suffixes) will always denote a tree. For the purposes of this paper, a *homeomorphism* of T into T' is a function $\phi: V(T) \to V(T')$ such that, for every $\xi \in V(T)$, the images under ϕ of the successors of ξ follow *distinct* successors of $\phi(\xi)$. The set of all trees will be quasi-ordered by the rule that $T \leqslant T'$ if and only if there exists a homeomorphism of T into T'. This paper presents a new and shorter proof of the following theorem of Kruskal (2).

THEOREM 1. *The set of all trees is wqo.*

If A, B are subsets of Q, a mapping $f: A \to B$ is *non-descending* if $a \leqslant f(a)$ for every $a \in A$. The class of finite subsets of Q will be denoted by SQ, and will be quasi-ordered by the rule that $A \leqslant B$ if and only if there exists a one-to-one non-descending mapping of A into B, where A, B denote members of SQ. The Cartesian product $Q \times Q'$ will be quasi-ordered by the rule that $(q_1, q_1') \leqslant (q_2, q_2')$ if and only if $q_1 \leqslant q_2$ and $q_1' \leqslant q_2'$. The cardinal number of a set A will be denoted by $|A|$.

The following two lemmas are well known (see (1)), but for the reader's convenience their proofs are given here.

LEMMA 1. *If Q, Q' are wqo, then $Q \times Q'$ is wqo.*

Proof. We must prove an arbitrary infinite sequence $(q_1, q_1'), (q_2, q_2'), \dots$ of elements of $Q \times Q'$ to be good. Call q_m *terminal* if there is no $n > m$ such that $q_m \leqslant q_n$. The number

53-3

of q_m which are terminal must be finite, since otherwise they would form a bad subsequence of $q_1, q_2, \ldots$. Therefore there is an N such that q_r is not terminal if $r > N$. We can therefore select a positive integer $f(1) > N$, then an $f(2) > f(1)$ such that $q_{f(1)} \leqslant q_{f(2)}$, then an $f(3) > f(2)$ such that $q_{f(2)} \leqslant q_{f(3)}$ and so on. Since Q' is wqo, there exist i, j such that $i < j$ and $q'_{f(i)} \leqslant q'_{f(j)}$, whence $(q_{f(i)}, q'_{f(i)}) \leqslant (q_{f(j)}, q'_{f(j)})$ and therefore our original sequence is good.

Lemma 2. *If Q is wqo, then SQ is wqo.*

Proof. Assume that the lemma is false. Select an $A_1 \in SQ$ such that A_1 is the first term of a bad sequence of members of SQ and $|A_1|$ is as small as possible. Then select an A_2 such that A_1, A_2 (in that order) are the first two terms of a bad sequence of members of SQ and $|A_2|$ is as small as possible. Then select an A_3 such that A_1, A_2, A_3 (in that order) are the first three terms of a bad sequence of members of SQ and $|A_3|$ is as small as possible. Assuming the Axiom of Choice, this process yields a bad sequence $A_1, A_2, A_3, \ldots$. Since this sequence is bad, no A_i is empty: therefore we can select an element a_i from each A_i. Let $B_i = A_i - \{a_i\}$. If there existed a bad sequence $B_{f(1)}, B_{f(2)}, \ldots$ such that $f(1) \leqslant f(i)$ for all i, the sequence

$$A_1,\ A_2,\ \ldots,\ A_{f(1)-1},\ B_{f(1)},\ B_{f(2)},\ \ldots$$

would be bad (since $A_i \leqslant B_j$ entails $A_i \leqslant A_j$ and is therefore impossible if $i < j$). Since this would contradict the definition of $A_{f(1)}$, there can be no bad sequence $B_{f(1)}, B_{f(2)}, \ldots$ such that $f(1) \leqslant f(i)$ for all i. It follows that the class ($\mathfrak{B}$, say) of sets B_i is wqo, since any bad sequence of sets B_i would have a (bad) infinite subsequence in which no suffix was less than the first. Therefore, by Lemma 1, $Q \times \mathfrak{B}$ is wqo. Therefore there exist i, j such that $i < j$ and $(a_i, B_i) \leqslant (a_j, B_j)$, which implies that $A_i \leqslant A_j$ and thus contradicts the badness of $A_1, A_2, \ldots$. This contradiction proves the lemma.

The *branch* of T at a vertex ξ is the tree R such that $V(R)$ is the set of those vertices of T which follow ξ and

$$E(R) = E(T) \cap (V(R) \times V(R)).$$

Proof of Theorem 1. Assume that the theorem is false. Select a tree T_1 such that T_1 is the first term of a bad sequence of trees and $|V(T_1)|$ is as small as possible. Then select a T_2 such that T_1, T_2 (in that order) are the first two terms of a bad sequence of trees and $|V(T_2)|$ is as small as possible. Continuing this process as in the proof of Lemma 2 yields a bad sequence $T_1, T_2, \ldots$. Let B_i be the set of branches of T_i at the successors of its root, and let $B = B_1 \cup B_2 \cup \ldots$. If there existed a bad sequence $R_1, R_2, \ldots$ such that $R_i \in B_{f(i)}$ and $f(1) \leqslant f(i)$ for every i, the sequence

$$T_1,\ T_2,\ \ldots,\ T_{f(1)-1},\ R_1,\ R_2,\ \ldots$$

would be bad (since $T_i \leqslant R \in B_j$ entails $T_i \leqslant T_j$ and is therefore impossible if $i < j$). Since this would contradict the definition of $T_{f(1)}$, there can be no bad sequence $R_1, R_2, \ldots$ such that $R_i \in B_{f(i)}$ and $f(1) \leqslant f(i)$ for every i. Since any bad sequence of elements of B would have a bad subsequence of this form, it follows that no sequence of elements of B is bad. Therefore B is wqo and hence, by Lemma 2, SB is wqo. Therefore

$B_i \leqslant B_j$ for some pair i, j such that $i < j$. Therefore there is a one-to-one non-descending mapping ϕ: $B_i \to B_j$. For each $R \in B_i$, $R \leqslant \phi(R)$ and so there exists a homeomorphism h_R of R into $\phi(R)$. A homeomorphism h of T_i into T_j may thus be defined by writing $h(\rho(T_i)) = \rho(T_j)$ and making h coincide with h_R on the vertices of each $R \in B_i$. Therefore $T_i \leqslant T_j$, which contradicts the badness of T_1, T_2, ... and thus proves Theorem 1.

The Tree Theorem of (2) is stronger than Theorem 1 of the present paper, but the above proof of Theorem 1 can easily be adapted to prove the Tree Theorem by considering $X \times F(B)$ in place of SB (where X, F have the meanings stated in (2)). Because the necessary changes are easy to make, I have sacrificed this much generality in the interests of readability.

Note added 10 *August* 1963. It has been brought to the author's notice that Kruskal's proof of the Tree Theorem (2) anticipated a somewhat similar proof obtained independently by S. Tarkowski (*Bull. Acad. Polon. Sci. Sér. Sci. Math. Astr. Phys.* **8** (1960), 39–41).

REFERENCES

(1) HIGMAN, G. Ordering by divisibility in abstract algebras. *Proc. London Math. Soc.* (3), **2** (1952), 326–336.

(2) KRUSKAL, J. B. Well-quasi-ordering, the tree theorem, and Vázsonyi's conjecture. *Trans. American Math. Soc.* **95** (1960), 210–225.

Reprinted from
Proc. Cambridge Phil. Soc. **59** (1963), 833–835

Z. Wahrscheinlichkeitstheorie 2, 340–368 (1964)

On the Foundations of Combinatorial Theory
I. Theory of Möbius Functions

By

GIAN-CARLO ROTA

Contents

1. Introduction

One of the most useful principles of enumeration in discrete probability and combinatorial theory is the celebrated *principle of inclusion-exclusion* (cf. FELLER*, FRÉCHET, RIORDAN, RYSER). When skillfully applied, this principle has yielded the solution to many a combinatorial problem. Its mathematical foundations were thoroughly investigated not long ago in a monograph by FRÉCHET, and it might at first appear that, after such exhaustive work, little else could be said on the subject.

One frequently notices, however, a wide gap between the bare statement of the principle and the skill required in recognizing that it applies to a particular combinatorial problem. It has often taken the combined efforts of many a combinatorial analyst over long periods to recognize an inclusion-exclusion pattern. For example, for the ménage problem it took fifty-five years, since CAYLEY's attempts, before JACQUES TOUCHARD in 1934 could recognize a pattern, and thence readily obtain the solution as an explicit binomial formula. The situation becomes bewildering in problems requiring an enumeration of any of the numerous collections of combinatorial objects which are nowadays coming to the fore. The counting of trees, graphs, partially ordered sets, complexes, finite sets on which groups act, not to mention more difficult problems relating to permutations with restricted position, such as Latin squares and the coloring of maps, seem to lie beyond present-day methods of enumeration. The lack of a systematic theory is hardly matched by the consummate skill of a few individuals with a natural gift for enumeration.

This work begins the study of a very general principle of enumeration, of which the inclusion-exclusion principle is the simplest, but also the typical case. It often happens that a set of objects to be counted possesses a natural ordering, in general only a partial order. It may be unnatural to fit the enumeration of such a set into a linear order such as the integers: instead, it turns out in a great many cases that a more effective technique is to work with the natural order of the set. One is led in this way to set up a "difference calculus" relative to an arbitrary partially ordered set.

Looked at in this way, a surprising variety of problems of enumeration reveal themselves to be instances of the general problem of inverting an "indefinite sum" ranging over a partially ordered set. The inversion can be carried out by defining an analog of the "difference operator" relative to a partial ordering. Such an operator is the Möbius function, and the analog of the "fundamental theorem of the calculus" thus obtained is the Möbius inversion formula on a partially ordered set. This formula is here expressed in a language close to that of number theory, where it appears as the well-known inverse relation between the Riemann zeta function and the Dirichlet generating function of the classical Möbius function. In fact, the algebra of formal Dirichlet series turns out to be the simplest non-trivial instance of such a "difference calculus", relative to the order relation of divisibility.

Once the importance of the Möbius function in enumeration problems is realized, interest will naturally center upon relating the properties of this function to the structure of the ordering. This is the subject of the first paper of this series; we hope to have at least begun the systematic study of the remarkable properties of this most natural invariant of an order relation.

We begin in Section 3 with a brief study of the incidence algebra of a locally finite partially ordered set and of the invariants associated with it: the zeta function, Möbius function, incidence function, and Euler characteristic. The language of number theory is kept, rather than that of the calculus of finite differences, and the results here are quite simple.

The next section contains the main theorems: Theorem 1 relates the Möbius functions of two sets related by a Galois connection. By suitably varying one of the sets while keeping the other fixed one can derive much information. Theorem 2 of this section is suggested by a technique that apparently goes back to RAMANUJAN. These two basic results are applied in the next section to a variety of special cases; although a number of applications and special cases have been left out, we hope thereby to have given an idea of the techniques involved.

The results of Section 6 stem from an "Ideenkreis" that can be traced back to Whitney's early work on linear graphs. Theorem 3 relates the Möbius function to certain very simple invariants of "cross-cuts" of a finite lattice, and the analogy with the Euler characteristic of combinatorial topology is inevitable. Pursuing this analogy, we were led to set up a series of homology theories, whose Euler characteristic does indeed coincide with the Euler characteristic which we had introduced by purely combinatorial devices.

Some of the work in lattice theory that was carried out in the thirties is useful in this investigation; it turns out, however, that modular lattices are not combinatorially as interesting as a type of structure first studied by WHITNEY, which we have called geometric lattices following BIRKHOFF and the French school. The remarkable property of such lattices is that their Möbius function alternates in sign (Section 7).

To prevent the length of this paper from growing beyond bounds, we have omitted applications of the theory. Some elementary but typical applications will be found in the author's expository paper in the American Mathematical Monthly. Towards the end, however, the temptation to give some typical examples became irresistible, and Sections 9 and 10 were added. These by no means exhaust the range of applications, it is our conviction that the Möbius inversion formula on a partially ordered set is a fundamental principle of enumeration, and we hope to implement this conviction in the successive papers of this series. One of them will deal with structures in which the Möbius function is multiplicative, —-that is, has the analog of the number-theoretic property $\mu(mn) = \mu(m)\,\mu(n)$ if m and n are coprime — and another will give a systematic development of the Ideenkreis centering around POLYA's Hauptsatz, which can be significantly extended by a suitable Möbius inversion.

A few words about the history of the subject. The statement of the Möbius inversion formula does not appear here for the first time: the first coherent version—with some redundant assumptions—is due to WEISNER, and was independently rediscovered shortly afterwards by PHILIP HALL. Ward gave the statement in full generality. Strangely enough, however, these authors did not pursue the combinatorial implications of their work; nor was an attempt made to systematically investigate the properties of Möbius functions. Aside from HALL's applications to p-groups, and from some applications to statistical mechanics by M. S. GREEN and NETTLETON, little has been done; we give a hopefully complete bibliography at the end.

It is a pleasure to acknowledge the encouragement of G. BIRKHOFF and A. GLEASON, who spotted an error in the definition of a cross-cut, as well as of SEYMOUR SHERMAN and KAI-LAI CHUNG. My colleagues D. KAN, G. WHITEHEAD, and especially F. PETERSON gave me essential help in setting up the homological interpretation of the cross-cut theorem.

2. Preliminaries

Little knowledge is required to read this work. The two notions we shall not define are those of a *partially ordered set* (whose order relation is denoted by $\leq$) and a *lattice*, which is a partially ordered set where max and min of two elements (we call them join and meet, as usual, and write them $\vee$ and $\wedge$) are defined. We shall use instead the symbols $\cup$ and $\cap$ to denote union and intersection *of sets* only. A *segment* $[x, y]$, for x and y in a partially ordered set P, is the set of all elements z between x and y, that is, such that $x \leq z \leq y$. We shall occasionally use open or half-open segments such as $[x, y)$, where one of the endpoints is to be omitted. A segment is endowed with the induced order structure; thus, a segment of a lattice is again a lattice. A partially ordered set is *locally finite* if every segment is finite. We shall only deal with locally finite partially ordered sets.

The *product* $P \times Q$ of partially ordered sets P and Q is the set of all ordered pairs (p, q), where $p \in P$ and $q \in Q$, endowed with the order $(p, q) \geqq (r, s)$ whenever $p \geqq r$ and $q \geqq s$. The product of any number of partially ordered sets is defined similarly. The *cardinal power* Hom (P, Q) is the set of all monotonic functions from P to Q, endowed with the partial order structure $f \geqq g$ whenever $f(p) \geqq g(p)$ for every p in P.

In a partially ordered set, an element p *covers* an element q when the segment $[q, p]$ contains two elements. An *atom* in P is an element that covers a minimal element, and a *dual atom* is an element that is covered by a maximal element.

If P is a partially ordered set, we shall denote by P^* the partially ordered set obtained from P by inverting the order relation.

A *closure relation* in a partially ordered set P is a function $p \to \bar{p}$ of P into itself with the properties (1) $\bar{p} \geqq p$; (2) $\bar{\bar{p}} = \bar{p}$; (3) $p \geqq q$ implies $\bar{p} \geqq \bar{q}$. An element is *closed* if $p = \bar{p}$. If P is a finite Boolean algebra of sets, then a closure relation on P defines a lattice structure on the closed elements by the rules $p \wedge q = p \cap q$ and $p \vee q = \overline{p \cup q}$, and it is easy to see that every finite lattice is isomorphic to one that is obtained in this way. A *Galois connection* (cf. ORE, p. 182ff.) between two partially ordered sets P and Q is a pair of functions $\zeta : P \to Q$ and $\pi : Q \to P$ with the properties: (1) both ζ and π are order-inverting; (2) for p in P, $\pi(\zeta(p)) \geqq p$, and for q in Q, $\zeta(\pi(q)) \geqq q$. Under these circumstances the mappings $p \to \pi(\zeta(p))$ and $q \to \zeta(\pi(q))$ are closure relations, and the two partially ordered sets formed by the closed sets are isomorphic.

In Section 7, the notion of a closure relation with the *Mac Lane-Steinitz exchange property* will be used. Such a closure relation is defined on the Boolean algebra P of subsets of a finite set E and satisfies the following property: if p and q are points of E, and S a subset of E, and if $p \notin \bar{S}$ but $p \in \overline{S \cup q}$, then $q \in \overline{S \cup p}$. Such a closure relation can be made the basis of WHITNEY's theory of independence, as well as of the theory of geometric lattices. The closed sets of a closure relation satisfying the MACLANE-STEINITZ exchange property where every point is a closed set form a geometric (= matroid) lattice in the sense of BIRKHOFF (Lattice Theory, Chapter IX).

A partially ordered set P is said to have a 0 or a I if it has a unique minimal or maximal element. We shall always assume $0 \neq I$. A partially ordered set P having a 0 and a I satisfies the *chain condition* (also called the JORDAN-DEDEKIND chain condition) when all totally ordered subsets of P having a maximal number of elements have the same number of elements. Under these circumstances one introduces the *rank* $r(p)$ of an element p of P as the length of a maximal chain in the segment $[o, p]$, minus one. The rank of 0 is 0, and the rank of an atom is 1. The height of P is the rank of any maximal element, plus one.

Let P be a finite partially ordered set satisfying the chain condition and of height $n + 1$. The *characteristic polynomial* of P is the polynomial $\sum_{x \in P} \mu(0, x) \lambda^{n-r(x)}$, where r is the rank function (see the def. of μ below).

If A is a finite set, we shall write $n(A)$ for the number of elements of A.

3. The incidence algebra

Let P be a locally finite partially ordered set. The *incidence algebra* of P is defined as follows. Consider the set of all real-valued functions of two variables $f(x, y)$, defined for x and y ranging over P, and with the property that $f(x, y) = 0$ if $x \nleqq y$. The sum of two such functions f and g, as well as multiplication by scalars, are defined as usual. The product $h = fg$ is defined as follows:

$$h(x, y) = \sum_{x \leqq z \leqq y} f(x, z)\, g(z, y).$$

In view of the assumption that P is locally finite, the sum on the right is well-defined. It is immediately verified that this is an associative algebra over the real field (any other associative ring could do). The incidence algebra has an identity element which we write $\delta(x, y)$, the Kronecker delta.

The *zeta function* $\zeta(x, y)$ of the partially ordered set P is the element of the incidence algebra of P such that $\zeta(x, y) = 1$ if $x \leqq y$ and $\zeta(x, y) = 0$ otherwise. The function $n(x, y) = \zeta(x, y) - \delta(x, y)$ is called the *incidence function*.

The idea of the incidence algebra is not new. The incidence algebra is a special case of a semigroup algebra relative to a semigroup which is easily associated with the partially ordered set. The idea of taking "interval functions" goes back to DEDEKIND and E. T. BELL; see also WARD.

Proposition 1. *The zeta function of a locally finite partially ordered set is invertible in the incidence algebra.*

Proof. We define the inverse $\mu(x, y)$ of the zeta function by induction over the number of elements in the segment $[x, y]$. First, set $\mu(x, x) = 1$ for all x in P. Suppose now that $\mu(x, z)$ has been defined for all z in the open segment $[x, y)$. Then set

$$\mu(x, y) = -\sum_{x \leqq z < y} \mu(x, z).$$

Clearly μ is an inverse of ζ.

The function μ, inverse to ζ, is called the *Möbius function* of the partially ordered set P.

The following result, simple though it is, is fundamental:

Proposition 2. (Möbius inversion formula). *Let $f(x)$ be a real-valued function, defined for x ranging in a locally finite partially ordered set P. Let an element p exist with the property that $f(x) = 0$ unless $x \geqq p$.*

Suppose that

(*) $$g(x) = \sum_{y \leqq x} f(y).$$

Then

(**) $$f(x) = \sum_{y \geqq x} g(y)\, \mu(y, x).$$

Proof. The function g is well-defined. Indeed, the sum on the right can be written as $\sum_{p \leqq y \leqq x} f(y)$, which is finite for a locally finite ordered set.

Substituting the right side of (*) into the right side of (**) and simplifying, we get

$$\sum_{y \leq x} g(y)\mu(y, x) = \sum_{y \leq x} \sum_{z \leq y} f(z)\mu(y, x) = \sum_{y \leq x} \sum_{z} f(z)\zeta(z, y)\mu(y, x).$$

Interchanging the order of summation, this becomes

$$\sum_{z} f(z) \sum_{y \leq x} \zeta(z, y)\mu(y, x) = \sum_{z} f(z)\delta(z, x) = f(x), \quad \text{q. e. d.}$$

Corollary 1. *Let $r(x)$ be a function defined for x in P. Suppose there is an element q such that $r(x)$ vanishes unless $x \leq q$. Suppose that*

$$s(x) = \sum_{y \geq x} r(y).$$

Then

$$r(x) = \sum_{y \geq x} \mu(x, y)s(y).$$

The proof is analogous to the above and is omitted.

Proposition 3. (Duality). *Let P^* be the partially ordered set obtained by inverting the order of a locally finite partially ordered set P, and let μ^* and μ be the Möbius functions of P^* and P. Then $\mu^*(x, y) = \mu(y, x)$.*

Proof. We have, in virtue of Proposition 2 and Corollary 1,

$$\sum_{x \geq^* y \geq^* z} \mu^*(x, y) = \delta(x, z).$$

Letting $q(x, y) = \mu^*(y, x)$, it follows that q is an inverse of ζ in the incidence algebra of P. Since the inverse is unique, $q = \mu$, q. e. d.

Proposition 4. *The Möbius function of any segment $[x, y]$ of P equals the restriction to $[x, y]$ of the Möbius function of P.*

The proof is omitted.

Proposition 5. *Let $P \times Q$ be the direct product of locally finite partially ordered sets P and Q. The Möbius function of $P \times Q$ is given by*

$$\mu((x, y), (u, v)) = \mu(x, u)\mu(y, v), \; x, u \in P; \; y, v \in Q.$$

The proof is immediate and is omitted.

The same letter μ has been used for the Möbius functions of three partially ordered sets, and we shall take this liberty whenever it will not cause confusion.

Corollary (Principle of Inclusion-Exclusion). *Let P be the Boolean algebra of all subsets of a finite set of n elements. Then, for x and y in P,*

$$\mu(x, y) = (-1)^{n(y)-n(x)}, \qquad y \geq x,$$

where $n(x)$ denotes the number of elements of the set x.

Indeed, a Boolean algebra is isomorphic to the product of n chains of two elements, and every segment $[x, y]$ in a Boolean algebra is isomorphic to a Boolean algebra.

Aside of the simple result of Proposition 5, little can be said in general about how the Möbius function varies by taking subsets and homomorphic images of a partially ordered set. We shall see that more sophisticated notions will be required to relate the Möbius functions of two partially ordered sets.

Let P be a finite partially ordered set with 0 and I. The *Euler characteristic* E of P is defined as

$$E = 1 + \mu(0,1).$$

The simplest result relating to the computation of the Euler characteristic was proved by PHILIP HALL by combinatorial methods. We reprove it below with a very simple proof which shows one of the uses of the incidence algebra:

Proposition 6. *Let P be a finite partially ordered set with 0 and I. For every k, let C_k be the number of chains with k elements stretched between 0 and I. Then*

$$E = 1 - C_2 + C_3 - C_4 + \cdots.$$

Proof. $\mu = \zeta^{-1} = (\delta + n)^{-1} = \delta - n + n^2 \ldots$. It is easily verified that $n^{k-1}(x, y)$ equals the number of chains of k elements stretched between x and y. Letting $x = 0$ and $y = I$, the result follows at once.

It will be seen in section 6 that the Euler characteristic of a partially ordered set can be related to the classical Euler characteristic in suitable homology theories built on the partially ordered set.

Proposition 6 is a typical application of the incidence algebra. Several other results relating the number of chains and subsets with specified properties can often be expressed in terms of identities for functions in the incidence algebra. In this way, one obtains generalizations to an arbitrary partially ordered set of some classical identities for binomial coefficients. We shall not pursue this line here further, since it lies out of the track of the present work.

Example 1. The classical Möbius function $\mu(n)$ is defined as $(-1)^k$ if n is the product of k distinct primes, and 0 otherwise. The classical inversion formula first derived by Möbius in 1832 is:

$$g(m) = \sum_{n|m} f(n); \quad f(m) = \sum_{n|m} g(n)\, \mu\left(\frac{m}{n}\right).$$

It is easy to see (and will follow trivally from later results) that $\mu\left(\frac{m}{n}\right)$ is the Möbius function of the set of positive integers, with divisibility as the partial order. In this case the incidence algebra has a distinguished subalgebra, formed by all functions $f(n, m)$ of the form $f(n, m) = G\left(\frac{m}{n}\right)$. The product $H = FG$ of two functions in this subalgebra can be written in the simpler form

$$(*) \qquad H(m) = \sum_{kn=m} F(k)\, G(n).$$

If we associate with the element F of this subalgebra the *formal Dirichlet series* $\hat{F}(s) = \sum_{n=1}^{\infty} F(n)/n^s$, then the product (*) corresponds to the product of two formal Dirichlet series considered as functions of s, $\hat{H}(s) = \hat{F}(s)\, \hat{G}(s)$. Under this representation, the zeta function of the partially ordered set is the classical *Riemann zeta function* $\zeta(s) = \sum_{n=1}^{\infty} 1/n^s$, and the statement that the Möbius function is the inverse of the zeta function reduces to the classical identity $1/\zeta(s) = \sum_{n=1}^{\infty} \mu(n)/n^s$. It is hoped this example justifies much of the terminology introduced above.

Example 2. If P is the set of ordinary integers, then $\mu(m, n) = -1$ if $m = n - 1$, $\mu(m, m) = 1$, and $\mu(m, n) = 0$ otherwise. The Möbius inversion formula reduces to a well known formula of the calculus of finite differences, which is the discrete analog of the fundamental theorem of calculus.

The Möbius function of a partially ordered set can be viewed as the analog of the classical difference operator $\Delta f(n) = f(n+1) - f(n)$, and the incidence algebra serves as a calculus of finite differences on an arbitrary partially ordered set.

4. Main results

It turns out that the Möbius functions of two partially ordered sets can be compared, when the sets are related by a Galois connection. By keeping one of the sets fixed, and varying the other from among sets with a simpler structure, such as Boolean algebras, subspaces of a finite vector space, partitions, etc., one can derive much information about a Möbius function. This is the program we shall develop. The basic result is the following:

Theorem 1. *Let P and Q be finite partially ordered sets, where P has a 0 and Q has a 0 and a 1. Let μ_P and μ be their Möbius functions. Let*

$$\pi : Q \to P; \quad \varrho : P \to Q$$

be a Galois connection such that

$$\pi(x) = 0 \quad \text{if and only if} \quad x = 1. \tag{1}$$

$$\varrho(0) = 1. \tag{2}$$

Then

$$\mu(0, 1) = \sum_{a>0} \mu_P(0, a)\zeta(\varrho(a), 0) = \sum_{[a:\varrho(a)=0]} \mu_P(0, a).$$

One gets a significant summand on the right for every $a > 0$ in P which is mapped into 0 by ϱ. One therefore expects the right side to contain "few" terms. In general, μ_P is a known function and μ is the function to be determined.

Proof. We shall first establish the identity

$$\sum_{a \geqq b} \delta(\pi(x), a) = \zeta(x, \varrho(b)) \tag{*}$$

for every b in P. Here ζ on the right stands for the zeta function of Q. Equation (*) is equivalent to the following statement: $\pi(x) \geqq b$ if and only if $x \leqq \varrho(b)$. But this latter statement is immediate from the properties of a Galois connection. Indeed, if $\pi(x) \geqq b$, then $\varrho(\pi(x)) \leqq \varrho(b)$, but $x \leqq \varrho(\pi(x))$, hence $x \leqq \varrho(b)$, and similarly for the converse implication.

To identity (*) we apply the Möbius inversion formula relative to P, thereby obtaining the identity

$$\delta(\pi(x), 0) = \sum_{a \geqq 0} \mu_P(0, a)\zeta(x, \varrho(a)). \tag{**}$$

Now, $\delta(\pi(x), 0)$ takes the value 1 if and only if $\pi(x) = 0$, that is, in view of assumption (1), if and only if $x = 1$. For all other values of x, we have $\delta(\pi(x), 0) = 0$. Therefore,

$$\delta(\pi(x), 0) = 1 - n(x, 1).$$

We can now rewrite equation (**) in the form

$$1 - n(x, 1) = \zeta(x, \varrho(0)) + \sum_{a>0} \mu_P(0, a)\zeta(x, \varrho(a))$$

However, in view of assumption (2), $\zeta(x, \varrho(0)) = \zeta(x, 1)$, and this is identically one for all x in Q. Therefore, simplifying,

$$-n(x, 1) = \sum_{a>0} \mu_P(0, a)\zeta(x, \varrho(a)).$$

Now, since $\zeta = \delta + n$, we have $\mu = \delta - \mu n$, hence, recalling that $0 \neq 1$,

$$\mu(0, 1) = -\sum_{0 \leqq x \leqq 1} \mu(0, x) n(x, 1) = \sum_{0 \leqq x \leqq 1} \sum_{a>0} \mu_P(0, a)\mu(0, x)\zeta(x, \varrho(a)).$$

Interchanging the order of summation, we get

$$\mu(0, 1) = \sum_{a>0} \mu_P(0, a) \sum_{0 \leqq x \leqq 1} \mu(0, x)\zeta(x, \varrho(a)).$$

The last sum on the right equals $\delta(0, \varrho(a))$, and this equals $\zeta(\varrho(a), 0)$. The proof is therefore complete.

For simplicity of application, we restate Theorem 1 inverting the order of P.

Corollary. *Let $p: Q \to P$; $q: P \to Q$ be order preserving functions between P and Q such that*

(1) If $p(x) = 1$ then $x = 1$, and conversely.

(2) $q(1) = 1$.

(3) $p(q(x)) \leqq x$ and $q(p(x)) \geqq x$.

Then

$$\mu(0, 1) = \sum_{a<1} \mu_P(a, 1)\zeta(q(a), 0) = \sum_{[a:q(a)=0]} \mu_P(a, 1)$$

where μ is the Möbius function of Q.

The second result is suggested by a technique which apparently goes back to RAMANUJAN (cf. HARDY, RAMANUJAN, page 139).

Theorem 2. *Let Q be a finite partially ordered set with 0, and let P be a partially ordered set with 0. Let $p: Q \to P$ be a monotonic function of Q onto P. Assume that the inverse image of every interval $[0, a]$ in P is an interval $[0, x]$ in Q, and that the inverse image of 0 contains at least two points.*

Then

$$\sum_{[x:p(x)=a]} \mu(0, x) = 0$$

for every a in P.

The *proof* is by induction over the set P. Since $[0, 0]$ is an interval and its inverse image is an interval $[0, q]$ with $q > 0$, we have

$$\sum_{[x:p(x)=0]} \mu(0, x) = \sum_{0 \leqq x \leqq q} \mu(0, x) = 0.$$

Suppose now the statement is true for all b such that $b < a$ in P. Then

$$\sum_{b<a} \sum_{[x:p(x)=b]} \mu(0, x) = 0.$$

It follows that

$$\sum_{[x:p(x)=a]} \mu(0, x) = \sum_{b \leqq a} \sum_{[x:p(x)=b]} \mu(0, x).$$

The last sum equals the sum over some interval $[0, r]$ which is the inverse image of the segment $[0, a]$, that is

$$\sum_{b \leqq a} \sum_{[x:p(x)=b]} \mu(0, x) = \sum_{0 \leqq x \leqq r} \mu(0, x) = \delta(0, r).$$

But $r > 0$ because a is strictly greater than 0. Hence $\delta(r, 0) = 0$, and this concludes the proof.

5. Applications

The simplest (and typical) application of Theorem 1 is the following:

Proposition 1. *Let R be a subset of a finite lattice L with the following properties: $1 \notin R$, and for every x of L, except $x = 1$, there is an element y of R such that $y \geqq x$.*

For $k \geqq 2$, let q_k be the number of subsets of R containing k elements whose meet is 0. Then $\mu(0, 1) = q_2 - q_3 + q_4 + \cdots$.

Proof. Let $B(R)$ be the Boolean algebra of subsets of R. We take $P = B(R)$ and $Q = L$ in Theorem 1, and establish a Galois connection as follows. For x in L, let $\pi(x)$ be the set of elements of R which dominate x. In particular, $\pi(1)$ is the empty set. For A in $B(R)$, set $\varrho(A) = \wedge A$, namely, the meet of all elements of A, an empty meet giving as usual the element 1. This is evidently a Galois connection. Conditions (1) and (2) of the Theorem are obviously satisfied.

The function μ_P is given by the Corollary of Proposition 5 of Section 3, and hence the conclusion is immediate.

Two noteworthy special cases are obtained by taking R to be the set of dual atoms of Q, or the set of all elements < 1 (cf. also Weisner).

Closure relations. A useful application of Theorem 1 is the following:

Proposition 2. *Let $x \to \bar{x}$ be a closure relation on a partially ordered set Q having 1, with the property that $\bar{x} = 1$ only if $x = 1$. Let P be the partially ordered subset of all closed elements of Q. Then:* (a) *If $\bar{x} > x$, then $\mu(x, 1) = 0$;* (b) *If $\bar{x} = x$, then $\mu(x, 1) = \mu_P(x, 1)$, where μ_P is the Möbius function of P.*

Proof. Considering $[x, 1]$, it may be assumed that P has a 0 and $x = 0$. We apply Corollary 1 of Theorem 1, setting $p(x) = \bar{x}$ and letting q be the injection map of P into Q. It is then clear that the assumptions of the Corollary are satisfied, and the set of all a in P such that $q(a) = 0$ is either the empty set or the single element 0, q. e. d.

Corollary (Ph. Hall). *If 0 is not the meet of dual atoms of a finite lattice L, or if 1 is not the join of atoms, then $\mu(0, 1) = 0$.*

Proof. Set $\bar{x} = \wedge A(x)$, where $A(x)$ is the set of dual atoms of Q dominating x, and apply the preceding result. The second assertion is obtained by inverting the order.

Example 1. *Distributive lattices.* Let L be a locally finite distributive lattice. Using Proposition 2, we can easily compute its Möbius function. Taking an interval

$[x, y]$ and applying Proposition 4 of Section 3, we can assume that L is finite. For $a \in L$, define $\bar{a}$ to be the join of all atoms which a dominates. Then $a \to \bar{a}$ is a closure relation in the inverted lattice L^*. Furthermore, the subset of closed elements is easily seen to be isomorphic to a finite Boolean algebra (cf. BIRKHOFF Lattice Theory, Ch. IX) Applying Proposition 5 of Section 3, we find: $\mu(x, y) = 0$ if y is not the join of elements covering x, and $\mu(x, y) = (-1)^n$ if y is the join of n distinct elements covering x.

In the special case of the integers ordered by divisibility, we find the formula for the classical Möbius function (cf. Example 1 of Section 3.).

The Möbius function of cardinal products. Let P and Q be finite partially ordered sets. We shall determine the Möbius function of the partially ordered set Hom(P, Q) of monotonic functions from P to Q, in terms of the Möbius function of Q. It turns out that very little information is needed about P.

A few preliminaries are required for the statement.

Let R be a subset of a partially ordered set Q with 0, and let $\bar{R}$ be the ideal generated by R, that is, the set of all elements x in Q which are below ($<$) some element of R. We denote by Q/R the partially ordered set obtained by removing off all the elements of $\bar{R}$, and leaving the rest of the order relation unchanged. There is a natural order-preserving transformation of Q onto Q/R which is one-to-one for elements of Q not in $\bar{R}$. We shall call Q/R the *quotient* of Q by the ideal generated by R.

Lemma. *Let $f: P \to Q$ be monotonic with range $R \subset Q$. Then the segment $[f, 1]$ in* Hom (P, Q) *is isomorphic with* Hom $(P, Q/R)$.

Proof. For g in $[f, 1]$, set $g'(x) = g(x)$ to obtain a mapping $g \to g'$ of $[f, 1]$ to Hom $(P, Q/R)$. Since $g \geq f$, the range of g lies above R, so the map is an isomorphism.

Proposition 3. *The Möbius function μ of the cardinal product* Hom (P, Q) *of the finite partially ordered set P with the partially ordered set Q with 0 and 1 is determined as follows:*

(a) *If $f(p) \neq 0$ for some element p of P which is not maximal, then $\mu(0, f) = 0$.*

(b) *In all other cases,*

$$\mu(0, f) = \prod_m \mu(0, f(m)), \qquad f \in P,$$

where the product ranges over all maximal elements of P, and where μ on the right stands for the Möbius function of Q.

(c) *For $f \leq g$, $\mu(f, g) = \mu(0, g')$, where g' is the image of g under the canomial map of $[f, 1]$ onto Hom $(P, Q/R)$, provided Q/R has a 0.*

Proof. Define a closure relation in $[0, f]^*$, namely the segment $[0, f]$ with the inverted order relation, as follows. Set $\bar{g}(m) = g(m)$ if m is a maximal element of P, and $\bar{g}(a) = 0$ if a is not a maximal element of P. If $\bar{g} = 0$, then $g(m) = 0$ for all maximal elements m, hence $g(a) = 0$ for all $a <$ some maximal element, since g is monotonic. Hence $g = 0$, and the assumption of Proposition 2 is satisfied. The set of closed elements is isomorphic to Hom (M, P), where M is a set of as many elements as there are maximal elements in P. Conclusion (a) now follows from Proposition 2, and conclusion (b) from Proposition 5 of Section 3. Conclusion (c) follows at once from the Lemma.

We pass now to some applications of Theorem 2.

Proposition 4. *Let $a \to \bar{a}$ be a closure relation on a finite lattice Q, with the property that $\overline{a \vee b} = \bar{a} \vee \bar{b}$ and $\bar{0} > 0$. Then for all $a \in Q$,*

$$\sum_{[x:\bar{x}=a]} \mu(0, x) = 0 .$$

Proof. Let P be a partially ordered set isomorphic to the set of closed elements of L. We define $p(x)$, for x in Q, to be the element of P corresponding to the closed element $\bar{x}$. Since $\bar{0} > 0$, any x between 0 and $\bar{0}$ is mapped into $\bar{0}$. Hence the inverse image of 0 in P under the homomorphism p is the nontrival interval $[0, \bar{0}]$.

Now consider an interval $[0, a]$ in P. Then $p^{-1}([0, a]) = [0, \bar{x}]$, where $\bar{x}$ is the closed element of L corresponding to a. Indeed, if $0 \leqq y \leqq \bar{x}$ then $\bar{y} \leqq \bar{\bar{x}} = \bar{x}$, hence $p(y) \leqq a$. Conversely, if $p(y) \leqq a$, then $\bar{y} \leqq \bar{x}$ but $y \leqq \bar{y}$, hence $y \leqq \bar{x}$. Therefore the condition of Theorem 2 is satisfied, and the conclusion follows at once.

Corollary (Weisner).

(a) *Let $a > 0$ in a finite lattice L. Then, for any b in L,*

$$\sum_{x \vee a = b} \mu(0, x) = 0$$

(b) *Let $a < 1$ in L. Then, for any b in L,*

$$\sum_{x \wedge a = b} \mu(x, 1) = 0 .$$

Proof. Take $\bar{x} = x \vee a$. Part (b) is obtained by inverting the order.

Example 2. Let V be a finite-dimensional vector space of dimension n over a finite field with q elements. We denote by $L(V)$ the lattice of subspaces of V. We shall use Proposition 4 to compute the Möbius function of $L(V)$.

In the lattice $L(V)$, every segment $[x, y]$, for $x \leqq y$, is isomorphic to the lattice $L(W)$, where W is the quotient space of the subspace y by the subspace x. If we denote by $\mu_n = \mu_n(q)$ the value of $\mu(0, 1)$ for $L(V)$, it follows that $\mu(x, y) = \mu_j$, when j is the dimension of the quotient space W. Therefore once μ_n is known for for every n, the entire Möbius function is known.

To determine μ_n, consider a subspace a of dimension $n - 1$. In view of the preceding Corollary, we have for all $a < 1$ (where 1 stands for the entire space V):

$$\sum_{x \wedge a = 0} \mu(x, 1) = 0$$

where 0 stands of course for the 0-subspace. Let a be a dual atom of $L(V)$, that is, a subspace of dimension $n - 1$. Which subspaces x have the property that $x \wedge a = 0$? x must be a line in V, and such a line must be disjoint except for 0 from a. A subspace of dimension $n - 1$ contains q^{n-1} distinct points, so there will be $q^n - q^{n-1}$ points outside of a. However, every line contains exactly $q - 1$ points. Therefore, for each subspace a of dimension $n - 1$ there are

$$\frac{q^n - q^{n-1}}{q - 1} = q^{n-1}$$

distinct lines x such that $x \wedge a = 0$. Since each interval $[x, 1]$ is isomorphic to a space of dimension $n-1$, we obtain

$$\mu_n = \mu(0,1) = -\sum_{\substack{x\wedge a=0\\ x\neq 0}} \mu(x,1) = -q^{n-1}\mu_{n-1}.$$

This is a difference equation for μ_n which is easily solved by iteration. We obtain the result, first established by PHILIP HALL (see also WEISNER and S. DELSARTE):

$$\mu_n(q) = (-1)^n q^{n(n-1)/2} = (-1)^n q^{\binom{n}{2}}.$$

6. The Euler characteristic

Sharper results relating $\mu(0,1)$ to combinatorial invariants of a finite lattice can be obtained by application of Theorem 1, when the "comparison set" P remains a Boolean algebra.

A *cross-cut* C of a finite lattice L is a subset of L with the following properties:

(a) C does not contain 0 or 1.

(b) no two elements of C are comparable (that is, if x and y belong to C, then neither $x < y$ nor $x > y$ holds).

(c) Any maximal chain stretched between 0 and 1 meets the set C.

A *spanning subset* S of L is a subset such that $\vee S = 1$ and $\wedge S = 0$.

The main result is the following *Cross-cut Theorem*:

Theorem 3. *Let μ be the Möbius function and E the Euler characteristic of a non-trivial finite lattice L, and let C be a cross-cut of L. For every integer $k \geqq 2$, let q_k denote the number of spanning subsets of C containing k distinct elements. Then*

$$E-1 = \mu(0,1) = q_2 - q_3 + q_4 - q_5 + \cdots$$

The *proof* is by induction over the distance of a cross-cut C from the element 1.

Define the distance $d(x)$ of an element x from the element 1 as the maximum length of a chain stretched between x and 1. For example, the distance of a dual atom is two. If C is a cross-cut of L, define the distance $d(C)$ as $\max d(x)$ as x ranges over C. Thus, the distance of the cross-cut consisting of all dual atoms is two, and conversely, this is the only cross-cut having distance two.

It follows from Proposition 1 of Section 5 that the result holds when $d(C) = 2$ (take $R = C$ in the assertion of the Proposition). Thus, we shall assume the truth of the statement for all cross-cuts whose distance is less than n, and prove it for a cross-cut with $d(C) = n$.

If C is a subset of L, we shall write $x > C$ or $x \leqq C$ to mean that there is an element y or C such that $x > y$, or that there is an element y of C such that $x \leqq y$. For a general C, these possibilities may not be mutually exclusive; they are mutually exclusive when C is a cross-cut. We shall repeatedly make use of this remark below.

Define a modified lattice L' as follows. Let L' contain all the elements x such that $x \leqq C$ in the same order. On top of C, add an element 1 covering all the elements of C, but no others; this defines L'.

In L', consider the cross-cut C and apply Proposition 1 of section 5 again. If μ' is the Möbius function of L', then

$$\mu'(0,1) = p_2 - p_3 + p_4 \cdots,$$

where p_k is the number of all subsets $A \subset C \subset L'$ of k elements, such that $\wedge A = 0$.

Comparing the lattices L and L', we have

$$0 = \sum_{x \leqq C} \mu(0, x) + \sum_{x > C} \mu(0, x) = \sum_{x \leqq C} \mu'(0, x) + \mu'(0, 1).$$

However, for $x \leqq C$, we have $\mu'(0, x) = \mu(0, x)$ by construction of L'. Hence

$$\sum_{x \leqq C} \mu(0, x) = -p_2 + p_3 - p_4 + \cdots$$

Since the sets $(x/x \leqq C)$ and $(x/x > C)$ are disjoint, we can write

$$\mu(0, 1) = -\sum_{x<1} \mu(0, x) = -[\sum_{x \leqq C} \mu(0, x) + \sum_{1>x>C} \mu(0, x)].$$

We now simplify the first summation on the right:

(*)
$$\mu(0, 1) = p_2 - p_3 + p_4 \cdots - \sum_{1>x>C} \mu(0, x).$$

Now let $q_k(x)$ be the number of subsets of C having k elements, whose meet is 0 and whose join is x. In particular, $q_k(1) = q_k$. Then clearly·

$$p_k = \sum_{x>C} q_k(x), \quad k \geqq 2,$$

the summation in (*) can be simplified to

(**)
$$\mu(0, 1) = (q_2 - q_3 + q_4 - \cdots) - \sum_{1>x>C} [-q_2(x) + q_3(x) - q_4(x) + + \cdots + \mu(0, x)].$$

For x above C and unequal to 1, consider the segment $[0, x]$. We prove that $C(x) = C \cap [0, x]$ is a cross-cut of the lattice $[0, x]$ such that $d(C(x)) < d(C)$. Once this is done, it follows by the induction hypothesis that every term in brackets on the right of (**) vanishes, and the proof will be complete.

Conditions (a) and (b) in the definition of a cross-cut are trivially satisfied by $C(x)$, and condition (c) is verified as follows. Suppose Q is a maximal chain in $[0, x]$ which does not meet $C(x)$. Choose a maximal chain R in the segment $[x, 1]$; then the chain $Q \cup R$ is maximal in L, and does not intersect C.

It remains to verify that $d(C(x)) < d(C)$, and this is quite simple. There is a chain Q stretched between C and x whose length is $d(C(x))$. Then $d(C)$ exceeds the length of the chain $Q \cup R$, and since $x < 1$, R has length at least 2, hence the length of $Q \cup R$ exceeds that of Q by at least one. The proof is therefore complete.

Theorem 3 gives a relation between the value $\mu(0, 1)$ and the width of narrow cross-cuts or *bottlenecks* of a lattice. The proof of the following statement is immediate.

Corollary 1. (a) *If L has a cross-cut with one element, then* $\mu(0, 1) = 0$.

(b) *If L has a cross-cut with two elements, then the only two possible values of* $\mu(0, 1)$ *are* 0 *and* 1.

(c) *If L has a cross-cut having three elements, then the only possible values of* $\mu(0, 1)$ *are* 2, 1, 0 *and* -1.

In this connection, an interesting combinatorial problem is to determine all possible values of $\mu(0, 1)$, given that L has a cross-cut with n elements.

Reduction of the main formula. In several applications of the cross-cut theorem, the computation of the number q_k of spanning sets may be long, and systematic procedures have to be devised. One such procedure is the following:

Proposition 1. *Let C be a cross-cut of a finite lattice L. For every integer $k \geqq 0$, and for every subset $A \subset C$, let $q(A)$ be the number of spanning sets containing A, and let $S_k = \sum_A q(A)$, where A ranges over all subsets of C having k elements. Set S_0 to be the number of elements of C. Then*

$$\mu(0, 1) = S_0 - 2S_1 + 2^2 S_2 - 2^3 S_3 + \cdots.$$

Proof. For every subset $B \subset C$, set $p(B) = 1$ if B is a spanning set, and $p(B) = 0$ otherwise. Then

$$q(A) = \sum_{C \supseteq B \supseteq A} p(B).$$

Applying the Möbius inversion formula on the Boolean algebra of subsets of C, we get

$$p(A) = \sum_{B \supseteq A} q(B)\mu(A, B),$$

where μ is the Möbius function of the Boolean algebra. Summing over all subsets $A \subset C$ having exactly k elements,

$$q_k = \sum_{n(A)=k} p(A) = \sum_{n(A)=k} \sum_{B \supseteq A} q(B)\mu(A, B).$$

Interchanging the order of summation on the right, recalling Proposition 5 of Section 3 and the fact that a set of $k + l$ elements possesses $\binom{k+l}{l}$ subsets of k elements, we obtain

$$q_k = S_k - \binom{k+1}{1} S_{k+1} + \binom{k+2}{2} S_{k+2} \cdots + (-1)^{n-k} \binom{n}{k} S_n.$$

A convenient way of recasting this expression in a form suitable for computation is the following. Let V be the vector space of all polynomials in the variable x. over the real field. The polynomials $1, x, x^2, \ldots$, are linearly independent in V. Hence there exists a linear functional L in V such that

$$L(x^k) = S_k, \quad k = 0, 1, 2, \ldots.$$

Formula (*) can now be rewritten in the concise form

$$q_k = L(x^k - (k+1)x^{k+1} + \binom{k+2}{2} x^{k+2} - \cdots) = L\left(\frac{x^k}{(1+x)^{k+1}}\right).$$

Upon applying the cross-cut theorem, we find the expression (where q_0 and q_1 are also given by (*), but turn out to be 0)

$$\begin{aligned}\mu(0, 1) &= L\left(\frac{1}{1+x} - \frac{x}{(1+x)^2} + \frac{x^2}{(1+x)^3} - \cdots\right)\\ &= L\left(\frac{1}{1+2x}\right) = L(1 - 2x + 4x^2 - 8x^3 + \cdots)\\ &= S_0 - 2S_1 + 4S_2 - \cdots, \quad \text{q.e.d.}\end{aligned}$$

The cross-cut theorem can be applied to study which alterations of the order relation of a lattice preserve the Euler characteristic. Every alteration which preserves meets and joins of the spanning subsets of some cross-cut will preserve the Euler characteristic. There is a great variety of such changes, and we shall not develop a systematic theory here. The following is a simple case.

Following BIRKHOFF and JÓNSSON and TARSKI we define the *ordinal sum* of lattices as follows. Given a lattice L and a function assigning to every element x of L a lattice $L(x)$, (all the $L(x)$ are distinct) the *ordinal sum* $P = \sum_L L(x)$ of the lattices $L(x)$ *over* the lattice L is the partially ordered set P consisting of the set $\bigcup_{x \in L} L(x)$, where $u \leq v$ if $u \in L(x)$ and $v \in L(x)$ and $u \leq v$ in $L(x)$, or if $u \in L(x)$ and $v \in L(y)$ and $x < y$. It is clear that P is a lattice if all the $L(x)$ are finite lattices.

Proposition 2. *If the finite lattice P is the ordinal sum of the lattices $L(x)$ over the non-trivial lattice L, and μ_P, μ_x and μ_L are the corresponding Möbius functions, then:*
If $L(0)$ is the one element lattice, then $\mu_P(0, 1) = \mu_L(0, 1)$.

Proof. The atoms of P are in one-to-one correspondence with the atoms of L and the spanning subsets are the same. Hence the result follows by applying the cross-cut theorem to the atoms.

In virtue of a theorem of JÓNSSON and TARSKI, every lattice P has a unique maximal decomposition into an ordinal sum over a "skeleton" L. This can be used in connection with the preceding Corollary to further simplify the computation of $\mu(0, n)$ as n ranges through P.

Homological interpretation. The alternating sums in the Cross-Cut Theorem suggest that the Euler characteristic of a lattice be interpreted as the Euler characteristic in a suitable homology theory. This is indeed the case. We now define* a *homology theory* $H(C)$ relative to an arbitrary cross-cut C of a finite lattice L. For the homological notions, we refer to Eilenberg-Steenrod.

Order the elements of C, say $a_1, a_2, \ldots, a_n$. For $k \geqq 0$, let a *k-simplex* σ be any subset of C of $k + 1$ elements which does *not* span. Let C_k be the free abelian group generated by the k-simplices. We let $C_{-1} = 0$; for a given simplex σ, let σ_i be the set obtained by omitting the $(i + 1)$-st element of σ, when the elements of σ are ordered according to the given ordering of C. The boundary of a k-simplex is defined as usual as $\partial_k \sigma = \sum_{i=0}^{k} (-1)^i \sigma_i$, and is extended by linearity to all of C_k, giving a linear mapping of C_k into C_{k-1}. The k-th homology group H_k is defined as the abelian group obtained by taking the quotient of the kernel of ∂_k by the image of ∂_{k+1}. The rank b_k of the abelian group H_k, that is, the number of independent generators of infinite cyclic subgroups of H_k, is the k-th *Betti number*.

Let α_k be the rank of C_k, that is, the number of k-simplices. The *Euler characteristic* of the homology $H(C)$ is defined in homology theory as

$$E(C) = \sum_{k=0}^{\infty} (-1)^k \alpha_k .$$

* This definition was obtained jointly with D. KAN, F. PETERSON and G. WHITEHEAD, whom I now wish to thank.

It follows from well-known results in homology theory that

$$E(C) = \sum_{k=0}^{\infty} (-1)^k b_k .$$

Let q_k be the number of spanning subsets with k elements as in Theorem 3. Then $q_{k+1} + \alpha_k$ is the total number of subsets of C having $k+1$ elements; if C has N elements, then $\alpha_k = \binom{N}{k+1} - q_{k+1}$. It follows from the Cross-Cut Theorem that

$$E(C) = \sum_{k=0}^{\infty} (-1)^k \binom{N}{k+1} - \sum_{k=0}^{\infty} (-1)^k q_{k+1}$$
$$= \sum_{k=0}^{\infty} (-1)^k \binom{N}{k+1} + \mu(0,1).$$

We have however

$$\sum_{k=0}^{\infty} (-1)^k \binom{N}{k+1} = -\sum_{i=1}^{\infty} (-1)^i \binom{N}{i} = 1 - \sum_{i=0}^{\infty} (-1)^i \binom{N}{i} = 1 - (1-1)^N = 1,$$

and hence

$$E(C) = 1 + \mu(0,1) = E;$$

in other words:

Proposition 3. *In a finite lattice, the Euler characteristic of the homology of any cross-cut C equals the Euler characteristic of the lattice.*

This result can sometimes be used to compute the Möbius functions of "large" lattices. In general, the numbers q_k are rather redundant, since any spanning subset of k elements gives rise to several spanning subsets with more than k elements. A method for eliminating redundant spanning sets is then called for. One such method consists precisely in the determination of the Betti numbers b_k.

We conjecture that the Betti numbers of $H(C)$ are themselves independent of the cross-cut C, and are also "invariants" of the lattice L, like the Euler characteristic $E(C)$. In the special case of lattices of height 4 satisfying the chain condition, this conjecture has been proved (in a different language) by DOWKER.

Example 1. *The Betti numbers of a Boolean algebra.* We take the cross-cut C of all atoms. If the height of the Boolean algebra is $n+1$, then every k-cycle, for $k < n-2$, bounds, so that $b_0 = 1$ and $b_k = 0$ for $0 < k < n-2$. On the other hand, there is only one cycle in dimension $n-2$. Hence $b_{n-2} = 1$ and we find $E = 1 + (-1)^{n-2}$, which agrees with Proposition 5 of Section 3.

A notion of Euler characteristic for *distributive* lattices has been recently introduced by HADWIGER and KLEE. For finite distributive lattices, KLEE's Euler characteristic is related to the one introduced in this work. We refer to KLEE's paper for details.

7. Geometric lattices

An ordered structure of very frequent occurrence in combinatorial theory is the one that has been variously called matroid (WHITNEY), matroid lattice (BIRKHOFF), closure relation with the exchange property (MACLANE), geometric lattice

(Birkhoff), abstract linear dependence relation (Bleicher and Preston). Roughly speaking, these structures arise in the study of combinatorial objects that are obtained by piecing together smaller objects with a particularly simple structure. The typical such case is a linear graph, which is obtained by piecing together edges. Several counting problems associated with such structures can often be attacked by Möbius inversion, and one finds that the Möbius functions involved have particularly simple properties.

We briefly summarize the needed facts out of the theory of such structures, referring to any of the works of the above authors for the proofs.

A finite lattice L is a *geometric lattice* when every element of L is the join of atoms, and whenever if a and b in L cover $a \wedge b$, then $a \vee b$ covers both a and b. Equivalently, a geometric lattice is characterized by the existence of a rank function satisfying $r(a \wedge b) + r(a \vee b) \leq r(a) + r(b)$. Notice that this implies the chain condition. In particular if a is an atom, then $r(a \vee c) = r(c)$ or $r(c) + 1$. If M is a semimodular lattice, then the partially ordered subset of all elements which are joins of atoms is a geometric sublattice.

Geometric lattices are most often obtained from a closure relation on a finite set which satisfies the MacLane-Steinitz exchange property. The lattice L of closed sets in such a closure relation is a geometric lattice whenever every one-element set is closed. Conversely, every geometric lattice can be obtained in this way by defining one such closure relation on the set of its atoms.

The fundamental property of the Möbius function of geometric lattices is the following:

Theorem 4. *Let μ be the Möbius function of a finite geometric lattice L. Then:*

(a) *$\mu(x, y) \neq 0$ for any pair x, y in L, provided $x \leq y$.*

(b) *If y covers z, then $\mu(x, y)$ and $\mu(x, z)$ have opposite signs.*

Proof. Any segment $[x, y]$ of a geometric lattice is also a geometric lattice. It will therefore suffice to assume that $x = 0$, $y = 1$ and that z is a dual atom of L.

We proceed by induction. The theorem is certainly true for lattices of height 2, where $\mu(0, 1) = -1$. Assume it is true for all lattices of height $n - 1$, and let L be a lattice of height n. By the Corollary to Proposition 4 of Section 5, with $b = 1$, and a an atom of L, we have

$$\mu(0, 1) = -\sum_{\substack{x \vee a = 1 \\ x \neq 1}} \mu(0, x).$$

Now from the subadditive inequality

$$r(x \wedge a) + r(x \vee a) \leq r(x) + r(a)$$

we infer that if $x \vee a = 1$, then $n \leq \dim x + \dim a$, hence $\dim x \geq n - 1$. The element x must therefore be a dual atom. It follows from the induction assumption and from the fact that L satisfies the chain condition, that all the $\mu(0, x)$ in the sum on the right have the same sign, and none of them is zero. Therefore, $\mu(0, 1)$ is not zero, and its sign is the opposite of that $\mu(0, x)$ for any dual atom x. This concludes the proof.

Corollary. *The coefficients of the characteristic polynomial of a geometric lattice alternate in sign.*

We next derive a combinatorial interpretation of the Euler characteristic of a geometric lattice, which generalizes a technique first used by WHITNEY in the study of linear graphs.

A subset $\{a, b, \ldots, c\}$ of a geometric lattice L is *independent* when

$$r(a \vee b \vee \cdots \vee c) = r(a) + r(b) + \cdots + r(c).$$

Let C_k be the cross-cut of L of all elements of rank $k > 0$. A maximal independent subset $\{a, b, \ldots, c\} \subset C_k$ is a *basis* of C_k. All bases of C_k have the same number of elements, namely, $n - k$ if the lattice has height n. A subset $A \subset C_k$ is a *circuit* (WHITNEY) when it is not independent but every proper subset is independent. A set is independent if and only if it contains no circuits.

Order the elements of L of rank k in a linear order, say $a_1, a_2, \ldots, a_l$. This ordering induces a lexicographic ordering of the circuits of C_k.

If the subset $\{a_{i_1}, a_{i_2}, \ldots, a_{i_j}\}$ $(i_1 < i_2 < \cdots < i_j)$ is a circuit, the subset $a_{i_1}, a_{i_2}, \ldots, a_{i_{j-1}}$ will be called a *broken circuit.*

Proposition 1. *Let L be a geometric lattice of height $n + 1$, and let C_k be the cross-cut of all elements of rank k. Then $\mu(0, 1) = (-1)^n m_k$, where m_k is the number of subsets of C_k whose meet is 0, containing $n - k + 1$ elements each, and not containing all the arcs of any broken circuit.*

Again, the assertion implies that $m_1 = m_2 = m_3 = \cdots$.

Proof. Let the lexicographically ordered broken circuits be $P_1, P_2, \ldots, P_\sigma$, and let S_i be the family of all spanning subsets of C_k containing P_i but not $P_1, P_2, \ldots,$ or P_{i-1}. In particular, $S_{\sigma+1}$ is the family of all those spanning subsets not containing all the arcs of any broken circuit. Let q_j^i be the number of spanning subsets of j elements and not belonging to S_i. We shall prove that for each $i \geqq 1$

$$(*) \qquad \mu(0, 1) = q_2^i - q_3^i + q_4^i \cdots.$$

First, set $i = 1$. The set S_1 contains all spanning subsets containing the broken circuit P_1. Let $\bar{P}_1$ be the cicuit obtained by completing the broken circuit P_1. — A spanning set contained in S_1 contains either $\bar{P}_1$ or else P_1 but not $\bar{P}_1$; call these two families of spanning subsets A and B, and let q_j^A and q_j^B be defined accordingly. Then $q_j = q_j^1 + q_j^A + q_j^B$, and

$$\mu(0, 1) = q_2 - q_3 + q_4 \cdots = q_2^1 - q_3^1 + \cdots +$$
$$+ q_2^A + (q_2^B - q_3^A) - (q_3^B - q_4^A) + \cdots.$$

Now, $q_2^A = 0$, because no circuit can contain two elements; there is a one-to-one correspondence between the elements of A and those of B, obtained by completing the broken circuit P_1. Thus, all terms in parentheses cancel and the identity (*) holds for $i = 1$.

To prove (*) for $i > 1$, remark that the element c_i of C_k, which is dropped from a circuit to obtain the broken circuit P_i, does not occur in any of the previous circuits, because of the lexicographic ordering of the circuits. Hence the induction can be continued up to $i = \sigma + 1$.

Any set belonging to $S_{\sigma+1}$ does not contain any circuit. Hence, it is an independent set. Since it is a spanning set, it must contain $n-k+1$ elements. Thus, all the integers $q_{\sigma+1}$ vanish except $q^{\sigma+1}_{n-k+1}$ and the statement follows from (*), q. e. d.

Corollary 1. *Let* $q(\lambda)=\lambda^n+m_1\lambda^{n-1}+m_2\lambda^{n-2}+\cdots+m_n$ *be the characteristic polynomial of a geometric lattice of height* $n+1$. *Then* $(-1)^k m_k$ *is a positive integer for* $1\leq k\leq n$, *equal to the number of independent subsets of* k *atoms not containing any broken circuit.*

The *proof* is immediate: take $k=1$ in the preceding Proposition.

The homology of a geometric lattice is simpler than that of a general lattice:

Proposition 2. *In the homology relative to the cross-cut* C_k *of all elements of rank* $k=1$, *the Betti numbers* $b_1, b_2, \ldots, b_{k-2}$ *vanish.*

The *proof* is not difficult.

Example 1. *Partitions of a set.*

Let S be a finite set of n elements. A *partition* π of S is a family of disjoint subsets $B_1, B_2, \ldots, B_k$, called *blocks*, whose union is S. There is a (well-known) natural ordering of partitions, which is defined as follows: $\pi\leq\sigma$ whenever every block of π is contained in a block of partition σ. In particular, 0 is the partition having n blocks, and I is the partition having one block. In this ordering, the partially ordered set of partitions is a geometric lattice (cf. BIRKHOFF).

The Möbius function for the lattice of partitions was first determined by SCHÜTZENBERGER and independently by ROBERTO FRUCHT and the author. We give a new proof which uses a recursion. If π is a partition, the *class* of π is the (finite) sequence $(k_1, k_2, \ldots)$, where k_i is the number of blocks with i elements.

Lemma. *Let* L_n *be the lattice of partitions of a set with* n *elements. If* $\pi\in L_n$ *is of rank* k, *then the segment* $[\pi, 1]$ *is isomorphic to* L_{n-k}. *If* π *is of class* $(k_1, k_2, \ldots)$, *then the segment* $[0, \pi]$ *is isomorphic to the direct product of* k_1 *lattices isomorphic to* L_1, k_2 *lattices isomorphic to* L_2, etc.

The *proof* is immediate.

It follows from the Lemma that if $[x, y]$ is a segment of L_n, then it is isomorphic to a product of k_i lattices isomorphic to L_i, $i=1,2,\ldots$. We call the sequence $(k_1, k_2, \ldots)$ the *class* of the *segment* $[x, y]$.

Proposition 3. *Let* $\mu_n=\mu(0,1)$ *for the lattice of partitions of a set with* n *elements. Then* $\mu_n=(-1)^{n-1}(n-1)!$.

Proof. By the Corollary to Proposition 4 of Section 5, $\sum_{x\wedge a=0}\mu(x,1)=0$. Let a be the dual atom consisting of a block C_1 containing $n-1$ points, and a second block C_2 containing one point. Which non-zero partitions x have the property that $x\wedge a=0$? Let the blocks of such a partition x be $B_1, \ldots, B_k$. None of the blocks B_i can contain two distinct points of the block C_1, otherwise the two points would still belong to the same block in the intersection. Furthermore, only one of the B_i can contain the block C_2. Hence, all the B_i contain one point, except one, which contains C_2 and an extra point. We conclude that x must be an atom, and there are $n-1$ such atoms. Hence, $\mu_n=\mu(0,1)=-\sum_x\mu(x,1)$, where x ranges over a set of $n-1$ atoms. By the Lemma, the segment $[x, 1]$ is isomorphic to the lattice of partitions of a set with $n-1$ elements, hence $\mu_n = -(n-1)\mu_{n-1}$. Since $\mu_2 = -1$, the conclusion follows.

Corollary. *If the segment $[x, y]$ is of class $(k_1, k_2, \ldots, k_n)$, then*

$$\mu(x, y) = \mu_1^{k_1} \mu_2^{k_2} \ldots \mu_n^{k_n} = (-1)^{k_1+k_2+\cdots+k_n-n} (2!)^{k_3} (3!)^{k_4} \ldots ((n-1)!)^{k_n}.$$

The Möbius inversion formula on the partitions of a set has several combinatorial applications; see the author's expository paper on the subject.

8. Representations

There is, as is well known, a close analogy between combinatorial results relating to Boolean algebras and those relating to the lattice of subspaces of a vector space. This analogy is displayed for example in the theory of q-difference equations developed by F. H. JACKSON, and can be noticed in many number-theoretic investigations. In view of it, we are led to surmise that a result analogous to Proposition 1 of Section 5 exists, in which the Boolean algebra of subsets of R is replaced by a lattice of subspaces of a vector space over a finite field. Such a result does indeed exist; in order to establish it a preliminary definition is needed.

Let L be a finite lattice, and let V be a finite-dimensional vector space over a finite field with q elements. A *representation* of L over V is a monotonic map p of L into the lattice M of subspaces of V, having the following properties:

(1) $p(0) = 0$.

(2) $p(a \vee b) = p(a) \vee p(b)$.

(3) Each atom of L is mapped to a line of the vector space V, and the set of lines thus obtained spans the entire space V.

A representation is *faithful* when the mapping p is one-to-one. We shall see in Section 9 that a great many ordered structures arising in combinatorial problems admit faithful representations. Given a representation $p: L \to M$, one defines the *conjugate map* $q: M \to L$ as follows.

Let K be the set of atoms of M (namely, lines of V), and let A be the image under p of the set of atoms of L. For $s \in M$, let $K(s)$ be the set of atoms of M dominated by s, and let $B(s)$ be a minimal subset of A which spans (in the vector space sense) every element of $K(s)$. Let $A(s)$ be the subset of A which is spanned by $B(s)$. A simple vector-space argument, which is here omitted, shows that the set $A(s)$ is well defined, that is, that it does not depend upon the choice of $B(s)$, but only upon the choice of s.

Let $C(s)$ be the set of atoms of L which are mapped by p onto $A(s)$. Set $q(s) = \vee C(s)$ in the lattice L; this defines the map q. It is obviously a monotonic function.

Lemma. *Let $p: L \to M$ be a faithful representation and let $q: M \to L$ be the conjugate map. Assume that every element of L is a join of atoms. Then $p(q(s)) \geqq s$ and $q(p(x)) \leqq x$.*

Proof. By definition, $q(s) = \vee C(s)$, where $C(s)$ is the inverse image of $A(s)$ under p. By property (2) of a representation,

$$p(q(s)) = p(\vee C(s)) = \vee p(C(s)) = \vee A(s).$$

But this join of the set of lines $A(s)$ in the lattice M is the same as their span in the vector space V. Hence $\bigvee A(s) \geqq s$, and we conclude that $p(q(s)) \geqq s$.

To prove that $q(p(x)) \leqq x$, it suffices to show that $A(p(x)) = B$, where B is the set of atoms in A dominated by $p(x)$. Clearly $B \subset A(p(x))$, and it will suffice to establish the converse implication. By (2), and by the fact that x is a join of atoms, we have $p(x) = \bigvee B$. Therefore every line l dominated by $p(x)$ is spanned by a subset of B. If in addition $l \in A$, then $l \leqq \bigvee C$ for some subset $C \subset B$, hence $l \in B$. This shows $B \supset A(p(x))$, q. e. d.

Theorem 5. *Let L be a finite lattice, where every element is a join of atoms, let $p: L \to M$ be a faithful representation of L into the lattice M of subspaces of a vector space V over a finite field with q elements, and let $q: M \to L$ be the conjugate map. For every $k \geqq 2$, let m_k be the number of k-dimensional subspaces s of V such that $q(s) = I$. Then*

$$(*) \qquad \mu(0, 1) = q^{\binom{2}{2}} m_2 - q^{\binom{3}{2}} m_3 + q^{\binom{4}{2}} m_4 - \cdots,$$

where μ is the Möbius function of L.

Proof. Let $Q = L^*$, let $c: L \to Q$ and $c^*: Q \to L$ be the canonical isomorphisms between L and Q. Define $\pi: Q \to M$ as $\pi = pc^*$, and $\varrho: M \to Q$ as $\varrho = cq$. We verify that π and ϱ give a Galois connection between Q and M satisfying the hypothesis of Theorem 1. If $\pi(x) = 0$, then there is a $y \in L$ such that $y = c^*(x)$ and $p(y) = 0$. It follows from the definition of a representation that $y = 0$. Hence $x = c(y) = 1$. Furthermore, $\varrho(0) = c(q(0)) = 1$. It follows from the preceding Lemma that π and ϱ are a Galois connection. Applying Theorem 1 and the result of Example 2 of Section 5, formula (*) follows at once.

Remark. It is easy to see that every lattice having a faithful representation is a geometric lattice. The converse is however not true, as an example of T. Lazarson shows.

A reduction similar to that of Proposition 1 of Section 7 can be carried out with Theorem 5 and representations, and another combinatorial property of the Euler characteristic is obtained.

9. The coloring of graphs

By way of illustration of the preceding theory, we give some applications to the classic problem of coloring of graphs, and to the problem of constructing flows in networks with specified properties. Our results extend previous work of G. D. Birkhoff, D. C. Lewis, W. T. Tutte and H. Whitney.

A *linear graph* $G = (V, E)$ is a structure consisting of a finite set V, whose elements are called vertices, together with a family E of two-element subsets of V, called edges. Two vertices a and b are adjacent when the set (a, b) is an edge; the vertices a and b are called the endpoints of (a, b). Alternately, one calls the vertices *regions* and calls the graph a *map*, and we use the two terms interchangeably, considering them as two words for the same object. If S is a set of edges, the *vertex set* $V(S)$ consists of all vertices which are incident to some edge in S.

A set of edges S is *connected* when in any partition $S = A \cup B$ into disjoint non-empty sets A and B, the vertex sets $V(A)$ and $V(B)$ are not disjoint. Every set of edges is the union of disjoint connected *blocks*.

The *bond closure* on a graph $G = (V, E)$ is a closure relation defined on the set E of edges as follows. If $S \subset E$, let $\bar{S}$ be the set of all edges both of whose endpoints belong to one and the same block of S. Every set consisting of a single edge is closed, and these are the only minimal non-empty closed sets.

Lemma 1. *The bond closure $S \to \bar{S}$ has the exchange property.*

Proof. Suppose e and f are edges, $S \subset E$, and $e \in \overline{S \cup f}$ but $e \notin \bar{S}$. Then every endpoint of e which is not in $V(S)$ is an endpoint of f; on the other hand, S and f have at least one point in common, otherwise $e \in \bar{S}$. Thus both e and f either connect the same two blocks of S, or else they have one endpoint in S and one common endpoint; hence $f \in \overline{S \cup e}$, q.e.d.

The lattice $L = L(G)$ of bond-closed subsets of E is called the *bond lattice* of the graph G. Suppose that E has n blocks and $p(\lambda)$ is the characteristic polynomial of L, then the polynomial $\lambda^n p(\lambda)$ is the *chromatic polynomial* of the graph G, first studied by G. D. BIRKHOFF. From Theorem 4 we infer at once the theorem of WHITNEY that the coefficients of the chromatic polynomial alternate in sign.

The chromatic polynomial has the following combinatorial interpretation. Let C be a set of n elements, called colors. A function $f: V \to C$ is a *proper coloring* of the graph, when no two adjacent vertices are assigned the same color. To every coloring f — not necessarily proper — there corresponds a subset of E, the *bond* of f, defined as the set of all edges whose endpoints are assigned the same color by f. The bond of f is a closed set of edges. For every closed set S, let $p(\lambda, S)$ be the number of colorings whose bond is S. Then we shall prove that $p(\lambda, S) = \lambda^n q(\lambda, S)$, where $q(\lambda, S)$ is the characteristic polynomial of the segment $[S, I]$ in the lattice L. Since every coloring has a bond $\sum_{T \geq S} p(\lambda, T)$ equals the total number of colorings having some bond $T \geq S$. But this number is evidently $\lambda^{k-r(s)}$, where k is the number of vertices of the graph and $r(S)$ is the rank of S in L. Applying the Möbius inversion formula on the bond-lattice, we get

$$(*) \qquad p(\lambda) = p(\lambda, 0) = \sum_{T \in L} \lambda^{k-r(T)} \mu(0, T).$$

But the number of colorings whose bond is the null set 0 is exactly the number of proper colorings.

WHITNEY's evaluation (cf. A logical expansion in Mathematics) of the chromatic polynomials of a graph in terms of the number of subgraphs of s edges and p connected components is an immediate consequence of the cross-cut theorem applied to the atoms of the bond-lattice of G. This result of WHITNEY's can now be sharpened in two directions: first, a cross-cut other than that of the atoms can be taken; secondly, the computation of the coefficients of the chromatic polynomial can be simplified by Proposition 1 of Section 8. The cross-cut of all elements of rank 2 is particularly suited for computation, and can be programmed. The interested reader may wish to explicitly translate the cross-cut theorem and the results of Section 8 into the geometric language of graphs.

Example 1. For a *complete graph* on n vertices, where every two-element subset is an edge, the bond-lattice is isomorphic to the lattice of partitions of a set with n elements. The chromatic polynomial is evidently $(\lambda)_n = \lambda(\lambda - 1) \ldots (\lambda - n + 1)$, and the coefficients $s(n, k)$ are the *Stirling numbers of the first kind*.

Thus, $\sum_{r(\pi)=k} \mu(0, \pi) = s(n, k)$. This gives a combinatorial interpretation to the Stirling numbers of the first kind.

For a map $\mathfrak{m}$ embedded in the plane, where regions and boundaries have their natural meaning and no region bounds with itself, one obtains an interesting geometric result by applying the cross-cut theorem to the dual atoms of the bond lattice $L(\mathfrak{m})$.

Let $\mathfrak{m}$ be a connected map in the plane; without loss of generality we can assume: (a) that all the regions of $\mathfrak{m}$, except one which is unbounded, lie inside a convex polygon, the outer boundary of $\mathfrak{m}$; (b) that all boundaries are segments of straight lines. The *dual graph* of $\mathfrak{m}$ is the linear graph made up of the boundaries of $\mathfrak{m}$. A *circuit* in a linear graph is defined as a simple closed curve contained in the graph. We give an expression of the polynomial $P(\lambda, \mathfrak{m})$ in terms of the circuits of the dual graph. The outer boundary is always a circuit.

A set of circuits of a map $\mathfrak{m}$ in the plane *spans*, when their union — in the set-theoretic sense — is the entire boundary of $\mathfrak{m}$.

Proposition 1. *For every integer $k \geqq 1$, let C_k be the number of spanning sets of k distinct circuits of a map $\mathfrak{m}$ in the plane. Then*

$$\mu_{\mathfrak{m}}(0, 1) = -C_1 + C_2 - C_3 + C_4 - \cdots$$

Proof. If the map has two regions, then $C_1 = 1$ and all other $C_c = 0$, so the result is trivial. Assume now that $\mathfrak{m}$ has at least 3 regions. Then $C_1 = 0$. All we have to prove is that the integers C_k are the integers q_k of Theorem 3, relative to the cross-cut of $L(\mathfrak{m})$ consisting of all the dual atoms.

By the Jordan curve theorem, every circuit divides the plane into two regions; this gives a one-to-one correspondence of the circuits with the dual atoms of $L(\mathfrak{m})$. Conversely, because we can assume that the map is of the special type described above, every dual atom in $L(\mathfrak{m})$ is a map with two connected regions, and so must have as a boundary a simple closed curve, q.e.d.

It has been shown by Richard Rado (p. 312) that the bond-lattice $L(G)$ of any linear graph G has a faithful representation. Accordingly, Theorem 5 can also be applied to obtain expression for $\mu(0,1)$. These expressions usually give sharper bounds than similar expressions based upon the cross-cut of atoms.

Farther-reaching techniques for the computation of the Möbius function of $L(G)$ are obtained by applying Theorem 1 to situations where P and Q are both bond-lattices of graphs. This we shall now do. A *monomorphism* of a graph G into a graph H is a one-to-one function f of the vertices of G *onto* the vertices of H, which induces a map $\bar{f}$ of the edges of G *into* the edges of H. Every monomorphism $f: G \to H$ induces a monotonic map $p: L(G) \to L(H)$, where $p(S)$ is defined as the closure of the image $\bar{f}(S)$ in H. It also induces a monotonic map $q: L(H) \to$ $\to L(G)$, where $q(T)$ is defined as the set of edges of G whose image is in T.

Lemma 2. *$q(p(S)) = S$ for S in $L(G)$ and $p(q(T)) \leqq T$ for T in $L(H)$.*

Proof. Intuitively, $p(S)$ is obtained by "adding edges" to S, and $q(p(S))$ simply removes the added edges. Thus, the first statement is graphically clear. The second one can be seen as follows. $q(T)$ is obtained from T by removing a number of edges. Taking $p(q(T))$, some of the edges may be replaced, but in general not all. Thus, $p(q(T)) \leq T$.

Taking $M = L(H)^*$ and $c : L(H) \to M$ to be the canonical order-inverting map, we see that $\pi = cp$ and $\zeta = qc$ give a Galois connection between $L(G)$ and M. Now, $\pi(x) = 0$ is equivalent to $p(x) = 1$ for $x \in L(G)$. This can happen only if x has only one component, that is — since x is closed — only if $x = 1$ in $L(G)$. Thus $\pi(x) = 0$ if and only if $x = 1$. Secondly, $\varrho(0) = q(1) = 1$, evidently. We have verified all the hypotheses of Theorem 1, and we therefore obtain:

Proposition 2. *Let $f : G \to H$ be a monomorphism of a linear graph G into a linear graph H, and let μ_G and μ_H be the Möbius functions of the bond-lattices. Then*

$$\mu_G(0, 1) = \sum_{[a \in L(H);\, q(a) = 0]} \mu_H(a, 1),$$

where q is the map of $L(H)$ into $L(G)$ naturally associated with f, as above.

Proposition 1 can be used to derive a great many of the reductions of G. D. BIRKHOFF and D. C. LEWIS, and provides a systematic way of investigating the changes of Möbius functions — and hence of the chromatic polynomial — when edges of a graph are removed. It has a simple geometric interpretation.

An interesting application is obtained by taking H to be the complete lattice on n elements. We then obtain a formula for μ which completes the statements of Theorems 3 and 5. Let G be a linear graph on n vertices. Let C be the family of two-element subsets of G which are not edges of G. Let F be the family of all subsets of C which are closed sets in the bond-lattice of the complete graph on n vertices built on the vertices of G. Then,

Corollary.
$$\mu_G(0, 1) = \sum_{a \in F} \mu(a, 1),$$

where μ is the Möbius function of the lattice of partitions (cf. Example 5) of a set of n elements.

Stronger results can be obtained by considering "epimorphisms" rather than "monomorphisms" of graphs, relating μ_G to the Möbius function obtained from G by "coalescing" points. In this way, one makes contact with G. A. DIRAC's theory of critical graphs. We leave the development of this topic to a later work.

10. Flows in networks

A network $N = (V, E)$ is a finite set V of vertices, together with a set of ordered pairs of vertices, called edges.

We shall adopt for networks the same language as for linear graphs.

A *circuit* is a sequence of edges S such that every vertex in $V(S)$ belongs to exactly two edges of S. Every edge has a positive and a negative endpoint. Given a function Φ from E to the integers from 0 to $\lambda - 1$, let for each vertex v, $\overline{\Phi}(v)$ be defined as

$$\overline{\Phi}(v) = \sum_{e} \eta(e, v)\, \Phi(e),$$

where the sum ranges over all edges incident to v, and the function $\eta(e, v)$ takes the value $+1$ or -1 according as the positive or negative end of the edge e abuts at the vertex v, and the value zero otherwise. The function Φ is a *flow* (mod. λ) when $\bar{\Phi}(v) = 0$ (mod. λ) for every vertex v. The value $\Phi(e)$ for an edge e is called the *capacity* of the *flow* through e. The mod. λ restriction is inessential, but will be kept throughout.

A *proper flow* is one in which no edge is assigned zero capacity. TUTTE was the first to point out the importance of the problem of counting proper flows (cf. A contribution to the theory of chromatic polynomials) in combinatorial theory.

We shall reduce the solution of the problem to a Möbius inversion on a lattice associated with the network. This will give an expression for the number of proper flows as a polynomial in λ, whose coefficients are the values of a Möbius function.

Every flow through N is a proper flow of a suitable subnetwork of N, obtained by removing those edges which are assigned capacity 0. However, the converse of this assertion is not true: given a subnetwork S of N, it may not be possible to find a flow which is proper on the complement of N. This happens because every flow which assigns capacity zero to each edge of S may assign capacity zero to some further edges. We are therefore led to define a closure relation on the set of all subgraphs as follows: $\bar{S}$ shall be the set of all edges which necessarily are assigned capacity zero, in any flow of N which assigns capacity zero to every edge of S. In other words, if $e \notin \bar{S}$, then there is a flow in N which assigns capacity $\neq 0$ to the edge e, but which assigns capacity zero to all the edges of $\bar{S}$. It is immediately verified that $S \to \bar{S}$ is a closure relation. We call it the *circuit closure* of S. The circuit closure *has the exchange property*: if $e \in \overline{S \cup p}$ but $e \notin \bar{S}$, then $p \in \overline{S \cup e}$. Before verifying it, we first derive a geometric characterization of the circuit closure. A set S is circuit closed ($S = \bar{S}$) if and only if through every edge e not in S there passes a circuit which is disjoint from S. For if S is closed and $e \notin S$, then there is a flow through e and disjoint from S. But this can happen only if there is a circuit through e.

If there is a circuit through the edge p disjoint from $\overline{S \cup e}$, and a circuit through e disjoint from $\bar{S}$ and containing p, then there is — as has been observed by WHITNEY — also a circuit through e not containing $\bar{S} \cup p$. This implies that e is not in the closure of $\bar{S} \cup p$ and verifies the exchange property.

The lattice $C(N)$ of closed subsets of edges of the network N is the *circuit lattice of N*. An atom in this lattice is not necessarily a single edge.

Proposition 1. *The number of proper flows,* (mod. λ) *on a network N with v vertices, e edges and p connected components is a polynomial $p(\lambda)$ of degree $e - v + p$. This polynomial is the characteristic polynomial of the circuit lattice of N. The coefficients alternate in sign.*

Proof. The last statement is an immediate consequence of Theorem 4 of Section 8.

The total number of flows on N (not necessarily proper) is determined as follows. Assume for simplicity that N is connected. Remove a set D of $v - 1$ edges from N, one adjacent to each but one of the vertices.

Every flow on N can be obtained by first assigning to each of the edges not in D an arbitrary capacity, between 0 and $\lambda - 1$, and then filling in capacities for the edges in D to match the requirement of zero capacity through each vertex. There are λ^{e-v+1} ways of doing this, and this is therefore the total number of flows mod. λ. If the network is in p connected components, the same argument gives λ^{e-v+p}. Now, every flow on G is a proper flow on a unique closed subset $\bar{S}$, obtained by removing all edges having capacity zero.

Hence

$$\lambda^{e-v+p} = \sum_{\bar{S} \in \underline{C}(G)} p(\bar{S}, \lambda),$$

where $p(\bar{S}, \lambda)$ is the characteristic polynomial of the closed subgraph $\bar{S}$. Setting $n(s) = e(s) - v(s) + p(s)$, the number of edges, vertices and components of s, and applying the inversion formula, we get

$$p(G, \lambda) = \sum_{S \in \underline{C}(G)} \lambda^{n(s)} \mu(S, G), \quad \text{q.e.d.}$$

In the course of the proof we have also shown that $n(s)$ is the *rank* of S in the circuit lattice of G. The rank of the null subgraph is one.

The four-color problem is equivalent to the statement that every planar network without an isthmus has a proper flow mod 5. (An isthmus is an edge that disconnects a component of the network when removed.)

Most of the results of the preceding section extend to circuit lattices of a network, and give techniques for computation of the flow polynomials of networks. We shall not write down their translation into the geometric language of networks.

References

AUSLANDER, L., and H. M. TRENT: Incidence matrices and linear graphs. J. Math. Mech. 8, 827 — 835 (1959).

BELL, E. T.: Algebraic Arithmetic. New York: Amer. Math. Soc. (1927),

— Exponential polynomials. Ann. of Math., II. Ser. **35**, 258—277 (1934).

BERGE, C.: Théorie des graphes et ses applications. Paris: Dounod 1958.

BIRKHOFF, GARRETT: Lattice Theory, third preliminary edition. Harvard University, 1963.

— Lattice Theory, revised edition. American Mathematical Society, 1948.

BIRKHOFF, G. D.: A determinant formula for the number of ways of coloring a map. Ann. of Math., II. Ser. **14**, 42—46 (1913).

—, and D. C. LEWIS: Chromatic polynomials. Trans. Amer. math. Soc. **60**, 355—451 (1946).

BLEICHER, M. N., and G. B. PRESTON: Abstract linear dependence relations. Publ. Math., Debrecen **8**, 55—63 (1961).

BOUGAYEV, N. V.: Theory of numerical derivatives. Moscow, 1870—1873, pp. 1—222.

BRUIJN, N. G. DE: Generalization of Polya's fundamental theorem in enumerative combinatorial analysis. Indagationes math. **21**, 59—69 (1959).

CHUNG, K.-L., and L. T. C. HSU: A combinatorial formula with its application to the theory of probability of arbitrary events. Ann. math. Statistics **16**, 91—95 (1945).

DEDEKIND, R.: Gesammelte Mathematische Werke, volls. I—II—III. Hamburg: Deutsche Math. Verein. (1930).

DELSARTE, S.: Fonctions de Möbius sur les groupes abéliens finis. Ann. of Math., II. Ser. **49**, 600—609 (1948).

DILWORTH, R. P.: Proof of a conjecture on finite modular lattices. Ann. of Math., II. Ser. **60**, 359—364 (1954).

DIRAC, G. A.: On the four-color conjecture. Proc. London math. Society, III. Ser. **13**, 193 to 218 (1963).

DOWKER, C. H.: Homology groups of relations. Ann. of Math., II. Ser. **56**, 84—95 (1952).

DUBREIL-JACOTIN, M.-L., L. LESIEUR et R. CROISOT: Leçons sur la théorie des treilles des structures algebriques ordonnées et des treilles géometriques. Paris: Gauthier-Villars 1953.

EILENBERG, S., and N. STEENROD: Foundations of algebraic topology. Princeton: University Press 1952.

FARY, I.: On straight-line representation of planar graphs. Acta Sci. math. Szeged **11**, 229--233 (1948).

FELLER, W.: An introduction to probability theory and its applications, second edition. New York: Wiley 1960.

FRANKLIN, P.: The four-color problem. Amer. J. Math. **44**, 225—236 (1922).

FRÉCHET, M.: Les probabilités associées à un système d'évenements compatibles et dépendants. Actualitées scientifiques et industrielles, nos. 859 et 942. Paris: Hermann 1940 et 1943.

FRONTERA MARQUÉS, B.: Una función numérica en los retículos finitos que se anula para los retículos reducibles. Actas de la 2a, Reunión de matemáticos españoles. Zaragoza 103—111 1962.

FRUCHT, R., and G.-C. ROTA: La función de Möbius para el retículo di particiones de un conjunto finito. To appear in Scientia (Chile).

GOLDBERG, K., M. S. GREEN and R. E. NETTLETON: Dense subgraphs and connectivity. Canadian J. Math. **11** (1959).

GOLOMB, S. W.: A mathematical theory of discrete classification. Fourth Symposium in Information Theory, London, 1961.

GREEN, M. S., and R. E. NETTLETON: Möbius function on the lattice of dense subgraphs. J. Res. nat. Bur. Standards **64B**, 41—47 (1962).

— — Expression in terms of modular distribution functions for the entropy density in an infinite system. J. Chemical Physisc **29**, 1365—1370 (1958).

HADWIGER, H.: Eulers Charakteristik und kombinatorische Geometrie. J. reine angew. Math. **194**, 101--110 (1955).

HALL, PHILIP: A contribution to the theory of groups of prime power order. Proc. London math. Soc., II. Ser. **36**, 39--95 (1932).

— The Eulerian functions of a group. Quart. J. Math. Oxford Ser. 134--151, 1936.

HARARY, F.: Unsolved problems in the enumeration of graphs. Publ. math. Inst. Hungar Acad. Sci. **5**, 63--95 (1960).

HARDY, G. H.: Ramanujan. Cambridge: University Press 1940.

—, and E. M. WRIGHT: An introduction to the theory of numbers. Oxford: University Press 1954.

HARTMANIS, J.: Lattice theory of generalized partitions. Canadian J. Math. **11**, 97—106 (1959).

HILLE, E.: The inversion problems of Möbius. Duke math. J. **3**, 549--568 (1937).

HSU, L. T. C.: Abstract theory of inversion of iterated summation. Duke math. J. **14**, 465 to 473 (1947).

— On Romanov's device of orthogonalization. Sci. Rep. Nat. Tsing Hua Univ. **5**, 1—12 (1948).

— Note on an abstract inversion principle. Proc. Edinburgh math. Soc. (2) **9**, 71—73 (1954).

JACKSON, F. H.: Series connected with the enumeration of partitions. Proc. London math. Soc., II. Ser. **1**, 63--88 (1904).

— The q-form of Taylor's theorem. Messenger of Mathematics **38**, 57—61 (1909).

JÓNSSON, B.: Lattice-theoretic approach to projective and affine geometry. Symposium on the Axiomatic Method. Amsterdam, North-Holland Publishing Company, 1959, 188—205.

—, and A. TARSKI: Direct decomposition of finite algebraic systems. Notre Dame Mathematical lectures, no. 5. Indiana: Notre Dame 1947.

KAC, M., and J. C. WARD: A combinatorial solution of the two-dimensional Ising model. Phys. Review **88**, 1332—1337 (1952).

KAPLANSKI, I., and J. RIORDAN: The problème des ménages. Scripta math. **12**, 113—124 (1946).

KLEE, V.: The Euler characteristic in combinatorial geometry. Amer. math. Monthly **70**, 119—127 (1963).

LAZARSON, T.: The representation problem for independence functions. J. London math. Soc. **33**, 21–25 (1958).
MACLANE, S.: A lattice formulation of transcendence degrees and p-bases. Duke math. J. **4**, 455–468 (1938).
MACMILLAN, B.: Absolutely monotone functions. Ann. of Math., II. Ser. **60**, 467–501 (1954).
MÖBIUS, A. F.: Über eine besondere Art von Umkehrung der Reihen. J. reine angew. Math. **9**, 105–123 (1832).
ORE, O.: Theory of graphs. Providence: American Mathematical Society 1962.
POLYA, G.: Kombinatorische Anzahlbestimmungen für Gruppen, Graphen und chemische Verbindungen. Acta math. **68**, 145–253 (1937).
RADO, R.: Note on independence functions. Proc. London math. Soc., III. Ser. **7**, 300–320 (1957).
READ, R. C.: The enumeration of locally restricted graphs, I. J. London math. Soc. **34**, 417 to 436 (1959).
REDFIELD, J. H.: The theory of group-reduced distributions. Amer. J. Math. **49**, 433–455 (1927).
REVUZ, ANDRÉ: Fonctions croissantes et mesures sur les espaces topologiques ordonnés. Ann. Inst. Fourier **6** 187–268 (1955).
RIORDAN, J.: An introduction to combinatorial analysis. New York: Wiley 1958.
ROMANOV, N. P.: On a special orthonormal system and its connection with the theory of primes. Math. Sbornik, N. S. **16**, 353–364 (1945).
ROTA, G.-C.: Combinatorial theory and Möbius functions. To appear in Amer. math. Monthly.
-- The number of partitions of a set. To appear in Amer. math. Monthly.
RYSER, H. J.: Combinatorial Mathematics. Buffalo: Mathematical Association of America 1963.
SCHÜTZENBERGER, M. P.: Contribution aux applications statistiques de la théorie de l'information. Publ. Inst. Stat. Univ. Paris, **3**, 5–117 (1954).
TARSKI, A.: Ordinal algebras. Amsterdam: North-Holland Publishing Company 1956.
TOUCHARD, J.: Sur un problème de permutations. C. r. Acad. Sci., Paris, **198**, 631–633 (1934).
TUTTE, W. T.: A contribution to the theory of chromatic polynomials. Canadian J. Math. **6**, 80–91 (1953).
A class of Abelian group. Canadian J. Math. **8**, 13–28 (1956).
A homotopy theorem for matroids. I. and II. Trans. Amer. math. Soc. **88**, 144–140 (1958).
Matroids and graphs. Trans. Amer. math. Soc. **90**, 527–552 (1959).
WARD, M.: The algebra of lattice functions. Duke math. J. **5**, 357–371 (1939).
WEISNER, L.: Abstract theory of inversion of finite series. Trans. Amer. math. Soc. **38**, 474–484 (1935).
-- Some properties of prime-power groups. Trans. Amer. math. Soc. **38**, 485–492 (1935).
WHITNEY, H.: A logical expansion in mathematics. Bull. Amer. math. Soc. **38**, 572–579 (1932).
— Characteristic functions and the algebra of logic. Ann. of Math., II. Ser. **34**, 405–414 (1933).
— The abstract properties of linear dependence. Amer. J. Math. **57**, 507–533 (1935).
WIELANDT, H.: Beziehungen zwischen den Fixpunktzahlen von Automorphismengruppen einer endlichen Gruppe. Math. Z. **73**, 146–158 (1960).
WINTNER, A.: Eratosthenian Averages. Baltimore (privately printed) 1943.

Department of Mathematics
Massachusetts Institute of Technology
Cambridge 39, Massachusetts

(Received September 2, 1963)

Reprinted by TRUEXpress Oxford England

PATHS, TREES, AND FLOWERS

JACK EDMONDS

1. Introduction. A *graph G* for purposes here is a finite set of elements called *vertices* and a finite set of elements called *edges* such that each edge *meets* exactly two vertices, called the *end-points* of the edge. An edge is said to *join* its end-points.

A *matching* in G is a subset of its edges such that no two meet the same vertex. We describe an efficient algorithm for finding in a given graph a matching of maximum cardinality. This problem was posed and partly solved by C. Berge; see Sections 3.7 and 3.8.

Maximum matching is an aspect of a topic, treated in books on graph theory, which has developed during the last 75 years through the work of about a dozen authors. In particular, W. T. Tutte **(8)** characterized graphs which do not contain a *perfect* matching, or *1-factor* as he calls it—that is a set of edges with exactly one member meeting each vertex. His theorem prompted attempts at finding an efficient construction for perfect matchings.

This and our two subsequent papers will be closely related to other work on the topic. Most of the known theorems follow nicely from our treatment, though for the most part they are not treated explicitly. Our treatment is independent and so no background reading is necessary.

Section 2 is a philosophical digression on the meaning of "efficient algorithm." Section 3 discusses ideas of Berge, Norman, and Rabin with a new proof of Berge's theorem. Section 4 presents the bulk of the matching algorithm. Section 7 discusses some refinements of it.

There is an extensive combinatorial-linear theory related on the one hand to matchings in bipartite graphs and on the other hand to linear programming. It is surveyed, from different viewpoints, by Ford and Fulkerson in **(5)** and by A. J. Hoffman in **(6)**. They mention the problem of extending this relationship to non-bipartite graphs. Section 5 does this, or at least begins to do it. There, the König theorem is generalized to a matching-duality theorem for arbitrary graphs. This theorem immediately suggests a polyhedron which in a subsequent paper **(4)** is shown to be the convex hull of the vectors associated with the matchings in a graph.

Maximum matching in non-bipartite graphs is at present unusual among combinatorial extremum problems in that it is very tractable and yet not of the "unimodular" type described in **(5** and **6)**.

Received November 22, 1963. Supported by the O.N.R. Logistics Project at Princeton University and the A.R.O.D. Combinatorial Mathematics Project at N.B.S.

Section 6 presents a certain invariance property of the dual to maximum matching.

In paper **(4)**, the algorithm is extended from maximizing the cardinality of a matching to maximizing for matchings the sum of weights attached to the edges. At another time, the algorithm will be extended from a capacity of one edge at each vertex to a capacity of d_i edges at vertex v_i.

This paper is based on investigations begun with G. B. Dantzig while at the RAND Combinatorial Symposium during the summer of 1961. I am indebted to many people, at the Symposium and at the National Bureau of Standards, who have taken an interest in the matching problem. There has been much animated discussion on possible versions of an algorithm.

2. Digression. An explanation is due on the use of the words "efficient algorithm." First, what I present is a conceptual description of an algorithm and not a particular formalized algorithm or "code."

For practical purposes computational details are vital. However, my purpose is only to show as attractively as I can that there is an efficient algorithm. According to the dictionary, "efficient" means "adequate in operation or performance." This is roughly the meaning I want—in the sense that it is conceivable for maximum matching to have no efficient algorithm. Perhaps a better word is "good."

I am claiming, as a mathematical result, the existence of a *good* algorithm for finding a maximum cardinality matching in a graph.

There is an obvious finite algorithm, but that algorithm increases in difficulty exponentially with the size of the graph. It is by no means obvious whether *or not* there exists an algorithm whose difficulty increases only algebraically with the size of the graph.

The mathematical significance of this paper rests largely on the assumption that the two preceding sentences have mathematical meaning. I am not prepared to set up the machinery necessary to give them formal meaning, nor is the present context appropriate for doing this, but I should like to explain the idea a little further informally. It may be that since one is customarily concerned with existence, convergence, finiteness, and so forth, one is not inclined to take seriously the question of the existence of a *better-than-finite* algorithm.

The relative cost, in time or whatever, of the various applications of a particular algorithm is a fairly clear notion, at least as a natural phenomenon. Presumably, the notion can be formalized. Here "algorithm" is used in the strict sense to mean the idealization of some physical machinery which gives a definite output, consisting of cost plus the desired result, for each member of a specified domain of inputs, the individual problems.

The problem-domain of applicability for an algorithm often suggests for itself possible measures of size for the individual problems—for maximum matching, for example, the number of edges or the number of vertices in the graph. Once a measure of problem-size is chosen, we can define $F_A(N)$ to be the least upper bound on the cost of applying algorithm A to problems of size N.

When the measure of problem-size is reasonable and when the sizes assume values arbitrarily large, an asymptotic estimate of $F_A(N)$ (let us call it *the order of difficulty of algorithm A*) is theoretically important. It cannot be rigged by making the algorithm artificially difficult for smaller sizes. It is one criterion showing how good the algorithm is—not merely in comparison with other given algorithms for the same class of problems, but also on the whole how good in comparison with itself. There are, of course, other equally valuable criteria. And in practice this one is rough, one reason being that the size of a problem which would every be considered is bounded.

It is plausible to assume that any algorithm is equivalent, both in the problems to which it applies and in the costs of its applications, to a "normal algorithm" which decomposes into elemental steps of certain prescribed types, so that the costs of the steps of all normal algorithms are comparable. That is, we may use something like Church's thesis in logic. Then, it is possible to ask: Does there or does there not exist an algorithm of given order of difficulty for a given class of problems?

One can find many classes of problems, besides maximum matching and its generalizations, which have algorithms of exponential order but seemingly none better. An example known to organic chemists is that of deciding whether two given graphs are isomorphic. For practical purposes the difference between algebraic and exponential order is often more crucial than the difference between finite and non-finite.

It would be unfortunate for any rigid criterion to inhibit the practical development of algorithms which are either not known or known not to conform nicely to the criterion. Many of the best algorithmic ideas known today would suffer by such theoretical pedantry. In fact, an outstanding open question is, essentially: "how good" is a particular algorithm for linear programming, the simplex method? And, on the other hand, many important algorithmic ideas in electrical switching theory are obviously not "good" in our sense.

However, if only to motivate the search for good, practical algorithms, it is important to realize that it is mathematically sensible even to question their existence. For one thing the task can then be described in terms of concrete conjectures.

Fortunately, in the case of maximum matching the results are positive. But possibly this favourable position is very seldom the case. Perhaps the twoness of edges makes the algebraic order for matching rather special in comparison with the order of difficulty for more general combinatorial extremum problems (cf. **3**).

An upper bound on the order of difficulty of the matching algorithm is n^4, where n is the number of vertices in the graph. The algorithm consists of "growing" a number of trees in the graph—at most n—until they augment or become Hungarian. A tree is grown by branching from a vertex in the tree to an edge-vertex pair not yet in the tree—at most n times. Such a branching may give rise to a back-tracing through at most n edge-vertex pairs in the tree in order to relabel some of them as forming a blossom or an augmenting path. At each of these three levels there may be other labelling work involved—but it is majorized by the work already cited. The work of identifying and labelling the vertex at the other end of some edge to a given vertex need not increase more than linearly with n.

An upper bound on the order of magnitude of memory needed for the algorithm is n^2—the same order of magnitude of memory used to store the graph itself.

3. Alternating paths.

3.0. A *subgraph* of graph G is a graph consisting of a subset of vertices in G and a subset of edges in G under the same incidences which hold for them in G. A non-empty graph G is called *connected* if there is no pair of non-empty subgraphs of G such that each vertex of G and each edge of G is contained in exactly one of the subgraphs. The vertices and edges of any graph partition uniquely into zero or more connected subgraphs, called its *components*.

Maximum, *minimum*, and *odd* will refer to cardinality unless otherwise stated.

3.1. The graph E, *formed* from a set E of edges in G, is the subgraph of G consisting of edges E and their end-points. Any graph H, unless it has a single-vertex component, is formed by its edges. Thus in some contexts it causes no confusion to make no explicit distinction between a graph and its edge-set. In particular, a matching in G may be thought of as a subgraph of G whose components are distinct edges. The sum of two sets D and E is commonly defined as $D + E = (D - E) \cup (E - D)$. The sum $D + E$ of two graphs D and E, formed by edge-sets D and E, is defined to be the graph formed by the edge-set $D + E$.

3.2. There are two other kinds of subtraction for graphs besides the set-theoretic difference used above. With these we must distinguish between a subgraph and the edges which form it. Where G is a graph and E is a set of edges, $G - E$ is the subgraph of G consisting of all the vertices of G and the edges of G not in E. For two graphs G and H, $G - H$ is the subgraph of G consisting of the vertices of G not in H and the edges of G not meeting vertices of H.

Graph $G \cup H$ (graph $G \cap H$) consists of the union (intersection) of the vertex-sets and the edge-sets of graphs G and H, with incidences in $G \cup H$ (graph $G \cap H$) the same as in G and H. We may also take the intersection or union of a graph with a set of edges to get, respectively, a set of edges or a graph. In the latter case the end-points of the edges being adjoined to the graph must be specified. We shall have occasion to give the same edge different end-points in different graphs.

3.3. A *circuit* B in graph G is a connected subgraph in which each vertex of B meets exactly two edges of B. A (simple) *path* P in G is either a single vertex (joining itself to itself) or else a connected subgraph whose two *end-points* each meet one edge, an *end-edge*, of P and whose other vertices each meet two edges of P. A path is said to *join* its end-points.

3.4. For the pair (G, M), where M is a matching in G, a vertex is called *exposed* if it meets no edge of M. Let $\bar{M}$ denote the edges of G not in M. Define an *alternating* path or *alternating* circuit, P, in (G, M) to be such that one edge in $M \cap P$ and one edge in $\bar{M} \cap P$ meets each vertex of P, except the end-points in the case of a path. Several authors, beginning with J. Peterson in 1891, have used alternating paths to prove the existence of "factors" in certain kinds of graphs.

3.5. *For any two matchings M_1 and M_2 in G, the components of the subgraph formed by $M_1 + M_2$ are paths and circuits which are alternating for (G, M_1) and for (G, M_2). Each path end-point is exposed for either M_1 or M_2.*

A vertex of G meets no more than one edge, each, of M_1 and M_2—and thus no more than two edges of $M_1 + M_2$, one in $M_1 \cap \bar{M}_2$ and one in $M_2 \cap \bar{M}_1$. An end-point v of a path in graph $M_1 + M_2$, meeting an end-edge in $M_1 \cap \bar{M}_2$, say, meets no other edge of M_1. Hence, if an edge of M_2 meets v, it does not belong to M_1 and so it *does* belong to $M_1 + M_2$. But then v is not an end-point. Therefore v is exposed for M_2. This completes the proof.

3.6. An alternating path A in (G, M) joining two exposed vertices contains one more edge of $\bar{M}$ than of M. $M + A$ is a matching of G larger than M by one. Such a path is called *augmenting*. Thus matching M is not maximum if (G, M) contains an augmenting path. The converse also holds:

3.7 (Berge, **1**). *A matching M in G is not of maximum cardinality if and only if (G, M) contains an alternating path joining two exposed vertices of M.*

If M_2 is a larger matching than M, some component of graph $M + M_2$ must contain more M_2-edges than M. By 3.5, such a component is an augmenting path for (G, M).

3.8. Berge proposed searching for augmenting paths as an algorithm for maximum matching. In fact, he proposed to trace out an alternating path from an exposed vertex until it must stop and, then, if it is not augmenting, to back up a little and try again, thereby exhausting possibilities.

His idea is an important improvement over the completely naive algorithm. However, depending on what further directions are given, the task can still be one of exponential order, requiring an equally large memory to know when it is done.

Norman and Rabin **(7)** present a similar method for finding in G a minimum *cover-by-edges*, C, a minimum cardinality set of edges in G which meets every vertex in G. The Berge-Norman-Rabin theorem **(2)** is generalized in **(3)**, but a corresponding generalization of the algorithm presented here in Section 4 is unknown.

3.9. Norman and Rabin also show that the maximum matching problem and the minimum cover-by-edges problem are equivalent.

Assuming every vertex meets an edge, the minimum cardinality of a cover of the vertices in G by a set of edges equals the minimum cardinality of a cover of the vertices in G by a set of edges and vertices, where a vertex is regarded as covering itself. By replacing edges by vertices or vice versa, one can go back or forth between a minimum cover by a set of edges and a minimum cover by a set of edges and vertices, where the latter set consists of a maximum matching together with its exposed vertices.

4. Trees and flowers.

4.0. A *tree* may be defined as (1) a graph T every pair of whose vertices is joined by exactly one path in T; (2) inductively, as either a single vertex or else the union of two disjoint trees together with an edge which has one end-point in each; (3) as a connected graph with one more vertex than edges; and so on.

4.1. An *alternating tree* J is a tree each of whose edges joins an *inner* vertex to an *outer* vertex so that each inner vertex of J meets exactly two edges of J. *An alternating tree contains one more outer vertex than inner vertices.* This follows from the third definition of tree by regarding each inner vertex with its two edges as a single edge joining two outer vertices.

4.2. *For each outer vertex v of an alternating tree J there is a unique maximum matching of J which leaves v exposed and the only exposed vertex in J. Every maximum matching of J is one of these.*

Definition (2) of tree can be strengthened to the statement that a tree minus any one of its edges is two trees. Thus J minus any one of its inner vertices, say u, is two alternating trees. One of these, J_1, contains v as an outer vertex. Assume inductively that J_1 can be matched uniquely so only v is exposed and that J_2, the other subtree, can be matched uniquely so only the vertex v_2, joined in J to u by edge e_2, is exposed. Then the union of e_2 and these two matchings is a matching of J which leaves only v exposed. Since every edge of J has one inner and one outer end-point, every maximum matching leaves only an outer vertex exposed.

4.3. A *planted tree*, $J = J(M)$, of G for matching M is an alternating tree in G such that $M \cap J$ is a maximum matching of J and such that the vertex r

in J which is exposed for $M \cap J$ is also exposed for M. That is, all matching edges which meet J are in J. Vertex r is called the *root* of $J(M)$.

In planted tree $J(M)$ every alternating path $P(M)$, which has outer vertex v and the matching edge to v at one of its ends, is a subpath of the alternating path $P_v(M)$ in $J(M)$ which joins v to the root r.

For $k \geqslant 1$, assume that P_{2k-1} is the unique path $P(M)$ which contains $2k - 1$ edges and assume that at its non-v end it has an inner vertex u_k and a matching edge. Then P_{2k}, consisting of P_{2k-1} together with the unique non-matching edge in J which meets u_k, is the unique path $P(M)$ with $2k$ edges. It has outer vertices v_k and v at its two ends. If $v_k \neq r$, then P_{2k+1}, consisting of P_{2k} together with the unique matching edge which meets v_k, is the unique path $P(M)$ with $2k + 1$ edges. It has an inner vertex u_{k+1} and a matching edge at its non-v end. Since our assumption is true for $k = 1$ and since k cannot become infinite, the theorem follows by induction.

We define a *stem* in (G, M) as either an exposed vertex or an alternating path with an exposed vertex at one end and a matching edge at the other end. The exposed vertex and the vertex at the other end are, respectively, the *root* and the *tip* of the stem. The preceding theorem tells us that (1) no trial-and-error search is required to find the path in J from any of its vertices back to the root and (2) the path P_v in J joining any outer vertex v to the root of J is a stem.

4.4. An *augmenting tree*, $J_A = J_A(M)$, in (G, M) is a planted tree $J(M)$ plus an edge e of G such that one end-point of e is an outer vertex v_1 of J and the other end-point v_2 is exposed and not in J. *The path in J_A which joins v_2 to the root of J is an augmenting path.* This follows immediately from (4.3).

4.5. For each vertex b of an odd circuit B there is a unique maximum matching of B which leaves b exposed. A *blossom*, $B = B(M)$, in (G, M) is an odd circuit in G for which $M \cap B$ is a maximum matching in B with say vertex b exposed for $M \cap B$. A *flower*, $F = F(M)$, consists of a blossom and a stem which intersect only at the tip of the stem (the vertex b).

A *flowered tree*, J_F, in (G, M) is a planted tree J plus an edge e of G which joins a pair of outer vertices of J. *The union of e and the two paths which join its outer-vertex end-points to the root of J is a flower, F.*

Let v_1 and v_2 be these outer vertices, and P_1 and P_2 be the paths in J joining them to r. We have seen that P_1 and P_2 are stems (which are easily recovered from J). Since they intersect in at least r and since the path in J joining r to any other vertex is unique, $P_b = P_1 \cap P_2$ is an alternating path with an end at r. If its other end-point, say b, were inner, it would be distinct from r, v_1, and v_2. Thus r would be distinct from v_1 and v_2, and b would meet three different edges of J, one in P_b, one in P_1 not in P_b, and one in P_2 not in P_b. But an inner vertex meets only two edges in the tree. Therefore b is outer and P_b is a stem. Thus $P_1' = P_1 - (P_b - b)$ and $P_2' = P_2 - (P_b - b)$, unless one is a vertex $v_1 = b$ or $v_2 = b$, have non-matching edges at their b-ends and matching edges at their outer ends. It follows in any case that $B = P_1' \cup P_2' \cup e$ is a circuit with only b exposed for $M \cap B$, and thus B is a blossom with p_b as its stem.

4.6. A *Hungarian tree* H in a graph G is an alternating tree whose outer vertices are joined by edges of G only to its inner vertices.

4.7. *For a matching M in a graph G, an exposed vertex is a planted tree. Any planted tree $J(M)$ in G can be extended either to an augmenting tree, or to a flowered tree, or to a Hungarian tree* (merely by looking at most once at each of the edges in G which join vertices of the final tree).

An exposed vertex satisfies the definition of planted tree. Suppose we are given a planted tree J and a set D (perhaps empty) of edges in G which are not in J but which join outer to inner vertices of J. (1) If no outer vertex of J meets an edge not in $D \cup J$, then J is Hungarian. Suppose outer vertex v_1 meets an edge e not in $D \cup J$, whose other end-point is, say, v_2. (2) If v_2 is an inner vertex of J, we can enlarge D by adjoining e. (3) If v_2 is an outer vertex of J, then $e \cup J$ is a flowered tree. (4) If v_2 is exposed and not in J, then $e \cup J$ is an augmenting tree. (5) Finally, if v_2 is not exposed and not in J, then the M-edge e_2 which meets v_2 is not in J, and thus v_3, the other end-point of e_2, is not in J by the definition of planted tree. Therefore, in this case we can extend J to a larger planted tree with new inner vertex v_2 and new outer vertex v_3 by adjoining edges e and e_2. For any J and D, one of the five cases holds. Therefore by looking at any edge in G at most once, we can reach one of the three cases described in the theorem, because the other two cases, (2) and (5), consume edges and G is finite.

4.8. The algorithm which is being constructed is efficient because it does not require tracing many various combinations of the same edges in order to find an augmenting path or to determine that there are none. In fact we accomplish one or the other without ever looking again at the edges encountered in process (4.7), except to pick out from the tree the blossom or the augmenting path when case (3) or (4) occurs. We see from (4.3) and (4.5) how easy it is to retrieve the blossom or the path. When flowers arise we "shrink" the blossoms, and so if an augmenting path arises later, it will be in a "reduced" graph. However, only one other very simple kind of task translates the augmentation to (G, M) itself. That task is to expand a shrunken blossom to an odd circuit and find the maximum matching of the odd circuit which leaves a certain vertex exposed. Actually, we shall find in (7.3) that it is desirable to leave odd circuits shrunk while looking in the reduced graph for as many successive augmentations as possible since they are all reflected in augmentations of (G, M).

4.9. For H, a subgraph of G, G is the disjoint union $(G - H) \cup \delta H \cup H^+$, where δH is the set of the edges with one end-point in H and one end-point in

$G - H$, and where $H^+ = G - (G - H)$ is the subgraph consisting of H and all edges of G with both end-points in H. When H is connected, *shrinking H* means constructing the new graph $G/H = (G - H) \cup \delta H \cup h$ by regarding H^+ as a single new vertex, $h = H/H$, which meets the edges $\delta H = \delta h$. The end-points in $G - H$ of the edges δH do not change.

4.10. If B is an odd circuit in G, then $b' = B/B$ is called a *pseudovertex* of G/B. To *expand* b' means to recover G from G/B. The algorithm, after it expands a pseudovertex b', will make use of the circuit B. In general, finding a "Hamiltonian" circuit in a graph B^+ is difficult. Therefore, when the algorithm shrinks B to form G/B, it should remember circuit B as having effected the shrinking. Thus we call circuit B in G (rather than B^+) the *expansion of* b'. In formal calculation shrinking B in G is an easy operation. Essentially, just assign all the vertices and edges of B a label, b', and then, until b' is expanded by erasing these labels, ignore any distinction between vertices labelled b' and ignore edges joining them to each other.

Where M is a matching set of edges in G, M/B is defined as $M \cap (G/B)$. Clearly, if B is a blossom for (G, M), then M/B is a matching of G/B.

4.11. Let $G_0 = G$, $G_i = G_{i-1}/B_i$, and $b_i = B_i/B_i$ for $i = 1, \ldots, n$, where B_i is an odd circuit in graph G_{i-1}. We inductively define the *pseudovertices* (with respect to G) of G_k $(k = 1, \ldots, n)$ to be b_k together with the pseudovertices in $G_{k-1} - B_k = G_k - b_k$. Of course not every b_i, $i \leqslant k$, will be a pseudovertex of G_k because some will have been absorbed into others. The order in which the pseudovertices of a G_k arise is immaterial. That is, the order in which the odd circuits B_i are shrunk is immaterial except in so far as one shrunken B_i is a vertex in another B_i. Thus we can expand any pseudovertex b_j of G_k to obtain a graph G_{kj} for which $G_{kj}/B_j = G_k$. The pseudovertices of G_{kj} (with respect to G) are the pseudovertices in B_j together with the pseudovertices in $G_k - b_j$; that is, graph G_{kj} can be obtained from G by shrinking in a proper order the odd circuits which were absorbed into these pseudovertices. On the other hand, we do not expand a vertex b_h in B_k until vertex b_k is expanded.

There is a partial order on the b_i's defined by the transitive completion of the relation $b_h < b_k$ where b_h is a vertex of B_k. (It is a special kind of partial ordering because each b_h is a vertex of at most one B_k.) There is a partial order on the sets, $S_\alpha, S_\beta, \ldots$, of mutually incomparable b_i's, where $S_\alpha \leqslant S_\beta$ when every member of S_α is less than or equal to some member of S_β. Evidently there is a unique family of graphs, $G_\alpha, G_\beta, \ldots$, which include the G_i's and G. They correspond 1–1 to the sets, $S_\alpha, S_\beta, \ldots$, so that the pseudovertices of G_α are S_α, etc. We have $S_\alpha \leqslant S_\beta$ if and only if G_β can be obtained from G_α by shrinking certain B_i, those for which b_i is less than or equal to some member of S_β and not less than or equal to any member of S_α. Graph G corresponds to the empty set and G_n corresponds to the set of b_i's which are maximal with respect to their partial order.

The *complete expansion* of a pseudovertex b_i is the subgraph

$$U^+ = G - (G - U) \subset G$$

where U consists of all vertices of G absorbed into b_i by shrinking.

4.12. *Where B is the blossom of a flower F for (G, M), M is a maximum matching of G if and only if M/B is a maximum matching of G/B.*

4.13. *Where blossom B is in J_F, a planted flowered tree for (G, M), J_F/B is a planted tree for $(G/B, M/B)$. It contains B/B as an outer vertex. Its other outer and inner vertices are respectively those of J_F which are not in B.*

Theorem (4.13) follows easily from (4.5). We separate the two converse statements of Theorem (4.12) into slightly stronger statements, (4.14) and (4.15).

4.14. *Where B is any odd circuit in G, for every matching M_1 of G/B there exists a maximum matching M_B of B such that $M = M_1 \cup M_B$ is a matching for G.*

Since any matching M_1 of G/B contains at most one edge meeting B/B, the edges M_1 in G meet at most one vertex, say b_1, of B. Therefore the desired M_B is the maximum matching of B which leaves b_1 exposed. Since the cardinality $|M_B|$ of M_B is constant, any augmentation of M_1 yields a corresponding augmentation of M. Therefore, the "only if" part of (4.12) is proved.

Applying the above matching operation to successive expansions of pseudovertices into odd circuits we have:

Where P is the complete expansion of a pseudovertex p in G_2, where G_1 is the graph obtained from G_2 by completely expanding p, and where M_2 is any matching of G_2, there exists a matching M_P of P leaving exactly one exposed vertex in P such that $M_P \cup M_2$ is a matching of G_1. Thus since $|M_P|$ is constant, any augmentation in G_2 yields a corresponding augmentation in G_1.

4.15. *For (G, M), let P be a subgraph such that* (1) *$M \cap P$ leaves exactly one exposed vertex in P,* (2) *M/P is a maximum matching of G/P, and* (3) *$p = P/P$ is the tip of a stem S_p for $(G/P, M/P)$. Then M is a maximum matching of G.*

The edges of S_p form in G a stem, S, for (G, M). (In case S_p has no edges, take S to be the vertex in P exposed for M.) Compare $M' = M + S$ and M'/P with M and M/P. The definition of stem implies that M' is a matching of G with $|M'| = |M|$ and that the exposure of the root of S is changed to the exposure of the tip of S. Similarly $M'/P = M/P + S_p$ is a matching of G/P with $|M'/P| = |M/P|$ and with vertex p exposed. Because the cardinalities do not change, it is sufficient to show that M' is maximum in G if M'/P is maximum in G/P.

Using (3.7), if M' is not maximum, G contains an augmenting path $A = A(M')$. If A contains no vertices of P, then it is also an augmenting path for M'/P in G/P. Otherwise, because P contains only one exposed vertex for M', at least one of the ends of A is at an exposed vertex u_1 not in P. There is a unique subpath A_1 of A with one end-point at u_1 and containing only one vertex p_1 of P, at its other end. The only difference between A_1 and $A_1/P = (A_1 \cup P)/P$ is that p_1 is replaced by p, which is exposed for M'/P. Thus A_1/P is an augmenting path for M'/P and so M'/P is not maximum. The theorem is proved.

The theorem extends as follows:

For (G, M), let $P_1, \ldots, P_n$ be a family of disjoint subgraphs in G such that (1) $M \cap P_i$ leaves exactly one exposed vertex in P_i, (2) $M_n = M \cap G_n$ is a maximum matching of $G_n = G/P_1/\ldots/P_n$, and (3) vertices P_i/P_i of G_n are outer vertices in a planted tree J_n for (G_n, M_n). Then M is a maximum matching of G.

We may assume that the indices order the P_i/P_i's so that (for $k = 1, \ldots, n-1$) those from 1 through k are contained in a planted subtree J_k of J_n not containing those from $k+1$ through n. Hence the theorem follows by induction after proving that $M_{n-1} = M \cap G_{n-1}$ is a maximum matching of $G_{n-1} = G/P_1/\ldots/P_{n-1}$. Since every outer vertex of J_n is the tip of a stem in G_n, this follows from the last theorem.

4.16. Theorems (4.7) and (4.13) show how by branching a planted tree out from an exposed vertex of (G, M) and shrinking blossoms B_i when they are encountered, we eventually obtain in a graph $G_k = G/B_1/\ldots/B_k$ either a tree with an augmenting path or a Hungarian tree. An augmenting path admits an augmentation of matching $M_k = M \cap G_k$ according to (3.7), and (4.14) shows how this induces an augmentation of matching $M_{k-1} = M \cap G_{k-1}$ and so on back through M. On the other hand, when a Hungarian tree J is obtained, submatching $(J \cup B_k \cup \ldots \cup B_1) \cap M$ of (G, M) cannot be improved and so this part of G is freed from further consideration. This follows immediately from (4.15) and the next theorem, (4.17), where G_k is denoted simply as G.

4.17. *Let J be a Hungarian tree in a graph G. A matching M_1 of $G - J$ is maximum in $G - J$ if and only if M_1 together with any maximum matching M_J of J is a maximum matching of G.*

Since J and $G - J$ are disjoint, if there exists a matching M_1' of $G - J$ which is larger than M_1, then $M_1' \cup M_J$ is a larger matching of G than $M_1 \cup M_J$. Conversely, suppose M_1 is maximum for $G - J$. Let

$$M' = M_1' \cup M_I \cup M_J'$$

be an arbitrary matching of G where $M_1' \subset G - J$, where $M_J' \subset J$, and where $M_I \cap ((G - J) \cup J)$ is empty. Then $|M_1'| \leqslant |M_1|$. Every edge in M_I meets at least one inner vertex of J; that is, where $I' \subset I(J)$ is the set of the inner vertices met by M_I, $|M_I| \leqslant |I'|$. The graph $J - I'$ consists of $|I'| + 1$

disjoint alternating trees whose inner vertices together are $I(J) - I'$. Therefore, since the maximum matching cardinality of an alternating tree equals the number of its inner vertices, $|M_J'| \leqslant |I(J) - I'|$. Adding the three inequalities gives $|M'| \leqslant |M_1| + |I(J)| = |M_1 \cup M_J|$. So the theorem is proved.

4.18. The matching M of $G = G^0$, to begin with, may be empty. If it leaves any exposed vertices, then the process (4.16) operates with respect to one of them. Either it produces an augmentation of M by one edge, thus disposing of two exposed vertices, or it reduces the possible domain for augmenting M to a subgraph $G^1 = G_k - J$ of G, containing one less exposed vertex and containing only edges and vertices not previously considered. Successive application of (4.16) may reduce the consideration of M to a subgraph G^i of G and reveal there an augmentation of M. After augmenting in G^i, obtaining a larger M for G with two less exposed vertices in G^i, (4.16) operates again in G^i, never returning to the matching in the rest of G.

4.19. Repeated application of (4.18) reduces the domain in question to a G^n containing no exposed vertices. Then we know that we have a maximum matching; let us still call it M, with n exposed vertices in G. Thus the construction of an algorithm for finding a maximum cardinality matching in a graph is complete. Often the last application of (4.18) is unnecessary. For verifying maximality, the algorithm may as well stop when it reduces the domain to a G^{n-1} containing one exposed vertex, since two exposed vertices are necessary in order to augment. However, for theoretical purposes it is convenient to have the algorithm grow a tree from each exposed vertex of the final, maximum matching.

4.20. We may define an *alternating forest* to be a family of disjoint alternating trees and a *planted forest* in (G, M) to be a family of disjoint planted trees in (G, M). A *dense* planted forest is one which contains all the exposed vertices of (G, M). The family of exposed vertices, itself, is a dense planted forest. The algorithm works as well by growing a dense planted forest all at once, rather than one tree at a time.

It is appropriate then to define *augmenting forest* (*flowered forest*) to be a planted forest plus an edge e of G whose end-points are outer vertices of different trees (of the same tree) of the planted forest.

A *Hungarian forest* in G is defined similarly to Hungarian tree, replacing the word "tree" by "forest." Notice that the trees of a Hungarian forest are not necessarily Hungarian trees—an outer vertex of one tree may be joined by an edge of G to an inner vertex of another tree in the forest.

The theorems on trees presented in this section are essentially the same for forests.

5. The dual to matching.

5.0. A *bipartite* graph K is one in which every circuit contains an even number of edges. This condition, that K contains no odd circuits, is equivalent to being able to partition the vertices of K into two parts so that each edge of K meets exactly one vertex in each part. The well-known König theorem states:

For a bipartite graph K, the maximum cardinality of a matching in K equals the minimum number of vertices which together meet all the edges of K.

5.1. The linear programming duality theorem states: *If*
(1) $x \geqslant 0$, $Ax \leqslant c$ *and*
(2) $y \geqslant 0$, $A^T y \geqslant b$,
for given real vectors b and c and real matrix A, then for real vectors x and y,

$$\max_x(b, x) = \min_y(c, y)$$

when such extrema exist.

The problems of finding a maximizing vector x and a minimizing vector y are called *linear programmes*, *dual* to each other.

5.2. The König theorem is now widely recognized as the instance of (5.1) where b and c consist of all ones and $A = A_K$ is the zero-one incidence matrix of edges (columns) versus vertices (rows) in a bipartite graph K. In view of Theorem (5.1) the König theorem is equivalent to the remarkable fact that, with b, c, and A as just described, the two linear programmes of (5.1) have solutions x and y whose components are zeros and ones whether or not this condition is imposed. An elegant theory centres on this phenomenon.

Graph-theoretic algorithms are well known for so-called assignment, transportation, and network flow problems **(5)**. These are linear programmes which have constraint matrices A that are essentially A_K.

5.3. For a linear programme with an arbitrary matrix A of integers, or even of zeros and ones, we cannot say that the extreme values will be assumed, as when $A = A_K$, by vectors with integer components. Therefore, in general when we impose the condition of integrality on x, the equality of the two extrema no longer holds.

In particular, when the maximum matching problem is extended from bipartite to general graphs G, a genuine integrality difficulty is introduced. Our matching algorithm met it by the device of shrinking blossoms.

5.4. The matching algorithm yields a generalization of the König theorem to maximum matchings in G. The new matching duality theorem, in the form "maximum cardinality of a matching in G equals minimum of something else," is also an instance of linear programming duality.

It is reasonable to hope for a theorem of this kind because any problem which involves maximizing a linear form by one of a discrete set of non-negative vectors has associated with it a dual problem in the following sense. The discrete set of vectors has a convex hull which is the intersection of a discrete set of half-spaces. The value of the linear form is as large for some vector of the discrete set as it is for any other vector in the convex hull. Therefore, the discrete problem is equivalent to an ordinary linear programme whose constraints, together with non-negativity, are given by the half-spaces. The dual (more precisely, a dual) of the discrete problem is the dual of this ordinary linear programme.

For a class of discrete problems, formulated in a natural way, one may hope then that equivalent linear constraints are pleasant even though they are not explicit in the discrete formulation.

5.5. Arising from the definition of a matching—no more than one matching edge to each vertex—are the obvious linear constraints that for each vertex $v \in G$ the sum of the x's corresponding to edges which meet v is less than one. To obtain a maximum cardinality matching, we want to maximize the sum of all the x's, corresponding to edges of G, subject to the additional condition that each x is zero or one.

It turns out that maximum matching can be turned into linear programming by substituting for the zero-one condition the additional constraints that the x's are non-negative and that for any set R of $2k + 1$ vertices in $G(k = 1, 2, \ldots)$ the sum of the x's which correspond to edges with both end-points in R is no greater than k. The former condition on the x's obviously implies the latter since for no matching in G do more than k matching edges have both ends in R.

The converse—that subject only to the linear constraints, $\sum x_i$ can be maximized by zeros and ones—is not so obvious, but in view of (5.1) it follows from (5.6), the generalized König theorem.

Actually the stronger converse holds—that subject only to these same linear constraints, $\sum c_i x_i$, for any real numbers c_i, can be maximized by zeros and ones. In other words, the polyhedron described by the constraints is, indeed, the convex hull of the zero-one vectors which correspond to matchings in G. We shall not prove this until we take up maximum weight-sum matching in paper **(4)**. Although the convex-hull notion suggested trying to generalize the König theorem, and although the generalization found does suggest the true convex hull, the success of the first suggestion does not necessarily validate the second.

5.6. A set consisting of one vertex in G is said to *cover* an edge e in G if e meets the vertex. The *capacity* of this set is one. A set consisting of $2k + 1$ vertices in $G(k = 1, 2, \ldots)$ is said to *cover* an edge e in G if both end-points of e are in the set. The *capacity* of this set is k. An *odd-set cover* of a graph G is a family of odd sets of vertices such that each edge in G is covered by a member of the family.

MATCHING-DUALITY THEOREM. *The maximum cardinality of a matching in G equals the minimum capacity-sum of an odd-set cover in G.*

It is obvious that the capacity-sum of any odd-set cover in G is at least as large as the cardinality of any matching in G, so we have only to prove the existence in G of an odd-set cover and a matching for which the numbers are equal.

5.7. The theorem holds for a graph which has a perfect matching M—that is, with no exposed vertices—since the odd-set cover consisting of two sets, one set containing one of the vertices and the other set containing all the other vertices, has capacity-sum equal to $|M|$. It also holds for a graph which has a matching with one exposed vertex. Here the odd-set cover may be taken as consisting of one member, the set containing all vertices of the graph. For the case of one exposed vertex, an odd-set cover may also be constructed as in (5.8) by applying the algorithm to construct a Hungarian tree even though it obviously will not result in augmentation.

5.8. Applying the algorithm to (G, M), where $|M|$ is maximum, using some exposed vertex as root, we obtain a graph G' containing a maximally matched Hungarian tree J, a number of whose outer vertices are pseudo. Let S_J consist of all odd sets of the following two types: sets each consisting of one inner vertex in J, and sets each consisting of the vertices in the complete expansion of one pseudovertex of J.

The number of edges of M which a member of S_J covers is *equal* to the capacity of the member. Every edge of M not in $G' - J$ is covered by *exactly* one member of S_J. An edge of G is covered by a member of S_J if and only if it is not in $G' - J$.

Matching $M \cap (G' - J)$ is a maximum matching of $G' - J$ with one less exposed vertex than (G, M). Assuming that $|M \cap (G' - J)|$ equals the capacity of an odd-set cover, say S'_J, of $G' - J$, we have that $|M|$ equals the capacity of $S_J \cup S'_J$, an odd set cover of G. Theorem (5.6) follows by induction on the number of exposed vertices.

5.9. It is evident from the proof that we may require the minimum odd-set cover to have certain other structure—in particular, that each member with more than one vertex contain the vertices of at least one odd circuit in G. With the latter restriction the theorem becomes a strict generalization of the König theorem.

6. Invariance of the dual.

6.0. For any particular application of the algorithm (4) to G, yielding, say, the maximum matching M, we may skip the augmentation steps in (4.16) by regarding the augmented matching as being the one already at hand. This gives a particular application of (4) to G starting with maximum matching M. In the application of the algorithm to (G, M), we can regard all the branchings and blossom shrinkings as taking place without subtracting the trees J_i as they arise. Thus we obtain from (G, M) a graph G^* with a number of pseudovertices which are outer vertices in a sequence $\{J_i\}$ $(i = 1, \ldots, n)$ of disjoint planted trees in G^*, one corresponding to each exposed vertex of (G, M). By expanding all the pseudovertices of G^* completely, we recover the graph G.

6.1. The tree J_i is Hungarian in $G^* - J_1 \ldots - J_{i-1}$, but usually not Hungarian in G^* because an outer vertex of J_i might be joined to an inner vertex of any other tree with a lower index. Hence the partition of the outer and inner vertices into trees J_i depends on the order of their construction. Also non-matching edges which can occur in each tree are not unique. In general, joining outer to inner vertices of a J_i are many other $\bar{M}$ edges which would do as well. The particular blossoms which led to the pseudovertices are also fairly arbitrary. And, finally, the maximum matching is far from unique. However, (6.2) will show that the graph G^* is uniquely determined by G alone.

6.2. *For a* (G, M) *where* M *is any maximum matching, let* G^* *and* $\{J_i\}$ *be obtained from* (G, M) *by* (6.0).

(*a*) *The non-pseudo outer vertices of the* J_i*'s and the vertices of the pseudovertex complete expansions, all called the outer vertices* $O(G)$ *of* G*, are precisely the vertices of* G *which are left exposed by some maximum matching of* G.

(*b*) *The inner vertices of the* J_i*'s, called the inner vertices* $I(G)$ *of* G*, are precisely those vertices of* G *not in* $O(G)$ *but joined to vertices in* $O(G)$.

(*c*) G^* *is obtained from* G *by shrinking the connected components of* $O(G)^+$*, the subgraph of* G *consisting of vertices* $O(G)$ *and all edges of* G *joining them.*

6.3. We have defined vertex families $O(G)$ and $I(G)$ in terms of particular J_i. The theorem yields definitions dependent only on G itself.

Clearly $O(G^*)$ and $I(G^*)$, defined in terms of the J_i in G^*, are respectively the outer and the inner vertices of the J_i. Notice that the early definitions of inner and outer, for vertices in an alternating tree, are consistent with the definitions for a general graph.

6.4. *Proof of* (6.2), (b) *and* (c). Let the vertex v^* of G^* be joined in G^* to some outer vertex u^* of J_i. Then v^* is a vertex in some $J_h (h \leqslant i)$, since J_i is Hungarian in $G^* - J_1 - \ldots - J_{i-1}$. But v^* cannot be an outer vertex of J_h since u^* is not inner and since J_h is Hungarian in $G^* - J_1 - \ldots - J_{h-1}$. Therefore v^* is inner. It follows that each outer vertex u of G is joined only to inner vertices and to other vertices in the complete expansion of its image u^*. By construction, each inner vertex is joined to an outer vertex of G. Hence, (b) is true.

Since by construction the complete expansion of each outer vertex of G^* is connected, it also follows that the connected components of $O(G)^+$ correspond precisely to outer vertices of G^*. Hence, (c) is true.

6.5. An outer vertex u of G, by definition, either is identical with or is contained in the complete expansion of some outer vertex u^* of, say, J_i. For any maximum matching M_i of alternating tree J_i, $M_i \cup [M \cap (G^* - J_i)]$ is a maximum matching of G^*, which by (4.14) induces a maximum matching M' of G. Let M_i be the one which leaves u^* exposed. If u^* is pseudo, then by (4.14) M' can be chosen so that u is exposed in the expansion. This proves half of (6.2), (a).

6.6. C. Witzgall suggested the following simplified proof of the converse, viz. that only the outer vertices are ever exposed for a maximum matching. A non-outer vertex v meets an edge e of M. Deleting v and its adjoining edges, $\bigcup J_i - v$ is a Hungarian forest in $G^* - v$.

If v is inner, then the forest is dense in $G^* - v$. Otherwise it is dense in $G^* - v$ except for one exposed vertex, the other end of e. In either case it follows that $M - e$ is a maximum matching of $G - v$.

Assume that M' is a maximum matching of G which leaves v exposed. Then M' is also a matching of $G - v$. Since M' is larger than $M - e$, we have a contradiction. This completes the proof of (6.2).

6.7. The definition of odd-set cover may be expanded (more than necessary for Theorem (5.6)) to include the possibility of members which are even sets of vertices in G. A set of $2k$-vertices has capacity k and covers the edges which have both end-points in the set. Then, clearly, Theorem (5.6) still holds for this kind of cover.

With this definition of cover, it follows from the uniqueness of G^* that there is a unique *preferred* minimum cover, S^*, for any graph G. The one-vertex members of S^* are the inner vertices of G^*, the other odd members of S^* correspond to the pseudovertices of G^*, and the one even member of S^* consists of the non-inner, non-outer vertices of G^*.

7. Refinement of the algorithm.

7.0. Several possibilities for refining the algorithm suggest themselves.

We could remember an old tree, uprooted by an augmentation, so that when a new rooted tree takes on a vertex in it, we can immediately adjoin a piece of it to the new tree. This appears not worth doing. A tree is easy to grow, easier than selecting from an old tree the piece which may be grafted.

7.1. A quite useful refinement is to leave the pseudovertices of the old tree shrunk until their expansion is necessary. We see from (4.14) that any further augmentation of a matching M' in a graph G' with pseudovertices yields a further augmentation in G just as easily as the first. On the other hand, a maximum matching in G', reached after one or more augmentations, does not necessarily yield a maximum matching of G. The sufficiency part of (4.12) depends on the blossom being part of a flower, whereas the first augmentation in G' uproots the stem.

7.2. However, we may easily observe the circumstance arising in the application of the algorithm to (G', M') where the shrinkage might hide a possible augmentation in G. It is where a pseudovertex, say b', becomes an inner vertex of the planted tree, say $J' = J'(M')$.

In this case, we obtain a graph G'' from G' by expanding b' to an odd circuit B. The edges of J' form in G'' a subgraph which we still call J'. The set M' is also a matching in G''. One edge of $J' \cap M'$ has an end-point, say b_1, in B. One edge of $J' \cap \bar{M}'$ has an end-point, say b_2, in B. The maximum matching M_B of B which is compatible with M' in G'' leaves b_1 exposed. The vertices b_1 and b_2 partition B into two paths, P_2 even and P_1 odd, which join b_1 and b_2. The graph $J'' = P_2 \cup J'$ is a planted tree in G'' for the matching $M_B \cup M'$.

Unless b_1 and b_2 coincide, P_2 will contain outer vertices of J''. These may be joined to vertices not in J'' which admit an extension of J'', not possible for $J''/B = J' \subset G'$, to a planted tree with an augmenting path.

7.3. If $J' \subset G'$ can be extended in G' to a tree with an augmenting path, it does not matter that some of the inner vertices are pseudo because a further augmentation for G is thus determined. If J' with pseudo inner vertex b' can be extended in (G', M') to a flowered tree whose blossom B' contains b', then b' loses its distinction as an inner vertex. It might as well stay shrunk and be absorbed into the new pseudovertex B'/B' of G'/B'. In fact, Theorems (4.15) and (4.17), together, tell us that any pseudo outer vertex might as well be left pseudo during the algorithm.

Therefore a pseudo inner vertex should be retained until a planted Hungarian tree J_H is obtained. If no inner vertices of J_H are pseudo, then (4.17) is applicable. Otherwise, at this point, a pseudo inner vertex should be expanded according to (7.2).

7.4. One of the main operations of the algorithm is described in (4.3). That is back-tracing along paths in a tree already constructed, either to obtain an augmentation as in (4.4) or to delineate a new blossom as in (4.5). The back-tracing takes place in an alternating tree only because blossoms have been shrunk to pseudovertices. A pseudovertex may be compounded from many earlier blossom shrinkings and may thus encompass a complicated subgraph of G. After shrinking, back-tracing entirely bypasses the internal structure of a pseudovertex.

A possible alternative to actually shrinking is some method for tracing through the internal structure of a pseudovertex. Witzgall and Zahn (**9**) have designed a variation of the algorithm which does that. Their result is attractive and deceptively non-trivial.

References

1. C. Berge, *Two theorems in graph theory,* Proc. Natl. Acad. Sci. U.S., *43* (1957), 842–4.

2. ——— *The theory of graphs and its applications* (London, 1962).

3. J. Edmonds, *Covers and packings in a family of sets,* Bull. Amer. Math. Soc., *68* (1962), 494–9.

4. ——— *Maximum matching and a polyhedron with* (0, 1) *vertices*, appearing in J. Res. Natl. Bureau Standards *69*B (1965).
5. L. R. Ford, Jr. and D. R. Fulkerson, *Flows in networks* (Princeton, 1962).
6. A. J. Hoffman, *Some recent applications of the theory of linear inequalities to extremal combinatorial analysis*, Proc. Symp. on Appl. Math., *10* (1960), 113–27.
7. R. Z. Norman and M. O. Rabin, *An algorithm for a minimum cover of a graph*, Proc. Amer. Math. Soc., *10* (1959), 315–19.
8. W. T. Tutte, *The factorization of linear graphs*, J. London Math. Soc., *22* (1947), 107–11.
9. C. Witzgall and C. T. Zahn, Jr., *Modification of Edmonds' algorithm for maximum matching of graphs*, appearing in J. Res. Natl. Bureau Standards *69*B (1965).

National Bureau of Standards and
Princeton University

Reprinted from
Canad. J. Math. **17** (1965), 449–467

A THEOREM OF FINITE SETS

by

G. KATONA

Mathematical Institute of the Hungarian Academy of Sciences
Budapest, Hungary

§ 1. Introduction

Let $A_1, \ldots, A_n$ be a system of different subsets of a finite set H, where $|H| = h$ and $|A_i| = l$ $(1 \leq i \leq n)$ ($|A|$ denotes the number of elements of A). We ask for a system $A_1, \ldots, A_n$ (for given h, l, n) for which the number of sets B satisfying $|B| = l - 1$ and $B \subset A_i$ for some i is minimum. The first lower estimation for this minimum is given by SPERNER ([1], Hilfssatz). His estimation is $\dfrac{n \cdot l}{h - l + 1}$. This depends on h. However, if $n = \binom{N}{l}$, it is expected that the minimizing system is the system of all l-tuples chosen from a subset of N elements of H. In this case the number of B's is $\binom{N}{l-1}$ which does not depend on h. A. HAJNAL proved this statement in the case of $l = 3$ (unpublished). In this paper I prove for all cases that this is, indeed, the minimum, and find the (more complicated) minimum also for arbitrary n. The theorem is probably useful in proofs by induction over the maximal number of elements of the subsets in a system, as was SPERNER's lemma in his paper [1].

KLEITMAN told me in Tihany (Hungary) that he thought I could solve the following problem of ERDŐS by the aid of the above theorem and the "marriage problem": Let $A_1, \ldots, A_n$ be subsets of H, where $|H| = 2h$ and $|A_i| = h$. For what n's is it always possible to construct a system $B_1, \ldots, B_n$ with the properties $B_i \subset A_i$, $|B_i| = h - 1$ $(1 \leq i \leq n)$. § 3 contains the solution of this problem in a more general form.

§ 2. The main result

Before the exact formulation of the theorem we need the following simple but interesting

LEMMA 1. *If n and l are natural numbers, we can write the number n uniquely in the form*

$$(1) \qquad n = \binom{a_l(n, l)}{l} + \binom{a_{l-1}(n, l)}{l-1} + \ldots + \binom{a_{t(n,l)}(n, l)}{t(n, l)},$$

where $t(n, l) \geq 1$, $a_l > a_{l-1} > \ldots > a_{t(n,l)}$ are natural numbers and $a_i(n, l) \geq$ $\geq i$ $(i = t(n, l), t(n, l) + 1, \ldots, l)$.

187

PROOF. The existence of form (1) is proved by induction over l. For $l = 1$ the statement is trivial. Assume that for $l = k - 1$ it is true also and prove for $l = k$. Let a_k be the maximal integer satisfying the inequality $\binom{a_k}{k} \leq n$. If here equality holds, we are ready. If it does not, using the induction hypothesis we have for the number $n - \binom{a_k}{k}$ the following expression:

$$n - \binom{a_k}{k} = \binom{a_{k-1}}{k-1} + \cdots + \binom{a_t}{t}, \tag{2}$$

where $t \geq 1$, $a_{k-1} > \ldots > a_t$, $a_i \geq i$ $(i = t, t+1, \ldots, k-1)$. (2) gives an expression for n, we have to verify only $a_k > a_{k-1}$ and $a_k \geq k$. If $a_k \leq \leq a_{k-1}$ held, then

$$n \geq \binom{a_k}{k} + \binom{a_{k-1}}{k-1} \geq \binom{a_k}{k} + \binom{a_k}{k-1} = \binom{a_k+1}{k}$$

would hold also, which contradicts choosing of a_k. On the other hand, $a_k \geq k$ follows from $a_k > a_{k-1}$ and $a_{k-1} \geq k - 1$.

The unicity of Form (1) is proved also by induction over l. For $l = 1$ the statement is trivial. Assume that for $l = k - 1$ it is also true and prove for $l = k$. If, on the contrary, there exist two forms:

$$n = \binom{a_k}{k} + \binom{a_{k-1}}{k-1} + \cdots + \binom{a_t}{t} = \binom{a'_k}{k} + \binom{a'_{k-1}}{k-1} + \cdots + \binom{a'_r}{r}, \tag{3}$$

we may separate two different cases. If $a_k = a'_k$, we can obtain two different forms of $n - \binom{a_k}{k}$, which contradict our induction hypothesis. If $a_k < a'_k$, the contradiction follows from

$$n \leq \binom{a_k}{k} + \binom{a_k - 1}{k-1} + \cdots + \binom{a_k - k + 1}{1} = \binom{a_k+1}{k} - 1 < \binom{a_k+1}{k} \leq$$

$$\leq \binom{a'_k}{k} \leq \binom{a'_k}{k} + \cdots + \binom{a'_r}{r}.$$

Thus we proved the lemma.

In the future we will use the following two notations:

$$E_l(n) = \binom{a_l(n,l) - 1}{l-1} + \binom{a_{l-1}(n,l) - 1}{l-2} + \cdots + \binom{a_{t(n,l)}(n,l) - 1}{t(n,l) - 1}$$

and

$$F_l(n) = \binom{a_l(n,l)}{l-1} + \binom{a_{l-1}(n,l)}{l-2} + \cdots + \binom{a_{t(n,l)}(n,l)}{t(n,l) - 1}.$$

These numbers are uniquely determined by Lemma 1.

Let us consider now the problem. Let H be a finite set with h elements, and

$$\mathcal{A} = \{A_1, \dots, A_n\}$$

a system of different subsets of H, where the number of elements of A_i is

$$|A_i| = l \qquad (1 \leq i \leq n).$$

Obviously, l is a fixed integer between 1 and h. Let $c(\mathcal{A})$ denote the following system

$$c(\mathcal{A}) = \{B : |B| = l-1 \text{ and } B \subset A_j \text{ for at least one } j\}.$$

The problem is to determine the minimum of $|c(\mathcal{A})|$, if h, n and l are given. Theorem 1 gives the exact solution of this problem.

THEOREM 1. *Let h, n and l be given integers with the properties*

$$h \geq 1, \qquad 1 \leq l \leq h \quad \textit{and} \quad 1 \leq n \leq \binom{h}{l}.$$

If H is a set of h elements, and

$$\mathcal{A} = \{A_1, \dots, A_n\}, \quad |A_i| = l \qquad (i = 1, \dots, n)$$

a system of different subsets of H, then

$$\min |c(\mathcal{A})| = F_l(n),$$

where the minimum runs over all such systems $\mathcal{A}$.

REMARK. It is interesting, that $\min|c(\mathcal{A})|$ does not depend on h. For example, SPERNER's estimation [1]:

$$c(\mathcal{A}) \geq \frac{n \cdot l}{h - l + 1}$$

depends on h.

Before the proof we shall give another theorem. We will prove them together.

THEOREM 2. *Let h, n and l be given integers with the properties*

$$h \geq 1, \qquad 1 \leq l \leq h \quad \text{and} \quad \binom{h}{l} \leq n \leq 2\binom{h}{l}.$$

Further G and H are disjoint sets of h elements. If

$$\mathcal{A} = \{A_1, \dots, A_n\}$$

is a system of A_i 's, where

$$A_i \subset G \quad \text{or} \quad A_i \subset H \qquad (1 \leq i \leq n)$$

and

$$|A_i| = l \qquad (1 \leq i \leq n),$$

then

$$\min |c(\mathcal{A})| = \binom{h}{l-1} + F_l\left(n - \binom{h}{l}\right).$$

PROOF. 1. First we construct the minimizing system of Theorem 1. Denote this system by $\mathcal{M}(h, n, l)$. Obviously, it is sufficient to construct the system $\mathcal{M}(a_l^*(n), n, l)$, where $a_l^*(n)$ is the least integer satisfying

$$\binom{a_l^*(n)}{l} \geq n .$$

The construction will be carried out by induction over l. If $l = 1$, $a_1^*(n) = n$ and $\mathcal{M}(a_1^*(n), n, 1)$ consists of all the sets of one element. Assume we constructed already the system $\mathcal{M}(a_{l-1}^*(n), n, l-1)$ for all n. Construct now $\mathcal{M}(a_l^*(n), n, l)$. If $n = \binom{a_l(n, l)}{l}$, then the minimizing system consists of all the subsets having l elements. If $n > \binom{a_l(n, l)}{l}$, let H be a set of $a_l^*(n) = a_l(n, l)+1$ elements, and e an element of H. Since $a_l > a_{l-1}$, we can construct the system $\mathcal{M}\left(a_l(n, l), n - \binom{a_l(n, l)}{l}, l-1\right)$ on $H - \{e\}$ by the induction hypothesis. Define the system $\mathcal{N}$ in the following manner:

$$\mathcal{N} = \left\{N \cup \{e\} : N \in \mathcal{M}\left(a_l(n, l), n - \binom{a_l(n, l)}{l}, l-1\right)\right\}.$$

If $\mathcal{P}$ denotes the system of all subsets of $H - \{e\}$, having l elements, then $\mathcal{P}$ and $\mathcal{N}$ form together the system $\mathcal{M}(a_l^*(n), n, l)$. Indeed, the number of sets is $\binom{a_l(n, l)}{l} + n - \binom{a_l(n, l)}{l} = n$ and we have only to verify

$$|c(\mathcal{M}(a_l^*(n), n, l))| = \binom{a_l(n, l)}{l-1} + \dots + \binom{a_{t(n,l)}(n, l)}{t(n, l)-1} = F_l(n) . \tag{4}$$

However, it is easy to see, that

$$|c(\mathcal{M}(a_l^*(n), n, l))| = \binom{a_l(n, l)}{l-1} + \left|c\left(\mathcal{M}\left(a_l(n, l), n - \binom{a_l(n, l)}{l}, l-1\right)\right)\right|$$

and by the induction hypothesis

$$\left|c\left(\mathcal{M}\left(a_l(n, l), n - \binom{a_l(n, l)}{l}, l-1\right)\right)\right| = \binom{a_{l-1}(n, l)}{l-2} + \dots + \binom{a_{t(n,l)}(n, l)}{t(n, l)-1},$$

which proves (4).

2. The minimizing system of Theorem 2 consists of a complete system in G, and $\mathcal{M}\left(h, n - \binom{h}{l}, l\right)$ in H.

3. In the previous two points we showed that in the case of Theorem 1

$$\min |c(\mathcal{A})| \leq F_l(n) ,$$

and in the case of Theorem 2

$$\min|c(\mathcal{A})| \leq \binom{h}{l-1} + F_l\left(n - \binom{h}{l}\right).$$

Thus, it is sufficient to verify

$$|c(\mathcal{A})| \geq F_l(n) \tag{5}$$

and

$$|c(\mathcal{A})| \geq \binom{h}{l-1} + F_l\left(n - \binom{h}{l}\right), \tag{6}$$

respectively. These statements will be proved by induction over l. If $l = 1$, both statements are trivial. Assume we have proved for all numbers $< l$ and prove for l.

4. First we prove the inequality

$$F_l(n) \leq F_l(n_1) + F_{l-1}(n_2), \tag{7}$$

if

$$n = n_1 + n_2, \qquad n_1 \geq 0, \quad n_2 \geq 0 \tag{8}$$

are integers, and

$$n_2 \leq E_l(n). \tag{9}$$

The statement will be proved for fixed l and for every n, n_1, n_2 using the induction hypothesis for $l - 1$. For the sake of simplicity we use the following notations:

$$\begin{array}{llll} t = t(n, l) & a_i = a_i(n, l) & (t \leq i \leq l) & a_l^* = a_l^*(n), \\ r = t(n_1, l) & b_i = a_i(n_1, l) & (r \leq i \leq l) & b_l^* = a_l^*(n_1), \\ s = t(n_2, l-1) & c_i = a_i(n_2, l-1) & (s \leq i \leq l-1) & c_{l-1}^* = a_{l-1}^*(n_2). \end{array}$$

It follows from (8) and (9) that

$$n_1 \geq n - E_l(n) = \binom{a_l - 1}{l} + \ldots + \binom{a_t - 1}{t}. \tag{10}$$

Because of (10)

$$b_l \geq a_l - 1 \tag{11}$$

must hold, since in the contrary case it would be

$$n_1 \leq \binom{a_l - 2}{l} + \binom{a_l - 3}{l-1} + \ldots + \binom{a_l - (l+1)}{1} = \binom{a_l - 1}{l} - 1,$$

what contradicts (10). On the other hand

$$a_l \geq b_l \tag{12}$$

because of (8). Applying (11) and (12) we can distinguish two different cases: (a) $b_l = a_l$ and (b) $b_l = a_l - 1$.

(a) In this case (7) has the form

$$\binom{a_l}{l-1} + \binom{a_{l-1}}{l-2} + \ldots + \binom{a_t}{t-1} \leq$$

$$\leq \binom{a_l}{l-1} + \binom{b_{l-1}}{l-2} + \ldots + \binom{b_r}{r-1} + F_{l-1}(n_2)\,.$$

Decreasing both sides by $\binom{a_l}{l-1}$ we have

$$F_{l-1}\left(n - \binom{a_l}{l}\right) \leq F_{l-1}\left(n_1 - \binom{a_l}{l}\right) + F_{l-1}(n_2)\,. \tag{13}$$

Let H_1 and H_2 be disjoint sets. Construct the system $\mathcal{M}\left(b_{l-1}+1, n_1 - \binom{a_l}{l}, l-1\right)$ on H_1 and the system $\mathcal{M}(c^*_{l-1}, n_2, l-1)$ on H_2. In this manner we obtain a system $\mathcal{N}$ on $H_1 \cup H_2$. Applying the induction hypothesis (Point **3**. (5)) for $\mathcal{N}$ and $l-1$ we have

$$F_{l-1}\left(n - \binom{a_l}{l}\right) \leq |c(\mathcal{N})| =$$

$$= \left|c\left(\mathcal{M}\left(b_{l-1}+1, n_1 - \binom{a_l}{l}, l-1\right)\right)\right| + |c(\mathcal{M}(c^*_{l-1}, n_2, l-1))|\,. \tag{14}$$

However, we know (Point **1**. (4)) that

$$\left|c\left(\mathcal{M}\left(b_{l-1}+1, n_1 - \binom{a_l}{l}, l-1\right)\right)\right| = F_{l-1}\left(n_1 - \binom{a_l}{l}\right) \tag{15}$$

and

$$|c(\mathcal{M}(c^*_{l-1}, n_2, l-1))| = F_{l-1}(n_2)\,. \tag{16}$$

Finally, (13) follows from (14), (15), and (16).

(b) $b_l = a_l - 1$. We separate this case into two subcases:

$$\text{(ba) } n_2 \geq \binom{a_l - 1}{l-1}, \qquad \text{(bb) } n_2 < \binom{a_l - 1}{l-1}.$$

(ba) In this case (7) has the form

$$\binom{a_l}{l-1}+\binom{a_{l-1}}{l-2}+\ldots+\binom{a_t}{t-1}\le\binom{a_l-1}{l-1}+\binom{b_{l-1}}{l-2}+\ldots+\binom{b_r}{r-1}+$$
$$+\binom{a_l-1}{l-2}+\binom{c_{l-2}}{l-3}+\ldots+\binom{c_s}{s-1},$$

since $c_{l-1}=a_l-1$, because of (9) and the supposition (ba). Decreasing both sides by $\binom{a_l}{l-1}=\binom{a_l-1}{l-1}+\binom{a_l-1}{l-2}$ we have

$$F_{l-1}\left(n-\binom{a_l}{l}\right)\le F_{l-1}\left(n_1-\binom{a_l-1}{l}\right)+F_{l-2}\left(n_2-\binom{a_l-1}{l-1}\right). \tag{17}$$

We can prove (17) by using of the induction hypothesis if

$$n_2-\binom{a_l-1}{l-1}\le E_{l-1}\left(n-\binom{a_l}{l}\right) \tag{18}$$

holds. However (9) gives

$$\binom{a_l-1}{l-1}+\binom{c_{l-2}}{l-2}+\ldots+\binom{c_s}{s}\le\binom{a_l-1}{l-1}+ \tag{19}$$
$$+\binom{a_{l-1}-1}{l-2}+\ldots+\binom{a_t-1}{t-1}.$$

Decreasing both sides by $\binom{a_l-1}{l-1}$ we obtain

$$\binom{c_{l-2}}{l-2}+\ldots+\binom{c_s}{s}\le\binom{a_{l-1}-1}{l-2}+\ldots+\binom{a_t-1}{t-1} \tag{20}$$

and (20) is equivalent to (18).

(bb) In this case (7) has the form

$$\binom{a_l}{l-1}+\binom{a_{l-1}}{l-2}+\ldots+\binom{a_t}{t-1}\le\binom{a_l-1}{l-1}+$$
$$+\binom{b_{l-1}}{l-2}+\ldots+\binom{b_r}{r-1}+F_{l-1}(n_2).$$

Decreasing both sides by $\binom{a_l-1}{l-1}$ we have

$$\binom{a_l-1}{l-2}+F_{l-1}\left(n-\binom{a_l}{l}\right)\le F_{l-1}\left(n_1-\binom{a_l-1}{l}\right)+F_{l-1}(n_2). \tag{21}$$

Let G and H be two disjoint sets of $a_l - 1$ elements. Construct the system $\mathcal{M}\left(a_l - 1, n_1 - \binom{a_l - 1}{l}, l - 1\right)$. We can it construct if $n_1 - \binom{a_l - 1}{l} \leq \binom{a_l - 1}{l - 1}$. But this follows from $a_l - 1 = b_l > b_{l-1}$, since

$$n_1 - \binom{a_l - 1}{l} = \binom{b_{l-1}}{l - 1} + \ldots + \binom{b_r}{r}.$$

Construct further the system $\mathcal{M}(c^*_{l-1}, n_2, l - 1)$ on H. The possibility of this construction follows from the assumption (bb). In this manner we obtain a system $\mathcal{N}$ on $G \cup H$. Applying the induction hypothesis (Point **3**. (6)) for $\mathcal{N}$ and $l - 1$ we have

$$\binom{a_l - 1}{l - 2} + F_{l-1}\left(n - \binom{a_l}{l}\right) \leq \left|c\left(\mathcal{M}\left(a_l - 1, n_1 - \binom{a_l - 1}{l}, l - 1\right)\right)\right| + \tag{22}$$
$$+ |c(\mathcal{M}(c^*_{l-1}, n_2, l - 1))| .$$

However, we know (Point **1**. (4)) that

$$\left|c\left(\mathcal{M}\left(a_l - 1, n_1 - \binom{a_l - 1}{l}, l - 1\right)\right)\right| = F_{l-1}\left(n_1 - \binom{a_l - 1}{l}\right) \tag{23}$$

and

$$|c(\mathcal{M}(c^*_{l-1}, n_2, l - 1))| = F_{l-1}(n_2) , \tag{24}$$

further, (21) follows from (22), (23) and (24). Thus we proved the inequality for l.

5. However, we need (7) under the condition

$$n_2 \leq \frac{n \cdot l}{a^*_l} \tag{25}$$

instead of (9). Thus we are going now to prove the inequality

$$\frac{n \cdot l}{a^*_l} \leq E_l(n) . \tag{26}$$

We prove (26) by induction over l, but we should like to mention that the proof of (26) is independent from the whole proof of the theorems. For $l = 1$ the statement is trivial. Assume we proved it for the integers $< l$, and prove for l. If $n = \binom{a_l}{l}$ then $a^*_l = a_l$ and $E_l(n) = \binom{a_l - 1}{l - 1}$, thus (26) holds with equality. We may assume $a^*_l = a_l + 1$. Obviously

$$\frac{l}{a_l + 1}\binom{a_l}{l} \leq \binom{a_l - 1}{l - 1} \tag{27}$$

and by the induction hypothesis

$$(28)\quad \frac{l-1}{a_{l-1}+1}\left[\binom{a_{l-1}}{l-1}+\dots+\binom{a_r}{r}\right]\le\binom{a_{l-1}-1}{l-2}+\dots+\binom{a_r-1}{r-1}.$$

If $\frac{l}{a_l+1}\le\frac{l-1}{a_{l-1}+1}$, summarizing (27) and (28) we obtain (26). In the contrary case

$$(29)\quad \frac{l}{a_l+1}>\frac{l-1}{a_l}$$

holds because of $a_l\ge a_{l-1}+1$. Let us set out from the identity

$$\binom{a_l}{l-1}\left(\frac{l}{a_l+1}-\frac{l-1}{a_l}\right)=\binom{a_l-1}{l-1}-\frac{l}{a_l+1}\binom{a_l}{l}.$$

The expression in the bracket is positive because of (29), thus we can write

$$\left[\binom{a_{l-1}}{l-1}+\binom{a_{l-2}}{l-2}+\dots+\binom{a_r}{r}\right]\left(\frac{l}{a_l+1}-\frac{l-1}{a_l}\right)<\binom{a_l-1}{l-1}-\frac{l}{a_l+1}\binom{a_l}{l},$$

since $a_l>a_{l-1}$. Write $\frac{l-1}{a_{l-1}+1}$ instead of $\frac{l-1}{a_l}$, and reorder the inequality

$$\frac{l}{a_l+1}\left[\binom{a_l}{l}+\binom{a_{l-1}}{l-1}+\dots+\binom{a_r}{r}\right]<\binom{a_l-1}{l-1}+$$
$$+\frac{l-1}{a_{l-1}+1}\left[\binom{a_{l-1}}{l-1}+\dots+\binom{a_r}{r}\right].$$

Finally, from the above inequality (26) follows by (28).

6. Now let us prove statement (5) for l by induction over h if $h=l$ is trivial. Assume we have proved (5) for all sets $|H|<h$, and prove for h. There exists an element e of H, contained by at most $\frac{n\cdot l}{h}$ sets A_i. We define the following systems:

$$\mathscr{B}=\{A:A\in\mathscr{A},\ e\notin A\}$$

and

$$\mathscr{C}=\{A-\{e\}:A\in\mathscr{A},\ e\in\mathscr{A}\}$$

where

$$(30)\quad n_2=|\mathscr{C}|\le\frac{n\cdot l}{h}\le\frac{n\cdot l}{a_l^*(n)}.$$

Naturally,

$$c(\mathscr{B})\subset c(\mathscr{A})$$

and

$$c(\mathscr{C})\,(\cup)\,e\subset c(\mathscr{A}),$$

where $\mathcal{D}(\cup)a$ denotes in general the system $\{D \cup \{a\} : D \in \mathcal{D}\}$. Thus the inequality

$$|c(\mathcal{A})| \geq |c(\mathcal{B})| + |c(\mathcal{C})| \tag{31}$$

holds. However, $\mathcal{B}$ is a system in $H - \{e\}$, we may apply the induction hypothesis for $h - 1$

$$|c(\mathcal{B})| \geq F_l(n - n_2). \tag{32}$$

Further, applying the induction hypothesis for $l - 1$ we obtain

$$|c(\mathcal{C})| \geq F_{l-1}(n_2). \tag{33}$$

It follows from (31), (32) and (33) that

$$F_l(n - n_2) + F_{l-1}(n_2) \leq |c(\mathcal{A})|. \tag{34}$$

Using the result of Point **5**, inequality (5) follows from (34) and (7) by (30), since (7) is proved already for l.

7. Now prove statement (6) for l by induction over h. If $h = l$, it is trivial. Assume we have proved (6) for all sets $|G| = |H| < h$, and prove for h. The proof will be similar to the proof of the previous point.

Let $\mathcal{A}_1$ and $\mathcal{A}_2$ be given by

$$\mathcal{A}_1 = \{A : A \in \mathcal{A}, A \subset G\},$$

and

$$\mathcal{A}_2 = \{A : A \in \mathcal{A}, A \subset H\}.$$

If $|\mathcal{A}_1| = r$ and $|\mathcal{A}_2| = s$, there are two elements $e \in G$ and $f \in H$, such that e is contained by at most $\frac{r \cdot l}{h}$, and f is contained by at most $\frac{s \cdot l}{h}$ sets A_l. Define the following systems:

$$\mathcal{B} = \{A : A \in \mathcal{A}, e \notin A, f \notin A\},$$

$$\mathcal{C}_1 = \{A - \{e\} : A \in \mathcal{A}_1, e \in A\}$$

and

$$\mathcal{C}_2 = \{A - \{f\} : A \in \mathcal{A}_2, f \in A\},$$

where

$$r_2 = |\mathcal{C}_1| \leq \frac{r \cdot l}{h} \tag{35}$$

and

$$s_2 = |\mathcal{C}_2| \leq \frac{s \cdot l}{h}. \tag{36}$$

Naturally,

$$c(\mathcal{B}) \subset c(\mathcal{A}),$$

$$c(\mathcal{C}_1)(\cup)e \subset c(\mathcal{A})$$

and

$$c(\mathcal{C}_2)(\cup)f \subset c(\mathcal{A}).$$

Thus the inequality

$$(37)\qquad |c(\mathscr{A})| \geq |c(\mathscr{B})| + |c(\mathscr{C}_1)| + |c(\mathscr{C}_2)| = |c(\mathscr{B})| + |c(\mathscr{C}_1 \cup \mathscr{C}_2)|$$

holds. However $\mathscr{B}$ is a system in $G \cup H - \{e\} - \{f\}$, we may apply our induction hypothesis for $h - 1$:

$$(38)\qquad |c(\mathscr{B})| \geq \binom{h-1}{l-1} + F_l\left(n - r_2 - s_2 - \binom{h-1}{l}\right).$$

Further, applying the induction hypothesis for $l - 1$ we obtain

$$(39)\qquad |c(\mathscr{C}_1 \cup \mathscr{C}_2)| \geq \binom{h-1}{l-2} + F_{l-1}\left(r_2 + s_2 - \binom{h-1}{l-1}\right).$$

It follows from (37), (38) and (39) that

$$(40)\qquad \binom{h}{l-1} + F_l\left(n - r_2 - s_2 - \binom{h-1}{l}\right) + F_{l-1}\left(r_2 + s_2 - \binom{h-1}{l-1}\right) \leq |c(\mathscr{A})|.$$

Now we should like to use inequality (7) which is valid under condition (25) (Point **5**). For this reason we have to verify only

$$(41)\qquad r_2 + s_2 - \binom{h-1}{l-1} \leq \frac{\left[n - r_2 - s_2 - \binom{h-1}{l} + r_2 + s_2 - \binom{h-1}{l-1}\right] l}{a_l^*\left(n - \binom{h}{l}\right)} = \frac{\left[n - \binom{h}{l}\right] \cdot l}{a_l^*\left(n - \binom{h}{l}\right)}.$$

However

$$(42)\qquad r_2 + s_2 - \binom{h-1}{l-1} \leq \frac{\left[r + s - \binom{h}{l}\right] \cdot l}{h} = \frac{\left[n - \binom{h}{l}\right] \cdot l}{h}$$

is an immediate consequence of (35) and (36). Since $n \leq 2\binom{h}{l}$ is a condition of Theorem 2, $a_l^*\left(n - \binom{h}{l}\right) \leq h$ holds and (42) results (41). Thus we can use

(7) for this case:

$$(43)\quad F_l\left(n-\binom{h}{l}\right)\leq F_l\left(n-r_2-s_2-\binom{h-1}{l}\right)+F_{l-1}\left(r_2+s_2-\binom{h-1}{l-1}\right).$$

Finally, (40) and (43) gives the desired inequality, and the whole proof is finished.

Now we consider a natural generalization of the problem of Theorem 1. The problem is to determine the minimum of $|c^k(\mathcal{A})|$, where $1\leq k\leq l$, $c^k(\mathcal{A})=c(c^{k-1}(\mathcal{A}))$ and $c^1(\mathcal{A})=c(\mathcal{A})$. It is not difficult to conjecture what is the result. To the theorem we need the following notation:

$$F_l^k(n)=\binom{a_l(n,l)}{l-k}+\binom{a_{l-1}(n,l)}{l-1-k}+\ldots+\binom{a_{t(n,l)}(n,l)}{t(n,l)-k}\qquad(1\leq k\leq l),$$

where $\binom{a}{b}=0$ if $b<0$.

THEOREM 3. *Let h, n, l and k be given integers with the properties*

$$h\geq 1,\ 1\leq k\leq l\leq h\quad\text{and}\quad 1\leq n\leq\binom{h}{l}.$$

If H is a set of h elements and

$$\mathcal{A}=\{A_1,\ldots,A_n\},\qquad |A_i|=l\qquad(i=1,\ldots,n)$$

a system of different subsets of H, then

$$\min|c^k(\mathcal{A})|=F_l^k(n),$$

where the minimum runs over all such systems $\mathcal{A}$.

PROOF. It is easy to see by induction over l, that $|c^k(\mathcal{M}(h,n,l))|=F_l^k(n)$. Thus, we have to prove only

$$(44)\qquad |c^k(\mathcal{A})|\geq F_l^k(n).$$

This will be proved by induction over k. For $k=1$ Theorem 3 gives Theorem 1. Assume now (44) is true for values smaller than k, and prove for k. Obviously,

$$c^k(\mathcal{A})=c(c^{k-1}(\mathcal{A}))$$

holds and using the induction hypothesis and Theorem 1 we obtain

$$(45)\qquad |c^k(\mathcal{A})|\geq F_{l-(k-1)}(F_l^{k-1}(n)).$$

(a) If $t(n,l)-(k-1)>0$, then

$$F_l^{k-1}(n)=\binom{a_l(n,l)}{l-(k-1)}+\ldots+\binom{a_{t(n,l)}(n,l)}{t(n,l)-(k-1)}$$

is an expression of type (1). That is

$$(46)\quad \begin{aligned} &t\left(F_l^{k-1}(n), l-k+1\right) = t(n,l) - k + 1 \\ &a_i\left(F_l^{k-1}(n), l-k+1\right) = a_{i+k-1}(n,l) \quad \left(t(n,l) - k + 1 \le i \le l - k + 1\right) \end{aligned}$$

and

$$(47)\quad \begin{aligned} F_{l-k+1}\left(F_l^{k-1}(n)\right) &= \sum_{i=t(n,l)-k+1}^{l-k+1} \binom{a_i\left(F_l^{k-1}(n), l-k+1\right)}{i-1} = \\ &= \sum_{i=t(n,l)-k+1}^{l-k+1} \binom{a_{i+k-1}(n,l)}{i-1} = \sum_{j=t(n,l)}^{l} \binom{a_j(n,l)}{j-k} = F_l^k(n), \end{aligned}$$

which proves (44) and (45).

(b) If $t(n, l) - k + 1 \le 0$, then (46) does not hold. However in this case

$$F_l^{k-1}(n) - 1 = \binom{a_l(n,l)}{l-k+1} + \dots + \binom{a_k(n,l)}{1}$$

and

$$\begin{aligned} &t\left(F_l^{k-1}(n) - 1, l-k+1)\right) = 1 \\ &a_i\left(F_l^{k-1}(n) - 1, l-k+1\right) = a_{i+k-1}(n,l) \quad (1 \le i \le l-k+1) \end{aligned}$$

hold. Further, the equation

$$(48)\quad \begin{aligned} F_{l-k+1}\left(F_l^{k-1}(n) - 1\right) &= \sum_{i=1}^{l-k+1} \binom{a_i\left(F_l^{k-1}(n) - 1, l-k+1\right)}{i-1} = \\ &= \sum_{i=1}^{l-k+1} \binom{a_{i+k-1}(n,l)}{i-1} = \sum_{j=k}^{l} \binom{a_j(n,l)}{j-k} = \sum_{j=t(n,l)}^{l} \binom{a_j(n)}{j-k} = F_l^k(n) \end{aligned}$$

is true in this case instead of (47). If we prove

$$(49)\quad F_{l-k+1}\left(F_l^{k-1}(n)\right) = F_{l-k+1}\left(F_l^{k-1}(n) - 1\right),$$

then (44) follows from (45), (49) and (48). (49) will be proved by the following simple lemma.

LEMMA 2. *If $t(m, r) = 1$, then*

$$F_r(m+1) = F_r(m).$$

PROOF. Let s be the least index such that $a_s(m, r) > a_{s-1}(m, r) + 1$ $(2 \le s \le r)$. If there is not such s, let s be equal to $r + 1$. Thus, we can write

$$\begin{aligned} m = \binom{a_r(m,r)}{r} + \dots + \binom{a_{s-1}(m,r)}{s-1} + \binom{a_{s-1}(m,r) - 1}{s-2} + \dots + \\ + \dots + \binom{a_{s-1}(m,r) - (s-2)}{1} \end{aligned}$$

and

$$m+1=\binom{a_r(m,r)}{r}+\ldots+\binom{a_{s-1}(m,r)+1}{s-1}.$$

Now it is not difficult to see, that

$$F_r(m)=\binom{a_r(m,r)}{r-1}+\ldots+\binom{a_{s-1}(m,r)}{s-2}+\binom{a_{s-1}(m,r)-1}{s-3}+\ldots+$$

$$+\ldots+\binom{a_{s-1}(m,r)-(s-2)}{0}=\binom{a_r(m,r)}{r-1}+\ldots+\binom{a_{s-1}(m,r)+1}{s-2}=$$

$$=F_r(m+1),$$

which proves the lemma and Theorem 3.

§ 3. Solution of an Erdős-problem

Let H be a finite set of h elements, and $\mathscr{A}$ a system of subsets of H:

$$\mathscr{A}=\{A_1, A_2, \ldots, A_n\},\ A_i\subset H,\ |A_i|=l \qquad (1\leq i\leq n).$$

Erdős proposed the following problem. For which numbers n can we construct a system $\mathscr{B}$ with the properties

$$\mathscr{B}=\{B_1, B_2, \ldots, B_n\},\ B_i\subset A_i,\ |B_i|=l-k \qquad (1\leq i\leq n).$$

In the solution we use the well-known marriage problem. It is clear in this connection, that it is a very important question, in which cases does $F_l^k(n)<n$, $F_l^k(n)=n$ or $F_l^k(n)>n$ hold. The following sequence of lemmas deals with this problem.

Lemma 3. *If $1\leq k\leq l$ and x are positive integers, then*

$$f(x)=\binom{x}{l-k}-\binom{x}{l}$$

is a monotone increasing function between l and $2l-k-2$ but it is a monotone decreasing function from $2l-k-1$. The values $f(2l-k-2)$ and $f(2l-k-1)$ are equal.

Proof. Let $0\leq a<b\leq x$ be integers. It is easy to see that $\binom{x}{a}-\binom{x}{b}<0$, $\binom{x}{a}-\binom{x}{b}=0$ and $\binom{x}{a}-\binom{x}{b}>0$, respectively, if $a+b<x$, $a+b=x$ and $a+b>x$, respectively.

Consider the difference $f(x+1)-f(x)=\binom{x}{l-k-1}-\binom{x}{l-1}$. Using the above remark we obtain that

$$f(x+1)-f(x)<0 \quad \text{if} \quad 2l-k-2<x,$$

$$f(x+1)-f(x)=0 \quad \text{if} \quad 2l-k-2=x,$$

and finally,

$$f(x+1)-f(x)>0 \quad \text{if} \quad 2l-k-2>x.$$

This completes the proof.

The following two lemmas are immediate consequences of Lemma 3.

LEMMA 3a. *If $1\leq k\leq l$ and x are positive integers, then*

$$\binom{x}{l-k}-\binom{x}{l}\leq\binom{2l-k-1}{l-k}-\binom{2l-k-1}{l}.$$

LEMMA 3b. *If $1\leq k\leq l$ and $x\geq 2l-k+1$ are positive integers, then*

$$\binom{x}{l-k}-\binom{x}{l}\leq\binom{2l-k+1}{l-k}-\binom{2l-k+1}{l}.$$

LEMMA 4. *If $1\leq k\leq m$, then*

$$\sum_{i=k}^{m-1}\left[\binom{2i-k-1}{i-k}-\binom{2i-k-1}{i}\right]\leq\binom{2m-k-1}{m-k}-\binom{2m-k-1}{m}.$$

PROOF. Let a and b be positive integers, where $\frac{a}{2}\leq b\leq a-1$. Then

$$\binom{a}{b}-\binom{a}{b+1}=\binom{a}{b}\left[1-\frac{a-b}{b+1}\right]=\binom{a}{b}\left[\frac{2b-a+1}{b+1}\right], \tag{50}$$

and similarly

$$\binom{a+2}{b+1}-\binom{a+2}{b+2}=\binom{a+2}{b+1}\left[1-\frac{a-b+1}{b+2}\right]=\binom{a+2}{b+1}\left[\frac{2b-a+1}{b+2}\right]. \tag{51}$$

Further

$$\binom{a+2}{b+1}\left[\frac{2b-a+1}{b+2}\right]=\binom{a}{b}\left[\frac{2b-a+1}{b+1}\right]\cdot\left[\frac{(a+2)(a+1)}{(b+2)(a-b+1)}\right], \tag{52}$$

where

$$\frac{a+1}{b+2}\geq 1,$$

and

$$\frac{a+2}{a-b+1}\geq\frac{a+2}{\frac{a}{2}+1}=2.$$

That is

$$\binom{a+2}{b+1} - \binom{a+2}{b+2} \geq 2\left[\binom{a}{b} - \binom{a}{b+1}\right] \tag{53}$$

follows from (50), (51) and (52).

Applying (53) for $a = 2i = k - 1$, and $b = i - 1$, $(1 \leq k \leq i)$, we obtain

$$\binom{2(i+1)-k-1}{i} - \binom{2(i+1)-k-1}{i+1} \geq 2\left[\binom{2i-k-1}{i-1} - \binom{2i-k-1}{i}\right],$$

or

$$\binom{2(i+1)-k-1}{i+1-k} - \binom{2(i+1)-k-1}{i+1} \geq 2\left[\binom{2i-k-1}{i-k} - \binom{2i-k-1}{i}\right]. \tag{54}$$

Prove now the lemma by induction over m. If $m = k$, the statement is trivial. Let the lemma be true for m and prove it for $m + 1$.

$$\sum_{i=k}^{m}\left[\binom{2i-k-1}{i-k} - \binom{2i-k-1}{i}\right] = \sum_{i=k}^{m-1}\left[\binom{2i-k-1}{i-k} - \binom{2i-k-1}{i}\right] + \\ + \binom{2m-k-1}{m-k} - \binom{2m-k-1}{m}$$

and by induction hypothesis and (54)

$$\sum_{i=k}^{m}\left[\binom{2i-k-1}{i-k} - \binom{2i-k-1}{i}\right] \leq 2\left[\binom{2m-k-1}{m-k} - \binom{2m-k-1}{m}\right] \leq \\ \leq \binom{2(m+1)-k-1}{m+1-k} - \binom{2(m+1)-k-1}{m+1}$$

holds, which proves Lemma 4.

LEMMA 5. *If* $1 \leq k \leq l$ *and* $2l - k < a_l(n, l)$ *then*

$$F_l^k(n) < n. \tag{55}$$

PROOF. We may use Lemma 3b:

$$\binom{a_l(n,l)}{l-k} - \binom{a_l(n,l)}{l} \leq \binom{2l-k+1}{l-k} - \binom{2l-k+1}{l}. \tag{56}$$

On the other hand, by Lemma 3a

$$\binom{a_i(n,l)}{i-k} - \binom{a_i(n,l)}{i} \leq \binom{2i-k-1}{i-k} - \binom{2i-k-1}{i} \qquad (k \leq i \leq l-1)$$

holds and summarizing it we obtain

$$\sum_{i=k}^{l-1}\left[\binom{a_i(n,l)}{i-k}-\binom{a_i(n,l)}{i}\right]\le\sum_{i=k}^{l-1}\left[\binom{2i-k-1}{i-k}-\binom{2i-k-1}{i}\right].$$

Applying now Lemma 4 and (54):

$$\sum_{i=k}^{l-1}\left[\binom{a_i(n,l)}{i-k}-\binom{a_i(n,l)}{i}\right]\le\binom{2l-k-1}{l-k}-\binom{2l-k-1}{l}<$$

$$<\binom{2(l+1)-k-1}{l+1-k}-\binom{2(l+1)-k-1}{l+1}.$$

Obviously,

$$\sum_{i=1}^{l-1}\left[\binom{a_i(n,l)}{i-k}-\binom{a_i(n,l)}{i}\right]<\binom{2(l+1)-k-1}{l+1-k}-\binom{2(l+1)-k-1}{l+1} \tag{57}$$

also holds, since we added a nonpositive number to the left side. If we sum (56) and (57) the obtained inequality

$$F_l^k(n)-n<\binom{2l-k+1}{l-k}-\binom{2l-k+1}{l}+\binom{2(l+1)-k-1}{l+1-k}-$$

$$-\binom{2(l+1)-k-1}{l+1}=0$$

results (55).

Lemma 6. *If* $1\le k\le l$ *and* $2l-k>a_l(n,l)$ *then*

$$F_l^k(n)>n. \tag{58}$$

Proof. We know that

$$a_i(n,l)\le a_l(n,l)-(l-i). \tag{59}$$

If $l>i\ge a_l(n,l)-(l-k)$, then $a_l(n,l)-(l-i)\le 2i-k$ and by (59)

$$a_i(n,l)\le 2i-k$$

holds. In this case, obviously

$$\binom{a_i(n,l)}{i-k}-\binom{a_i(n,l)}{i}\ge 0 \tag{60}$$

follows. If $k\le i<a_l(n,l)-(l-k)$, then by (59) and Lemma 3

$$\binom{a_i(n,l)}{i-k}-\binom{a_i(n,l)}{i}\ge\binom{a_l(n,l)-(l-i)}{i-k}-\binom{a_l(n,l)-(l-i)}{i} \tag{61}$$

holds, but it is trivially true for $i<k$, too.

Sum (60) and (61)

$$\sum_{i=1}^{l-1}\left[\binom{a_l(n,l)}{i-k}-\binom{a_l(n,l)}{i}\right]\geq\sum_{i=1}^{a_l(n,l)-l+k-1}\left[\binom{a_l(n,l)-(l-i)}{i-k}-\right.$$

$$\left.-\binom{a_l(n,l)-(l-i)}{i}\right]=\binom{2a_l(n,l)-2l+k}{a_l(n,l)-l-1}-\binom{2a_l(n,l)-2l+k}{a_l(n,l)-l+k-1}+1.$$

That is

$$(62)\quad F_l^k(n)-n\geq\binom{2a_l(n,l)-2l+k}{a_l(n,l)-l-1}-\binom{2a_l(n,l)-2l+k}{a_l(n,l)-l+k-1}+1+$$

$$+\binom{a_l(n,l)}{l-k}-\binom{a_l(n,l)}{l}$$

s true. Here

$$(63)\quad \binom{2a_l(n,l)-2l+k}{a_l(n,l)-l-1}-\binom{2a_l(n,l)-2l+k}{a_l(n,l)-l+k-1}\geq\binom{a_l(n,l)-1}{a_l(n,l)-l-1}-$$

$$-\binom{a_l(n,l)-1}{a_l(n,l)-l+k-1}$$

because of Lemma 3. However we can write the right hand side of (63) in the form

$$(64)\quad \binom{a_l(n,l)-1}{a_l(n,l)-l-1}-\binom{a_l(n,l)-1}{a_l(n,l)-l+k-1}=$$

$$=\binom{a_l(n,l)-1}{l}-\binom{a_l(n,l)-1}{l-k}.$$

Here

$$\binom{a_l(n,l)-1}{l}-\binom{a_l(n,l)-1}{l-k}=\left[\binom{a_l(n,l)}{l}-\binom{a_l(n,l)}{l-k}\right]-\left[\binom{a_l(n,l)-1}{l-1}-\right.$$

$$\left.-\binom{a_l(n,l)-1}{l-k-1}\right]$$

and since $2l-k-1>a_l(n,l)-1$ by supposition of the lemma, thus

$$(65)\quad \binom{a_l(n,l)-1}{l}-\binom{a_l(n,l)-1}{l-k}\geq\binom{a_l(n,l)}{l}-\binom{a_l(n,l)}{l-k}.$$

Finally, (65), (64) (63) and (62) give

$$F_l^k(n)-n\geq 1$$

which proves our lemma.

LEMMA 7. *If*

$$n > \binom{2l-k}{l} + \binom{2(l-1)-k}{l-1} + \dots + \binom{k}{k}, \tag{66}$$

then

$$F_l^k(n) < n.$$

On the other hand, if

$$n \leq \binom{2l-k}{l} + \binom{2(l-1)-k}{l-1} + \dots + \binom{k}{k}, \tag{67}$$

then

$$F_l^k(n) \geq n$$

with equality only if

$$n = \binom{2l-k}{l} + \binom{2(l-1)-k}{l-1} + \dots + \binom{2s-k}{s} \tag{68}$$

for some s $(k \leq s \leq l)$.

PROOF. Consider first the case of (66). If $a_i(n, l) = 2i - k$ $(k \leq i \leq l)$, then $t(n, l) < k$ and

$$n = \binom{2l-k}{l} + \dots + \binom{k}{k} + \binom{k-1}{k-1} + \dots + \binom{t(n,l)}{t(n,l)}.$$

Obviously,

$$F_l^k(n) = \binom{2l-k}{l} + \dots + \binom{k}{k},$$

thus $F_l^k(n) < n$ holds.

In the contrary case

$$a_r(n, l) > 2r - k$$

$$a_i(n, l) = 2i - k \qquad (r < i \leq l)$$

hold for some r $(k \leq r < l)$. Since

$$n - F_l^k(n) = \left[\binom{a_r(n,l)}{r} + \dots\right] - \left[\binom{a_r(n,l)}{r-k} + \dots\right],$$

the statement follows by Lemma 5:

$$\binom{a_r(n,l)}{r} + \dots > F_r^k\left[\binom{a_r(n,l)}{r} + \dots\right].$$

The case (67) may occur in two different ways.

1. If (68) holds, then obviously $F_l^k(n) = n$
2. For some r $(k < r \leq l)$,

$$a_r(n, l) < 2r - k,$$

and

$$a_i(n, l) - 2i - k \qquad (r < i \leq l) .$$

Since

$$n - F_l^k(n) = \left[\binom{a_r(n, l)}{r} + \dots\right] - \left[\binom{a_r(n, l)}{r-k} + \dots\right],$$

the statement follows by Lemma 6:

$$\binom{a_r(n, l)}{r} + \dots < F_r^k\left[\binom{a_r(n, l)}{r} + \dots\right].$$

THEOREM 4. *Let* $1 \leq k \leq l \leq h$ *be positive integers,* H *a set of* h *elements and*

$$\mathcal{A} = \{A_1, \dots, A_n\}, \quad |A_i| = l \qquad (1 \leq i \leq n)$$

a system of subsets of H. *If*

$$n \leq \binom{2l-k}{l} + \binom{2(l-1)-k}{l-1} + \dots + \binom{k}{k}, \tag{69}$$

there exists a system

$$\mathcal{B} = \{B_1, \dots, B_n\}, \quad |B_i| = l - k, \quad B_i \subset A_i \quad (1 \leq i \leq n), \tag{70}$$

but in the case of

$$n > \binom{2l-k}{l} + \binom{2(l-1)-k}{l-1} + \dots + \binom{k}{k} \tag{71}$$

not necessarily .

PROOF. First we prove the latter case. If (71) holds then by Lemma 7 $F_l^k(n) < n$. We know (Theorem 1) that there exists a system $\mathcal{A}$ such that $|c^k(\mathcal{A})| = F_l^k(n)$. Thus, a system $\mathcal{B}$ satisfying (70) does not exist.

In the proof of the existence of $\mathcal{B}$ in the case of (69) we use the well-known marriage problem [2]:

THEOREM OF ORE. *Let* E *and* F *be disjoint sets and* G *a graph on* $E \cup F$ *Assume* G *has the property that for arbitrary* $D \subset E$ *there is a set* $H \subset F$ *such that every element of* H *is connected with at least one element of* D *and* $|H| \geq |D|$. *Then there exists a one-to-one mapping between* E *and a subset* K *of* F, *such that the associating vertices are connected in* G.

In our case $E = \mathcal{A}$, $F = c^k(\mathcal{A})$ and $A \in \mathcal{A}$, $B \in c^k(\mathcal{A})$ are connected if and only if $A \supset B$. Thus, it is sufficient to verify that for every subsystem

$$\mathcal{C} = \{A_{i_1}, \dots, A_{i_m}\} \subset \mathcal{A}$$

there are at least m sets in $c^k(\mathcal{A})$, which are contained in one of $A_{i_j} (1 \leq j \leq \leq m)$. However, $m \leq n$, thus by (69)

$$m \leq \binom{2l-k}{l} + \dots + \binom{k}{k}$$

and Lemma 7 gives

(72) $$F_l^k(m) \geq m.$$

Use now Theorem 1:

$$|c^k(\mathscr{C})| \geq F_l^k(m).$$

This and (72) results $|c^k(\mathscr{C})| \geq m$, which means that our graph has the property prescribed in the used theorem. Applying the theorem the obtained one-to-one mapping gives just the desired system $\mathscr{B}$.

COROLLARY. If $2l - k \geq h$, then (69) always holds and a system $\mathscr{B}$ satisfying (70) always exists.

This is an immediate consequence of the inequality

$$\binom{h}{l} \leq \binom{2l-k}{l} \leq \binom{2l-k}{l} + \dots + \binom{k}{k}$$

and the fact that $\mathscr{A}$ has at most $\binom{h}{l}$ elements.

REFERENCES

[1] SPERNER, E.: Ein Satz über Untermengen einer endlichen Menge, *Math. Z.* **27** (1928) 544—548.
[2] ORE, O.: Graphs and matching theorems, *Duke Math. J.* **22** (1955) 625—639.

Reprinted from
Proceeding of the Colloquium held at Tihany, Hungary, Sept. 1966
Academic Press and Akademiai Kiado, Budapest, 1968, pp. 187–207

Reprinted from JOURNAL OF COMBINATORIAL THEORY
Vol. 1, No. 2, September 1966

Printed in Italy

A Short Proof of Sperner's Lemma

Let S denote a set of N objects. By a *Sperner collection on S* we mean a collection of subsets of S such that no one contain another. In [1], Sperner showed that no such collection could have more than ${}_N C_{[N/2]}$ members. This follows immediately from the somewhat stronger

THEOREM. *Let Γ be a Sperner collection on S. Then*

$$\Sigma_{A \in \Gamma} {}_N C_{|A|}^{-1} \leq 1,$$

where $|A|$ denotes the cardinality of A.

PROOF. For each $A \subset S$, exactly $|A|!(N - |A|)!$ maximal chains of S (as a lattice under set inclusion) contain A. Since none of the $N!$ maximal chains of S meet Γ more than once, we have

$$\Sigma_{A \in \Gamma} |A|!(N - |A|)! \leq N!,$$

proving the theorem.

REFERENCE

1. E. SPERNER, Ein Satz über Untermenger einer endlichen Menge, *Math. Z.* 27 (1928), 544–548.

D. LUBELL
Systems Research Group, Inc.
1501 *Franklin Avenue*
Mineola, New York 11501

Sonderabdruck aus
ARCHIV DER MATHEMATIK
Vol. XIX, 1968 BIRKHÄUSER VERLAG, BASEL UND STUTTGART Fasc. 6

Möbius Inversion in Lattices

By

HENRY H. CRAPO[1])

1. Introduction. In the development of computational techniques for combinatorial theory, attention has lately centered on ROTA's theory of Möbius inversion [6]. The main theorem of ROTA's paper, concerning the computation of the Möbius invariant across a Galois connection, is a prerequisite to the use of lattice-theoretic methods in combinatorics.

By suitably combining ROTA's main theorem with a discrete analogue of integration-by-parts, we here obtain a perfectly general formulation of Möbius inversion across a Galois connection (theorem 3, below).

As immediate applications of this theory, we obtain a number of interesting computational results concerning finite lattices (section 3, 4) and combinatorial geometries (section 5).

2. Möbius Inversion across a Galois Connection. We begin with a restatement and a simplified proof of ROTA's main theorem. The proof turns on the essential fact that for any (locally finite) ordered set Q with least element 0, the recursion

$$\sum_{y \in Q} a(y)\,\zeta(y, z) = 0 \quad \text{for} \quad z \neq 0$$

has the unique solution $a(y) = 0$ with initial condition $a(0) = 0$, and has the unique solution $a(y) = \mu_Q(0, y)$ with initial condition $a(0) = 1$. Recall that the zeta function $\zeta(y, z)$ has value 1 if $y \leqq z$, and has value 0 otherwise.

Theorem 1. *If J is a closure operator on a finite lattice P, and $Q = P/J$ is the quotient lattice, consisting of the J-closed elements of P, then for all elements $x \in P$, and elements y closed in P, $x \leqq y$, the sum*

$$\sum_{t;\, x \leqq t \leqq J(t) = y} \mu(x, t)$$

has value $\mu_Q(x, y)$ if x is closed, and has value 0 otherwise.

[1]) We wish to express our gratitude to the National Research Council, Canada, for their support of this research (grant A-2994), to K. JACOBS, for his organization of the extraordinary conference "Kombinatorik" at Oberwolfach, and to D. KLEITMAN and J. GOLDMAN, for their organization of the combinatorics seminar at M.I.T., for which this material was prepared.

38*

Proof. Note that the theorem may be rewritten in the form

$$\delta(x, J(x))\,\mu_Q(J(x), y) = \sum_{t \in P} \mu(x, t)\,\delta(J(t), y)\,. \tag{1}$$

Without loss of generality, we assume $x = 0$ in P. For each element $y \in Q$, let $a(y) = \sum_{t \in P} \mu(0, t)\,\delta(J(t), y)$. Then

$$\sum_{y \in Q} a(y)\,\zeta(y, z) = \sum_{t, y \in P} \mu(0, t)\,\delta(J(t), y)\,\zeta(y, z) = \sum_{t \in P} \mu(0, t)\,\zeta(t, z) = \delta_P(0, z)\,.$$

If $0 < J(0)$, $\delta_P(0, z) = 0$ for all $z \in Q$, and $a(y) = 0$ for all $y \in Q$. If $0 = J(0)$, $\delta_P(0, z) = 1$ for $z = 0$, and $a(y) = \mu_Q(0, y)$. ∎

Given a function f from a finite lattice P into a ring with unit, associate the difference operators D, E

lower difference $$Df(x) = \sum_{y;\, y \le x} f(y)\,\mu(y, x)\,,$$

upper difference $$Ef(x) = \sum_{y;\, x \le y} \mu(x, y)\,f(y)\,.$$

Theorem 2 (Analogue of integration by parts). *If f, g are functions from a finite lattice P into a ring, then*

$$\sum_{x \in P} Df(x)\,g(x) = \sum_{x \in P} f(x)\,Eg(x)\,.$$

Proof. Both are equal to $\sum_{x, y} f(x)\,\mu(x, y)\,g(y)$. ∎

It is interesting to compare the proof of theorem 2 with the argument that cycles and coboundaries in a graph are orthogonal to one another. For each vertex p and edge x, let

$$\varepsilon(p, x) = \begin{cases} +1 & \text{if } p \text{ is the head of } x, \\ -1 & \text{if } p \text{ is the tail of } x, \\ 0 & \text{otherwise.} \end{cases}$$

Boundary and coboundary operators are defined by

$$\partial f(p) = \sum_x \varepsilon(p, x)\,f(x) \quad \text{for any 1-chain } f,$$
$$\delta g(x) = \sum_p g(p)\,\varepsilon(p, x) \quad \text{for any 0-chain } g.$$

If f is a 1-cycle ($\partial f = 0$) and h is a 1-coboundary ($h = \delta g$), then

$$\sum_x f(x)\,h(x) = \sum_{x, p} f(x)\,\varepsilon(p, x)\,g(p) = \sum_p \partial f(p)\,g(p) = \sum_p 0\,g(p) = 0\,.$$

If $\sigma: P \to L$ is a supremum-homomorphism from a complete lattice P into a complete lattice L, then $\sigma^{\Delta}: L \to P$ is an infimum-homomorphism, defined by

$$\sigma^{\Delta}(y) = \sup\{x;\, \sigma(x) \le y\}\,.$$

The pair σ, σ^{Δ} is a Galois connection, in the sense that $P/\sigma^{\Delta}(\sigma)$ is isomorphic to $L/\sigma(\sigma^{\Delta})$, where $\sigma^{\Delta}(\sigma)$ is a closure operator on P and $\sigma(\sigma^{\Delta})$ is a coclosure operator on L. All Galois connections between complete lattices arise in this fashion. In the special case where σ is onto L, $P/\sigma^{\Delta}(\sigma) \cong L$.

We now combine theorems 1 and 2 to establish a general theorem on Möbius inversion across Galois connections. This theorem is the discrete analogue of the change-of-variables formula for integration, and of Stokes' Theorem, $\int_{\partial S} \omega = \int_S d\omega$.

Theorem 3. *If σ: $P \to L$ is a sup-homomorphism from a finite lattice P into a finite lattice L, if f is a function from P and g is a function from L taking values in a ring, then*

$$\sum_{x \in P} D f(x)\, g(\sigma(x)) = \sum_{y \in L} f(\sigma^{\Delta}(y))\, E g(y)\,.$$

Proof. Let $Q = P/\sigma^{\Delta}\sigma \cong L/\sigma\sigma^{\Delta}$ be the common quotient lattice, and regard both f and g, restricted to closed elements in P and L, as functions on Q. Then

$$\begin{aligned} \sum_{x \in P} D f(x)\, g(\sigma(x)) &= \sum_{t, x \in P} f(t)\, \mu(t, x)\, g(\sigma(x)) = \sum_{t, x \in P, s \in Q} f(t)\, \mu_P(t, x)\, \delta(\sigma(x), s)\, g(s) = \\ &= \sum_{r, s \in Q} f(r)\, \mu_Q(r, s)\, g(s) = \sum_{r \in Q, z, y \in L} f(r)\, \delta(r, \sigma^{\Delta}(y))\, \mu_L(y, z)\, g(z) = \\ &= \sum_{y, z \in L} f(\sigma^{\Delta}(y))\, \mu_L(y, z)\, g(z) = \sum_{y \in L} f(\sigma^{\Delta}(y))\, E g(y)\,. \quad \blacksquare \end{aligned} \tag{2}$$

A number of related forms of Theorem 3 may be more convenient in applications. For instance, using the fact that the difference operators D and E are inverses to the summation operators S and T,

$$S f(x) = \sum_{y; y \leqq x} f(y); \quad T f(x) = \sum_{y; x \leqq y} f(y)\,,$$

we obtain corollary 1 by substituting Sf for f, Tg for g.

Corollary 1. *If σ: $P \to L$ is a sup-homomorphism from a finite lattice P into a finite lattice L, if f is a function on P and g is a function on L, then*

$$\sum_{x \in P} f(x)\, g(\sigma(x)) = \sum_{y \in L} S f(\sigma^{\Delta}(y))\, E g(y)\,,$$

$$\sum_{x \in P} D f(x)\, T g(\sigma(x)) = \sum_{y \in L} f(\sigma^{\Delta}(y))\, g(y)\,,$$

$$\sum_{x \in P} f(x)\, T g(\sigma(x)) = \sum_{y \in L} S f(\sigma^{\Delta}(y))\, g(y)\,. \quad \blacksquare$$

The symmetric intermediate form (2) appearing in the proof of Theorem 3 deserves special note:

Corollary 2. *If σ: $P \to L$ is a sup-homomorphism from a finite lattice P into a finite lattice L, if Q is the quotient lattice $P/\sigma^{\Delta}(\sigma) \cong L/\sigma(\sigma^{\Delta})$, and if functions f on P and g on L are defined on Q by restriction to closed elements of P, coclosed elements of L, respectively, then*

$$\sum_{x \in P} D f(x)\, g(\sigma(x)) = \sum_{r, s \in Q} f(r)\, \mu_Q(r, s)\, g(s) = \sum_{y \in L} f(\sigma^{\Delta}(y))\, E g(y)\,. \quad \blacksquare$$

For Galois connections given directly as a pair of order-inverting maps σ, τ whose composites $\sigma\tau$ and $\tau\sigma$ are increasing, it is more convenient to have Theorem 3 in the form obtained by inverting the lattice L, as follows.

Corollary 3. *If* $\sigma\colon P \to L$, $\tau\colon L \to P$ *is a Galois connection between finite lattices* P *and* L, *if* f *is a function on* P, *and* g *is a function on* L, *then*

$$\sum_{x\in P} Df(x)\, g(\sigma(x)) = \sum_{y\in L} f(\tau(y))\, Dg(y)\,. \ \blacksquare$$

Theorem 1, above, lacks the full symmetry of Galois connections because it operates between a lattice P and its quotient Q, rather than between two lattices P, L, with a common quotient Q. The symmetric form for Theorem 1, and thus for Rota's main theorems, is recoverable from Theorem 3 as follows.

Corollary 4. *If* σ *is a sup-homomorphism from a finite lattice* P *into a finite lattice* L, *then for any elements* $t \in P$, $z \in L$

$$\sum_{x\in P} \mu(t,x)\, \delta(\sigma(x), z) = \sum_{y\in L} \delta(t, \sigma^{\Delta}(y))\, \mu(y,z)\,.$$

This common value is clearly equal to 0 *unless* t *is closed in* P (ie: $t < x \Rightarrow \sigma(t) < \sigma(x)$) *and* z *is coclosed in* L (ie: $\exists x \in P;\ \sigma(x) = z$). *If* t *is closed in* P *and* z *is closed in* L, *both* t *and* z *correspond to elements of the common quotient lattice* Q, *and the common value of the summations is equal to* $\mu_Q(t, z)$.

Proof. Set $f(x) = \delta(t,x)$, $g(y) = \delta(y,z)$. Then $Df(x) = \mu_P(t,x)$ and $Eg(y) = \mu_L(y,z)$. The intermediate symmetric form (2) arising in the proof of Theorem 3 is in this case equal to

$$\sum_{r,s\in Q} \delta(t,r)\, \mu_Q(r,s)\, \delta(s,z)\,. \ \blacksquare$$

The theory of Möbius inversion across a composite of sup-homomorphisms develops directly from Theorem 3.

Corollary 5. *If* $\sigma_i\colon P_{i-1} \to P_i$, $i = 1, \ldots, k$, *is a sequence of sup-homomorphisms between finite lattices, if* f *is a function on* P_0 *and* g *is a function on* P_k, *then the sums*

$$\sum_{x,y\in P_i} f(\sigma_1^{\Delta}(\cdots \sigma_i^{\Delta}(x) \quad))\, \mu(x,y)\, g(\sigma_k(\cdots \sigma_{i+1}(y) \quad)) \tag{3}$$

are equal, for $i = 0, \ldots, k$. (Note that for $i = k$ the evaluation of g is at y.) $\blacksquare$

For computations involving composites such as $\sigma^{\Delta}(\tau^{\Delta})$ it should be borne in mind that Δ is a contravariant functor, ie: $\sigma^{\Delta}(\tau^{\Delta}) = (\tau(\sigma))^{\Delta}$.

The applicability of Corollary 5 is appreciably extended by the observation that composites of sup-homomorphisms give rise to commutative diagrams involving the intermediate quotients. For two sup-homomorphisms, diagram 1 applies.

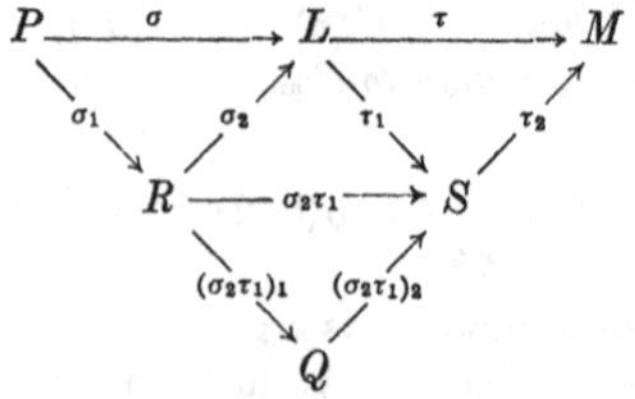

Diagram 1

For any sup-homomorphism β, the symbol β_1 indicates the map of each element to its closure, regarded as an element of the quotient lattice, while the symbol β_2 indicates the map of each element of the quotient, regarded as a closed element of the domain, to its image under β. Note that

$$(\sigma\tau)_1 = \sigma_1(\sigma_2\tau_1)_1 \quad\text{and}\quad (\sigma\tau)_2 = (\sigma_2\tau_1)_2\tau_2\,.$$

A result not obvious from previous forms of Theorem 3 derives from such consideration of quotients. Note that the lattices L and Q in the following corollary are not related by a Galois connection.

Corollary 6. *If σ: $P \to L$ and τ: $L \to M$ are sup-homomorphisms between finite lattices, if f is a function on P and g is a function on M, and if Q is the common quotient lattice of P and M relative to the composite $\tau(\sigma)$, then*

$$\sum_{x,y\in L} f(\sigma^{\Delta}(x))\,\mu_L(x,y)\,g(\tau(y)) = \sum_{r,s\in Q} f(r)\,\mu_Q(r,s)\,g(s)\,. \quad \blacksquare$$

The expressions $f(r)$, $g(s)$ in Corollary 6 refer as usual to $f((\tau(\sigma))_1^{\Delta}(r))$ and $g((\tau(\sigma))_2(s))$, the values of f and g at elements closed in L, coclosed in M, relative to $\tau(\sigma)$.

3. Enumerative Lattice Theory. To each binary relation ϱ: $X \to Y$ between finite sets X and Y there corresponds a Galois connection (σ, τ) between the Boolean algebras $\mathsf{B}(X)$, $\mathsf{B}(Y)$. For all $A \subseteq X$, $B \subseteq Y$,

$$\sigma(A) = \{y\in Y;\ x\in A \Rightarrow x\varrho y\}\,,$$
$$\tau(B) = \{x\in X;\ y\in B \Rightarrow x\varrho y\}.$$

(These definitions are simply $\sigma(A) = \bigcap A$, $\sigma(B) = \bigcap B$, if elements of X are viewed as subsets of Y, and elements of Y are viewed as subsets of X.)

Theorem 4. *If ϱ: $X \to Y$ is a relation between finite sets, if f is a function defined on subsets of X, and if g is a function defined on subsets of Y, then*

$$\sum_{A\subseteq X} Df(A)\,g\left(\bigcap A\right) = \sum_{B\subseteq Y} f\left(\bigcap B\right) Dg(B)\,.$$

Proof. Apply Corollary 3 of Theorem 3 to the Galois connection $\mathsf{B}(X) \underset{\tau}{\overset{\sigma}{\rightleftarrows}} \mathsf{B}(Y)$ defined by $\sigma(A) = \bigcap A \subseteq Y$ for $A \subseteq X$, $\tau(B) = \bigcap B \subseteq X$ for $B \subseteq Y$. $\blacksquare$

The elements of the common quotient lattice Q are precisely those pairs (A, B) $A \subseteq X$, $B \subseteq Y$, which are

1) totally related: $x\in A$, $y \in B \Rightarrow x\varrho y$
and maximal, in the sense that
2) $x \notin A \Rightarrow \exists y \in B$, $x\bar{\varrho}y$,
3) $y \notin B \Rightarrow \exists x \in A$, $x\bar{\varrho}y$,
where $\bar{\varrho}$ denotes negation of relation ϱ.

Each element $s \in Q$ thus has a cardinality $|s|_1$ as a subset of X and a cardinality $|s|_2$ as a subset of Y.

Corollary 1. *If (σ, τ) is the Galois connection defined by a finite relation $\varrho: X \to Y$,*

$$\sum_{A \subseteq X} (\varphi - 1)^{|A|} \nu^{|\sigma(A)|} = \sum_{B \subseteq Y} \varphi^{|\tau(B)|} (\nu - 1)^{|B|}.$$

This sum may in turn be calculated on the common quotient lattice Q, and is equal to

$$\sum_{r, s \in Q} \varphi^{|r|_1} \mu_Q(r, s) \nu^{|s|_2}.$$

Proof. Let $f(A) = \varphi^{|A|}$ for all $A \subseteqq X$, and $g(B) = \nu^{|B|}$, for all $B \subseteqq Y$. Noting that $Df(A) = \sum_{C \subseteq A} \varphi^{|C|} (-1)^{|A-C|} = (\varphi - 1)^{|A|}$, and similarly for Dg, the result follows directly from Theorem 4. ▌

Modulo a few redundancies, Corollary 1 to Theorem 4 is also the fundamental enumerative structure theorem for finite lattices. The redundancies arise, causing nonisomorphic relations to have isomorphic lattices, when $\sigma(x) = \sigma(A)$, for some subset $A \subseteq X$ and some element $x \notin A$ (also when this situation occurs for some element and subset of Y). When such redundancies do not occur in the relation ϱ, the supremum-irreducible elements of the lattice Q are precisely the pairs $(x, \sigma(x))$ for $x \in X$, the infimum-irreducible elements of Q are precisely the pairs $(\tau(y), y)$ for $y \in Y$, and the relation ϱ may be recovered from the lattice Q by

$$x \varrho y \Leftrightarrow x \leqq y \text{ in } Q. \tag{4}$$

Corollary 2. *Let Q be a finite lattice, with set X of supremum-irreducible elements and set Y of infimum-irreducible elements. For each element $z \in Q$, let $\alpha(z)$, $\beta(z)$ be the numbers of sup-irreducibles beneath z and inf-irreducibles above z, respectively. Then*

$$\sum_{A \subseteq X} (\varphi - 1)^{|A|} \nu^{\beta(\sup A)} = \sum_{B \subseteq Y} \varphi^{\alpha(\inf B)} (\nu - 1)^{|B|},$$

and both sums are equal to

$$\sum_{r, s \in Q} \varphi^{\alpha(r)} \mu(r, s) \nu^{\beta(s)}. \quad ▌$$

Redundancies can be reintroduced on the other side of the relation $\leftrightarrow$ lattice correspondence, with interesting results. Given a finite lattice L, and functions j and k from finite sets X and Y, respectively, into L, a binary relation ϱ is defined by

$$x \varrho y \Leftrightarrow j(x) \leqq k(y). \tag{5}$$

Let Q be the quotient lattice of the Galois connection determined by the relation ϱ. Then Q has two order-embeddings in L, neither of which is associated with a closure operator on L. Corollary 6 to Theorem 3 applies.

Theorem 5. *Let j and k be functions from finite sets X and Y into a finite lattice L, and let ϱ and Q be the relation and quotient lattice described above. Each element $z \in L$ has cardinalities $|z|_1$ and $|z|_2$ given by*

$$|z|_1 = |\{e \in X;\ j(e) \leqq z\}|, \quad |z|_2 = |\{e \in Y;\ z \leqq k(e)\}|.$$

Each element $s \in Q$ is realized as a pair (A, B) of subsets of X, Y, and thus has cardinalities $|s|_1 = |A|$, $|s|_2 = |B|$.

$$\sum_{x,y\in L} \varphi^{|x|_1} \mu_L(x, y)\, \nu^{|y|_2} = \sum_{r,s\in Q} \varphi^{|r|_1} \mu_Q(r, s)\, \nu^{|s|_2}.$$

In particular, $\mu_Q(0, 1) = \sum_{x,y\in L} \delta(0, |x|_1)\, \mu(x, y)\, \delta(|y|_2, 0)$.

Proof. The function j extends to a sup-homomorphism σ from the Boolean algebra $\mathsf{B}(X)$ into L by $\sigma(A) = \sup_L\{j(e);\, e \in A\}$. Similarly, k extends to an inf-homomorphism τ^Δ from the inverted Boolean algebra $\tilde{\mathsf{B}}(Y)$ into L (with opposite $\tau\colon L \to \tilde{\mathsf{B}}(Y)$), defined by $\tau^\Delta(B) = \inf_L\{k(e);\, e \in A\}$. Then

$$\tau(\sigma(A)) = \{b \in Y;\; a \in A \Rightarrow j(a) \leqq k(b)\},$$

and the quotient lattice with respect to the composite $\tau(\sigma)$ is equal to Q. The formula follows from Corollary 6 to Theorem 3, and the special case results from setting $\varphi = \gamma = 0$, realizing that $|r|_1 = 0 \Rightarrow r = 0 \in Q$ and $|s|_2 = 0 \Rightarrow s = 1 \in Q$. ∎

If, in the situation described above, $X = Y = L$, and if L is assumed to be an ordered set, not necessarily a lattice, then the resulting lattice Q is the MacNeille completion of the ordered set L.

4. Cross-cuts and Complementation. Rota's cross-cut theorem [6] and this author's complementation theorem [1] have in common a double application of Möbius inversion. Interesting sidelights on these theorems are obtainable by consideration of a lattice of the intervals of a finite lattice.

Theorem 6. *Given a finite lattice L, let $I(L)$ be the set consisting of the empty interval $\emptyset$, together with all intervals $[x, y]$, for $x \leqq y$ in L, ordered by containment. Then $\mu_{I(L)}([x, y], [w, z]) = \mu_L(w, x)\, \mu_L(y, z)$ if $w \leqq x \leqq y \leqq z$, and*

$$\mu_{I(L)}(\emptyset, [x, y]) = -\, \mu_L(x, y).$$

Proof. If $w \leqq x \leqq y \leqq z$, the interval from $[x, y]$ to $[w, z]$ is isomorphic to the cartesian product of the inverted interval $[\tilde{w}, \tilde{x}]$ in $\tilde{L}$ with the interval $[y, z]$ in L. But $\mu_{\tilde{L}}(\tilde{w}, \tilde{x}) = \mu_L(x, w)$, and the Möbius invariant is multiplicative on cartesian products. This establishes the product formula.

$$\mu_{I(L)}(\emptyset, [w, z]) = -\sum_{x,y;\, w\leqq x\leqq y\leqq z} \mu_{I(L)}([x, y], [w, z]) = -\sum_{x,y;\, w\leqq x\leqq y\leqq z} \mu_L(w, x)\, \mu_L(y, z) =$$
$$= -\sum_{x;\, w\leqq x\leqq z} \mu_L(w, x)\, \delta(x, z) = -\, \mu_L(w, z).\ ∎$$

Theorem 7. *Given a finite lattice L, an arbitrary subset $X \subseteq L$, a function f defined on subsets of X, and a function g defined on intervals of L, then*

$$\sum_{A\subseteq X} D f(A)\, g([\inf A, \sup A]) =$$
$$= f(\emptyset)\, g(\emptyset) - f(\emptyset) \sum_{w,z\in L} \mu(w, z)\, g([w, z]) + \sum_{w,x,y,z\in L} f(X \cap [x, y])\, \mu(w, x)\, \zeta(x, y)\, \mu(y, z)\, g([w, z]).$$

Proof. The map σ defined by $\sigma(A) = [\inf A, \sup A]$ is a sup-homomorphism from the Boolean algebra $\mathsf{B}(X)$ into the interval lattice $I(L)$, because

$$\sigma(A \cup B) = [\inf(A \cup B), \sup(A \cup B)] = [\inf A, \sup A] \vee [\inf B, \sup B].$$

Note that $\sigma^{\Delta}([x, y]) = X \cap [x, y]$. By Theorem 3,

$$\sum_{A \subseteq X} Df(A)\, g([\inf A, \sup A]) =$$
$$= \sum_{\emptyset \leqq [x,y] \leqq [w,z] \leqq [0,1]} f(X \cap [x, y])\, \mu_{I(L)}([x, y], [w, z])\, g([w, z]),$$

which reduces to the required form, by Theorem 6, once the summation is separated into three parts:

$$\emptyset = [x, y] = [w, z], \quad \emptyset = [x, y] < [w, z], \quad \text{and} \quad \emptyset < [x, y] \leqq [w, z]. \quad \blacksquare$$

Corollary 1. *If X and Y are arbitrary subsets of a finite lattice L, let q_k be the number of k-element subsets A of X disjoint from Y and spanning L* (ie: $\inf A = 0$, $\sup A = 1$). *Then*

$$q_0 - q_1 + q_2 - \cdots =$$
$$= \delta_L(0, 1) - \mu_L(0, 1) + \sum_{x, y \in L} \zeta(X \cap [x, y], Y)\, \mu(0, x)\, \zeta(x, y)\, \mu(y, 1).$$

Proof. Set $f(A) = \zeta(A, Y)$, so that

$$Df(A) = \sum_{B \subseteq A} (-1)^{|A| - |B|} \zeta(B, Y) = (-1)^{|A|} \delta(\emptyset, A \cap Y).$$

Set $g([w, z]) = \delta(0, w)\, \delta(z, 1)$. The sinister of the equation in Theorem 7 becomes

$$\sum_{A \subseteq X} (-1)^{|A|} \delta(\emptyset, A \cap Y)\, \delta(0, \inf A)\, \delta(\sup A, 1) = \sum_{k=0}^{\infty} (-1)^k q_k,$$

and the simplification of the dexter is obvious. $\blacksquare$

The cross-cut theorem, the complementation theorem, and, one may conjecture, other interesting facts about Möbius invariants of lattices are evaluations of Corollary 1 at particular sets X, Y.

Corollary 2 (The Cross-cut Theorem). *If X is a cross-cut of a finite lattice L, and if q_k is the number of k-element subsets of X which span L, then*

$$q_0 - q_1 + q_2 - \cdots = \mu_L(0, 1).$$

Proof. In Corollary 1 to Theorem 7, let X be the crosscut, and let $Y = \emptyset$. The condition $X \cap [x, y] = \emptyset$ is satisfied if and only if $x \leqq y < z$ for some $z \in X$, or $z < x \leqq y$ for some $z \in X$. These possibilities are mutually exclusive, and are indicated $y < X$ and $X < x$, respectively. Thus

$$\sum_{x, y \in L;\, X \cap [x, y] = \emptyset} \mu(0, x)\, \zeta(x, y)\, \mu(y, 1) =$$
$$= \sum_{x, y;\, y < X} \mu(0, x)\, \zeta(x, y)\, \mu(y, 1) + \sum_{x, y;\, X < x} \mu(0, x)\, \zeta(x, y)\, \mu(y, 1) =$$
$$= \sum_{y;\, y < X} \delta(0, y)\, \mu(y, 1) + \sum_{x;\, X < x} \mu(0, x)\, \delta(x, 1) = 2\mu(0, 1).$$

Substitution of this formula into that of Corollary 1 completes the proof. $\blacksquare$

Corollary 3 (The Complementation Theorem). *If s is any fixed element in a finite lattice L, then*

$$\mu(0,1) = \sum_{x,y \in s^{\perp}} \mu(0,x)\zeta(x,y)\mu(y,1)$$

where $s^{\perp}$ is the set of complements of s in L.

Proof. In Corollary 1 to Theorem 7, let $X = L$ and $Y = s^{\perp}$. Note that

$$\zeta(X \cap [x,y], Y) = \zeta([x,y], s^{\perp}) = 1$$

if and only if both x and y are complements of s. If $0 = 1$, $q_0 - q_1 + \cdots = q_0 = 1$. If $0 \neq 1$, then at most one of 0, 1 are in $s^{\perp}$. Assume w.l.o.g. that $0 \notin s^{\perp}$. Then a subset A disjoint from $s^{\perp} \cup \{0\}$ spans if and only if $A \cup \{0\}$ spans. A and $A \cup \{0\}$ have cardinalities of opposite parity, so $q_0 - q_1 + \cdots = 0$. Thus $q_0 - q_1 + \cdots = \delta(0,1)$, and the corollary follows. ▮

5. Combinatorial Geometry. A combinatorial geometry (or simply, a geometry) (e.g.: [4]) is most easily defined in terms of the lattice structure of its flats (closed subgeometries). Such lattices, which are called *geometric lattices*, have the distinguishing characteristic that, for all $x, y \in L$

$$y \text{ covers } x \Leftrightarrow \exists \text{ atom } p \text{ complementary to } x \text{ in } [0,y].$$

In this definition, "y covers x" means $x < y$ and $x < t \leqq y \Rightarrow t = y$. The complementarity condition requires $x \wedge p = 0$, $x \vee p = y$. We shall consider only finite geometric lattices here, so this single property will suffice for a definition. Geometric lattices are consequently relatively-complemented semimodular lattices, generated by atoms, generated by coatoms, and possessed of a well-defined rank $\lambda(x)$ = length of all maximal chains from 0 to x. (Note $\lambda(0) = 0$.)

The points of the associated *geometry* are the atoms of the geometric lattice. The lines, planes, ..., of the geometry are the sets of points beneath elements of rank 2, 3, ..., respectively, in the lattice.

Linear graphs give rise to geometries. If G is a linear graph with edge set X and vertex set H, the equivalence relation of path-connection along edges in a subset $A \subseteqq X$ yields a partition π_A of the vertex set H into A-path connected components. The map $\sigma: A \to \pi_A$ is a sup-homomorphism from the Boolean algebra $\mathsf{B}(X)$ into the partition lattice $\mathsf{P}(H)$. The Galois-closed edge sets, ie: the maximal sets A of each rank $\lambda(\sigma(A))$, form a geometric lattice $L(G)$.

The coboundary operator, defined parenthetically in section 2, maps ν^q 0-chains $f: H \to \{0, 1, \ldots, \nu - 1\}$ to each coboundary δf, where q is the number of connected components of G. Colorings, those 0-chains which have unequal values on the ends of any edge, correspond to coboundaries which take non-zero values on each edge. The *kernel* $\ker g$ of a coboundary is the set $g^{-1}(0)$, which is necessarily closed. We wish to calculate $P(x;\nu)$, the number of coboundaries with kernel x and values in the ring $\{0, 1, \ldots, \nu - 1\}$. The number of ν-colorings of the graph is $\nu^q p(0;\nu)$.

A coboundary is freely-determined by its values on any basis (spanning tree) for the graph, so there are $\nu^{\lambda(1)-\lambda(x)}$ ν-coboundaries with kernel $\geqq x$, for any $x \in L(G)$.

By Möbius inversion on L, there are

$$p(x;\nu) = \sum_{y;\,x \leqq y} \mu(x,y)\,\nu^{\lambda(1)-\lambda(y)} = E\,p(\ ;\nu)(x)$$

ν-coboundaries with kernel x.

The polynomial $p(0;\nu)$ is clearly well-defined for any finite Dedekind lattice L, ie: a lattice satisfying the chain condition, and thus having a rank function λ. $p(0;\nu)$ has been called the *Poincaré polynomial* of the lattice L. Recent unpublished work tends to establish a relation between Poincaré polynomials and general "coloring problems" on such lattices.

Poincaré polynomials may be considered as polynomial-valued elements of the incidence algebra of the lattice L. Let

$$p(x,z;\nu) = \sum_{y \in L} \mu(x,y)\,\zeta(y,z)\,\nu^{\lambda(z)-\lambda(y)},$$

so that $p(x;\nu) = p(x,1;\nu)$. Such polynomial-valued matrices have easily-calculable inverses in the incidence algebra.

Theorem 8. *Let functions a, b, c on a finite lattice L take values which are invertible elements of a ring. Then*

$$q(x,z) = \sum_{y \in L} a(x)\,\mu(x,y)\,b(y)\,\zeta(y,z)\,c(z)$$

has an inverse

$$q^{-1}(x,z) = \sum_{y \in L} \frac{1}{c(x)}\,\mu(x,y)\,\frac{1}{b(y)}\,\zeta(y,z)\,\frac{1}{a(z)}$$

in the incidence algebra of L. In particular, the Poincaré polynomial $p(x,z;\nu)$ has inverse

$$p^{-1}(x,z;\nu) = \sum_{y \in L} \nu^{\lambda(y)-\lambda(x)}\,\mu(x,y)\,\zeta(y,z). \quad \blacksquare$$

Every geometric lattice L may be realized in a number of ways as a quotient of other geometric lattices with respect to sup-homomorphisms which also preserve the relation covers-or-equals. Such maps are called *strong maps*, and map atoms either to atoms or to 0. There is a notion of orthogonality [3] with respect to any such realization $\sigma\colon M \to L$, giving rise to a strong map $\sigma^*\colon \tilde{M} \to L^*$, whenever the domain lattice M is also modular. The relation $\sigma^{**} = \sigma$ holds.

If the Poincaré polynomial is modified so as to have two numerical variables, it becomes possible to obtain from the polynomial for a Boolean representation, by simple substitution of variables, the corresponding polynomial for the orthogonal geometry. For graphs, this process converts coboundary enumeration to cycle enumeration[2]). The appropriate two-variable polynomial is the *coboundary polynomial* for any strong map $\sigma\colon P \to L$, defined by

(6) $$\tau(\sigma;\varphi,\nu) = \sum_{x \in L} \varphi^{\lambda(\sigma^{\Delta}(x))}\,p(x;\nu).$$

[2]) Cf. [2], [7], [8]. "A Ring in Graph Theory" is an important work, in which TUTTE calculates what has come to be known as the Grothendieck group, for a category of graphs.

Theorem 9. *If $\sigma\colon P \to L$ is a strong map between geometric lattices P, L*

$$\tau(\sigma; \varphi, \nu) = \sum_{x,y \in P} \varphi^{\lambda(x)} \mu(x, y) \nu^{\lambda(1) - \lambda(\sigma(x))}. \tag{7}$$

Proof. Directly from Theorem 3. ∎

The application of Theorem 9 to Boolean representations of a geometry, such as the map from subsets A of the set of atoms to sup A in L, is particularly useful.

Corollary 1. *If $\mathsf{B} = \mathsf{B}(X)$ is a finite Boolean algebra and $\sigma\colon \mathsf{B} \to L$ is a strong map into an geometric lattice L, then*

$$\tau(\sigma; \varphi, \nu) = \sum_{x \in L} \varphi^{|x|} p(x; \nu) = \sum_{A \subseteq X} (\varphi - 1)^{|A|} \nu^{\lambda(1) - \lambda(\sigma(A))}. \quad ∎$$

The definition of orthogonality relative to a strong map $\sigma\colon M \to L$ of a modular geometry onto a geometry L is as follows. The strong map $\sigma\colon M \to L$ determines a closure $J = \sigma^{\Delta}(\sigma)$ on M. There is a unique coclosure J^* on M satisfying, for all x, y in M such that y covers x

$$y \leqq J(x) \Leftrightarrow J^*(y) \nleqq x. \tag{8}$$

The coclosure J^* determines a quotient lattice P (with order induced by that on M), and a map $\tau\colon M \to P$ which is an inf-homomorphism preserving the relation covers-or-equals. The inverted lattice $\tilde{P}$ is a geometric lattice, so we set $L^* = \tilde{P}$. Let σ^* be the associated strong map from $\tilde{M}$ onto L^*.

The *rank generating function* ϱ of a strong map $\sigma\colon P \to L$ defined by

$$\varrho(\sigma; \xi, \eta) = \sum_{x \in P} \xi^{\lambda_L(1) - \lambda_L(\sigma(x))} \eta^{\lambda_P(x) - \lambda_L(\sigma(x))} \tag{9}$$

has symmetry [2] relative to orthogonality, whenever σ maps a modular geometry onto L.

Theorem 10. *$\varrho(\sigma^*; \xi, \eta) = \varrho(\sigma; \eta, \xi)$ for any pair $\sigma\colon M \to L$, $\sigma^*\colon \tilde{M} \to L^*$ of orthogonal maps of a modular geometry.*

Proof. The measurement $\lambda_L(1) - \lambda_L(\sigma(x))$ which provides the exponent of ξ in $\varrho(\sigma)$, enumerates the number of intervals $[y, z]$ of length 1 ("steps") in any maximal chain ("path") from x to 1 in M, for which $z \nleqq J(y)$, ie: $J^*(z) \leqq y$. But this is precisely the measurement $\lambda_{\tilde{M}}(\tilde{x}) - \lambda_{L^*}(\sigma^*(\tilde{x}))$, which provides the exponent of η in $\varrho(\sigma^*)$. Similarly,

$$\lambda_M(x) - \lambda_L(\sigma(x)) = \lambda_{L^*}(1) - \lambda_{L^*}(\sigma^*(\tilde{x}))$$

is the number of steps $[y, z]$ in any path from 0 to x in M for which $z \leqq J(y)$, ie: $J^*(z) \nleqq y$. ∎

Corollary 2. *If $\mathsf{B} = \mathsf{B}(X)$ is a finite Boolean algebra and $\sigma\colon \mathsf{B} \to L$ is a strong map into a geometric lattice L, then*

$$\tau(\sigma; \varphi, \nu) = (\varphi - 1)^{\lambda_L(1)} \varrho\left(\frac{\nu}{\varphi - 1}, \varphi - 1\right).$$

Proof. By Corollary 1,

$$\tau(\sigma;\varphi,\nu)=\sum_{A\subseteq X}(\varphi-1)^{|A|}\nu^{\lambda_L(1)-\lambda(\sigma(A))}=$$

$$=(\varphi-1)^{\lambda_L(1)}\sum_{A\subseteq X}\left(\frac{\nu}{\varphi-1}\right)^{\lambda_L(1)-\lambda(\sigma(A))}(\varphi-1)^{|A|-\lambda(\sigma(A))}. \quad ▮$$

We may now complete the calculation of the cycle polynomial $\tau(\sigma^*)$ from the coboundary polynomial $\tau(\sigma)$.

Corollary 3. *If* $\mathsf{B}=\mathsf{B}(X)$ *is a finite Boolean algebra and* $\sigma\colon \mathsf{B}\to L$ *is a strong map onto a geometric lattice* L, *with orthogonal* $\sigma^*\colon \tilde{\mathsf{B}}\to L^*$, *then*

$$\tau(\sigma^*;\varphi,\nu)=(\varphi-1)^{|1|}\nu^{-\lambda(1)}\tau\left(\sigma;\frac{\nu+\varphi-1}{\varphi-1},\nu\right).$$

Proof. From Corollary 2 we have $\varrho(\xi,\eta)=\eta^{-\lambda(1)}\tau(\sigma;\eta+1,\xi\eta)$. Also $\lambda^*(1)=$ $=|1|-\lambda(1)$. Thus

$$\tau(\sigma^*;\varphi,\nu)=(\varphi-1)^{|1|-\lambda(1)}\varrho^*\left(\frac{\nu}{\varphi-1},\varphi-1\right)=(\varphi-1)^{|1|-\lambda(1)}\varrho\left(\varphi-1,\frac{\nu}{\varphi-1}\right)=$$

$$=(\varphi-1)^{|1|-\lambda(1)}\nu^{-\lambda(1)}(\varphi-1)^{\lambda(1)}\tau\left(\sigma;\frac{\nu+\varphi-1}{\varphi-1},\nu\right). \quad ▮$$

So far we have dealt with representations of geometries as quotients of simpler geometries of higher rank. A few parallel results are available for embeddings of a given geometry as a subgeometry of various larger geometries, usually of equal rank.

Corollary 4. *If* $\sigma\colon P\to L$ *is a strong map between geometric lattices and if* $\iota\colon L\to N$ *is a* 1-1 *strong map from* L *into a geometric lattice* N, *then*

$$\tau(\iota(\sigma);\varphi,\nu)=\nu^{\lambda_N(1)-\lambda_N(\iota(1))}\tau(\sigma;\varphi,\nu).$$

Proof. $\tau(\iota(\sigma);\varphi,\nu)=\sum_{x,y\in L}\varphi^{\lambda_P(\sigma^\Delta(x))}\mu_L(x,y)\,\nu^{\lambda_N(1)-\lambda_N(\iota(y))}=\nu^{\lambda_N(1)-\lambda_N(\iota(1))}\tau(\sigma;\varphi,\nu)$ because $\lambda_N(\iota(y))=\lambda_L(y)$ for all $y\in L$. ▮

Corollary 5. *If* $\iota\colon L\to N$ *is a* 1-1 *strong map from a geometric lattice* L *into a geometric lattice* N, *then the relation*

$$\nu^{\lambda_N(1)-\lambda_L(1)}P_L(0,\nu)=\sum_{x\in N;\,\iota^\Delta(x)=0}P_N(x;\nu).$$

In particular, if N *is the lattice of all partitions of the set* H *of vertices of a graph* G *and* L *is the lattice* $L(G)$ *of closed subsets of the edge set* X *of* G, *then*

$$\nu^{|H|-1-\lambda(\iota(X))}P_L(0,\nu)=\sum_{k=1}^{\infty}\pi_k(\nu-1)(\nu-2)\cdots(\nu-k+1)$$

where π_k *is the number of* k*-part color-partitions of the vertex set* H *of* G.

Proof. Evaluate $\tau(\iota;\varphi,\nu)$, simplify by using Corollary 4, and set $\varphi=0$. ▮

Bibliography

[1] H. H. Crapo, The Möbius Function of a Lattice. J. Combinatorial Theory **1**, 126–131 (1966).
[2] H. H. Crapo, The Tutte Polynomial. Aequationes Math. (to appear).
[3] H. H. Crapo, Geometric Duality. Rend. Sem. Mat. Univ. Padova **38**, 23–26 (1967).
[4] D. A. Higgs, Strong Maps of Geometries. J. Combinatorial Theory **5** (1968) (to appear).
[5] O. Ore, Galois Connexions. Trans Amer. Math. Soc. **55**, 493–513 (1944).
[6] G.-C. Rota, On the Foundations of Combinatorial Theory I. Z. Wahrscheinlichkeitstheorie und verw. Gebiete **2**, 340–368 (1964).
[7] W. T. Tutte, A Ring in Graph Theory. Proc. Cambridge Philos. Soc. **43**, 26–40 (1947).
[8] W. T. Tutte, A Contribution to the Theory of Chromatic Polynomials. Canad. J. Math. **6**, 80–91 (1954).

Eingegangen am 9. 11. 1967

Anschrift des Autors:
Henry H. Crapo
Department of Mathematics, University of Waterloo, Waterloo, Ontario, Canada

Reprinted from JOURNAL OF COMBINATORIAL THEORY
Vo. 7, No. 3, November 1969

A Generalization of a Combinatorial Theorem of Macaulay

G. F. CLEMENTS AND B. LINDSTRÖM

University of Colorado, Boulder, Colorado 80302, and University of Stockholm, Sweden

Communicated by D. H. Younger

Received May 26, 1968

ABSTRACT

Let E denote the set of all vectors of dimension n ($n \geqslant 2$) with non-negative integral components. E is ordered in the lexicographic order. Let E_v denote the subset of all vectors in E with component sum v. If H_v denotes any subset of E_v let LH_v denote the set of the $| H_v |$ last elements in E_v, where $| H_v |$ is the number of elements of H_v. Let PH_v denote the set of all vectors of E_{v+1}, which are obtained by the addition of 1 to a component of a vector in H_v. In [3] Macaulay proved the inclusion $P(LH_v) \subset L(PH_v)$. Sperner gave a shorter proof in [4]. Let $k_1 \leqslant k_2 \leqslant \cdots \leqslant k_n$ be given positive integers and let F denote the set of all vectors $(a_1, ..., a_n)$ with integer components and $0 \leqslant a_i \leqslant k_i$ $i = 1, ..., n$. We shall prove Macaulay's inclusion for subsets H_v of F_v even if the operators P and L are restricted to operate in F. This will follow from our theorem. As another application we prove a generalization of the main result in [2]. By a different method Katona proved the theorem when $k_1 = k_2 = \cdots = k_n = 1$ (see [1, Theorem 1]).

1. INTRODUCTION AND STATEMENT OF THE RESULTS

Let the integers $k_1, k_2, ..., k_n$ be given such that $1 \leqslant k_1 \leqslant k_2 \leqslant \cdots \leqslant k_n$. The set of all vectors $(a_1, ..., a_n)$ of dimension n with integer components a_i for which $0 \leqslant a_i \leqslant k_i$, $i = 1, ..., n$, will be denoted by F. Write $(a_1, ..., a_n) < (b_1, ..., b_n)$ if $a_1 < b_1$ or if $a_1 = b_1, ..., a_{i-1} = b_{i-1}$, $a_i < b_i$ for some i, $2 \leqslant i \leqslant n$ (lexicographic order). Put $k_1 + \cdots + k_n = k$. Define

$$F_v = \{(a_1, ..., a_n) \mid a_1 + \cdots + a_n = v\} \cap F, \qquad v = 0, 1, ..., k.$$

If H is a subset of F, put $H_v = H \cap F_v$ and let $| H_v |$ denote the number of elements of H_v. The set of the $| H_v |$ first elements of F_v in the lexicographic order will be denoted by CH_v and is called the *compression* of H_v. The set of the last $| H_v |$ elements of F_v is denoted by LH_v. If H is any subset of F let

$$CH = \bigcup_{v=0}^{k} CH_v. \tag{1.1}$$

Let Γ be the multivalued function from F into F which associates with $(a_1, ..., a_n)$ the set

$$\{(a_1 - 1, a_2, ..., a_n), (a_1, a_2 - 1, a_3, ..., a_n), ..., (a_1, ..., a_{n-1}, a_n - 1)\} \cap F.$$

If $\Gamma(H) \subset H$ we say that H is *closed*.

Let P be the multivalued function from F to F which associates with $(a_1, ..., a_n)$ the set

$$\{(a_1 + 1, a_2, ..., a_n), (a_1, a_2 + 1, a_3, ..., a_n), ..., (a_1, ..., a_{n-1}, a_n + 1)\} \cap F.$$

Let S_m denote the set of the m first vectors of F. Put

$$\alpha(a_1, ..., a_n) = \sum_{i=1}^{n} a_i \qquad \text{and} \qquad \alpha(H) = \sum_{\mathbf{a}\in H} \alpha(\mathbf{a}).$$

We can now state our results.

THEOREM. *If $H_v \subset F_v$ then $\Gamma(CH_v) \subset C(\Gamma H_v)$, $v = 0, 1, ..., k$.*

COROLLARY 1. *If $H_v \subset F_v$ then $P(LH_v) \subset L(PH_v)$, $v = 0, 1, ..., k$.*

COROLLARY 2. *If $H \subset F$ and H is closed then CH is closed.*

COROLLARY 3. *If $H \subset F$, $|H| = m$ and H is closed then $\alpha(H) \leqslant \alpha(S_m)$.*

The notions in Corollary 2 were introduced by Lindström and Zetterström in solving a problem of k-adic integers [2]. They proved Corollary 2 in the special case $n = 2$ and all k_i equal (see [2, Lemma 1, p. 167]), but convinced themselves by an incorrect example that the corresponding result for $n = 3$ was wrong (they overlooked 102 in their example on page 169). Theorem 1 in [2] is a special case of Corollary 3 when all k_i are equal.

The theorem of Macaulay follows from Corollary 1 for fixed v when $k_1 \geqslant vn$.

The reader will perhaps find the following picture helpful to grasp the theorem. Let n, k_i for $i = 1, ..., n$ and v be fixed positive integers satisfying

$$1 \leqslant k_1 \leqslant k_2 \leqslant \cdots \leqslant k_n, \qquad 1 \leqslant v \leqslant k = \sum_{i=1}^{n} k_i .$$

Let the points $(a_1, ..., a_n)$ with integral coordinates a_i such that $0 \leqslant a_i \leqslant k_i$ be locations of buttons which turn on lights at those positions

$$(a_1 - 1, a_2, ..., a_n), ..., (a_1, ..., a_{n-1}, a_n - 1)$$

which have non-negative coordinates. Suppose one is required to press m of the buttons in the hyperplane $a_1 + a_2 + \cdots + a_n = v$. Which ones should be pressed so as to minimize the number of lights that go on? The minimum will be realized by using the first m buttons in the lexicographic order which lie in the hyperplane. For instance, if $n = 3$, $k_1 = k_2 = k_3 = 4$, and $v = 7$, the relation between the buttons, indicated by small solid circles, and lights, indicated by large open circles, are shown in Figure 1. A button turns on the lights connected to it by a straight line.

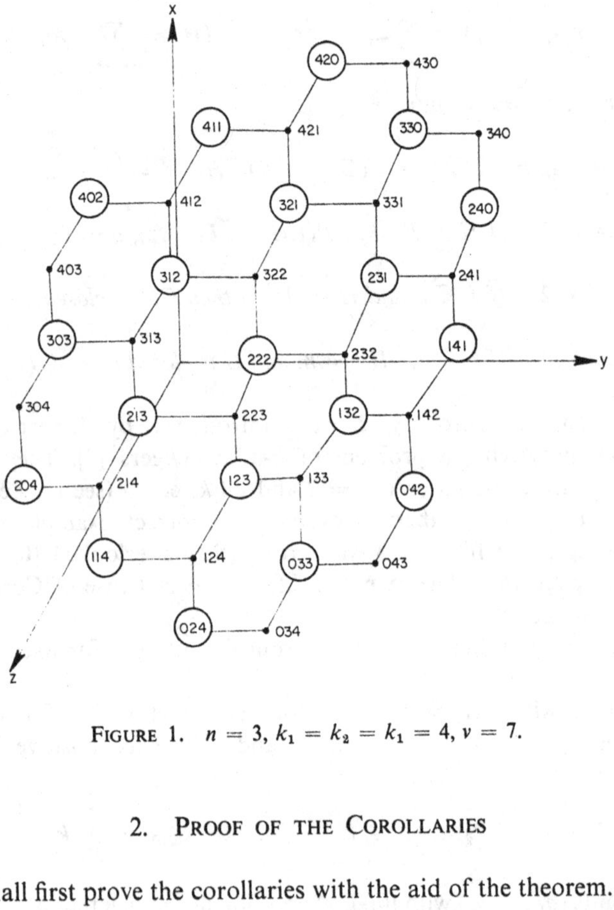

FIGURE 1. $n = 3$, $k_1 = k_2 = k_1 = 4$, $v = 7$.

2. PROOF OF THE COROLLARIES

We shall first prove the corollaries with the aid of the theorem.

PROOF OF COROLLARY 1: Apply the mapping of F onto F which maps $(a_1, \ldots, a_n)$ on $(k_1 - a_1, \ldots, k_n - a_n)$. F_v is then mapped on F_{k-v}. If H_v is mapped on H_{k-v} then CH_v is mapped on LH'_{k-v} and ΓH_v is mapped on PH'_{k-v}. The rest of the proof is now obvious.

PROOF OF COROLLARY 2: If $\Gamma(H) \subset H$ then $\Gamma(H_v) \subset H_{v-1}$ for $v = 1, \ldots, k$ From the theorem it follows $\Gamma(CH_v) \subset C(H_{v-1})$ and then

$$\Gamma(CH) = \Gamma\left(\bigcup_{v=0}^{k} CH_v\right) = \bigcup_{v=1}^{k} \Gamma(CH_v) \subset \bigcup_{v=1}^{k} C(H_{v-1}) \subset CH.$$

PROOF OF COROLLARY 3: Assume $CH \neq S_m$ and let $\mathbf{a} = (a_1, \ldots, a_n)$ be the first element of F which is not element in CH. Let $\mathbf{b} = (b_1, \ldots, b_n)$ be the last element of CH. It follows $\mathbf{a} < \mathbf{b}$, for $CH \neq S_m$. We shall next prove $\alpha(\mathbf{a}) > \alpha(\mathbf{b})$.

If $\alpha(\mathbf{a}) = \alpha(\mathbf{b})$, $\mathbf{a} \in CH$ follows from the definition (1.1) since $\mathbf{a} < \mathbf{b}$, and $\mathbf{a} \notin CH$ is contradicted. We now prove that $\alpha(\mathbf{a}) < \alpha(\mathbf{b})$ implies the same contradiction. Since $\mathbf{a} < \mathbf{b}$ let $a_1 = b_1, \ldots, a_{i-1} = b_{i-1}$, $a_i < b_i$. If $a_{i+1} = \cdots = a_n = 0$ we obtain $\mathbf{a}$ from $\mathbf{b}$ by the subtraction of $\alpha(\mathbf{b}) - \alpha(\mathbf{a})$ ones from $b_i, b_{i+1}, \ldots, b_n$. It then follows $\mathbf{a} \in CH$, for CH is closed by Corollary 2. If $a_j > 0$ for some $j > i$ put $d = \alpha(\mathbf{b}) - \alpha(\mathbf{a})$ and $d' = d - \min\{d, b_i - a_i - 1\}$. It follows $d' \leqslant b_{i+1} + \cdots + b_n$. We can now subtract d' ones from the integers $b_{i+1}, \ldots, b_n$ so that we obtain non-negative integers $c_{i+1}, \ldots, c_n$. Put $c_i = b_i - d + d'$ and $c_j = a_j$ for $j = 1, 2, \ldots, i-1$. We now have $\mathbf{c} = (c_1, \ldots, c_n) \in CH$, for $\mathbf{b} \in CH$ and CH is closed by Corollary 2. From $a_i < c_i$ it follows $\mathbf{a} < \mathbf{c}$. Then we get $\mathbf{a} \in CH$, for $c \in CH$ and $\alpha(\mathbf{c}) = \alpha(\mathbf{a})$. Thus $\alpha(\mathbf{a}) > \alpha(\mathbf{b})$ follows.

If we delete $\mathbf{b}$ from CH and adjoin $\mathbf{a}$ to CH, we obtain a set H' such that $\Gamma H' \subset H'$, $CH' = H'$ and $\alpha(H') > \alpha(CH)$. If $H' \neq S_m$ we can repeat the operation ′ and after m steps, at most, we have $H'^{\cdots\prime} = S_m$ and $\alpha(H'^{\cdots\prime}) > \alpha(CH) = \alpha(H)$, which is the result.

3. PROOF OF THE THEOREM

We shall first define some auxiliary notions. If $H \subset F$, put

$$H_{i:d} = \{(a_1, \ldots, a_n) \mid a_i = d\} \cap H, \qquad i = 1, 2, \ldots, n; \quad d = 0, 1, \ldots, k_i.$$

For subsets H_v of F_v let $(CH_v)_{i:d}$ denote the set of the $|(H_v)_{i:d}|$ first elements of $(F_v)_{i:d}$. We shall say that H_v is *i-compressed* if $(CH_v)_{i:d} = (H_v)_{i:d}$, $d = 0, 1, \ldots, k_i$. If $CH_v = H_v$ we say that H_v is *compressed.*

For any subset H_v of F_v we can define the sequence of sets $H_v^1, H_v^2, \ldots, H_v^j, \ldots$ by putting $H_v = H_v^1$ and

$$H_v^{j+1} = \bigcup_{d=0}^{k_i} (CH_v^j)_{i:d}, \qquad \text{where} \quad i \equiv j (\text{mod } n), \quad 1 \leqslant i \leqslant n. \tag{3.1}$$

We shall prove five lemmas.

LEMMA 1. *One can find p such that H_v^p is i-compressed for $i = 1, 2, \ldots, n$.*

PROOF: Enumerate the elements of F_v in the lexicographic order. If $\mathbf{a} \in F_v$ let $n(\mathbf{a})$ be $\mathbf{a}$'s number. For any subset H_v of F_v define $n(H_v)$ as the sum of numbers of its elements. It is evident that

$$n(H_v^1) \geqslant n(H_v^2) \geqslant \cdots \geqslant n(H_v^j) \geqslant \cdots$$

and

$$n(H_v^j) > n(H_v^{j+1}) \qquad \text{if} \quad H_v^{j+1} \neq H_v^j.$$

Since the sequence cannot decrease indefinitely, there must exist a p such that $H_v^p = H_v^{p+1} = \cdots$, i.e., H_v^p is i-compressed for $i = 1, 2, \ldots, n$.

LEMMA 2. *If the theorem is true in $n - 1$ dimensions and if $\Gamma H_v \subset H_{v-1}$ (n dimensions), it follows $\Gamma H_v^j \subset H_{v-1}^j$ for $j = 2, 3, \ldots$.*

PROOF: The proof is by induction from j to $j + 1$. From

$$\Gamma(H_v^j)_{i:d} \cap (F_{v-1})_{i:d} \subset (H_{v-1}^j)_{i:d}$$

it follows (in $n - 1$ dimensions)

$$\Gamma(CH_v^j)_{i:d} \cap (F_{v-1})_{i:d} \subset (CH_{v-1}^j)_{i:d}.$$

From $\Gamma H_v^j \subset H_{v-1}^j$, we obtain

$$|(H_v^j)_{i:d}| \leqslant |(H_{v-1}^j)_{i:d-1}, \qquad d \geqslant 1,$$

and then

$$\Gamma(CH_v^j)_{i:d} \cap (F_{v-1})_{i:d-1} \subset (CH_{v-1}^j)_{i:d-1}, \qquad d \geqslant 1, \tag{3.3}$$

for the left side is the first $|(CH_v^j)_{i:d}|$ elements of $(F_{v-1})_{i:d-1}$ and the right side is the first $|(CH_{v-1}^j)_{i:d-1}|$ elements of $(F_{v-1})_{i:d-1}$. From (3.2) and (3.3) we have

$$\Gamma(CH_v^j)_{i:0} \subset (CH_{v-1}^j)_{i:0}, \quad \Gamma(CH_v^j)_{i:d} \subset (CH_{v-1}^j)_{i:d} \cup (CH_{v-1}^j)_{i:d-1}, \quad d \geqslant 1.$$

If we take the union for $d = 0, 1, \ldots, k_i$, we obtain by (3.1) $\Gamma H_v^{j+1} \subset H_{v-1}^{j+1}$, and the lemma follows by induction since $\Gamma H_v^1 \subset H_{v-1}^1$.

LEMMA 3. *If H_v is compressed, then ΓH_v is compressed.*

PROOF: Assume $\mathbf{a} = (a_1, \ldots, a_n)$, $\mathbf{b} = (b_1, \ldots, b_n)$ are elements in F_{v-1} and $\mathbf{a} < \mathbf{b}$. Then, if

$$(b_1, \ldots, b_j + 1, \ldots, b_n) \in H_v, \qquad 1 \leqslant j \leqslant n,$$

we shall prove $\mathbf{a} \in \Gamma H_v$.

Let $a_1 = b_1, \ldots, a_{i-1} = b_{i-1}$, $a_i < b_i$. Then if $j \leqslant i$, we have

$$(a_1, \ldots, a_j + 1, \ldots, a_n) < (b_1, \ldots, b_j + 1, \ldots, b_n) \in H_v.$$

It follows $(a_1, \ldots, a_j + 1, \ldots, a_n) \in H_v$ since H_v is compressed. Hence $\mathbf{a} \in \Gamma H_v$. If $j > i$, we disregard the first $i - 1$ components and assume $a_1 < b_1$. If $a_v < k_v$ for some $v > 1$, or if $v = 1$ and $a_v < b_v - 1$, we have $(a_1, \ldots, a_v + 1, \ldots, a_n) < (b_1, \ldots, b_i + 1, \ldots, b_n) \in H_v$ and then $\mathbf{a} \in \Gamma H_v$. If $\mathbf{a} = (b_1 - 1, k_2, \ldots, k_n)$, we find that $\mathbf{b} = (b_1, k_2, \ldots, k_v - 1, \ldots, k_n)$ for some $v > 1$ and then $j = 1$ or v. Hence $(b_1 + 1, k_2, \ldots, k_v - 1, \ldots, k_n) \in H_v$ or $(b_1, k_2, \ldots, k_n) \in H_v$. It follows $(b_1, k_2, \ldots, k_n) \in H_v$ and $\mathbf{a} \in \Gamma H_v$.

LEMMA 4. *Let $n \geqslant 3$. Assume that $\mathbf{g} = (g_1, \ldots, g_n)$ and $\mathbf{h} = (h_1, \ldots, h_n)$ are elements in F_v, $\mathbf{g} < \mathbf{h}$ and $h_n = 0$ or $g_n = k_n$. Then if $\mathbf{h} \in S$, and S is i-compressed for $i = 1, 2, n$, it follows $\mathbf{g} \in S$.*

PROOF: The conclusion follows if we can find an increasing sequence of vectors in F_v beginning in $\mathbf{g}$ and ending in $\mathbf{h}$ such that any two consecutive vectors have an i-th component equal ($i = 1, 2$ or n).

First assume $g_n = k_n$. Consider three cases:

(1°) $g_1 = h_1$ is trivial.

(2°) If $g_1 < h_1$ and $g_i > 0$ for some i, $2 \leqslant i \leqslant n - 1$, we have

$$(g_1, g_2, \ldots, g_n) < (g_1 + 1, g_2', \ldots, g_{n-1}', g_n)$$

where

$$g_2' + \cdots + g_{n-1}' = g_2 + \cdots + g_{n-1} - 1 \qquad \text{and} \qquad (g_2', \ldots, g_{n-1}')$$

is as small as possible in the lexicographic order. We find then

$$(g_1 + 1, g_2', \ldots, g_{n-1}', g_n) \leqslant (h_1, h_2, \ldots, h_n), \quad \text{if} \quad g_1 + 1 = h_1. \quad (3.5)$$

The sequence follows from (3.4) and (3.5) if $g_1 + 1 = h_1$.

If $g_1 + 1 < h_1$ and $g_i' > 0$ for some $i \geqslant 2$ proceed as in (2°). If $g_1 + 1 < h_1$ and $g_2' = \cdots = g_{n-1} = 0$ proceed to (3°).

(3°) If $g_1 < h_1$ and $g_2 = g_3 = \cdots = g_{n-1} = 0$, we obtain the inequalities

$$(g_1, g_2, ..., g_n) < (h_1, g_2, ..., g_{n-1}, g_n - h_1 + g_1) \leqslant (h_1, ..., h_n). \quad (3.6)$$

The second vector belongs to F_v since $h_1 - g_1 \leqslant k_1 \leqslant k_n = g_n$.

We have proved that if $g_n = k_n$ one can find the desired sequence of vectors. Then assume $h_n = 0$. Apply the preceding result to the vectors $(k_1 - h_1, ..., k_n - h_n) < (k_1 - g_1, ..., k_n - g_n)$.

LEMMA 5. *The theorem is true when* $n = 2$.

PROOF: A subset $B = \{(a_1, a_2), (a_1 + 1, a_2 - 1), ..., (a_1 + c, a_2 - c)\}$ of F_v is called a *block*. A block which contains the first element of F_v is called an *initial block*. If B is any block and B_0 is an initial block, we easily obtain in all cases, since $k_1 \leqslant k_2$,

$$| \Gamma B | - | B | \geqslant | \Gamma B_0 | - | B_0 |.$$

(This is not true when $k_1 \geqslant v > k_2$). If $B_1, ..., B_r$ are all the maximal blocks which are subsets of H_v, it follows $\Gamma B_i \cap \Gamma B_j = \emptyset$ if $i \neq j$ and then

$$| \Gamma H_v | - | H_v | = \sum_{i=1}^{r} (| \Gamma B_i | - | B_i |) \geqslant | \Gamma B_0 | - | B_0 |$$

for any initial block B_0. In particular when $B_0 = CH_v$, we have $| \Gamma H_v | \geqslant | \Gamma(CH_v)|$ and $\Gamma(CH_v) \subset C(\Gamma H_v)$, since $\Gamma(CH_v)$ is compressed.

PROOF OF THE THEOREM: The theorem will be proved by induction from $n - 1$ to n. It is true for $n = 2$ by Lemma 5. Assume that the theorem is true in $n - 1$ dimensions. Let H_v be any subset of F_v and consider $H_v{}^j$ and $(\Gamma H_v)^j$ for $j = 2, 3, ...$. By Lemma 1 we can determine p such that $S = H_v{}^p$ is i-compressed for $i = 1, ..., n$. Put $(\Gamma H_v)^p = T$ for abbreviation. From Lemma 2 it follows

$$\Gamma S \subset T. \quad (3.7)$$

To complete the proof we show how to alter S to CH_v and T to a subset of $C(\Gamma H_v)$ in such a way that $\Gamma(CH_v) \subset C(\Gamma H_v)$ is obtained.

First, if $S = F_v$ then $| S | = | CH_v |$ shows that $CH_v = F_v$. Also $\Gamma(S) = \Gamma(F_v) = F_{v-1} \subset T$ implies $T = F_{v-1}$. Then since $| T | = | C(\Gamma H_v)|$, it follows $C(\Gamma H_v) = F_{v-1}$, and hence $\Gamma(CH)_v = \Gamma(F_v) = F_{v-1} = C(\Gamma H_v)$, and we are done.

If $S \neq F_v$ there is a first vector $\mathbf{g} = (g_1, ..., g_n)$ of F_v which is not in S,

Let $\mathbf{h} = (h_1, ..., h_n)$ denote the last vector of S. If $\mathbf{h} < \mathbf{g}$, then $S = CH_v$, and so it is no loss of generality to assume $\mathbf{h} > \mathbf{g}$.

It follows from Lemma 4 that $h_n > 0$, since S is i-compressed for $i = 1, 2, ..., n$. Define $\mathbf{h}^* = (h_1, ..., h_{n-1}, h_n - 1)$ and

$$\mathbf{g}^* = (g_1, ..., g_{n-1}, g_n - 1) \quad \text{if } g_n > 0.$$

Note that $\mathbf{h}^*, \mathbf{g}^* \in F_{v-1}$.

From (3.7) it follows $\mathbf{h}^* \in T$. If $\mathbf{x} \in S - \{\mathbf{h}\}$, then $\Gamma(\mathbf{x}) < \mathbf{x} < \mathbf{h}$ (where $\Gamma(\mathbf{x})$ denotes any image of $\mathbf{x}$). This shows that $\Gamma(\mathbf{x}) = \mathbf{h}^*$ for no $\mathbf{x} \in S - \{\mathbf{h}\}$. Now let

$$S' = (S - \{\mathbf{h}\}) \cup \{\mathbf{g}\},$$

$$T' = \begin{cases} (T - \{\mathbf{h}^*\}) \cup \{\mathbf{g}^*\} & \text{if } g_n > 0, \\ T & \text{if } g_n = 0. \end{cases}$$

We now show that $\Gamma S' \subset T'$. Since $\Gamma(\mathbf{x}) = \mathbf{h}^*$ for no $\mathbf{x} \in S - \{\mathbf{h}\}$, it suffices to show that $\Gamma(\mathbf{g}) \subset T'$. If $g_n > 0$ then $\mathbf{g}^*$ is an image of $\mathbf{g}$ under Γ, which is in T' by construction. Observe that $g_n < k_n$. This follows from Lemma 4 since S is i-compressed for $i = 1, 2, ..., n$. If $g_i > 0$ for some i, $1 \leqslant i \leqslant n - 1$, then $(g_1, ..., g_i - 1, ..., g_n)$ is an image of $\mathbf{g}$ under Γ. But it is also an image of $(g_1, ..., g_i - 1, ..., g_n + 1)$, which is in S because it precedes $\mathbf{g}$ and $\mathbf{g}$ was the smallest element of F_v not in S.

Since $\Gamma(S) \subset T$, it follows $(g_1, ..., g_i - 1, ..., g_n) \in T$. Also, because $h_1 > g_1$ (S is 1-compressed), we find $\mathbf{h}^* \neq (g_1, ..., g_i - 1, ..., g_n)$ so $(g_1, ..., g_i - 1, ..., g_n)$ is (still) in T'. Thus all images of $\mathbf{g}$ are in T', so $\Gamma(S') \subset T'$. Obviously, S' is i-compressed for $i = 1, 2, ..., n$.

After a finite number of applications of ′, we have $S'^{\cdots\prime} = CH_v$ and $\Gamma(CH_v) \subset T'^{\cdots\prime} = U$. Now $C(\Gamma H_v)$ is the first $|\Gamma H_v|$ elements of F_{v-1} while $\Gamma(CH_v)$ is the first $|\Gamma(CH_v)|$ elements of F_{v-1} by Lemma 3. But $|\Gamma(CH_v)| \leqslant |U| = |T| = |\Gamma H_v|$. It follows $\Gamma(CH_v) \subset C(\Gamma H_v)$, and the theorem is proved.

4. Concluding Remarks

We can show that $p = 4$ suffices in Lemma 1 when $n = 3$. The proof is rather long since the number of cases which one must consider is large.

We recently noticed that Katona [1] has proved our theorem when $k_1 = k_2 = \cdots = k_n = 1$. Katona puts his result in the language of set theory. He observes that "the theorem is probably useful in proofs by induction over the maximal number of elements of the subsets in a system, as was Sperner's lemma in his paper" [5].

References

1. G. Katona, A Theorem of Finite Sets, *Theory of Graphs* (Proceedings of the colloquium held at Tihany, Hungary September 1966), ed. by P. Erdös and G. Katona, Academic Press, New York and London, 1968.
2. B. Lindström and H.-O. Zetterström, A Combinatorial Problem in the k-adic Number System, *Proc. Amer. Math. Soc.* **18** (1967), 166–170.
3. F. S. Macaulay, Some Properties of Enumeration in the Theory of Modular Systems, *Proc. London Math. Soc.* **26** (1927), 531–555.
4. E. Sperner, Über einen kombinatorischen Satz von Macaulay und seine Anwendung auf die Theorie der Polynomideale, Abh. Math. Sem. Univ. Hamburg **7** (1930), 149–163.
5. E. Sperner, Ein Satz über Untermengen einer endlichen Menge, *Math. Z.* **27** (1928), 544–548.

PRINTED IN BRUGES, BELGIUM, BY THE ST. CATHERINE PRESS, LTD.

Reprinted from:
JOURNAL OF MATHEMATICAL PHYSICS VOLUME 11, NUMBER 6 JUNE 1970

Short Proof of a Conjecture by Dyson

I. J. GOOD
Department of Statistics, Virginia Polytechnic Institute, Blacksburg, Virginia

(Received 26 December 1969)

Dyson made a mathematical conjecture in his work on the distribution of energy levels in complex systems. A proof is given, which is much shorter than two that have been published before.

Let $G(\mathbf{a})$ denote the constant term in the expansion of

$$F(\mathbf{x};\mathbf{a}) = \prod_{i\neq j}\left(1-\frac{x_j}{x_i}\right)^{a_j}, \quad i,j = 1,2,\cdots,n,$$

where $a_1, a_2, \cdots, a_n$ are nonnegative integers and where $F(\mathbf{x};\mathbf{a})$ is expanded in positive and negative powers of $x_1, x_2, \cdots, x_n$. Dyson[1] conjectured that $G(\mathbf{a}) = M(\mathbf{a})$, where $M(\mathbf{a})$ is the multinomial coefficient $(a_1 + \cdots + a_n)!/(a_1! \cdots a_n!)$. This was proved by Gunson[2] and by Wilson.[3] A much shorter proof is given here.

By applying Lagrange's interpolation formula (see, for example, Kopal[4]) to the function of x that is identically equal to 1 and then putting $x = 0$, we see that

$$\sum_j \prod_i \left(1-\frac{x_j}{x_i}\right)^{-1} = 1, \; i \neq j.$$

By multiplying $F(\mathbf{x};\mathbf{a})$ by this function we see that, if $a_j \neq 0$, $j = 1, \cdots, n$, then

$$F(\mathbf{x};\mathbf{a}) = \sum_j F(\mathbf{x}; a_1, a_2, \cdots, a_{j-1}, a_j - 1, a_{j+1}, \cdots, a_n),$$

so that

$$G(\mathbf{a}) = \sum_j G(a_1, \cdots, a_{j-1}, a_j - 1, a_{j+1}, \cdots, a_n). \tag{1}$$

If $a_j = 0$, then x_j occurs only to negative powers in $F(\mathbf{x};\mathbf{a})$ so that $G(\mathbf{a})$ is then equal to the constant term in

$$F(x_1, \cdots, x_{j-1}, x_{j+1}, \cdots, x_n; a_1, \cdots, a_{j-1}, a_{j+1}, \cdots, a_n),$$

that is,

$$G(\mathbf{a}) = G(a_1, \cdots, a_{j-1}, a_{j+1}, \cdots, a_n), \quad \text{if } a_j = 0. \tag{2}$$

Also, of course,

$$G(\mathbf{0}) = 1. \tag{3}$$

Equations (1)–(3) clearly uniquely define $G(\mathbf{a})$ recursively. Moreover, they are satisfied by putting $G(\mathbf{a}) = M(\mathbf{a})$. Therefore $G(\mathbf{a}) = M(\mathbf{a})$, as conjectured by Dyson.

[1] F. J. Dyson, J. Math. Phys. 3, 140, 157, 166 (1962).
[2] J. Gunson, J. Math. Phys. 3, 752 (1962).
[3] K. G. Wilson, J. Math. Phys. 3, 1040 (1962).
[4] Z. Kopal, *Numerical Analysis* (Chapman and Hall, London, 1955), p. 21.

Reprinted from ADVANCES IN MATHEMATICS Vol. 5, No. 1, August 1970
Printed in Belgium

On a Lemma of Littlewood and Offord on the Distributions of Linear Combinations of Vectors*

DANIEL J. KLEITMAN

Department of Mathematics, Massachusetts Institute of Technology, Cambridge, Massachusetts 02139

In this paper we prove the following result:

THEOREM I. *Let $a_1, \ldots, a_n$ be vectors in a Hilbert space S, each with length at least unity. The number of their linear combinations with coefficients 0 or 1 that can lie in the union of any k regions $R_1, \ldots, R_k$ in S each of diameter less than ($<$) unity is not more than the sum of k largest binomial coefficients on N.*

This result is related to a problem of Littlewood and Offord [1] on the distribution of roots of algebraic equations. Various special cases have been obtained by several authors [2–4]. Theorem I settles a long standing conjecture of P. Erdös [2].

The method of proof can be extended straightforwardly to prove the following generalization.

THEOREM II. *Let $a_1, \ldots, a_n$ be vectors in a Hilbert space S each of length at least unity and let $m_1, M_1, \ldots, m_n, M_n$ be integers. Then the number of linear combinations $\sum_{n=1}^{N} c_i a_i$ with coefficients c_i integral and in $[m_i, M_i]$ that lie in the union of any k regions in S each of diameter less than ($<$) one is no more than the number of such linear combinations whose "weights" ($\sum c_i$) are among k "most populous" weights. This is the number of linear combinations satisfying*

$$\left[\frac{1}{2}\left(1 - k + \sum_{i=1}^{N} (M_i + m_i)\right)\right] \leqslant \sum_{i=1}^{N} c_i \leqslant \left[\frac{1}{2}\left(k - 1 + \sum_{i=1}^{N} (M_i + m_i)\right)\right].$$

* This work was supported in part by NSF GP-13778.

As this bound can be achieved by choosing all a's identical, the result is best possible.

We present the proof in detail for Theorem I only. Parallel steps for Theorem II are easily obtained. We can, without loss of generality, assume that our regions R_j are mutually disjoint. We do so below.

We first note a well-known property of binomial coefficients. The sum of k largest binomial coefficients on N is equal to the sum of $k+1$ and $k-1$ largest binomial coefficients on $N-1$. To prove this we note that k largest binomial coefficients on N are $\binom{N}{r}, \binom{N}{r+1}, \ldots, \binom{N}{s}$, where $r = [(N-k+1)/2]$ and $s = r + k - 1 = [(N+k-1)/2]$. Applying the recursion $\binom{N}{j} = \binom{N-1}{j} + \binom{N-1}{j-1}$ to each of these coefficient yield the desired relation, viz.,

$$\sum_{j=r}^{s} \binom{N}{j} = \sum_{j=r-1}^{s} \binom{N-1}{j} + \sum_{j=r}^{s-1} \binom{N-1}{j}.$$

We prove our theorem by induction on N. In light of the property just described we need only show that (0, 1)-linear combinations of $(a_1, \ldots, a_N)$ lying in k disjoint regions of diameter < 1 can be put in 1:1 correspondence with (0, 1)-linear combinations of $(a_1, \ldots, a_{N-1})$ lying either in $k+1$ or $k-1$ such regions.

Now (0, 1)-linear combinations of $(a_1, \ldots, a_N)$ have coefficient of a_N either zero or one. In the former case they may be considered as (0, 1)-linear combinations of $(a_1, \ldots, a_{N-1})$ as they stand. They must lie in our k disjoint regions as sums of $(a_1, \ldots, a_{N-1})$ if they are to do so as sums of $(a_1, \ldots, a_N)$. In the latter case, involving linear combinations explicitly containing a_N, the condition that they lie in our k disjoint regions is that their $(a_1, \ldots, a_{N-1})$ parts lie in the translation of these regions by $-a_N$.

If we show that the translation by $-a_N$ of at least one of our regions is disjoint from each of our k original regions we are done, since sums lying in our original k regions with a_N correspond to sums without a_N lying in these regions plus our translated region ($k+1$ disjoint regions) or lying in the $k-1$ other translated regions which are themselves mutually disjoint and of diameter < 1.

If we consider a hyperplane normal to a_N placed so that all regions $R_1 \cdots R_k$ lie on one side of it (the side on which $x \cdot a_N \geqslant 0$) and so that it just touches the closure of some R_j, the translation of R_j by $-a_N$ must lie on the other side of the hyperplane and hence must be disjoint from the regions $R_1 \cdots R_N$. This remark completes the proof.

If we were concerned with linear combinations as in Theorem II we could proceed in the same manner. If the coefficient of a_N can take on $M_N - m_N$ different values we obtain $(M_N - m_N)k$ regions in terms of $(a_1, \ldots, a_{N-1})$ to correspond to the k regions $R_1 \cdots R_k$. The desired recursion relation is obtained if these can be divided into disjoint families of regions of sizes $k + (M_N - m_N)$, $k + (M_N - m_N) - 2$, $k + (M_N - m_N) - 4, \ldots$. The hyperplane construction above permits us to find such families. That is if $m_N = 0$ the original k regions along with the translation of R_j as defined by $-a_N$, $-2a_N, \ldots, -M_N a_N$, form $k + M_N$ disjoint regions. Another disjoint family can be obtained by throwing away these and repeating the procedure just described. Iteration of this procedure produces disjoint families of regions which by induction yield the recusion satisfied by k most populous weights.

Acknowledgment

The author thanks R. Graham who suggested the geometric hyperplane interpretation of the argument described above.

References

1. J. E. Littlewood and C. Offord, On the number of real roots of a random algebric equation (III), *Mat. USSR Sb.* **12** (1943), 277–285.
2. P. Erdös, On a lemma of Littlewood and Offord, *Bull. Amer. Math. Soc.* **5** (1945), 898–902.
3. D. Kleitman, On a lemma of Littlewood and Offord on the distribution of certain sums, *Math. Z.* **90** (1965), 251–259.
4. G. Katona, On a conjecture of Erdös and a stronger form of Sperner's theorem, *Studia Sci. Math. Hungar.* **1** (1966), 59–63.

PRINTED IN BELGIUM BY THE ST. CATHERINE PRESS, TEMPELHOF 37, BRUGES, LTD.

Ramsey's Theorem for a Class of Categories

R. L. GRAHAM

Bell Telephone Laboratories, Incorporated, Murray Hill, New Jersey

K. LEEB

Universität Erlangen, Erlangen, Germany

AND

B. L. ROTHSCHILD

University of California, Los Angeles, California 90024

DEDICATED TO RICHARD RADO ON THE OCCASION OF HIS 65TH BIRTHDAY

1. INTRODUCTION AND BASIC TERMINOLOGY

In this paper we present a Ramsey theorem for certain categories which is sufficiently general to include as special cases the finite vector space analog to Ramsey's theorem (conjectured by Gian-Carlo Rota), the Ramsey theorem for n-parameter sets [2], as well as Ramsey's theorem itself [4, 6]. The Ramsey theorem for finite affine spaces is obtained here simultaneously with that for vector spaces. That these two are equivalent was already known [5, 1], and the arguments previously used to show that the affine theorem implies the projective theorem are also special cases of the results of this paper.

The argument used here to establish the main result is essentially the same as that used for n-parameter sets [2]. What we do here is to abstract the properties of n-parameter sets which suffice to allow the induction argument. In particular, the properties described for n-parameter sets in Remarks 1-3 of [2] are essential.

In order to state the Ramsey property for a category C we must have a notion of rank with which to index the objects and subobjects of the category. To this end, it is convenient to consider henceforth only categories C with the following property:

(a) The objects of C are the nonnegative integers $0, 1, 2, \ldots$, and if $l > k$, $C(l, k) = \varnothing$, where $C(l, k)$ is the set of all morphisms from l to k in C.

Using this property, we define a rank on subobjects of an object l in C. Namely, if $k \xrightarrow{f} l$ and $k' \xrightarrow{f'} l$ are representatives of the same subobject of l, then there must be isomorphisms $k \xrightarrow{\alpha} k'$ and $k' \xrightarrow{\beta} k$. But by (a), this means that $k = k'$. We define the rank of this subobject to be k, and we refer to it as a *k-subobject* of l. We denote by $C\begin{bmatrix} l \\ k \end{bmatrix}$ the class of subobjects of l in C of rank k. We make the convention that for $k < 0$, or $l < 0$,

$C\begin{bmatrix} l \\ k \end{bmatrix} = \varnothing$. In order to make our induction argument work, we need a finiteness condition. We assume in addition to (a) that all categories considered here satisfy:

(b) For each pair of integers there is an integer $y_{k,l}$ such that $C\begin{bmatrix} l \\ k \end{bmatrix}$ is a finite set with $y_{k,l}$ elements. In particular, $y_{0,0} = 1$.

For convenience, all categories we consider are assumed to satisfy

(c) All morphisms of C are monomorphisms.

If $k \xrightarrow{f} l$ is a morphism of C, we let $\bar{f}$ denote the induced mapping on subjects of l. That is, if $s \xrightarrow{g} k$ represents a subobject of k, then $\bar{f}$ takes this subobject into the subobject of l represented by the composition fg. This is clearly well defined, and $\bar{f}\colon C\begin{bmatrix} k \\ s \end{bmatrix} \to C\begin{bmatrix} l \\ s \end{bmatrix}$. An *r-coloring* of $C\begin{bmatrix} l \\ s \end{bmatrix}$ is a function $c\colon C\begin{bmatrix} l \\ s \end{bmatrix} \to \{1, \ldots, r\}$. We say that a subobject has color i if its image under c is i. An r-coloring c of $C\begin{bmatrix} l \\ s \end{bmatrix}$ induces an r-coloring on $C\begin{bmatrix} k \\ s \end{bmatrix}$ by the composition $c\bar{f}$, where $k \xrightarrow{f} l$ is in C. If the image of $c\bar{f}$ is only a single element, we say that c *has a monochromatic k-subobject*, namely, the k-subobject represented by $k \xrightarrow{f} l$.

We can now state the Ramsey property for a category C satisfying (a)-(c):

Given integers k, l, r, there exists a number n, depending only on k, l, r, so that for all $m \geqslant n$, every r-coloring of $C\begin{bmatrix} m \\ k \end{bmatrix}$ has a monochromatic l-subobject.

When C has morphisms $k \xrightarrow{f} l$ which are all the monomorphic functions from $\{1, \ldots, k\}$ to $\{1, \ldots, l\}$, then this is just the statement of Ramsey's Theorem. If C has morphisms $k \xrightarrow{f} l$ which are the linear monomorphisms from $V_k = \langle v_1, \ldots, v_k \rangle$ to $V_l = \langle v_1, \ldots, v_l \rangle$, where $v_1, v_2, \ldots$ form a basis for a vector space V over $GF(q)$, then this is the statement of Rota's conjecture. In this case, the k-subobjects of l correspond to the subspaces of V_l of dimension k. Other examples of special cases of the Ramsey property will be given later.

2. STATEMENT OF THE MAIN RESULT

In order to establish the Ramsey property for certain categories C, we consider a somewhat stronger version of it which makes the induction argument easier.

$C(k; l_1, \ldots, l_r)$: There is a number $N = N_c(k; r; l_1, \ldots, l_r)$ depending only on k, r, $l_1, \ldots, l_r$, such that for any $m \geqslant N$ and any r-coloring c of $C\begin{bmatrix} m \\ k \end{bmatrix}$, there is an i, $1 \leqslant i \leqslant r$, and a morphism $l_i \xrightarrow{f} m$ such that

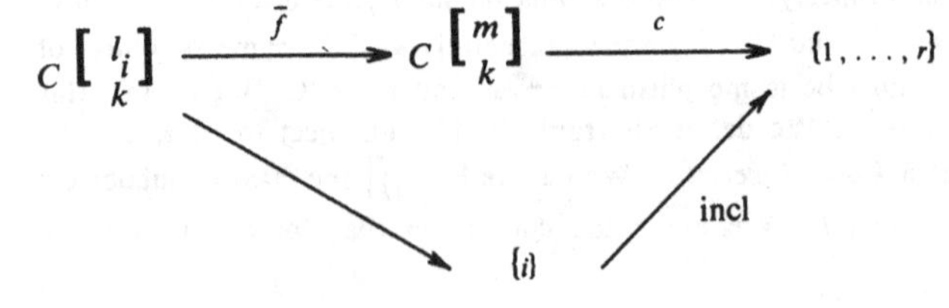

commutes, where incl$(i) = i$.

This statement always holds for $k < 0$, since $C\begin{bmatrix} l_i \\ k \end{bmatrix} = \emptyset$, by convention. If all the l_i are equal, this becomes the Ramsey property stated above.

Theorem 1 below provides the induction step in establishing $C(k; l_1, \ldots, l_r)$ for certain categories. It establishes $B(k+1; l_1, \ldots, l_r)$ if we know $A(k; l_1, \ldots, l_r)$ for all r and l_i provided the categories A and B are related in a special way. This relation is given by the conditions below. For a functor M from A to B with $M(x) = y$ for integers x and y, we denote by $\overline{M}$ the induced function from subobjects of x to subobjects of y. This is given by letting $\overline{M}$ take the subobject represented by $s \xrightarrow{f} x$ in A into the subobject represented by $M(s) \xrightarrow{M(f)} y$ in B.

Conditions on Categories A and B

There is a functor M from A to B with $M(l) = l+1$, $l = 0, 1, \ldots$, a functor P from B to A with $P(l) = l$, $l = 0, 1, \ldots$, an integer $t \geqslant 0$, and for each $l = 0, 1, \ldots t$ morphisms, $l \xrightarrow{\phi_{lj}} l+1$, $1 \leqslant j \leqslant t$, satisfying the following:

I. For each $k+1 = 0, 1, 2, \ldots$, the diagonal d in the following diagram is epic, where $\underline{\text{II}}$ (together with the indicated injections) is coproduct, and d is the unique map determined by the coproduct to make the diagram commute:

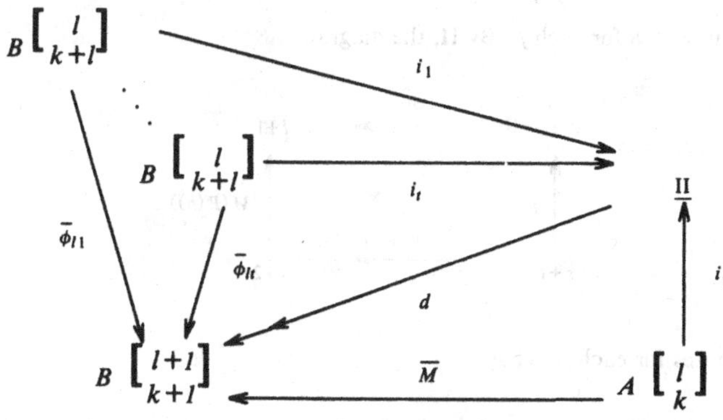

II. For each $s \xrightarrow{g} l$ in B and each $j = 1, \ldots, t$ the following diagram commutes:

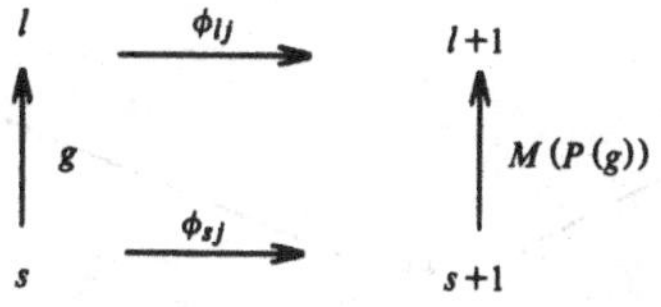

III. For some $l \xrightarrow{e} l+1$ in A, the following diagram commutes for all $j = 1, \ldots, t$:

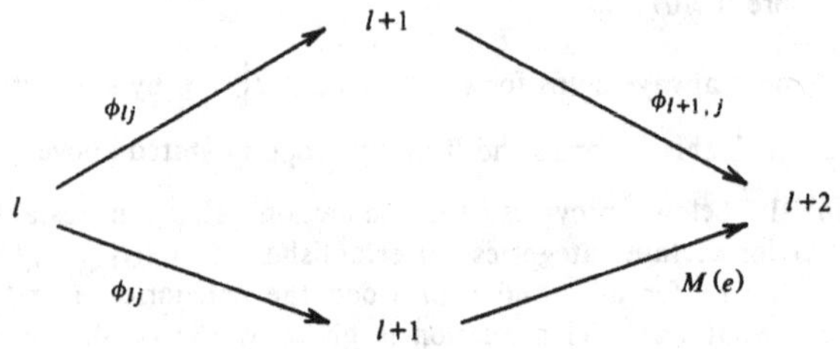

Remark. Let $s+l \xrightarrow{h} l$ in B. Then by III there is some $s \xrightarrow{e} s+1$ in A such that

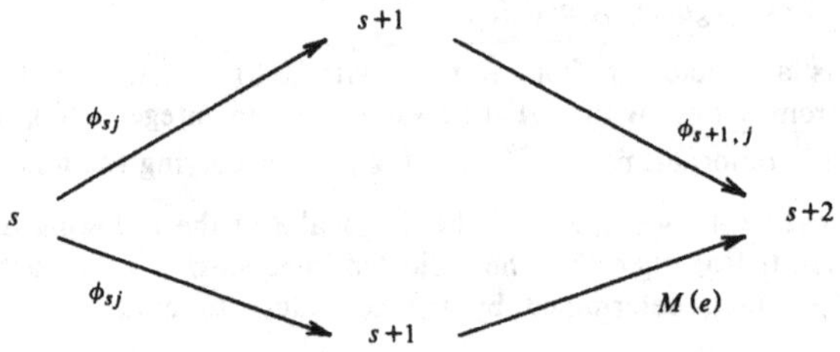

commutes in B for each j. By II, the diagram

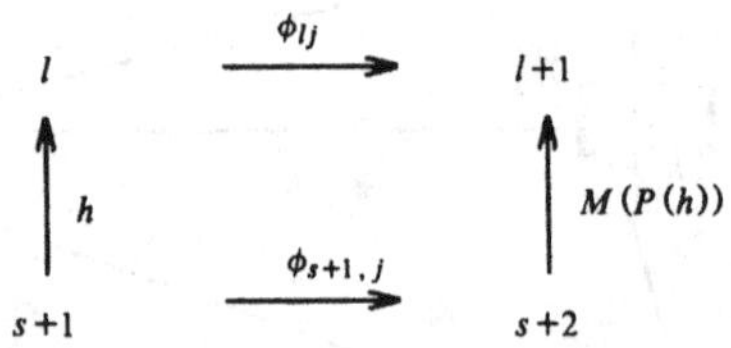

commutes for each j. Thus

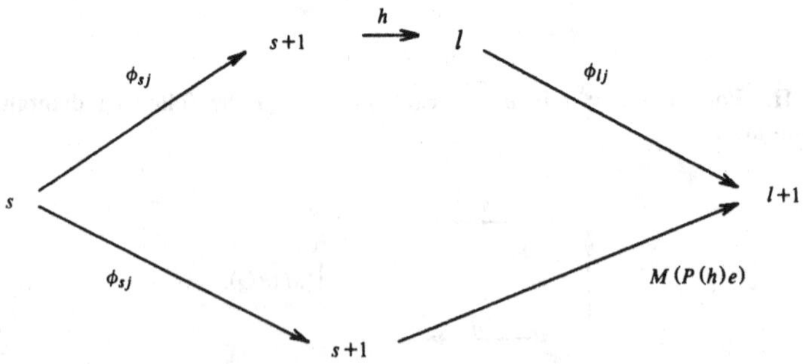

commutes for each j.

THEOREM 1. *Let A and B be two categories satisfying the conditions above. Assume* $A(k; l_1, \ldots, l_r)$ *holds for all* $l_1, \ldots, l_r$ *and* $r > 0$.

Then $B(k+1; l_1, \ldots, l_r)$ *holds for all* $l_1, \ldots, l_r$, *and* $r > 0$.

3. PROOF OF MAIN RESULT

We will eventually need a lemma about n-dimensional arrays of points. We state it now without proof. Proofs can be found in [3] and [2]. (It is a special case of Corollary 4 below, in fact.) We denote by A^n the set of n-tuples $(x_1, \ldots, x_n)$ of elements of x_i of a set A.

LEMMA 1. *Given integers* $r > 0$, $t \geqslant 0$, *there exists an integer* $N = N(r, t)$, *depending only on* r *and* t, *such that if* $n \geqslant N$, A *is a set of* t *elements, and* A^n *is* r*-colored in any way, then there exists a set of* t n*-tuples* $(x_1(j), \ldots, x_n(j))$, $1 \leqslant j \leqslant t$, *all the same color with the property that for each* i, $1 \leqslant i \leqslant n$, *either* $x_i(j) = j$ *for all* j, *or* $x_i(j) = a_i$ *for all* j *and some* $a_i \in A$.

Proof of Theorem 1. We use induction on $L = l_1 + \ldots + l_r$. $B(k+1; l_1, \ldots, l_r)$ holds vacuously if $l_i < k+1$ for any i or if $k+1 < 0$ and trivially if $t = 0$. So we assume $l_i \geqslant k+1 \geqslant 0$ and $t > 0$. If any $l_i = 0$, then $k+1 = 0$, and $B(k+1; l_1, \ldots, l_r)$ holds trivially, since $y_{0,0} = 1$. So we may assume all $l_i > 0$, and, in particular, that $L > 0$. Assume, then, that $B(k+1; l_1, \ldots, l_r)$ holds for $L-1$, and let $l_1 + \ldots + l_r = L$, $l_i > 0$.

DEFINITION. For $1 \leqslant h \leqslant m$, suppose $k+1 \xrightarrow{f} l+h$ is in B, and $f = M(f')$ for some $k \xrightarrow{f'} l+h-1$ in A. For any fixed choice of $j_h, j_{h+1}, \ldots, j_m - 1$, $1 \leqslant j_i \leqslant t$, let $\phi_i = \phi_{l+i,j}$. Then the $(k+1)$-subobject of $l+m$ represented by the composition

$$k+1 \xrightarrow{f} l+h \xrightarrow{\phi_h} l+h+1 \longrightarrow \cdots \longrightarrow l+m-1 \xrightarrow{\phi_{m-1}} l+m$$

is said to have *signature* $(h; j_m - 1, \ldots, j_h)$ with respect to l and m. (The signature need not be unique for a given subobject, nor must every subobject have a signature.) An r-coloring of $B\begin{bmatrix} l+m \\ k+1 \end{bmatrix}$ such that all $(k+1)$-subobjects with the same signature have the same color is called an (l, m)*-coloring.*

For integers l and m we define recursively some numbers needed to prove Lemma 2 below.

$$v_1 = N_A(k; r^{t^{m-1}}; l, \ldots, l)$$

$$v_2 = N_A(k; r^{t^{m-2}}; v_1 + 1, \ldots, v_1 + 1)$$

$$\vdots$$

$$v_m = N_A(k; r^{t^0}; v_m - 1 + 1, \ldots, v_m - 1 + 1).$$

The existence of these numbers is guaranteed by the hypothesis of Theorem 1.

LEMMA 2. *With the same assumptions as in Theorem 1, let* $l \geqslant 0$, $m \geqslant 1$ *be integers; let* $x \geqslant v_m + 1$; *and let* $B\begin{bmatrix} x \\ k+1 \end{bmatrix} \xrightarrow{c} \{1, \ldots, r\}$ *be an* r*-coloring. Then there exists* $l+m \xrightarrow{g} x$ *in* B *such that* $c\bar{g}$ *is an* (l, m)*-coloring of* $B\begin{bmatrix} l+m \\ k+1 \end{bmatrix}$.

Proof. We use induction on m. For $m = 1$ the lemma is trivially true. Assume for some $m \geqslant 2$ that it holds for $m - 1$. Then by induction, and by the choice of the v_i, there is some $v_1 + m \xrightarrow{g} x$ in B such that $B\begin{bmatrix} v_1+m \\ k+1 \end{bmatrix}$ is $(v_1 + 1, m - 1)$-colored by $c\bar{g}$.

We now color $B\begin{bmatrix} v_1+1 \\ k+1 \end{bmatrix}$ as follows: Two subobjects, represented by $k+1 \xrightarrow{f} v_1 + 1$ and $k + 1 \xrightarrow{f'} v_1 + 1$ have the same color if and only if for each choice of $j_m - 1, \ldots, j_1$, $1 \leqslant j_i \leqslant t$, the subobjects represented by the compositions

$$k+1 \xrightarrow{f} v_1 + 1 \xrightarrow{\phi_1} v_1 + 2 \longrightarrow \cdots \longrightarrow v_1 + m - 1 \xrightarrow{\phi_{m-1}} v_1 + m$$

and

$$k+1 \xrightarrow{f'} v_1 + 1 \xrightarrow{\phi_1} v_1 + 1 \longrightarrow \cdots \longrightarrow v_1 + m - 1 \xrightarrow{\phi_{m-1}} v_1 + m$$

have the same color, where $\phi_i = \phi_{v_1+i, j_i}$, $1 \leqslant i \leqslant m - 1$. This is an $r^{t^{m-1}}$-coloring of $B\begin{bmatrix} v_1+1 \\ k+1 \end{bmatrix}$; call it c'.

Next, we color $A\begin{bmatrix} v_1 \\ k \end{bmatrix}$ by the coloring induced by M. That is, a subobject in $A\begin{bmatrix} v_1 \\ k \end{bmatrix}$ is assigned the same color as its image under $\overline{M}$ in $B\begin{bmatrix} v_1+1 \\ k+1 \end{bmatrix}$. In other words, $c'\overline{M}$ is the coloring we use. By the choice of v_1, there is some i, $1 \leqslant i \leqslant r^{t^{m-1}}$, and some $l \xrightarrow{w} v_1$ in A such that the following diagram commutes:

$$\begin{array}{ccccc} A\begin{bmatrix} l \\ k \end{bmatrix} & \xrightarrow{\overline{w}} & A\begin{bmatrix} v_1 \\ k \end{bmatrix} & \xrightarrow{c'\overline{M}} & \{1, \ldots, r^{t^{m-1}}\} \\ & \searrow & & \nearrow \scriptstyle\text{incl} & \\ & & \{i\} & & \end{array}$$

Thus all the subobjects in $\overline{M}(A\begin{bmatrix} l \\ k \end{bmatrix})$ have the same color in $B\begin{bmatrix} l+1 \\ k+1 \end{bmatrix}$ colored by $c'\overline{M(w)}$.

Suppose $k+1 \xrightarrow{f} l + h$ is in B, $1 \leqslant h \leqslant m$, with $f = M(f')$ for some $k \xrightarrow{f'} l + h - 1$ in A. Consider the following diagram:

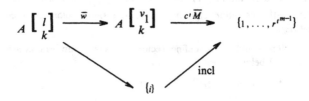

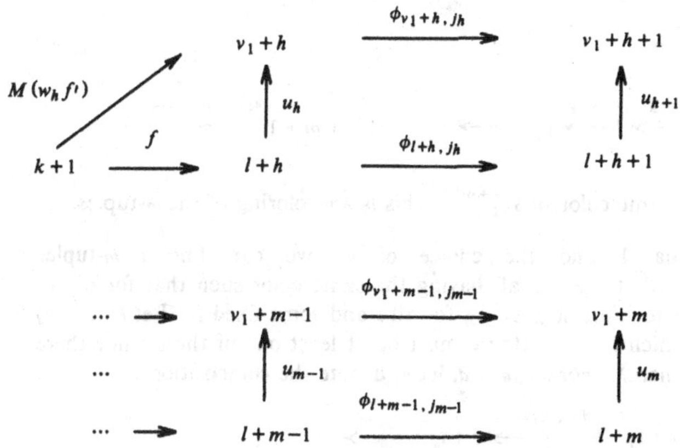

where $u_1 = M(w)$, $u_i = M(P(u_{i-1}))$, $i = 2, 3, \ldots, m$, and $w_1 = w$, $w_i = P(u_{i-1})$, $i = 2, 3, \ldots, m$. By condition II this commutes for each choice of $j_h, j_{h+1}, \ldots, j_m - 1$. Consider any subobject of $l + m$ with signature $(1; j_m - 1, \ldots, j_1)$ with respect to l and m. Let it be represented by $k + 1 \xrightarrow{e} l + m$, where e is the bottom row of the diagram above with $h = 1$. Then $u_m e$ represents a subobject of $v_1 + m$. By the definition of c' and the choice of w, all such subobjects with the same signature $(1; j_m - 1, \ldots, j_1)$ have the same color in $B\begin{bmatrix} v_1+m \\ k+1 \end{bmatrix}$, since the diagram above commutes. On the other hand, consider a subobject of $l + m$ with signature $(h; j_m - 1, \ldots, j_h)$, $h \geqslant 2$, and let it be represented by $k + 1 \xrightarrow{e} l + m$, where e is the bottom row of the diagram. By the commutativity of the diagram, $u_m e = bM(w_h f')$, where $v_1 + h \xrightarrow{b} v_1 + m$ is the top row of the diagram. This means that $u_m e$ has signature $(h - 1; j_m - 1, \ldots, j_h)$ with respect to $v_1 + 1$ and $m - 1$. Since $c\bar{g}$ was a $(v_1 + 1, m - 1)$-coloring of $B\begin{bmatrix} v_1+m \\ k+1 \end{bmatrix}$, the color of this subobject is determined only by the j_i. Thus the color of any subobject with signature $(h; j_m - 1, \ldots, j_h)$ with respect to l and m, $h \geqslant 1$, has its color under the coloring $c\bar{g}\bar{u}_m$ determined only by the j_i. So $c\bar{g}\bar{u}_m$ is an (l, m)-coloring, and the lemma is proved.

We may now proceed with the proof of Theorem 1. Let

$$l = \max_{1 \leqslant i \leqslant r} N_B(k + 1; r; l_1, \ldots, l_{i-1}, l_i - 1, l_{i+1}, \ldots, l_r),$$

a number which must exist by the induction hypothesis. Let $y = r^{y_{l,k+1}}$, where $y_{l,k+1}$ is the number given by property (b). Let $m = N(y, t)$, where $N(y, t)$ is the number given by Lemma 1. Let v_m be the number used in the hypothesis of Lemma 2 (depending on l and m), and let $x \geqslant v_m + 1$. Finally, let $B\begin{bmatrix} x \\ k+1 \end{bmatrix} \xrightarrow{c} \{1, \ldots, r\}$ be an r-coloring. By Lemma 2 there is some $l + m \xrightarrow{g} x$ in B such that $c\bar{g}$ is an (l, m)-coloring of $B\begin{bmatrix} l+m \\ k+1 \end{bmatrix}$. We now color the m-tuples $(j_1, \ldots, j_m)$, $1 \leqslant j_i \leqslant t$, by letting $(j_1, \ldots, j_m)$ and $(k_1, \ldots, k_m)$ have the same color if and only if for each $k + 1 \xrightarrow{h} l$ in B the subobjects represented by the compositions

$$k+1 \xrightarrow{h} l \xrightarrow{\phi_{l,j_1}} l+1 \longrightarrow \cdots \longrightarrow l+m-1 \xrightarrow{\phi_{l+m-1,j_m}} l+m$$

and

$$k+1 \xrightarrow{h} l \xrightarrow{\phi_{l,k}} l+1 \longrightarrow \cdots \longrightarrow l+m-1 \xrightarrow{\phi_{l+m-1,k_m}} l+m$$

both have the same color in $B\begin{bmatrix} l+m \\ k+1 \end{bmatrix}$. This is a y-coloring of the m-tuples.

By Lemma 1 and the choice of m, we can find t m-tuples $(j_1(z), \ldots, j_m(z))$, $1 \leqslant z \leqslant t$, all having the same color such that for each i either $j_i(z) = z$ for all z or $j_i(z) = j_i$ for all z and some fixed j_i. Let $i_1, \ldots, i_d$ be the i for which $j_i(z) = z$ (there must be at least one of these since there are t m-tuples here). For $0 \leqslant a \leqslant d$, let h_a denote the composition

$$l+i_a \xrightarrow{\phi_{l+i_a, j_{i_a+1}}} l+i_a+1 \longrightarrow \cdots$$

$$\cdots \longrightarrow l+i_{a+1}-2 \xrightarrow{\phi_{l+i_{a+1}-2, j_{i_{a+1}-1}}} l+i_{a+1}-1,$$

where we let $i_0 = 0$ and $i_{d+1} = m+1$. Consider the following diagram:

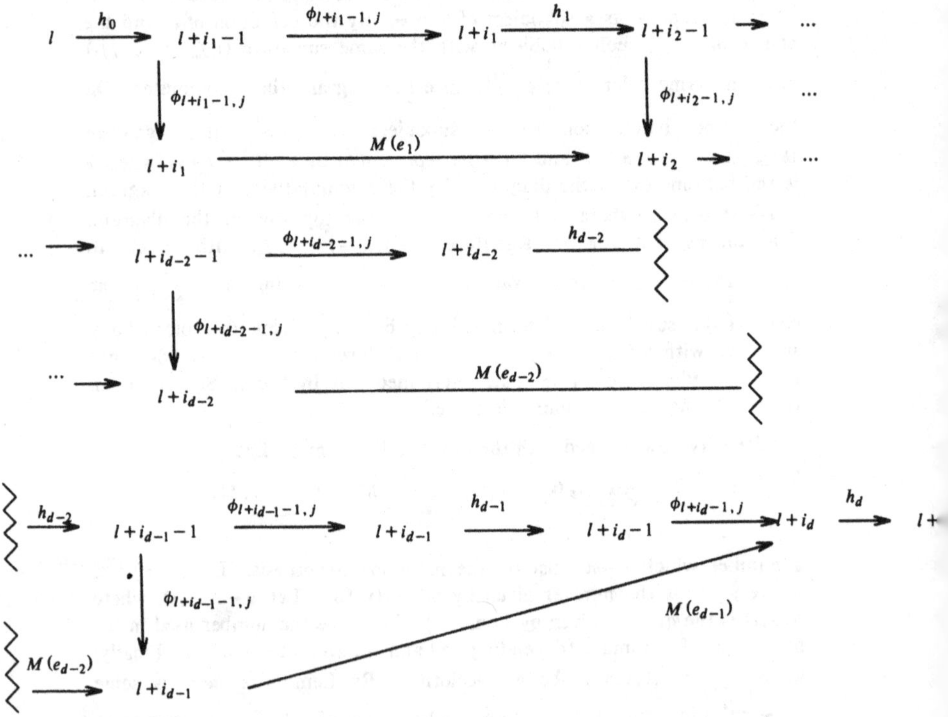

where the $l+i_d-s-1 \xrightarrow{e_{d-s}} l+i_{d-s+1}-1$ in A are those guaranteed by the Remark (following Condition III) to make this diagram commute for each $j = 1, 2, \ldots, t$.

By the choice of the h_a we have for any $k+1 \xrightarrow{h} l$ that the t subobjects represented by

$$k+1 \xrightarrow{h} l \xrightarrow{h_0} l+i_1-1 \xrightarrow{\phi_{l+i_1-1,j}} l+i_1 \longrightarrow \cdots$$

$$\cdots \xrightarrow{M(e_{d-1}e_{d-2}\cdots e_2 e_1)} l+i_d \xrightarrow{h_d} l+m \text{ in } B\begin{bmatrix} l+m \\ k+1 \end{bmatrix}, \quad 1 \leqslant j \leqslant t,$$

all have the same color. By Condition II, the following diagram commutes for all j:

$$\begin{array}{ccc} l+i_1-1 & \xrightarrow{\quad\phi_{l+i_1-1,j}\quad} & l+i_1 \\ \uparrow{\scriptstyle h_0} & & \uparrow{\scriptstyle M(P(h_0))} \\ l & \xrightarrow{\quad\phi_{l,j}\quad} & l+1 \end{array}$$

Then letting $\alpha = h_d M(e_{d-1} \cdots e_1 P(h_0))$ we see that for $k+1 \xrightarrow{h} l$ in B the subobjects represented by the t compositions

$$k+1 \xrightarrow{h} l \xrightarrow{\phi_{l,j}} l+1 \xrightarrow{\alpha} l+m, \quad 1 \leqslant j \leqslant t,$$

all have the same color. Thus $\overline{cg\alpha\phi_{l,j}}$ are equal for all $j = 1, 2, \dots, t$, on $B\begin{bmatrix} l \\ k+1 \end{bmatrix}$.

Now consider any subobject of $\overline{M}(A\begin{bmatrix} l \\ k \end{bmatrix})$ in $B\begin{bmatrix} l+1 \\ k+1 \end{bmatrix}$. Let it be represented by $k+l \xrightarrow{f} l+1$ in B, where $f = M(f')$, $k \xrightarrow{f'} l$ in A. Then the subobject represented by αf has signature $(i_d; j_m, \dots, j_{i_d+1})$ with respect to l and m, since αf is just $h_d M(e_{d-1} \cdots e_1 P(h_0) f')$. Since $l+m$ is (l, m)-colored by $c\,\bar{g}$, all subobjects of $l+m$ with this signature have the same color. Thus $c\,\bar{g}\,\alpha$ gives the same color to any subobject of $\overline{M}(A\begin{bmatrix} l \\ k \end{bmatrix})$, since the signature was independent of the choice of f. That is, $\overline{cg\alpha M}(A\begin{bmatrix} l \\ k \end{bmatrix}) = \{q\}$ for some q, $1 \leqslant q \leqslant r$.

Consider the coloring $\overline{cg\alpha\phi_{l,1}}$ on $B\begin{bmatrix} l \\ k+1 \end{bmatrix}$. By the choice of l, either there is some $l_p \xrightarrow{f_p} l$ in B such that

$$B\begin{bmatrix} l_p \\ k+1 \end{bmatrix} \xrightarrow{\overline{cg\alpha\phi_{l,1}f_p}} \{p\}, \quad p \neq q,$$

or there is some $l_q - 1 \xrightarrow{f_q} l$ in B such that

$$B\begin{bmatrix} l_q - 1 \\ k+1 \end{bmatrix} \xrightarrow{\overline{cg\alpha\phi_{l,1}f_q}} \{q\}.$$

In the former case, we have the desired monochromatic subobject, and the theorem is proved. Hence we may assume that

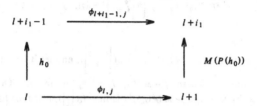

$$B\begin{bmatrix} l_q-1 \\ k+1 \end{bmatrix} \xrightarrow{\overline{cg\alpha\phi_{l,1}f_q}} \{q\} .$$

We recall that $\overline{cg\alpha\phi_{l,1}} = \overline{cg\alpha\phi_{l,j}}$ on $B\begin{bmatrix} l \\ k+1 \end{bmatrix}$ for all j. In particular,

$$\overline{cg\alpha\phi_{l,j}f_q}\left(B\begin{bmatrix} l_q-1 \\ k+1 \end{bmatrix}\right) = \{q\} \quad \text{for all } j .$$

By Condition II, $\phi_{l,j}f_q = M(P(f_q))\,\phi_{l_q-1,j}$, $j = 1,\ldots,t$. Thus

$$\overline{cg\alpha M(P(f_q))\,\phi_{l_q-1,j}}\left(B\begin{bmatrix} l_q-1 \\ k+1 \end{bmatrix}\right) = \{q\}, \quad j = 1,\ldots,t .$$

Now consider any subobject in $\overline{M}\left(A\begin{bmatrix} l_q-1 \\ k \end{bmatrix}\right)$, and let it be represented by $k+1 \xrightarrow{f} l_q$ in B, where $f = M(f')$, $k \xrightarrow{f'} l_q - 1$ in A. The subobject represented by $M(P(f_q))f = M(P(f_q)f')$ is in $\overline{M}\left(A\begin{bmatrix} l \\ k \end{bmatrix}\right)$, and thus has color q by the coloring $\overline{cg\alpha}$. So $\overline{cg\alpha M(P(f_q))}$ colors all subobjects in $M\left(A\begin{bmatrix} l_q-1 \\ k \end{bmatrix}\right)$ color q. We also saw above that $\overline{cg\alpha M(P(f_q))}$ colors all subobjects in $\phi_{l_q-1,j}\left(B\begin{bmatrix} l_q-1 \\ k+1 \end{bmatrix}\right)$ color q. But by Condition I, this accounts for all of $B\begin{bmatrix} l_q \\ k+1 \end{bmatrix}$, and hence $l_q \xrightarrow{g\alpha M(P(f_q))} x$ is the desired morphism, and the theorem is proved.

4. CONSEQUENCES

PROPOSITION 1. *Let $\mathscr{C}$ be a class of categories such that for each category B in $\mathscr{C}$ there is a category A in $\mathscr{C}$ such that A and B satisfy the conditions of Theorem 1. Then $B(k;l_1,\ldots,l_r)$ holds for all $k, l_1,\ldots,l_r$ and all B in $\mathscr{C}$.*

Proof. $B(-1;l_1,\ldots,l_r)$ holds vacuously for all $l_1,\ldots,l_r$, as observed at the beginning of the proof of Theorem 1. This holds for all B in $\mathscr{C}$. Thus for each B we can find a suitable A and apply Theorem 1 to obtain $B(0;l_1,\ldots,l_r)$ for all $l_1,\ldots,l_r$. Proceeding in this fashion from 0 to 1 to 2, etc., we obtain $B(k;l_1,\ldots,l_r)$ for all $k, l_1,\ldots,l_r$ and B in $\mathscr{C}$.

COROLLARY 1 (Ramsey). *Let C be the category with objects the nonnegative integers and morphisms $k \xrightarrow{f} l$ all the monomorphic functions from $\{1,\ldots,k\}$ into $\{1,\ldots,l\}$, where composition is just composition of functions. Then $C(k;l_1,\ldots,l_r)$ holds in general.*

Proof. We must find a class $\mathscr{C}$ containing C which satisfies the conditions of Proposition 1. For $\mathscr{C}$ choose the single category C itself. This clearly satisfies (a)-(c). So for A and B both equal to C, we must show that they satisfy the conditions of Theorem 1.

Let P be the identity functor on C. For any $k \xrightarrow{f} l$ in C, let $M(f)$ be the function $k+1 \xrightarrow{f'} l+1$ in C given by letting $f'(x) = f(x)$, $x \leqslant k$, and $f'(k+1) = l+1$. Let ϕ_l be the function from $\{1,\ldots,l\}$ to $\{1,\ldots,l+1\}$ which acts identically on $\{1,\ldots,l\}$. That is, $\phi_l(x) = x$ for $x \leqslant l$. Then we claim these choices, together with choosing $t = 1$ satisfy I-III.

Consider a subobject in $C\begin{bmatrix} l+1 \\ k+1 \end{bmatrix}$ represented by some $k+1 \xrightarrow{f} l+1$. First suppose $f(s) = l+1$ for some s. Then f represents the same subobject as $f\pi_{s,k+1}$, where $\pi_{s,k+1}$ is the permutation of $\{1, \ldots, k+1\}$ fixing everything except s and $k+1$, which it interchanges. $\pi_{s,k+1}$ is an isomorphism and is its own inverse. Let $k \xrightarrow{f'} l$ be defined by letting $f'(x) = f\pi_{s,k+1}(x)$, $1 \leqslant x \leqslant k$. Then clearly $M(f') = f\pi_{s,k+1}$. Thus the subobject we chose is in $\overline{M}(C\begin{bmatrix} l \\ k \end{bmatrix})$. The only other subobjects are represented by some $k+1 \xrightarrow{f} l+1$ where $f(\{1, \ldots, k+1\}) \subset \{1, \ldots, l\}$. Then letting $k+1 \xrightarrow{f'} l$ be defined by $f'(x) = f(x)$, $1 \leqslant x \leqslant k+1$, we have $f = \phi_l f'$, and the subobject is in $\overline{\phi_l}(C\begin{bmatrix} l \\ k+1 \end{bmatrix})$. This establishes I. II is clear from the definitions. III follows by taking e to be ϕ_l, since $M(\phi_l)(x) = x$ for $1 \leqslant x \leqslant l$. This establishes Corollary 1. We note that if one examines the argument used in the proof of Theorem 1 for this special case, the usual proof of Ramsey's Theorem emerges.

Let V be an infinite-dimensional vector space over $GF(q)$ with basis $v_1, v_2, \ldots$. For each $k = 0, 1, \ldots$, let $V_k = \langle v_1, \ldots, v_k \rangle$, $V_0 = \langle 0 \rangle$. Let C be the category which has objects $0, 1, \ldots$, and morphisms $k \xrightarrow{\phi} l$, where ϕ is a linear monomorphism from V_k to V_l. Composition is ordinary composition of mappings. C clearly satisfies (a)-(c).

COROLLARY 2 (Vector Space Analog). *For the category C described above, $C(k; l_1, \ldots, l_r)$ holds in general.*

Proof. We apply Proposition 1 to a class containing C. Let A be an infinite-dimensional vector space over $GF(q)$ with basis $a_1, a_2, \ldots$, and let $A_m = \langle a_1, \ldots, a_m \rangle$, $A_0 = \langle 0 \rangle$. For $m = 0, 1, 2, \ldots$, the category C_m is defined as follows: The objects of C_m are $0, 1, 2, \ldots$, and the morphisms $k \xrightarrow{(w,\phi)} l$ are all pairs (w, ϕ) where $w \in A_m \otimes V_l$ and ϕ is a linear monomorphism from V_k to V_l. Let $k \xrightarrow{(w,\phi)} l$, where $w = \sum_{i=1}^{m} a_i \otimes w_i$, $w_i \in V_l$, and $l \xrightarrow{(x,\psi)} n$ be morphisms in C_m. Then their composition is defined to be $k \xrightarrow{(y,\psi\phi)} n$, where $y = x + \sum_{i=1}^{m} a_i \otimes \psi(w_i)$. Thus we can think of these morphisms as certain special affine transformations from $A_m \otimes V_k$ into $A_m \otimes V_l$. (a)-(c) are satisfied for the C_m. We choose for our class $\mathscr{C}$ all the C_m. When $m = 0$, we get the category C of Corollary 2.

For each m, let $B = C_m$ and $A = C_{m+1}$. We show that these satisfy Theorem 1. To define M, consider a morphism $k \xrightarrow{(w,\phi)} l$ in C_{m+1}. Then $w \in A_{m+1} \otimes V_l$ can be written uniquely as $w = w' + a_{m+1} \otimes w_{m+1}$, where $w' \in A_m \otimes V_l$. Let $\phi': V_{k+1} \to V_{l+1}$ be determined by letting $\phi'(v_{k+1}) = v_{l+1} + w_{m+1}$, and $\phi' = \phi$ on V_k. Then define $M((w,\phi)) = (w', \phi')$, where $k+1 \xrightarrow{(w',\phi')} l+1$ is in C_m. One can verify by a direct check that M preserves composition. We next define P. Let $k \xrightarrow{(w,\phi)} l$ be in C_m. Then $P((w,\phi)) = (w'', \phi'')$, where $w'' = w + a_{m+1} \otimes 0$, and $\phi'' = \phi$. Clearly P preserves composition. Also, since the identity morphism for k in C_m is $(0, 1_k)$, where 1_k is the identity transformation on V_k, and similarly for C_{m+1}, we see that $M(l) = l+1$ and $P(l) = l$ for each l. Finally, let $t = |A_m| = q^m$, and for each element $a \in A_m$ and each l let $\phi_{la} = (a \otimes v_{l+1}, e_l)$ in C_m, where e_l is the map from V_l to V_{l+1} acting identically on V_l. Then these choices are sufficient to satisfy I-III.

To check I, let $k+1 \xrightarrow{(w',\phi')} l+1$ represent a $(k+1)$-subobject of $l+1$ in C_m. First suppose $\phi'(V_{k+1}) \not\subset V_l$. Then we can choose some isomorphism $\psi: V_{k+1} \to V_{k+1}$ such that $\phi'\psi(V_k) \subset V_l$ and $\phi'\psi(v_{k+1}) = v_{l+1} + v'$ for some

$v' \in V_l$. Furthermore, for a suitable choice of $v \in A_m \otimes V_{k+1}$ we have $(w', \phi')(v, \psi) = (\hat{w}', \phi'\psi)$, with $\hat{w}' \in A_m \otimes V_l$. Of course (w', ϕ') and $(\hat{w}', \phi'\psi)$ represent the same subobject since (v, ψ) is an isomorphism. Now let $k \xrightarrow{(w,\phi)} l$ be in C_{m+1}, where $\phi = \phi\psi'$ on V_k, and $w = \hat{w}' + a_{m+1} \otimes v'$. Then we have $M((w, \phi)) = (\hat{w}', \phi'\psi)$. Thus all subobjects represented by a (w', ϕ') with $\phi'(V_{k+1}) \not\subset V_l$ are in $\overline{M}(C_{m+1} \begin{bmatrix} l \\ k \end{bmatrix})$. On the other hand, if $\phi'(V_{k+1}) \subset V_l$, then $(w', \phi') = (w'' + a \otimes v_{l+1}, \phi')$ for some $a \in A_m$ and some $w'' \in A_m \otimes V_l$. But

$$(w'' + a \otimes v_{l+1}, \phi') = (a \otimes v_{l+1}, e_l)(w'', \phi'') = \phi_{l,a}(w'', \phi''),$$

where $\phi'' = \phi'$ on V_{k+1}, $\phi'': V_{k+1} \to V_l$. Thus the subobject is in $\overline{\phi_{la}}(C_m \begin{bmatrix} l \\ k+1 \end{bmatrix})$. This establishes I.

To check II, let $s \xrightarrow{(w,\phi)} l$ in C_m. Then $M(P((w, \phi))) = (w', \phi')$, $s + 1 \xrightarrow{(w',\phi')} l + 1$, where $w' = w$ and ϕ' is the mapping determined by letting $\phi' = \phi$ on V_s, and $\phi'(v_{k+1}) = v_{l+1}$. Clearly

$$(a \otimes v_{l+1}, e_l)(w, \phi) = (w', \phi')(a \otimes v_{s+1}, e_s).$$

This establishes II.

Finally, for III, consider in C_{m+1} the morphism

$$l \xrightarrow{(a_{m+1} \otimes v_{l+1}, e_l)} l + 1.$$

$M((a_{m+1} \otimes v_{l+1}, e_l)) = (0, \psi')$, where ψ' acts identically on V_l, and $\psi'(v_{l+1}) = v_{l+2} + v_{l+1}$. Now we have for each $a \in A_m$,

$$(a \otimes v_{l+2}, e_{l+1})(a \otimes v_{l+1}, e_l) = (0, \psi')(a \otimes v_{l+1}, e_l).$$

This establishes III.

Thus $C_m(k; l_1, \ldots, l_r)$ holds in general for all m by Proposition I. In particular, as noted above, if $m = 0$, this establishes Corollary 2. We note also that for $m = 1$ the subobjects of an object l can be considered to be affine subspaces of V_l. Thus we have also proved the affine version of Ramsey's Theorem, which we state below.

COROLLARY 3 (Affine Analog). *For $C = C_1$ as described above, $C(k; l_1, \ldots, l_r)$ is true in general.*

The application of Theorem 1 to the case $A = C_1$, $B = C_0$ is just the statement that the affine analog for k and all $l_1, \ldots, l_r$ implies the vector space analog for $k + 1$ and all $l_1, \ldots, l_r$. This result was already known [1, 5], and the previous proof is the same as the proof of Theorem 1 specialized to this case. There was another way given in [5] to show that Corollary 3 implies Corollary 2. Namely, it shows that $C_1(k; l_1, \ldots, l_r)$ implies $C_0(k; l_1, \ldots, l_r)$. This argument is also a special case of Theorem 1, and we can describe it here.

Actually, we replace C_0 with the equivalent C_0', defined by letting $k \xrightarrow{f} l$ in C_0' if and only if $k - 1 \xrightarrow{f} l - 1$ is in C_0. We also must adjoin an identity 1_0 to C_0'. If $k \xrightarrow{(w,\phi)} l$ is in C_1, then $M((w, \phi)) = (0, \phi)$ in C_0', where we recall that $k + 1 \xrightarrow{(0,\phi)} l + 1$ in C_0'. We let $t = 0$, thus making the choices of P and ϕ_{lj} unnecessary. Clearly $C_0' \begin{bmatrix} l+1 \\ k+1 \end{bmatrix} = \overline{M}(C_1 \begin{bmatrix} l \\ k \end{bmatrix})$, and I is satisfied. II is vacuously true as is III, since $t=0$. Hence by Theorem 1, if $C_1'(k;l_1,\ldots,l_r)$ holds for all $l_1,\ldots,l_r$, then $C_0'(k+1;l_1,\ldots,l_r)$ holds and this is just $C_0(k;l_1,\ldots,l_r)$, as desired.

Finally we obtain the Ramsey theorem for n-parameter sets. We refer the reader to [2] to see that the definitions used there are essentially the same as those we will use here. In particular, the categories corresponding to the notions in [2] are the quotient categories described in the last paragraph in this paper. That is, the partially ordered sets of subobjects are isomorphic.

Let G be a finite group, and let $A=\{a_1,\ldots,a_{t_0}\}$ be a finite set. Let $C(A,G)$ be the category with objects $0, 1, 2, \ldots$, and morphisms described as follows:

For each k and l, the morphisms $k \xrightarrow{(f,s)} l$ are diagrams

$$G \xleftarrow{s} \{1,\ldots,l\}\cup A \xrightarrow{f} \{1,\ldots,k\}\cup A,$$

where f is any epimorphic function which acts identically on A, and s is any function such that $s(a)=1\in G$ for $a\in A$. Composition of the morphisms $k\xrightarrow{(f,s)} l$ and $l\xrightarrow{(g,t)} m$ is given by $k\xrightarrow{(fg,\,sg\cdot t)} m$, where fg is ordinary composition of functions, and $sg\cdot t$ is defined by $s(g(x))\cdot t(x)=(sg\cdot t)(x)$ in G for $x\in\{1,\ldots,m\}\cup A$.

We note several things about this choice for $C(A,G)$. First there is no mention of the relationship of G to A. G need not be a permutation group on A, nor even act on it at all. This was a necessary assumption for part of the proof in [2]. Second, we allow $|A|<2$ here, where in [2], $|A|\geqslant 2$ was required. Actually, in the situation in [2] where the n-parameter sets under consideration had constant set $B\subset A$, we did not need $|B|\geqslant 2$. But this took a separate argument. What we have there is the general result for n-parameter sets for arbitrary sets of constants B.

COROLLARY 4 (n-Parameter Sets). *If* $C=C(A,G)$, *then* $C(k;l_1,\ldots,l_r)$ *holds in general.*

Proof. Again, we consider a class $\mathscr{C}$ containing $C(A,G)$ for which Proposition 1 holds. There is more than one possibility here. We will give the proof in detail for one class $\mathscr{C}$. Then we will describe another class but omit the detailed verification of I-III. It is this second class $\mathscr{C}$ which provides a more direct translation of the proof in [2]. The first $\mathscr{C}$ we describe now is somewhat different.

Let $\{a, a_2, a_3,\ldots\}$ be an infinite set. For each $t=1,2,3,\ldots$, let $A_t=\{a_1,\ldots,a_t\}$, and let $C_t=C(A_t,G)$. Thus $C(A,G)$ above is C_{t_0} here. We claim that $A=C_{m+1}$ and $B=C_m$ satisfy Theorem 1, for all $m\geqslant 1$.

To see this we first define M. Let $k\xrightarrow{(f,s)} l$ be in C_{m+1}. Then $M((f,s))=(f',s')$, where $k-1\xrightarrow{(f',s')} l+1$ in C_m is defined as follows. For $x\in A_m\cup\{1,\ldots,l\}$, $f'(x)=f(x)$ if $f(x)\in A_m\cup\{1,\ldots,k\}$, $f'(x)=k+1$ if $f(x)=a_{m+1}$, and $f'(l+1)=k+1$. For $x\in A_m\cup\{1,\ldots,l\}$, $s'(x)=s(x)$, and $s'(l+1)=1$. One can check that M does preserve composition. For the identity map $(e_l,1)$, l in C_{m+1}, where e_l acts identically on l and $1(x)=1\in G$, $x\in A_{m+1}\cup\{1,\ldots,l\}$, we see that $M((e_l,1))=(e_{l+1},1)$ in C_m, so $M(l)=l+1$.

Next we define P. Let $k\xrightarrow{(h,r)} l$ be in C_m. Then $P((h,r))=(h'',r'')$, where $k\xrightarrow{(h'',r'')} l$ in C_{m+1} is defined by letting $h''(x)=h(x)$ and $u''(x)=u(x)$

for $x \in A_m \cup \{1, \ldots, l\}$, and $h''(a_{m+1}) = a_{m+1}$, $r''(a_{m+1}) = 1 \in G$. P clearly preserves composition, and $P(l) = l$ for all l.

Finally, for each l and any $g \in G$ and any j, $1 \leq j \leq m$, let $\phi_{l,(j,g)} = (d_{jl}, 1_{gl})$, or just $(j,g)_l$ for short, where $d_{jl}(x) = x$ for $x \in \{1, \ldots, l\} \cup A_m$, $d_{jl}(l+1) = a_j$, and $1_{gl}(x) = 1 \in G$ for $x \in \{1, \ldots, l\} \cup A_m$, $1_{gl}(l+1) = g$. These ϕ's are indexed by the pairs (j,g). We let $t = |A_m|\,|G| = m\,|G|$, and for the choices above we verify I-III.

Let $k+1 \xrightarrow{(f,s)} l+1$ represent a subobject in $C_m \begin{bmatrix} k+1 \\ l+1 \end{bmatrix}$. Suppose first that $f(l+1) \notin A_m$. Let π be a permutation on $\{1, \ldots, k+1\} \cup A_m$ fixing all $a \in A_m$ and such that πf takes $l+1$ onto $k+1$. Let $u = (s(l+1))^{-1}$, and let $(f', s') = (f, s)(\pi, 1_{uk}\pi)$, where as above, 1_{uk} maps $\{1, \ldots, k\} \cup A_m$ onto $1 \in G$, and $k+1$ onto u. Then $f' = \pi f$ and $s' = 1_{uk} f \cdot s$. In particular, since $(\pi, 1_{uk}\pi)$ is isomorphism in C_m (its inverse is $(\pi^{-1}, 1_{u^{-1}k})$), we see that (f,s) and (f', s') represent the same subobject of $l+1$. Now let $k \xrightarrow{(f'', s'')} l$ be defined in C_{m+1} as follows. For $x \in A_m \cup \{1, \ldots, l\}$, we let $f''(x) = f'(x)$ if $f'(x) \neq k+1$, and $f''(x) = a_{m+1}$ if $f'(x) = k+1$. We let $f''(a_{m+1}) = a_{m+1}$. For $x \in A_m \cup \{1, \ldots, l\}$, we let $s''(x) = s'(x)$, and $s''(a_{m+1}) = 1$. Then $M((f'', s'')) = (f', s')$. So the subobject represented by (f,s) is in $\overline{M}(C_{m+1} \begin{bmatrix} l \\ k \end{bmatrix})$. This is the case, then, for any (f,s) with $f(l+1) \notin A_m$. On the other hand, suppose $f(l+1) = a_j \in A_m$. Let $k+1 \xrightarrow{(f', s')} l$ in C_m be defined by $f'(x) = f(x)$ and $s'(x) = s(x)$ for $x \in \{1, \ldots, l\} \cup A_m$. Then $(f,s) = (j, s(l+1))_l (f', s')$ and (f,s) represents a subobject in $\overline{(j, s(l+1))_l}(C_m \begin{bmatrix} l \\ k+1 \end{bmatrix})$. This establishes I.

For II, we note that for $k \xrightarrow{(f,s)} l$ in C_m, $M(P((f,s)))$ is the morphism $k+1 \xrightarrow{(f', s')} l+1$ in C_m, where $f'(x) = f(x)$ and $s'(x) = s(x)$ for $x \in \{1, \ldots, l\} \cup A_m$, and $f'(l+1) = k+1$, $s'(l+1) = 1$. Then for each j and g we see that $(j,g)_l (f,s) = (f', s')(j,g)_k$, establishing II.

To verify III, we consider $(m+1, 1)_l$ in C_{m+1}. Then $M((m+1, 1)_l)$ is the morphism $l+1 \xrightarrow{(e', 1)} l+2$ in C_m where $1(x) = 1$ for all x in $\{1, \ldots, l+2\} \cup A_m$ and $e'(x) = x$ for $x \in \{1, \ldots, l\} \cup A_m$, and $e'(l+1) = l+1$, $e'(l+2) = l+1$. Then clearly $(j,g)_{l+1}(j,g)_l = (e', 1)(j,g)_l = M((m+1, 1)_l)(j,g)_l$. This establishes III and completes the proof of Corollary 4.

The alternate choice for the class $\mathscr{C}$ to prove Corollary 4 is as follows. For each $m = 0, 1, 2, \ldots$, let $A'_m = A \cup (\{1, \ldots, m\} \times G)$, and let $C'_m = C(A'_m, G)$. Then $C'_0 = C$. Let $\mathscr{C}$ be the class of all C'_m. For each m, C'_{m+1} and C'_m satisfy Theorem 1.

For $k \xrightarrow{(f,s)} l$ in C'_{m+1}, we let $M((f,s)) = (f', s')$, where for $x \in A'_m \cup \{1, \ldots, l\}$ we let

$$f'(x) = f(x) \text{ and } s'(x) = s(x) \text{ if } f(x) \in A'_m \cup \{1, \ldots, k\};$$

for $f(x) = (m+1, g)$, we let $f'(x) = k+1$, $s'(x) = g \cdot s(x)$; and $f'(l+1) = k+1$, $s'(l+1) = 1$. For $k \xrightarrow{(f,s)} l$ in C'_m we define $P((f,s)) = (f', s')$ in C'_{m+1} by letting $f'(x) = f(x)$, $s'(x) = s(x)$ if $x \in A'_m \cup \{1, \ldots, l\}$, and $f'((m+1, g)) = (m+1, g)$, $s'((m+1, g)) = 1$. For $a \in A'_m$ and $g \in G$, as before, we let $\phi_{l,(a,g)} = (d_{al}, 1_{gl})$. Then I, II and III can be verified, with $t = |A'_m|\,|G|$.

Now we still do not have an exact translation of the proof in [2]. In particular, we have taken no account of any action of G on A. To handle this we consider a set A and a group G acting on A, $a \to a^g \in A$ for $g \in G$. We consider the category $C(A,G)$ and obtain from it the category $\overline{C(A,G)}$ by identifying any two morphisms $k \to^{(f,s)} l$ and $k \to^{(g,u)} l$ for which $f(x) = g(x)$ and $s(x) = u(x)$ if $f(x) \in \{1,\ldots,k\}$, and $f(x)^{s(x)} = g(x)^{u(x)}$ otherwise. By considering G to act on $(\{1,\ldots,m\} \times G)$ by $(i,g)^h = (i,gh)$ for all $h \in G$, we obtain the categories $\overline{C'_m} = \overline{C(A'_m, G)}$. The categories $\overline{C'_{m+1}}$ and $\overline{C'_m}$ satisfy Theorem 1, where we take for M and P the functors determined by the M and P for C'_{m+1} and C'_m above by their action on classes of identified morphisms. For the ϕ's we use classes of identified $\phi_{l,(a,g)}$ from above. There are $|A'_m|$ of these, represented by the $\phi_{l,(a,1)}$. Thus we let $t = |A'_m|$ here. Letting $\mathscr{C}$ be the class consisting of all $\overline{C'_m}$, we can apply Proposition 1. This is the exact translation of the proof in [2].

REFERENCES

1. R. L. GRAHAM AND B. L. ROTHSCHILD, Rota's geometric analogue to Ramsey's theorem, *Proc. AMS Symp. in Pure Mathematics XIX Combinatorics AMS Providence* (1971), 101-104.
2. R. L. GRAHAM AND B. L. ROTHSCHILD, Ramsey's Theorem for n-parameter Sets, *Trans. Amer. Math. Soc.* **159** (1971), 257-292.
3. A. HALES AND R. I. JEWETT, Regularity and Positional games, *Trans. Amer. Math. Soc.* **106** (1963), 222-229.
4. F. P. RAMSEY, On a problem of formal logic, *Proc. London Math. Soc. 2nd Ser.* **30** (1930), 264-286.
5. B. L. ROTHSCHILD, A generalization of Ramsey's theorem and a conjecture of Rota, doctoral dissertation, Yale University, New Haven, CT, 1967.
6. H. J. RYSER, "Combinatorial Mathematics," Wiley, New York, 1963.

Reprinted from
Advances in Math. **8** (1972), 417–433

Reprinted from JOURNAL OF COMBINATORIAL THEORY Vol. 13, No. 2, October 1972
All Rights Reserved by Academic Press, New York and London Printed in Belgium

A Characterization of Perfect Graphs

L. LOVÁSZ

Eötvös L. University, Budapest, VIII. Muzeum krt. 6–8, Hungar

Communicated by W. T. Tutte

Received December 3, 1971

It is shown that a graph is perfect iff maximum clique · number of stability is not less than the number of vertices holds for each induced subgraph. The fact, conjectured by Berge and proved by the author, follows immediately that the complement of a perfect graph is perfect.

Throughout this note, graph means finite, undirected graph without loops and multiple edges. $\bar{G}$ and $|G|$ denote the complement and the number of vertices of G, respectively. Let $\mu(G)$ denote the maximum cardinality of a clique in the graph G, and let $\chi(G)$ be the chromatic number of G. Obviously

$$\chi(G) \geqslant \mu(G).$$

A graph G is called *perfect* if

$$\chi(G') = \mu(G')$$

for every induced subgraph G' of G. Berge [1] formulated two conjectures in connection with this notion:

(A) *A graph is perfect iff neither it nor its complement contains an odd circuit without diagonals.*

(B) *The complement of a perfect graph is perfect.*

Obviously, (A) is stronger than (B). In [3] (B) was proved. This result also follows from the theory of anti-blocking polyhedra, developed by Fulkerson [2].

In the present paper a theorem stronger than (B) but weaker than (A) is proved. This possibility of sharpening of (B) was raised by A. Hajnal.

THEOREM. *A graph G is perfect if and only if*

$$\mu(G')\,\mu(\bar{G}') \geqslant |\,G'\,|$$

for every induced subgraph G' of G.

Proof. Part "only if" is trivial. To prove part "if" we use induction on $|\,G\,|$. Thus we may assume that any proper induced subgraph of G, as well as its complement, is perfect.

Let *multiplication* of a vertex x by h ($h \geqslant 0$) mean substituting for it h independent vertices, joined to the same set of vertices as x. This notion is closely related to the notion of *pluperfection*, introduced by D. R. Fulkerson.

(I) As a first step of the proof we show that if G_0 arises from G by multiplication of its vertices then G_0 satisfies

$$\mu(G_0)\,\mu(\bar{G}_0) \geqslant |\,G_0\,|.$$

Assume this is not the case and consider a G_0 failing to have this property and with minimum number of vertices. Obviously, there is a vertex y of G which is multiplied by $h \geqslant 2$; let $y_1,\ldots,y_h$ be the corresponding vertices of G_0. Then

$$\mu(G_0 - y_1)\,\mu(\bar{G}_0 - y_1) \geqslant |\,G_0\,| - 1$$

by the minimality of G_0; hence

$$\mu(G_0) = \mu(G_0 - y_1) = p, \qquad \mu(\bar{G}_0) = \mu(\bar{G}_0 - y_1) = r$$

and

$$|\,G_0\,| = pr + 1.$$

Put $G_1 = G_0 - \{y_1,\ldots,y_h\}$. Then G_1 arises from $G - y$ by multiplication of its vertices, hence by [1, Theorem 1], $\bar{G}_1$ is perfect. Thus, $\bar{G}_1$ can be covered by $\mu(\bar{G}_1) \leqslant \mu(\bar{G}_0) = r$ disjoint cliques of G_1; let $C_1,\ldots,C_r$ be these cliques, $|\,C_1\,| \geqslant |\,C_2\,| \geqslant \cdots \geqslant |\,C_r\,|$.

Obviously, $h \leqslant r$. Since $|\,G_1\,| = |\,G_0\,| - h = pr + 1 - h$,

$$|\,C_1\,| = \cdots = |\,C_{r-h+1}\,| = p.$$

Let G_2 be the subgraph of G_0 induced by $C_1 \cup \cdots \cup C_{r-h+1} \cup \{y_1\}$, then

$$|\,G_2\,| = (r - h + 1)p + 1 < |\,G_0\,|;$$

thus, by the minimality of G_0,

$$\mu(G_2)\,\mu(\bar{G}_2) \geqslant |G_2|.$$

Since $\mu(G_2) \leqslant \mu(G_0) = p$, this implies

$$\mu(\bar{G}_2) \geqslant r - h + 2.$$

Let F be a stable set of $r - h + 2$ vertices of G_2; then $|F \cap C_i| \leqslant 1$ $(1 \leqslant i \leqslant r - h + 1)$, hence $y_1 \in F$. This implies that $F \cup \{y_2, \ldots, y_h\}$ is stable in G_0. On the other hand

$$|F \cup \{y_2, \ldots, y_h\}| = r + 1 > \mu(\bar{G}_0),$$

a contradiction.

(II) We show that $\chi(G) = \mu(G)$. It is enough to find a stable set F such that $\mu(G - F) < \mu(G)$ since then, by the induction hypothesis, $G - F$ can be colored by $\mu(G) - 1$ colors and, adding F as a further one, we obtain a $\mu(G)$-coloring of G.

Assume indirectly that $G - F$ contains a $\mu(G)$-clique C_F for any stable set F in G. Let, for $x \in G$, $h(x)$ denote the number of C_F's containing x. Let G_0 arise from G by multiplying each x by $h(x)$.

Then, by Part I above,

$$\mu(G_0)\,\mu(\bar{G}_0) \geqslant |G_0|.$$

On the other hand, obviously

$$|G_0| = \sum_x h(x) = \sum_F |C_F| = pf,$$

where f denotes the number of all stable sets in G_0, and

$$\mu(G_0) \leqslant \mu(G) = p,$$
$$\mu(\bar{G}_0) = \max_F \sum_{x \in F} h(x) = \max_F \sum_{F'} |F \cap C_{F'}| \leqslant \max_F \sum_{F' \neq F} 1 = f - 1,$$

a contradiction.

REMARK. The condition given in the theorem is strictly related to the max-max inequality given by Fulkerson [2]. Multiplication of a vertex is the same as what he calls *pluperfection*.

REFERENCES

1. C. BERGE, Färbung von Graphen, deren sämtliche bzw. deren ungerade Kreise starr sind, *Wiss. Z. Martin-Luther-Univ. Halle-Wittenberg Math.-Natur. Reihe* (1961), 114.
2. D. R. FULKERSON, Blocking and anti-blocking pairs of polyhedra, *7th International Programming Symposium*, The Hague, 1970.
3. L. LOVÁSZ, Normal hypergraphs and the perfect graph conjecture, *Discrete Math.*, in press.

Printed by the St Catherine Press Ltd., Tempelhof 37, Bruges, Belgium.

Reprinted from JOURNAL OF COMBINATORIAL THEORY
All Rights Reserved by Academic Press, New York and London
Vol. 13, No. 3, December 1972
Printed in Belgium

Note

A Note on the Line Reconstruction Problem

L. LOVÁSZ

Eötvös L. University, Budapest, Hungary

Communicated by W. T. Tutte

Received May 29, 1972

It is shown that if a graph has more lines than its complement does, then it can be reconstructed from its line-deleted subgraphs.

As in Harary's book [4], *graph* means finite, undirected graph without loops or multiple lines. $V(G)$ and $E(G)$ denote the sets of points and lines of G, respectively. Ulam [6] conjectured that, if two graphs G_1 and G_2 are such that $V(G_1) = \{v_1, ..., v_n\}$, $V(G_2) = \{w_1, ..., w_n\}$, $n \geqslant 3$, and $G_1 - v_i \cong G_2 - w_i$, for each i, then $G_1 \cong G_2$. In other words, every graph with at least three points can be uniquely reconstructed from its maximal induced subgraphs. It seems that this conjecture is particularly difficult, and it is solved for special cases only; see, e.g., [5].

An analogous conjecture, formulated by Harary [3], replaces "maximal induced subgraphs" by "maximal subgraphs". This conjecture is actually weaker than Ulam's conjecture (see [1]). In this note we prove it for graphs with "many" lines.

THEOREM. *Let G_1, G_2 be two graphs, $E(G_1) = \{e_1, ..., e_m\}$, $E(G_2) = \{f_1, ..., f_m\}$, and $|V(G_1)| = |V(G_2)| = n$. Assume that $G_1 - e_i \cong G_2 - f_i$ for each $1 \leqslant i \leqslant m$, and $m > \frac{1}{2}\binom{n}{2}$. Then $G_1 \cong G_2$.*

Proof. Let $G \to H$ denote the set of all monomorphisms of G into H. Then, by the sieve formula,

$$| G \to H | = \sum_{X \subseteq G} (-1)^{|E(X)|} | X \to \bar{H} |, \tag{1}$$

where $\bar{H}$ is the complement of H and X runs over all graphs with $V(X) = V(G)$, $E(X) \subseteq E(G)$. In effect, the right-hand side of (1) just counts all maps from the points of G to the points of $\bar{H}$, then takes away those maps sending (at least) one line of G to a line of $\bar{H}$, then adds those sending (at least) two lines to lines of $\bar{H}$, etc. Thus it counts exactly those maps which send no lines of G to lines of $\bar{H}$, so every line of G goes to a line of H.

Applying (1) to G_1 and G_2 we have

$$| G_1 \to G_2 | = \sum_{X \subseteq G_1} (-1)^{|E(X)|} | X \to \bar{G}_2 |, \tag{2}$$

and for G_2 and G_2 we have

$$| G_2 \to G_2 | = \sum_{X \subseteq G_2} (-1)^{|E(X)|} | X \to \bar{G}_2 |. \tag{3}$$

Since the hypothesis on maximal subgraphs assures that G_1 and G_2 have the same proper subgraphs (see [2, p. 92]), the terms in (2) and (3), with $X \neq G_1$ and $X \neq G_2$, are equal. Also, since $m > \frac{1}{2}\binom{n}{2}$, $| G_1 \to \bar{G}_2 | = | G_2 \to \bar{G}_2 | = 0$. Hence $| G_1 \to G_2 | = | G_2 \to G_2 | > 0$, which proves the theorem.

References

1. D. L. Greenwell, Reconstructing graphs, *Proc. Amer. Math. Soc.* **30** (1971), 431–433.
2. D. L. Greenwell and R. L. Hemminger, Reconstructing graphs, "The Many Facets of Graph Theory" (G. T. Chartrand and S. F. Kapoor, eds.), Springer-Verlag, New York, 1969.
3. F. Harary, On the reconstruction of a graph from a collection of subgraphs, "Theory of Graphs and Its Applications" (M. Fiedler, ed.), Czechoslovak Academy of Sciences, Prague/Academic Press, New York, 1965, pp. 47–52.
4. F. Harary, "Graph Theory," Addison-Wesley, Reading, Mass., 1969.
5. F. Harary and B. Manvel, The reconstruction conjecture for labeled graphs, "Combinatorial Structures and Their Applications" (R. K. Guy, ed.), Gordon & Breach, New York, 1969.
6. S. M. Ulam, "A Collection of Mathematical Problems," Wiley (Interscience), New York, 1960, p. 29.

Printed by the St Catherine Press Ltd., Tempelhof 37, Bruges, Belgium.

© DISCRETE MATHEMATICS 5 (1973) 171–178. North-Holland Publishing Company

ACYCLIC ORIENTATIONS OF GRAPHS*

Richard P. STANLEY
Department of Mathematics, University of California, Berkeley, Calif. 94720, USA

Received 1 June 1972

Abstract. Let G be a finite graph with p vertices and χ its chromatic polynomial. A combinatorial interpretation is given to the positive integer $(-1)^p \chi(-\lambda)$, where λ is a positive integer, in terms of acyclic orientations of G. In particular, $(-1)^p \chi(-1)$ is the number of acyclic orientations of G. An application is given to the enumeration of labeled acyclic digraphs. An algebra of full binomial type, in the sense of Doubilet–Rota–Stanley, is constructed which yields the generating functions which occur in the above context.

1. The chromatic polynomial with negative arguments

Let G be a finite graph, which we assume to be without loops or multiple edges. Let $V = V(G)$ denote the set of vertices of G and $X = X(G)$ the set of edges. An edge $e \in X$ is thought of as an unordered pair $\{u, v\}$ of two distinct vertices. The integers p and q denote the cardinalities of V and X, respectively. An *orientation* of G is an assignment of a direction to each edge $\{u, v\}$, denoted by $u \to v$ or $v \to u$, as the case may be. An orientation of G is said to be *acyclic* if it has no directed cycles.

Let $\chi(\lambda) = \chi(G, \lambda)$ denote the chromatic polynomial of G evaluated at $\lambda \in \mathbf{C}$. If λ is a non-negative integer, then $\chi(\lambda)$ has the following rather unorthodox interpretation.

Proposition 1.1. $\chi(\lambda)$ *is equal to the number of pairs* $(\sigma, \mathcal{O})$, *where* σ *is any map* $\sigma : V \to \{1, 2, ..., \lambda\}$ *and* $\mathcal{O}$ *is an orientation of* G, *subject to the two conditions*:

(a) *The orientation* $\mathcal{O}$ *is acyclic.*

(b) *If* $u \to v$ *in the orientation* $\mathcal{O}$, *then* $\sigma(u) > \sigma(v)$.

* The research was supported by a Miller Research Fellowship.

Proof. Condition (b) forces the map σ to be a proper coloring (i.e., if $\{u, v\} \in X$, then $\sigma(u) \neq \sigma(v)$). From (b), condition (a) follows automatically. Conversely, if σ is proper, then (b) defines a unique acyclic orientation of G. Hence, the number of allowed σ is just the number of proper colorings of G with the colors $1, 2, \ldots, \lambda$, which by definition is $\chi(\lambda)$.

Proposition 1.1 suggests the following modification of $\chi(\lambda)$. If λ is a non-negative integer, define $\bar{\chi}(\lambda)$ to be the number of pairs $(\sigma, \mathcal{O})$, where σ is any map $\sigma : V \to \{1, 2, \ldots, \lambda\}$ and $\mathcal{O}$ is an orientation of G, subject to the two conditions:

(a′) The orientation $\mathcal{O}$ is acyclic,

(b′) If $u \to v$ in the orientation $\mathcal{O}$, then $\sigma(u) \geqslant \sigma(v)$. We then say that σ is *compatible* with $\mathcal{O}$.

The relationship between χ and $\bar{\chi}$ is somewhat analogous to the relationship between combinations of n things taken k at a time without repetition, enumerated by $\binom{n}{k}$, and with repetition, enumerated by $\binom{n+k-1}{k} = (-1)^k \binom{-n}{k}$.

Theorem 1.2. *For all non-negative integers* λ,

$$\bar{\chi}(\lambda) = (-1)^p \chi(-\lambda).$$

Proof. Recall the well-known fact that the chromatic polynomial $\chi(G, \lambda)$ is uniquely determined by the three conditions:

(i) $\chi(G_0, \lambda) = \lambda$, where G_0 is the one-vertex graph.

(ii) $\chi(G + H, \lambda) = \chi(G, \lambda)\, \chi(H, \lambda)$, where $G + H$ is the disjoint union of G and H,

(iii) for all $e \in X$, $\chi(G, \lambda) = \chi(G \backslash e, \lambda) - \chi(G/e, \lambda)$, where $G \backslash e$ denotes G with the edge e deleted and G/e denotes G with the edge e contracted to a point.

Hence, it suffices to prove the following three properties of $\bar{\chi}$:

(i′) $\bar{\chi}(G_0, \lambda) = \lambda$, where G_0 is the one-vertex graph,

(ii′) $\bar{\chi}(G + H, \lambda) = \bar{\chi}(G, \lambda)\, \bar{\chi}(H, \lambda)$,

(iii′) $\bar{\chi}(G, \lambda) = \bar{\chi}(G \backslash e, \lambda) + \bar{\chi}(G/e, \lambda)$.

Properties (i′) and (ii′) are obvious, so we need only prove (iii′). Let $\sigma : V(G \backslash e) \to \{1, 2, \ldots, \lambda\}$ and let $\mathcal{O}$ be an acyclic orientation of $G \backslash e$ compatible with σ, where $e = \{u, v\} \in X$. Let $\mathcal{O}_1$ be the orientation of G obtained by adjoining $u \to v$ to $\mathcal{O}$, and $\mathcal{O}_2$ that obtained by adjoining $v \to u$. Observe that σ is defined on $V(G)$ since $V(G) = V(G \backslash e)$. We will show that for each pair (σ, O), exactly one of O_1 and O_2 is an acyclic orientation compatible with σ, except for $\bar{\chi}(G/e, \lambda)$ of these pairs, in which case both O_1 and O_2 are acyclic orientations compatible with σ. It then follows that $\bar{\chi}(G, \lambda) = \bar{\chi}(G\backslash e, \lambda) + \bar{\chi}(G/e, \lambda)$, so proving the theorem.

For each pair (σ, O), where $\sigma: G\backslash e \to \{1, 2, ..., \lambda\}$ and O is an acyclic orientation of $G\backslash e$ compatible with σ, one of the following three possibilities must hold.

Case 1: $\sigma(u) > \sigma(v)$. Clearly O_2 is not compatible with σ while O_1 is compatible. Moreover, O_1 is acyclic, since if $u \to v \to w_1 \to w_2 \to ... \to u$ were a directed cycle in O_1, we would have $\sigma(u) > \sigma(v) \geqslant \sigma(w_1) \geqslant \sigma(w_2) \geqslant ... \geqslant \sigma(u)$, which is impossible.

Case 2: $\sigma(u) < \sigma(v)$. Then symmetrically to Case 1, O_2 is acyclic and compatible with σ, while O_1 is not compatible.

Case 3: $\sigma(u) = \sigma(v)$. Both O_1 and O_2 are compatible with σ. We claim that at least one of them is acyclic. Suppose not. Then O_1 contains a directed cycle $u \to v \to w_1 \to w_2 \to ... \to u$ while O_2 contains a directed cycle $v \to u \to w_1' \to w_2' \to ... \to v$. Hence, O contains the directed cycle

$$u \to w_1' \to w_2' \to ... \to v \to w_1 \to w_2 \to ... \to u,$$

contradicting the assumption that O is acyclic.

It remains to prove that both O_1 and O_2 are acyclic for exactly $\bar{\chi}(G/e, \lambda)$ pairs (σ, O), with $\sigma(u) = \sigma(v)$. To do this we define a bijection $\Phi(\sigma, O) = (\sigma', O')$ between those pairs (σ, O) such that both O_1 and O_2 are acyclic (with $\sigma(u) = \sigma(v)$) and those pairs (σ', O') such that $\sigma': G/e \to \{1, 2, ..., \lambda\}$ and O' is an acyclic orientation of G/e compatible with σ'. Let z be the vertex of G/e obtained by identifying u and v, so

$$V(G/e) = V(G\backslash e) - \{u, v\} \cup \{z\}$$

and $X(G/e) = X(G\backslash e)$. Given (σ, O), define σ' by $\sigma'(w) = \sigma(w)$ for all $w \in V(G\backslash e) - \{z\}$ and $\sigma'(z) = \sigma(u) = \sigma(v)$. Define O' by $w_1 \to w_2$ in O' if and only if $w_1 \to w_2$ in O. It is easily seen that the map $\Phi(\sigma, O) = (\sigma', O')$ establishes the desired bijection, and we are through.

Theorem 1.2 provides a combinatorial interpretation of the positive integer $(-1)^p \chi(G, -\lambda)$, where λ is a positive integer. In particular, when $\lambda = 1$ every orientation of G is automatically compatible with every map $\sigma: G \to \{1\}$. We thus obtain the following corollary.

Corollary 1.3. *If G is a graph with p vertices, then $(-1)^p \chi(G, -1)$ is equal to the number of acyclic orientations of G.*

In [5], the following question was raised (for a special class of graphs). Let G be a p-vertex graph and let ω be a *labeling* of G, i.e., a bijection $\omega: V(G) \to \{1, 2, ..., p\}$. Define an equivalence relation $\sim$ on the set of all $p!$ labelings ω of G by the condition that $\omega \sim \omega'$ if whenever $\{u, v\} \in X(G)$, then $\omega(u) < \omega(v) \Leftrightarrow \omega'(u) < \omega'(v)$. How many equivalence classes of labelings of G are there? Clearly two labelings ω and ω' are equivalent if and only if the unique orientations $\mathcal{O}$ and $\mathcal{O}'$ compatible with ω and ω', respectively, are equal. Moreover, the orientations $\mathcal{O}$ which arise in this way are precisely the acyclic ones. Hence, by Corollary 1.3, the number of equivalence classes is $(-1)^p \chi(G, -1)$.

We conclude this section by discussing the relationship between the chromatic polynomial of a graph and the order polynomial [4;5;6] of a partially ordered set. If P is a p-element partially ordered set, define the *order polynomial* $\Omega(P, \lambda)$ (evaluated at the non-negative integer λ) to be the number of order-preserving maps $\sigma: P \to \{1, 2, ..., \lambda\}$. Define the *strict order polynomial* $\overline{\Omega}(P, \lambda)$ to be the number of *strict* order-preserving maps $\sigma: P \to \{1, 2, ..., \lambda\}$, i.e., if $x < y$ in P, then $\sigma(x) < \sigma(y)$. In [5], it was shown that Ω and $\overline{\Omega}$ are polynomials in λ related by $\overline{\Omega}(P, \lambda) = (-1)^p \Omega(P, -\lambda)$. This is the precise analogue of Theorem 1.2. We shall now clarify this analogy.

If $\mathcal{O}$ is an orientation of a graph G, regard $\mathcal{O}$ as a binary relation $\geqslant$ on $V(G)$ defined by $u \geqslant v$ if $u \to v$. If $\mathcal{O}$ is acyclic, then the transitive and reflexive closure $\overline{\mathcal{O}}$ of $\mathcal{O}$ is a partial ordering of $V(G)$. Moreover, a map $\sigma: V(G) \to \{1, 2, ..., \lambda\}$ is compatible with $\mathcal{O}$ if and only if σ is order-preserving when considered as a map from $\overline{\mathcal{O}}$. Hence the number of σ compatible with $\mathcal{O}$ is just $\Omega(\overline{\mathcal{O}}, \lambda)$ and we conclude that

$$\overline{\chi}(G, \lambda) = \sum_{\mathcal{O}} \Omega(\overline{\mathcal{O}}, \lambda),$$

where the sum is over all acyclic orientations $\mathcal{O}$ of G. In the same way, using Proposition 1.1, we deduce

$$\chi(G, \lambda) = \sum_{\mathcal{O}} \overline{\Omega}(\overline{\mathcal{O}}, \lambda). \tag{1}$$

Hence, Theorem 1.2 follows from the known result $\bar{\Omega}(P, \lambda) = (-1)^p \Omega(P, -\lambda)$, but we thought a direct proof to be more illuminating. Equation (1) strengthens the claim made in [4] that the strict order polynomial $\bar{\Omega}$ is a partially-ordered set analogue of the chromatic polynomial χ.

2. Enumeration of labeled acyclic digraphs

Corollary 1.3, when combined with a result of Read (also obtained by Bender and Goldman), yields an immediate solution to the problem of enumerating labeled acyclic digraphs with n vertices. The same result was obtained by R.W. Robinson (to be published), who applies it to the unlabeled case.

Proposition 2.1. *Let $f(n)$ be the number of labeled acyclic digraphs with n vertices. Then*

$$\sum_{n=0}^{\infty} f(n)\, x^n/n!\, 2^{\binom{n}{2}} = \left(\sum_{n=0}^{\infty} (-1)^n x^n/n!\, 2^{\binom{n}{2}} \right)^{-1}.$$

Proof. By Corollary 1.3,

$$f(n) = (-1)^n \sum_G \chi(G, -1), \tag{2}$$

where the sum is over all labeled graphs G with n vertices. Now, Read [3] (see also [1]) has shown that if

$$M_n(k) = \sum_G \chi(G, k)$$

(where the sum has the same range as in (2)), then

$$\sum_{n=0}^{\infty} M_n(k)\, x^n/n!\, 2^{\binom{n}{2}} = \left(\sum_{n=0}^{\infty} x^n/n!\, 2^{\binom{n}{2}} \right)^k. \tag{3}$$

Actually, the above papers have $2^{n^2/2}$ where we have $2^{\binom{n}{2}}$ – this amounts to the transformation $x' = 2^{1/2}x$. One advantage of our 'normalization' is that the numbers $n!\,2^{\binom{n}{2}}$ are integers; a second is that the function

$$F(x) = \sum_{n=0}^{\infty} x^n/n!\,2^{\binom{n}{2}}$$

satisfies the functional relation $F'(x) = F(\frac{1}{2}x)$. A third advantage is mentioned in the next section. Thus setting $k = -1$ and changing x to $-x$ in (3) yields the desired result.

By analyzing the behavior of the function $F(x) = \Sigma_{n=0}^{\infty} x^n/n!\,2^{\binom{n}{2}}$, we obtain estimates for $f(n)$. For instance, Rouché's theorem can be used to show that $F(x)$ has a unique zero $\alpha \approx -1.488$ satisfying $|\alpha| \leqslant 2$. Standard techniques yield the asymptotic formula

$$f(n) \sim C\,2^{\binom{n}{2}}\,n!(-\alpha)^{-n},$$

where α is as above and $1.741 \approx C = 1/\alpha F(\frac{1}{2}\alpha)$. A more careful analysis of $F(x)$ will yield more precise estimates for $f(n)$.

3. An algebra of binomial type

The existence of a combinatorial interpretation of the coefficients $M_n(k)$ in the expansion

$$\left(\sum_{n=0}^{\infty} x^n/2^{\binom{n}{2}}\,n!\right)^k = \sum_{n=0}^{\infty} M_n(k)\,x^n/2^{\binom{n}{2}}\,n!$$

suggests the existence of an algebra of full binomial type with structure constants $B(n) = 2^{\binom{n}{2}}\,n!$ in the sense of [2]. This is equivalent to finding a locally finite partially ordered set P (said to be of *full binomial type*), satisfying the following conditions:

(a) In any segment $[x, y] = \{z \mid x \leqslant z \leqslant y\}$ of P (where $x \leqslant y$ in P), every maximal chain has the same length n. We call $[x, y]$ an *n-segment.*

(b) There exists an n-segment for every integer $n \geqslant 0$ and the number of maximal chains in any n-segment is $B(n) = 2^{\binom{n}{2}}\,n!$, (In particular, $B(1)$ must equal 1, further explaining the normalization $x' = 2^{1/2}x$ of Section 2.)

If such a partially ordered set P exists, then by [2] the value of $\zeta^k(x, y)$, where ζ is the zeta function of P, k is any integer and $[x, y]$ is any n-segment, depends only on k and n. We write $\zeta^k(x, y) = \zeta^k(n)$. Then again from [2],

$$\sum_{n=0}^{\infty} \zeta^k(n) x^n / B(n) = \left(\sum_{n=0}^{\infty} x^n / B(n) \right)^k .$$

Hence $\zeta^k(n) = M_n(k)$. In particular, the cardinality of any n-segment $[x, y]$ is $M_n(2)$, the number of labeled two-colored graphs with n vertices; while $\mu(x, y) = (-1)^n f(n)$, where μ is the Möbius function of P and $f(n)$ is the number of labeled acyclic digraphs with n vertices. The general theory developed in [2] provides a combinatorial interpretation of the coefficients of various other generating functions, such as $(\Sigma_{n=1}^{\infty} x^n/B(n))^k$ and $(2 - \Sigma_{n=0}^{\infty} x^n/B(n))^{-1}$.

Since $M_n(2)$ is the cardinality of an n-segment, this suggests taking elements of P to be properly two-colored graphs. We consider a somewhat more general situation.

Proposition 3.1. *Let V be an infinite vertex set, let q be a positive integer and let P_q be the set of all pairs (G, σ), where G is a function from all 2-sets $\{u, v\} \subseteq V$ $(u \neq v)$ into $\{0, 1, \ldots, q-1\}$ such that all but finitely many values of G are 0, and where $\sigma: V \to \{0, 1\}$ is a map satisfying the condition that if $G(\{u, v\}) > 0$ then $\sigma(u) \neq \sigma(v)$ and that $\Sigma_{u \in v} \sigma(u) < \infty$.*

If (G, σ) and (H, τ) are in P_q, define $(G, \sigma) \leqslant (H, \tau)$ if:

(a) *$\sigma(u) \leqslant \tau(u)$ for all $u \in V$, and*

(b) *If $\sigma(u) = \tau(u)$ and $\sigma(v) = \tau(v)$, then $G(\{u, v\}) = H(\{u, v\})$.*

Then P_q is a partially ordered set of full binomial type with structure constants $B(n) = n!\, q^{\binom{n}{2}}$.

Proof. If (H, τ) *covers* (G, σ) in P (i.e., if $(H, \tau) > (G, \sigma)$ and no (G', σ') satisfies $(H, \tau) > (G', \sigma') > (G, \sigma)$), then

$$\sum_{u \in V} \tau(u) = 1 + \sum_{u \in V} \sigma(u).$$

From this it follows that in every segment of P, all maximal chains have the same length.

In order to prove that an n-segment $S = [(G, \sigma), (H, \tau)]$ has $n!\,q^{\binom{n}{2}}$ maximal chains, it suffices to prove that (H, τ) covers exactly nq^{n-1} elements of S, for then the number of maximal chains in S will be $(nq^{n-1})((n-1)q^{n-2}) \ldots (2q^1) \cdot 1 = n!q^{\binom{n}{2}}$. Since S is an n-segment, there are precisely n vertices $v_1, v_2, \ldots, v_n \in V$ such that $\sigma(v_i) = 0 < 1 = \tau(v_i)$. Suppose (H, τ) covers $(H', \tau') \in S$. Then τ' and τ agree on every $v \in V$ except for one v_i, say v_1, so $\tau'(v_1) = 0$, $\tau(v_1) = 1$. Suppose now $H'(\{u, v\}) > 0$, where we can assume $\tau'(u) = 0$, $\tau'(v) = 1$. If v is not some v_i, then $\sigma(u) = 0$, $\sigma(v) = 1$, so $H'(\{u, v\}) = G(\{u, v\})$. If $v = v_i$ $(2 \leqslant i \leqslant n)$ and u is not v_1, then $\tau(u) = 0$, $\tau(v) = 1$, so $H'(\{u, v\}) = H(\{u, v\})$. Hence $H'(\{u, v\})$ is completely determined unless $u = v_1$ and $v = v_i$, $2 \leqslant i \leqslant n$. In this case, each $H'(\{v_1, v_i\})$ can have any one of q values. Thus, there are n choices of v_1 and q choices for each $H'(\{v_1, v_i\})$, $2 \leqslant i \leqslant n$, giving a total of nq^{n-1} elements $(H', \tau') \in S$ covered by (H, τ).

Observe that when $q = 1$, condition (b) is vacuous, so P_1 is isomorphic to the lattice of finite subsets of V. When $q = 2$, we may think of $G(\{u, v\}) = 0$ or 1 depending on whether $\{u, v\}$ is not or is an edge of a graph on the vertex set V. Then σ is just a proper two-coloring of v with the colors 0 and 1, and the elements of P_2 consist of all properly two-colored graphs with vertex set V, finitely many edges and finitely many vertices colored 1. We remark that P_q is not a lattice unless $q = 1$.

References

[1] E.A. Bender and J. Goldman, Enumerative uses of generating functions, Indiana Univ. Math. J. 20 (1971) 753–765.

[2] P. Doubilet, G.-C. Rota and R. Stanley, On the foundations of combinatorial theory: The idea of generating function, in: Sixth Berkeley symposium on mathematical statistics and probability (1972) 267–318.

[3] R. Read, The number of k-colored graphs on labelled nodes, Canad. J. Math. 12 (1960) 410–414.

[4] R. Stanley, A chromatic-like polynomial for ordered sets, in: Proc. second Chapel Hill conference on combinatorial mathematics and its applications (1970) 421–427.

[5] R. Stanley, Ordered structures and partitions, Mem. Am. Math. Soc. 119 (1972).

[6] R. Stanley, A Brylawski decomposition for finite ordered sets, Discrete Math. 4 (1973) 77–82.

Sonderabdruck aus
ARCHIV DER MATHEMATIK
Vol. XXIV, 1973 BIRKHÄUSER VERLAG, BASEL UND STUTTGART Fasc. 3

Valuations on Distributive Lattices

By
LADNOR GEISSINGER

Valuations on Distributive Lattices I

By

LADNOR GEISSINGER

Introduction. We continue the study, begun by G.-C. Rota, of the valuation ring of a distributive lattice. This ring is the representing object for all valuations on the lattice. In the locally finite case Rota established a connection with the incidence algebra of the set of join-irreducible elements, from which he derived interesting results about the Euler characteristic and Möbius function associated with some geometric objects. In this paper we give new proofs of some of his results, and extend others. In part I we discuss general properties of the valuation module and ring of a lattice, and determine their structure for a finite geometric lattice. We then describe the duality between maps of finite distributive lattices and of finite posets. This makes it easy to characterize finite projective distributive lattices, construct the free distributive lattice on a finite poset, and determine what properties of lattice homomorphisms correspond to strict and residuated maps of posets. We also use the valuation ring to give a construction for the coproduct of distributive lattices. In part II we will determine the structure and mapping properties of valuation rings and Möbius algebras. We use these to prove some theorems of Rota on Möbius functions, an identity due to Klee, and theorems on extending finitely additive measures.

1. The Valuation Module of a Lattice. A function f from a lattice L into an abelian group is *modular* if $f(a \vee b) + f(a \wedge b) = f(a) + f(b)$ for all $a, b \in L$. Following Rota [19], we call any such modular function a *valuation*. (Birkhoff [2] reserves the term valuation for real-valued modular functions.) In the free abelian group $Z^{(L)}$ on L, let $M(L)$ be the subgroup generated by all elements of the form $a \vee b + a \wedge b - a - b$ with $a, b \in L$. Then $V(L) = Z^{(L)}/M(L)$ is the *valuation module of* L introduced by Rota [19]. Let $i\colon L \to V(L)$ be the natural induced map. The following characteristic property of $(V(L), i)$ is an immediate consequence of its construction.

Proposition 1. *The function $i\colon L \to V(L)$ is the universal valuation on L, that is, i is a valuation and every valuation on L into an abelian group A factors uniquely as i followed by a group homomorphism from $V(L)$ into A.*

Thus the additive group of valuations on L into A can be identified with $\mathrm{Hom}(V(L), A)$. The functorial properties of $V(L)$ also follow easily from the construction.

Proposition 2. *A lattice homomorphism $\varphi\colon L_1 \to L_2$ induces a unique group homomorphism $\varphi'\colon V(L_1) \to V(L_2)$ such that $\varphi' i_1 = i_2 \varphi$.*

The existence of simple types of valuations implies certain structural properties of $V(L)$ and $i(L)$. For example, since every constant function from L into any abelian group is a valuation, by Proposition 1 the elements of $i(L)$ must be non-zero and of infinite order. More useful information about the linear independence of subsets of $i(L)$ can be derived from consideration of 2-valued valuations, or equivalently, of prime ideals of L and their complements, prime filters.

Proposition 3. *For any prime ideal or prime filter F of a lattice L, its characteristic function $c_F: L \to Z$, which is 1 on F and 0 on $L \backslash F$, is a valuation. Each element of $i(F)$ is linearly independent of the elements of $i(L \backslash F)$ and vice versa, though neither of these sets is necessarily independent.*

Proof. It is easy to verify directly the first statement. The second statement follows from the first by Proposition 1.

Proposition 4. *The map i is an injection iff L is distributive. When L is distributive, if $\{a_1, \ldots, a_r, b\} \subset L$ and if b is not in the interval $[\bigwedge a_i, \bigvee a_i]$, then b is linearly independent of $\{a_1, \ldots, a_r\}$ in $V(L)$.*

Proof. A well-known theorem of Stone states that any two elements of a distributive lattice can be separated by a prime filter [2, 17], hence they are independent in $V(L)$ by Proposition 3. If L is not distributive it contains distinct elements c, x, y with $c \wedge x = c \wedge y$ and $c \vee x = c \vee y$ from which $i(x) = i(y)$. The condition on b holds iff there is a prime filter separating b from $\{a_1, \ldots, a_r\}$.

Later we will give an elementary proof of this proposition which does not depend on Stone's theorem. Whenever we deal with distributive lattices we identify L and $i(L)$.

Now L with either of the operations $\vee$ or $\wedge$ is a semigroup, so $Z^{(L)}$ may be considered a semigroup algebra using either $\vee$ or $\wedge$ as multiplication.

Proposition 5. *If L is a distributive lattice, $M(L)$ is an ideal of the semigroup algebra $Z^{(L)}$ for both $\vee$ and $\wedge$ and so $V(L)$ is a commutative ring with either (the induced) $\vee$ or $\wedge$ as product. Moreover, for any homomorphism $\varphi: L_1 \to L_2$ of distributive lattices, the extended map $\varphi: V(L_1) \to V(L_2)$ is a homomorphism for both $\vee$ and $\wedge$.*

Proof. $(x \vee y + x \wedge y - x - y) \wedge t = (x \wedge t) \vee (y \wedge t) + (x \wedge t) \wedge (y \wedge t) - x \wedge t - y \wedge t$ and similarly for $\vee$.

Corollary. *The ring $(V(L), \wedge)$ and the map i are characterized by the following universal property. For any commutative ring A and any map $\beta: L \to A$ for which $\beta(x \wedge y) = \beta(x) \cdot \beta(y)$ and $\beta(x \vee y) = \beta(x) + \beta(y) - \beta(x \wedge y)$, there is a unique ring homomorphism $\alpha: (V(L), \wedge) \to (A, \cdot)$ such that $\alpha \cdot i = \beta$.*

Suppose L is a distributive lattice. If L does not have a unit (maximal element) u or a zero (minimal element) z we can adjoin such elements to L and the enlarged lattice is still distributive. Since this merely adds onto $V(L)$ one or two copies of Z as direct summands, whenever it is convenient we assume $u, z \in L$. Then u and z are the identities for $\wedge$ and $\vee$ in $V(L)$. The usual augmentation map $\varepsilon: Z^{(L)} \to Z$ given by $\varepsilon(\sum c_i x_i) = \sum c_i$ is a homomorphism for both $\wedge$ and $\vee$ and its kernel I, which is generated by all $x - y$ for all $x, y \in L$, contains $M(L)$. Thus $V(L)$ with the induced homomorphism $\varepsilon: V(L) \to Z$ is an augmented algebra relative to both multiplications $\vee$ and $\wedge$. Moreover, in Proposition 5 the ring homomorphism φ commutes with the augmentation homomorphisms ε and carries I_1 into I_2. In a sense the augmentation ideal I is the most important part of $V(L)$. Namely, for any $x \in L$, $V(L) = I \oplus Zx$ so that a homomorphism $f: I \to A$ extends uniquely to a valuation on L when an arbitrary value $f(x)$ in A is assigned to x. It is often convenient to take $x = z$ the zero of L so that $I \approx V(L)/Zz = \bar{V}(L)$ naturally represents valuations normalized to take value 0 on z and at the same time $\bar{V}(L)$ is again a ring for the $\wedge$-multiplication.

In the theory of Boolean algebras there is a duality principle which comes from complementation. For a general distributive lattice L with u and z there is no complementation process in L, however there is an endomorphism of $V(L)$ which can be used in much the same way. Namely, the map $\tau(x) = u + z - x$ for all $x \in L$ when extended to an endomorphism of $Z^{(L)}$ carries $M(L)$ into itself and so induces a homomorphism $\tau: V(L) \to V(L)$. τ could also be described by saying that $\tau(c) = -c$ for all c in the augmentation ideal and that $\tau(z) = u$ or $\tau(u) = z$. The following proposition is the substitute for De Morgan's laws and justifies our subsequent practice of usually ignoring $\vee$ and considering $V(L)$ only with multiplication $\wedge$.

Proposition 6. *$\tau: (V(L), \wedge) \to (V(L), \vee)$ is an isomorphism of augmented algebras, and $\tau^2 = \mathrm{id}$. Hence $\tau: (V(L), \vee) \to (V(L), \wedge)$ is also an isomorphism.*

Proof. $z + u - x \wedge y = z + u + x \vee y - x - y = (z + u - x) \vee (z + u - y)$ so that $\tau(x \wedge y) = \tau(x) \vee \tau(y)$. Clearly $\tau^2 = \mathrm{id}$ and $\varepsilon\tau = \varepsilon$.

As a further check that τ is the correct algebraic analogue of complementation, note that if $x \in L$ has a complement x' then

$$x \vee x' + x \wedge x' - x - x' = 0 = u + z - x - x' \text{ so that } \tau(x) = x'.$$

In any case, in $V(L)$ we always have $x \vee \tau(x) = u$ and $x \wedge \tau(x) = z$ for every $x \in L$. More generally, for any element x in an interval $[v, w]$ of L, the element $v + w - x$ in $V(L)$ acts as the relative complement of x.

A more useful and more familiar form of Proposition 6, again for distributive lattices, is the following.

Proposition 7. *For all $x_1, \ldots, x_n$ in L, $u - \bigvee x_i = \bigwedge (u - x_i)$ in $V(L)$, that is,*

$$\bigvee x_i = \sum x_i - \sum (x_i \wedge x_j) + \sum (x_i \wedge x_j \wedge x_k) - \cdots \quad (i < j < k < \cdots).$$

Proof. $\tau(\bigvee x_i) = z + u - \bigvee x_i = \bigwedge \tau(x_i) = \bigwedge (z + u - x_i) = z + \bigwedge (u - x_i)$ since $z \wedge (u - x_i) = 0$. For a direct proof, use induction beginning with either

$$u - x \vee y = u + x \wedge y - x - y = (u - x) \wedge (u - y) \text{ or } x \vee y = x + y - x \wedge y.$$

Note that in the direct proof u can be replaced by any $v \geqq \bigvee x_i$.

Corollary. *For any valuation f on L,*

$$f(\bigvee x_i) = \sum f(x_i) - \sum f(x_i \wedge x_j) + \sum f(x_i \wedge x_j \wedge x_k) - \cdots \quad (i < j < k < \cdots).$$

This is well known; in particular, when f is an additive set function on $L = 2^X$ this is the classical inclusion-exclusion formula.

It is somewhat unusual to have to consider two natural ring structures on the same abelian group $V(L)$. The relation between them, derived from $x \vee y = x + y - x \wedge y$ for all $x, y \in L$, is given by $a \vee b = \varepsilon(a) b + \varepsilon(b) a - a \wedge b$ for all $a, b \in V(L)$. But now it is easily checked that if $\varepsilon: A \to k$ is an augmentation of any k-algebra $(A, \cdot)$ there is another multiplication on A given by $a * b = \varepsilon(b) a + \varepsilon(a) b - a \cdot b$ for which $(A, *, \varepsilon)$ is an augmented algebra. Moreover, it is clear that the endomorphism $\tau(a) = -a$ of the augmentation ideal carries $a \cdot b$ into $a * b = \tau(a) * \tau(b)$. To see if τ extends to all of A, note the following. If $(A, \cdot)$ has a unit u, then $b * u = u * b = \varepsilon(b) u$, and conversely if z is a unit for $(A, *)$ and if $\varepsilon(z) = 1$ then $a \cdot z = z \cdot a = \varepsilon(a) z$. An element z with this property in an augmented algebra $(A, \cdot, \varepsilon)$ has been called an (invariant) integral [13]. If $(A, \cdot, \varepsilon)$ is an augmented algebra with unit u then τ can be extended to an isomorphism $(A, \cdot) \to (A, *)$ iff there is such an integral in A by letting $\tau(u) = z$ and $\tau(z) = u$. For any commutative ring A with unit, the set E of all idempotents forms a Boolean algebra with $a \wedge b = a \cdot b$ and $a \vee b = a + b - a \cdot b$ for all $a, b \in E$. Thus Proposition 7 holds for idempotents in A. What we have just shown is essentially that when A has an augmentation ε, the set $\{a \in E \mid \varepsilon(a) = 1\}$ (which is always a sublattice of E) is a Boolean algebra iff A has an integral z. Finally note that if K is any multiplicatively closed subset of E, then the sublattice generated by K consists of all elements of the form $\sum a_i - \sum (a_i \wedge a_j) + \sum (a_i \wedge a_j \wedge a_k) \cdots$ for all finite indexed families (a_i) of elements of K.

A *generalized Boolean algebra*, that is, a distributive lattice L which contains z and is relatively complemented, along with the usual symmetric difference operation $\triangle$ is a group. Moreover, $M(L)$ is then an ideal in the group algebra $(Z^{(L)}, \triangle)$ so $V(L)$ with the induced operation $\triangle$ is again a ring. In $V(L)$, $x \triangle y = x \vee y - x \wedge y + z$, for all $x, y \in L$, and more generally for all $b, c \in V(L)$, $b \triangle c = b \vee c - b \wedge c + \varepsilon(b)\varepsilon(c) z$. But it is easily checked that this formula for the operation $\triangle$, which makes sense for any distributive lattice L, always yields a third ring structure on $V(L)$, even when $\triangle$ is not defined on L or $Z^{(L)}$. Of course if $u \in L$ then $V(L)$ is also a ring with product the operation complementary to $\triangle$, that is, $x \triangle' y = x \wedge y - x \vee y + u$. We shall not pursue these other operations further; instead we turn to nondistributive lattices.

For a nondistributive lattice L, $M(L)$ need not be an ideal in $(Z^{(L)}, \wedge)$. The condition for $M(L)$ to be a $\wedge$-ideal is that in $V(L)$ for all $t, x, y \in L$,

$$0 = t \wedge (x \vee y) + t \wedge (x \wedge y) - t \wedge x - t \wedge y = t \wedge (x \vee y) - (t \wedge x) \vee (t \wedge y).$$

But in any case,

$$\begin{aligned} t \wedge (x \vee y) - (t \wedge x) \vee (t \wedge y) &= t + x \vee y - t \vee x \vee y + t \wedge x \wedge y - t \wedge x - t \wedge y = \\ &= t + x + y - x \wedge y - t \wedge x - t \wedge y - t \vee x \vee y + t \wedge x \wedge y = \\ &= -t - x - y + x \vee y + t \vee x + t \vee y - t \vee x \vee y + t \wedge x \wedge y = \\ &= (t \vee x) \wedge (t \vee y) - t \vee (x \wedge y). \end{aligned}$$

(We have suppressed the i's in $i(t)$, etc..) Thus $M(L)$ is a $\wedge$-ideal iff it is a $\vee$-ideal. Birkhoff [2] calls a valuation f on any lattice *distributive* if

$$f(t \vee x \vee y) - f(t \wedge x \wedge y) = f(x \vee y) + f(t \vee x) + f(t \vee y) - f(t) - f(x) - f(y) =$$
$$= f(x) + f(y) + f(t) - f(t \wedge x) - f(t \wedge y) - f(x \wedge y).$$

Thus $M(L)$ is a $\wedge$-ideal iff i is a distributive valuation. By Proposition 1, i is distributive iff every valuation on L is distributive.

For any lattice L, let $\tilde{M}(L)$ be the subgroup of $Z^{(L)}$ generated by $M(L)$ and all elements of the form $t \wedge [x \vee y + x \wedge y - x - y]$ for all $t, x, y \in L$. Then $\tilde{M}(L)$ is the $\wedge$-ideal generated by $M(L)$ and from our computation above it follows that $\tilde{M}(L)$ is also the $\vee$-ideal generated by $M(L)$. Thus $\tilde{V}(L) = Z^{(L)}/\tilde{M}(L)$ is a ring for each of the products $\wedge$ and $\vee$ and since $\tilde{M}(L)$ is contained in the augmentation ideal, $\tilde{V}(L)$ is an augmented algebra. The induced map $j: L \to \tilde{V}(L)$ is a homomorphism for both $\wedge$ and $\vee$; thus $j(L)$ is closed under both these operations. Hence $j(L)$ is a lattice and j is a lattice homomorphism. Moreover, by construction $j(t \wedge (x \vee y)) = j((t \wedge x) \vee (t \wedge y))$ so $j(L)$ is a distributive lattice. Also j as a function into $\tilde{V}(L)$ is a distributive valuation, and $\tilde{V}(L)$ is the valuation ring of $j(L)$.

Proposition 8. *The function $j: L \to \tilde{V}(L)$ is the universal distributive valuation on L and $j: L \to j(L)$ is the universal lattice homomorphism from L into distributive lattices.*

Proof. The first part is a consequence of the previously mentioned properties of $\tilde{M}(L)$. For the second, suppose φ is a lattice homomorphism of L into a distributive lattice L', and $i': L' \to V(L')$ is the natural injection. Then $i' \circ \varphi$ is a distributive valuation and so factors uniquely as $\alpha \circ j$ where $\alpha: \tilde{V}(L) \to V(L')$ is a group homomorphism. But since i', φ and j preserve $\wedge$ and $\vee$ and $i' \circ \varphi = \varphi \circ j$ then α preserves $\wedge$ and $\vee$ on $j(L)$, and hence on all of $\tilde{V}(L)$. Thus α is both a lattice and ring homomorphism.

The result stated in the Corollary to Proposition 7 is now seen to hold more generally for distributive valuations on any lattice.

Examples. For the modular 5-element lattice M_5 it is easily checked that $V(M_5)$ is free of rank 2 while $\tilde{V}(M_5)$ is free of rank 1. For the nonmodular 5-element lattice N_5, $V(N_5) = \tilde{V}(N_5)$ and the rank is 3.

If L_1, L_2 are lattices with minimal elements z_1, z_2 respectively, a valuation f on $L_1 \times L_2$ is determined by $f(z_1, z_2)$ and the normalized valuations $f_1(x) = f(x, z_2) - f(z_1, z_2)$ on L_1 and $f_2(y) = f(z_1, y) - f(z_1, z_2)$ on L_2. Conversely, given normalized valuations f_i on L_i into an abelian group A and an element c in A, the function $f(x, y) = c + f_1(x) + f_2(y)$ is a valuation on $L_1 \times L_2$ into A. Thus the augmentation subgroup $I(L_1 \times L_2)$ is the direct sum $I(L_1) \oplus I(L_2)$ and

$$V(L_1 \times L_2) \approx I(L_1) \oplus I(L_2) \oplus Z(z_1, z_2) \approx [V(L_1) \oplus V(L_2)]/Z(z_2 - z_1).$$

Similarly, $\tilde{V}(L_1 \times L_2) \approx [\tilde{V}(L_1) \oplus \tilde{V}(L_2)]/Z(z_2 - z_1)$.

Application. Let L be a finite geometric lattice which is connected [3], that is, which is not isomorphic to a direct product of two geometric lattices. Then for any two copoints (= coatoms) x, y there is by the path theorem [3] a connected path $x = x_0, x_1, \ldots, x_r = y$ of copoints from x to y, where connected path means that, for each i, $x_{i-1} \wedge x_i$ is a coline above which is at least one copoint t_i different from x_{i-1} and x_i. Thus x_{i-1}, t_i, x_i generate a copy of M_5 and so any valuation f on L must take the same value on all x_i and t_i, hence on all copoints. Now for any element y of rank k there are elements x, t with x of rank $k+1$ and t a copoint such that $y = x \wedge t$ and $x \vee t = u$. Thus if all flats of rank $k+1$ have the same f value, then the same is true of all flats of rank k and so by induction downward f is constant on the flats of any given rank, in particular, on the points. Since a flat x of rank k is the join of k independent points it is easy to see that $f(x) - f(z) = k(f(p) - f(z))$ for any point p. That is, the valuation $g(x) = f(x) - f(z)$ is given by $g(x) = [\mathrm{rk}(x)]\, g(p)$. The lattice L is modular iff the rank function is a valuation. Hence if L is modular geometric then $V(L)$ is free abelian on two generators p, z and the universal valuation $i: L \to V(L)$ is given by $i(x) = [\mathrm{rk}(x)]\,(p - z) + z$. When L is nonmodular there are flats x, y, t such that $x \vee t = y \vee t$, $x \wedge t = y \wedge t$, $x < y$ and $\mathrm{rk}(y) = \mathrm{rk}(x) + 1$. Hence for the valuation g above $[\mathrm{rk}(x)]\, g(p) = g(x) = g(y) = [\mathrm{rk}(y) + 1]\, g(p)$ and so $g(p) = 0$. Thus for nonmodular L all valuations are constant so that $V(L)$ is free abelian on one generator z and $i(x) = z$ for all x in L. Clearly then $V(L) = \hat{V}(L)$ in the nonmodular case. For modular L, since L is connected and atomic it cannot be distributive unless it contains just one point in which case L is the two element lattice and $V(L) = \hat{V}(L)$. So when L is modular and not distributive there are at least two points (atoms) $\{p, q\}$ and since $i(p) = i(q)$ then $j(p) = j(q) = j(p \wedge q) = j(p \vee q)$. That is, in the group homomorphism from $V(L)$ to $\hat{V}(L)$ the point p must be identified with z so that $\hat{V}(L)$ is free on the single generator z. Finally, for any finite geometric lattice L, if L is expressed as the product of connected geometric lattices, then $V(L)$ and $\hat{V}(L)$ are free abelian groups and rank $V(L) = 1 + (\#$ of modular connected components) while rank $\hat{V}(L) = 1 + (\#$ of 1-point components [isthmuses]).

2. Finite Posets and Distributive Lattices. We collect together here some facts we shall need about maps of finite posets and distributive lattices. With only slight modification most of the results hold also for infinite posets and distributive lattices which have a zero and are locally finite, that is, in which every interval is a finite set. A subset J (possibly empty) of an ordered set (poset) $(P, \leqq)$ is called an *order ideal* (*order filter*) if $y \in J$ and $x \leqq y$ $(y \leqq x)$ imply $x \in J$. The set $J(P)$ of all order ideals of P is a sublattice of 2^P with unit $u = P$ and zero $z = \emptyset$. An element x of a join-semilattice L is *join-irreducible* if $x = r \vee s$ implies $x = r$ or $x = s$. The set $P(L)$ of all (including z) join-irreducible elements of L is a poset with the induced order. If a poset P has a zero z, the poset $P \backslash z$ will be denoted by $\hat{P}$. For any finite distributive lattice L, the map $x \to \sigma(x) = \{p \in \hat{P}(L) \mid p \leqq x\}$ is a lattice isomorphism of L onto $J(\hat{P}(L))$ [2, 7]. Dually, for any finite poset M, the principal order ideals $(m] = \{k \in M \mid k \leqq m\}$ are precisely the nonzero elements of $P(J(M))$ so that $m \to (m]$ is an order isomorphism of M onto $\hat{P}(J(M))$. In a distributive lattice L which is finite or just satisfies DCC, a *filter* (order filter closed under $\wedge$) is *prime* (its complement is closed under $\vee$, i.e. is an *ideal*) iff it is principal and its minimal element is join-irreducible. Hence $\sigma(x)$ may be identified with the set of all prime filters containing x or all prime ideals not containing x, and σ is then the Stone representation of the distributive lattice L by a lattice of sets [2, 7].

By a (u, z)-*homomorphism* between lattices we mean a map which preserves $(u, z, \vee, \wedge)$. The following equivalences of pairs of categories will be much used.

Proposition 9. *The category of finite distributive lattices and (u, z)-homomorphisms (u-homomorphisms) is equivalent to the dual of the category of finite posets (with z) and order homomorphisms (preserving z).*

Proof. We prove both equivalences simultaneously. If $\beta: M_1 \to M_2$ is an order homomorphism of finite posets then taking inverse images we get a map $\beta': J(M_2) \to \to J(M_1)$ which is a (u, z)-homomorphism of finite distributive lattices. Moreover, if the M_i both have zeros then the $\hat{J}(M_i)$ are still distributive lattices and β' maps $\hat{J}(M_2)$ into $\hat{J}(M_1)$ iff $\beta(z_1) = z_2$. In this case $\beta': \hat{J}(M_2) \to \hat{J}(M_1)$ is a u-homomorphism. The association $M \to J(M)$ (or $M \to \hat{J}(M)$) and $\beta \to \beta'$ is a contravariant functor from the poset category to the distributive lattice category. Note that for $A \in J(M_2)$, $\beta'(A) = \bigvee \{(t] \mid (\beta(t)] \leqq A\}$. In the opposite direction we have the contravariant functor P which associates to each u-homomorphism $\lambda: L_1 \to L_2$ of finite distributive lattices the order homomorphism $\lambda^*: P(L_2) \to P(L_1)$ given by $\lambda^*(p_2) = = \bigwedge \{x \in L_1 \mid \lambda(x) \geqq p_2\}$. Clearly $\lambda(z_1) = z_2$ iff λ^* maps $\hat{P}(L_2)$ into $\hat{P}(L_1)$. The conclusion now follows from easy computations involving composites and the isomorphisms $L \approx J(\hat{P}(L))$ and $M \approx \hat{P}(J(M))$ mentioned above.

For generalizations and related results see [6].

Using the concrete correspondences above it is easy to investigate the properties of special (u, z)-homomorphisms of finite distributive lattices. For example, λ is a monomorphism iff λ^* is an epimorphism and it is easy to check that an epi in the category of finite posets is just a surjective order homomorphism. But it is also obvious that λ is a monomorphism iff it is injective. For each $q \in \hat{P}(L)$ we let q^0 denote the unique maximal element in L which lies below q. Then for $\lambda: L_1 \to L_2$ as above, we have $\lambda^*(p) = q$ iff $p \leqq \lambda(q)$ and $p \nleqq \lambda(q^0)$. That is $q \in \lambda^*(\hat{P}_2(L))$ iff $\lambda(q) > \lambda(q^0)$ and λ^* is surjective iff $\lambda(q) > \lambda(q^0)$ for all $q \in \hat{P}(L_2)$. Dually, λ is an epimorphism iff λ^* is a monomorphism and clearly a monomorphism in the category of posets is just an injective order homomorphism. Among injective order homomorphisms $\beta: M_1 \to M_2$ are the *strict* maps, those for which $p \leqq q$ iff $\beta(p) \leqq \beta(q)$, i.e. M_1 is isomorphic to $\beta(M_1)$ with the order inherited from M_2. For any map β if $\beta(p) \leqq \beta(q)$ then for any $A \in J(M_2)$ such that $q \in \beta'(A)$ we have also $p \in \beta'(A)$. Consequently, if $p \nleqq q$ then $(q]$ is not in the image of β'. Thus β' is a surjective epimorphism iff β is a strict map. We will later derive two other conditions which are equivalent to λ or β' being an epimorphism.

It is now easy to prove a theorem of Balbes which characterizes "projective" distributive lattices [1, 7]. In the category of finite distributive lattices and (u, z)-homomorphisms L is *weakly projective* if for any $\alpha: L_1 \to L_2$ which is surjective and any $\beta: L \to L_2$, there is a $\gamma: L \to L_1$ such that $\alpha\gamma = \beta$. In the category of finite posets and order homomorphisms M is *weakly injective* if for any $\alpha: M_1 \to M_2$ which is strict and any $\beta: M_1 \to M$, there is a $\gamma: M_2 \to M$ such that $\gamma\alpha = \beta$.

Proposition 10 (Balbes). *For a finite distributive lattice L the following are equivalent:*

(i) *L is weakly projective,*

(ii) *$\hat{P}(L)$ is weakly injective,*

(iii) *$\hat{P}(L)$ is a lattice.*

Proof. From our remarks above (i) and (ii) are equivalent. If M is weakly injective and if in the defining property above we take $M_1 = M$, M_2 a lattice, α a strict embedding, and β the identity, then a retraction γ of M_2 onto M must exist. It follows that M must be a lattice, provided that there is such a map α. But the natural map $a \to (a]$ is a strict embedding of M into the lattice $J(M)$. Conversely, if M is a lattice, $\beta\colon M_1 \to M$ any order homomorphism, and $\alpha\colon M_1 \to M_2$ a strict map, define γ by $\gamma(m_2) = \bigvee\{\beta(m_1) \mid \alpha(m_1) \leqq m_2\}$. It is easy to see that γ preserves order and because α is strict also $\gamma\alpha = \beta$.

The referee has pointed out that the equivalence of (ii) and (iii) for an arbitrary poset in place of $\hat{P}(L)$ (and with "complete" inserted in (iii)) is due to Banaschewski and Bruns: Arch. Math. 18, 369–377 (1967).

Another useful result comes from the observation that for an order homomorphism $\beta\colon M_1 \to M_2$ the induced map $\beta'\colon J(M_2) \to J(M_1)$ takes $\hat{P}(J(M_2))$ into $\hat{P}(J(M_1))$ iff for each $m_2 \in M_2$ there is a $\alpha(m_2) \in M_1$ such that $\beta'((m_2]) = (\alpha(m_2)]$. In this case α, β constitute a Galois connection [18] between M_1 and M_2. (Each of α and β is said to be residuated.) Namely, $\alpha(m_2) = \sup\{m_1 \mid \beta(m_1) \leqq m_2\}$ and $\beta(m_1) = \inf\{m_2 \mid \alpha(m_2) \geqq m_1\}$ and $\alpha\beta$ is a closure operator on M_1 and $\beta\alpha$ is an interior operator on M_2. In terms of distributive lattices L_1 and L_2, the result above states that a map $\lambda\colon \hat{P}(L_2) \to \hat{P}(L_1)$ can be extended to a lattice (u, z)-homomorphism of L_2 into L_1 iff λ has the same properties as α above, that is, the λ-preimage of every principal order filter in $\hat{P}(L_1)$ is a principal order filter in $\hat{P}(L_2)$.

It is clear that the product $L_1 \times L_2$ of finite distributive lattices is the categorical product in both of the lattice categories under consideration and $\hat{P}(L_1 \times L_2)$ is isomorphic to the disjoint union $\hat{P}(L_1) \cup \hat{P}(L_2)$ and $P(L_1 \times L_2)$ to the one point join $P(L_1) \cup P(L_2)/\{z_1, z_2\}$ which are the coproducts of the $\hat{P}(L_i)$ and $P(L_i)$ in the poset categories. Dually the categorical product is the Cartesian product in both poset categories so the coproduct (free distributive product) of L_1 and L_2 must be isomorphic to $J(\hat{P}(L_1) \times \hat{P}(L_2))$ and $\hat{J}(P(L_1) \times P(L_2))$ respectively in the lattice categories. We will see shortly that $V(L_i)$ is free with rank $|P(L_i)|$ and so $V(\hat{J}(P(L_1) \times P(L_2)))$ has rank $|P_1(L_1)| \cdot |P(L_2)|$. This suggests that $V(L_1) \otimes V(L_2)$ might be used as a model for $V(\hat{J}(P(L_1) \times P(L_2)))$. Suppose L_1, L_2 are any distributive lattices with units (not necessarily finite), then considering the $V(L_i)$ as augmented algebras with $\wedge$-multiplication, their coproduct is $V(L_1) \otimes V(L_2)$ with the natural embeddings α_i given by $\alpha_1(y_1) = y_1 \otimes u_2$ and $\alpha_2(y_2) = u_1 \otimes y_2$. The multiplicative semigroup generated by $\alpha_1(L_1) \cup \alpha_2(L_2)$ is $\{y_1 \otimes y_2 \mid y_i \in L_i\}$ and these are idempotents in $\varepsilon^{-1}(1)$. The distributive lattice L generated by this semigroup consists, as we saw before, of all $\sum (y_i \otimes t_i) - \sum (y_i \wedge y_j \otimes t_i \wedge t_j) + \sum (y_i \wedge y_j \wedge y_k \otimes t_i \wedge t_j \wedge t_k) \ldots$ with $i < j < k < \cdots$ for all finite indexed families (y_i) in L_1 and (t_i) in L_2. Fortunately, we never need to use this expression for elements of L.

Proposition 11. *The coproduct of L_1 and L_2 in the category of u-homomorphisms of distributive lattices with unit is (L, α_1, α_2), and $V(L) \approx V(L_1) \otimes V(L_2)$.*

Proof. For any L_3, if $\beta_i\colon L_i \to L_3$ are u-homomorphisms, then the induced $\beta_i\colon V(L_i) \to V(L_3)$ are algebra homomorphisms, so there is a unique algebra map $\gamma\colon V(L_1) \otimes V(L_2) \to V(L_3)$ given by $\gamma(y \otimes t) = \gamma(y \otimes u_2)\,\gamma(u_1 \otimes t) = \beta_1(y) \wedge \beta_2(t)$

such that $\gamma\alpha_i = \beta_i$. For any $s, t \in L$ we know $\varepsilon(s) = \varepsilon(t) = 1 = \varepsilon(\gamma(s)) = \varepsilon(\gamma(t))$ so that $s \vee t = s + t - s \wedge t$ and $\gamma(s \wedge t) = \gamma(s \cdot t) = \gamma(s)\gamma(t) = \gamma(s) \wedge \gamma(t)$ and $\gamma(s \vee t) = \gamma(s) + \gamma(t) - \gamma(s \wedge t) = \gamma(s) \vee \gamma(t)$. Thus γ is a u-homomorphism of L and $\gamma(L) \subseteqq L_3$ because L_3 contains the image $\gamma(\alpha_i(L_i)) = \beta_i(L_i)$ of the generators $\alpha_i(L_i)$ of L. By the Corollary to Proposition 5, the inclusion of L into $V(L_1) \otimes V(L_2)$ extends uniquely to a ring homomorphism of $V(L)$ onto $V(L_1) \otimes V(L_2)$. On the other hand, the ring homomorphisms $\alpha_i: V(L_i) \to V(L)$ yield a ring homomorphism of $V(L_1) \otimes V(L_2)$ onto $V(L)$ which is the identity on L. Hence $V(L) \approx V(L_1) \otimes V(L_2)$.

Since J takes the category of finite posets into a subcategory of itself we may apply J twice to get a covariant functor which associates to each order homomorphism $\beta: M_1 \to M_2$ the (u, z)-homomorphism $\beta'': J^2(M_1) \to J^2(M_2)$ given, for principal ideals $(A]$, by $\beta''((A]) = \{B \in J(M_2) \mid \beta'(B) \subseteqq A\}$ for every $A \in J(M_1)$. For each finite poset M let $\gamma: M \to J^2(M)$ be given by $\gamma(m) = \{A \in J(M) \mid m \notin A\}$. Then γ provides a natural transformation from the identity to the functor J^2 since for $\beta: M_1 \to M_2$ we have

$$\begin{aligned}\beta''\gamma(m) = \beta''(\{A \in J(M_1) \mid m \notin A\}) &= \bigcup \{\beta''((A]) \mid m \notin A\} = \\ &= \{B \in J(M_2) \mid m \notin \beta'(B)\} = \\ &= \{B \in J(M_2) \mid \beta(m) \notin B\} = \gamma\beta(m).\end{aligned}$$

Lemma. *The sublattice $F(M)$ generated by $\gamma(M)$ contains all elements of $J^2(M)$ except for the unit and zero. If L is a distributive lattice, there is a unique lattice homomorphism $\pi: J^2(L) \to L$ such that $\pi\gamma = \mathrm{id}$.*

Proof. If $C \in J(M)$ and $C \neq M$, then

$$\begin{aligned}\bigcap \{\gamma(m) \mid m \notin C\} &= \{A \in J(M) \mid m \notin A \text{ for all } m \notin C\} = \\ &= \{A \in J(M) \mid A \subseteqq C\} = (C].\end{aligned}$$

Thus $F(M)$ contains all principal ideals in $J^2(M)$ except $(M]$, and doesn't contain $\emptyset$ or $(M]$ since $\emptyset \in \gamma(m)$ and $M \notin \gamma(m)$ for all $m \in M$.

If L is a distributive lattice, the map $\lambda: \hat{P}(L) \to J(L)$ given by $\lambda(p) = \{x \in L \mid p \nleqq x\}$ is order preserving. The induced lattice homomorphism $\lambda': J^2(L) \to J(\hat{P}(L))$ composed with the isomorphism $J(\hat{P}(L)) \approx L$ yields a lattice homomorphism $\pi: J^2(L) \to L$ given by $\pi((A]) = \bigvee \{p \in P(L) \mid \lambda(p) \subseteqq A\}$ for all $A \in J(L)$. Thus

$$\pi\gamma(y) = \bigvee \{p \in P(L) \mid \lambda(p) \subseteqq \lambda(y)\} = y \quad \text{for all } y \in L.$$

Proposition 12. *For any finite poset M, $(F(M), \gamma)$ is the free distributive lattice on M. That is, every order homomorphism β of M into a distributive lattice factors uniquely thru $F(M)$ as γ followed by a lattice homomorphism, namely the homomorphism $\pi\beta''$ in case the lattice is finite.*

Proof. For finite L, $\pi\beta''\gamma = \pi\gamma\beta = \beta$ by the lemma and the fact that γ is a natural transformation. $\pi\beta''$ is unique since $\gamma(M)$ generates $F(M)$. If L is infinite just replace L by a finite sublattice of L containing $\beta(M)$.

The preceding is a natural generalization of the construction due to Skolem of the free distributive lattice on a finite number of generators [2].

References

[1] R. Balbes, Projective and injective distributive lattices. Pacific J. Math. **21**, 405–420 (1967).
[2] G. Birkhoff, Lattice Theory. Third ed., Providence 1967.
[3] H. Crapo and G.-C. Rota, Combinatorial Geometries. Cambridge (Mass.) 1970.
[4] R. L. Davis, Order Algebras. Bull. Amer. Math. Soc. **76**, 83–87 (1970).
[5] J. Folkman, The homology groups of a lattice. J. Math. Mech. **15**, 631–636 (1966).
[6] L. Geissinger and W. Graves, The category of complete algebraic lattices. J. Combinatorial Theory (A) **13**, 332–338 (1972).
[7] G. Grätzer, Lattice Theory. San Francisco 1971.
[8] C. Greene, On the Möbius Algebra of a Partially Ordered Set. Proc. Conf. on Möbius Algebras, University of Waterloo 1971.
[9] P. R. Halmos, Measure Theory. Princeton 1950.
[10] P. Hilton and S. Wylie, Homology Theory. Cambridge 1960.
[11] V. Klee, The Euler characteristic in combinatorial geometry. Amer. Math. Monthly **70**, 119–127 (1963).
[12] A. Horn and A. Tarski, Measures in Boolean algebras. Trans. Amer. Math. Soc. **64**, 467–497 (1948).
[13] R. Larson and M. Sweedler, An associative orthogonal bilinear form for Hopf algebras. Amer. J. Math. **91**, 75–94 (1969).
[14] H. M. MacNeille, Partially ordered sets. Trans. Amer. Math. Soc. **42**, 416–460 (1937).
[15] B. Pettis, Remarks on the extension of lattice functionals. Bull. Amer. Math. Soc. **54**, 471 (1948).
[16] B. Pettis, On the extension of measures. Ann. of Math. **54**, 186–197 (1951).
[17] H. Rasiowa and R. Sikorski, The Mathematics of Metamathematics. Warsaw 1963.
[18] G.-C. Rota, On the foundations of combinatorial theory. I. Z. Wahrscheinlichkeitstheorie Verw. Gebiete **2**, 340–368 (1964).
[19] G.-C. Rota, On the combinatorics of the Euler characteristic. In: Studies in Pure Mathemathics, pp. 221–233. London 1971.
[20] L. Solomon, The Burnside algebra of a finite group. J. Combinatorial Theory **2**, 603–615 (1967).
[21] E. Spanier, Algebraic Topology. New York 1966.

Eingegangen am 24. 8. 1970 *)

Anschrift des Autors:
Ladnor Geissinger
Mathematics Department
University of North Carolina
Chapel Hill, North Carolina 27514, USA

*) Eine revidierte Fassung ging am 1. 9. 1972 ein.

Sonderabdruck aus
ARCHIV DER MATHEMATIK
Vol. XXIV, 1973 BIRKHÄUSER VERLAG, BASEL UND STUTTGART Fasc. 4

Valuations on Distributive Lattices II

By
LADNOR GEISSINGER

Introduction. We continue the study begun in part I [Arch. Math. 24, 230–239 (1973)] of the valuation ring of a finite distributive lattice. We show that it is the Möbius algebra of the set of join-irreducible elements and we derive Solomon's formula for idempotents. We use the duality between posets and distributive lattices given in part I to derive mapping properties of Möbius algebras. From this we get theorems on extending finitely additive measures, theorems of Rota concerning Möbius functions, an identity due to Klee, a factorization theorem of Stanley and Greene, and results on the characteristic valuation of a distributive lattice.

3. Finite Valuation Rings and Möbius Algebras. If L is a distributive lattice with DCC (minimum condition) and $P(L)$ is the set of all (including the zero z) join-irreducible elements of L, then every element of L can be uniquely expressed as a finite irredundant join of elements of $P(L)$. For each $x \in L$ the set

$$\sigma(x) = \{p \in P(L) \mid p \leqq x\}$$

is a finitely generated order ideal in $P(L)$ and $x = \bigvee \sigma(x)$. Also the prime dual ideals (filters) in L are precisely the principal filters generated by elements of $P(L)$.

Theorem 1. *For every distributive lattice L with* DCC, *the valuation ring $V(L)$ is a free abelian group with $P(L)$ as basis. That is, every valuation on L is determined by its values on $P(L)$ and these values can be assigned arbitrarily.*

Proof. For every $x \in L$ there are $\{p_1, \ldots, p_r\} \subset P(L)$ such that $x = \bigvee p_i$. Then by Prop. 7, in $V(L)$ we have $x = \bigvee p_i = \sum p_i - \sum p_i \wedge p_j + \sum p_i \wedge p_j \wedge p_k \ldots$. If $x \notin P(L)$ then each of the summands $p_i \wedge \cdots \wedge p_k$ is strictly below x in L. Hence by induction upward on L (DCC) we conclude that every x in L is a linear combination in $V(L)$ of a finite number of elements in $\sigma(x)$. Thus $P(L)$ generates $V(L)$. For any $\{p_1, \ldots, p_k\} \subset P(L)$, if p_r is maximal among them then the remaining p_i are in the complement of the prime filter generated by p_r and so p_r is independent of the rest by Prop. 3. Thus $P(L)$ is an independent set in $V(L)$.

Since this theorem is the principal result upon which the remainder of our discussion is based, we sketch alternative proofs. Perhaps the most natural procedure is to attempt to prove the statement about valuations directly without using the valuation ring. The chief difficulty is to show that any function v from $P(L)$ into an abelian group A can be extended to a valuation on L. One can easily define by induction on L an extension of the function v to L by using the unique representation of an element x as an irredundant join $\bigvee p_i$ of elements of $P(L)$ and then setting $v(x) = \sum v(p_i) - \sum v(p_i \wedge p_j) \ldots$. However, to show that this extension is a valuation requires a rather delicate induction argument. For another proof, when L is locally finite, we can use instead the valuations v_p of Proposition 3 for each $p \in P(L)$, which take the value 1 for all $x \geqq p$ and otherwise the value 0. Every function $f\colon P(L) \to A$ then determines a valuation v_f on L by $v_f(x) = \sum_p f(p)\, v_p(x) = \sum_{p \leqq x} f(p)$. By Möbius inversion over $P(L)$ we can choose f so that v_f and any given v agree on $P(L)$. One can also easily show that the v_p are independent. Hence if L is finite we get yet another proof by observing that $V(L)$ is generated by $P(L)$ and its dual has $|P(L)|$ independent functionals and so again $V(L)$ must be free on $P(L)$. See also the proof by Greene [8].

Corollary. *If $P(L)$ is a $\wedge$-semilattice, then it is a $\wedge$-subsemilattice of L and $(V(L), \wedge)$ is the semigroup algebra of $(P(L), \wedge)$ over Z.*

For distributive lattice L with DCC, the elements of L correspond to the finitely generated order ideals in $P(L)$ and these are closed under finite intersection. L is then locally finite iff $P(L)$ is locally finite, and iff for each p in $\hat{P}(L)$ there is a unique maximal element p^0 in L lying below p. When L is locally finite for each p in $\hat{P}(L)$ let $e_p = p - p^0$ (in $V(L)$) and for zero z let $e_z = z$. Parts of the following theorem appear in a paper by Davis [4] in which he showed that the valuation ring $V(L)$ is isomorphic to the Möbius algebra of $P(L)$ as defined by Solomon [20].

Theorem 2. *Let L be a locally finite distributive lattice with zero, and let μ be the Möbius function of $P(L)$. Then $\{e_p \mid p \in P(L)\}$ is a basis of $V(L)$ consisting of orthogonal idempotents, $x = \sum e_p$ $(p \leqq x)$ for each x in L, $e_p = \sum \mu(r, p)\, r$ $(r \leqq p$ and r in $P(L))$ for each p in $P(L)$, and $x = \sum \mu(r, p)\, r$ $(p, r$ in $P(L)$ and $p \leqq x)$ for each x in L.*

Proof. For $p \neq q$ in $P(L)$ it's easy to check that e_p and e_q are idempotent and $e_p \wedge e_q = 0$. For p in $P(L)$, $p = e_p + p^0$ and for any x in L which is not in $P(L)$, $x = b \vee c$ where b, c are less than x and in L. Also if $b = \sum e_q$ $(q \leqq b)$ and $c = \sum e_q$ $(q \leqq c)$ then $b \wedge c = \sum e_q$ $(q \leqq b \wedge c)$ since the e_q are orthogonal idempotents. Hence $b \vee c = b + c - b \wedge c = \sum e_q$ $(q \leqq b \vee c)$. Thus by induction upward, $x = \sum e_q$ $(q \leqq x)$ for all x in L. Hence the $\{e_q\}$ form a basis of $V(L)$. The embedding $i\colon L \to V(L)$ restricted to $P(L)$ and the map e from $P(L)$ into $V(L)$ thus satisfy $i(p) = \sum e_q$ $(q \leqq p)$ and so by Möbius inversion $e_p = \sum \mu(r, p)\, i(r) = \sum \mu(r, p) r$ $(r \leqq p)$. It follows that for every x in L, $x = \sum e_p = \sum \mu(r, p) r$ $(r \leqq p \leqq x$ and r, p in $P(L))$.

The expression for any x in terms of $P(L)$ was discovered before the other formulas above. A direct proof of this follows. By Theorem 1 for any x in L, $x = \sum d_p p$ where d_p is integral and $d_p = 0$ if $p \notin \sigma(x)$. For any q in $\sigma(x)$,

$$q = x \wedge q = \Big(\sum_{p \geqq q} d_p\Big) q + \sum_{q \not\leqq p} d_p (p \wedge q)$$

and since q is independent of the $p \wedge q < q$ in the second summand, $\sum_{p \geqq q} d_p = 1$. This is true for each q in $\sigma(x)$, so by Möbius inversion over the finite subset $\sigma(x)$ of $P(L)$ we get $d_r = \sum_{p \leq x} \mu(r, p)$.

Corollary. *For every q in $P(L)$, $q^0 = -\sum \mu(r, q)\, r \;(r < q)$.*

Solomon [20] defined the Möbius algebra $M(Q)$, for any poset Q in which every principal ideal $(q]$ is finite, as the free abelian group on Q with multiplication given by $q \wedge r = \sum \mu(s, t)\, s \;(s \leq t, t \leq q$ and $t \leq r)$. If Q has a zero z, then Q is isomorphic to the poset of join-irreducible elements of the lattice $\hat{J}(Q)$ of nonempty order ideals of Q, and by Theorem 2 then $M(Q) \approx V(\hat{J}(Q))$. If Q does not have a zero, letting z denote the zero (= empty set) of $J(Q)$, it follows that z generates a 1-dimensional ideal in $V(J(Q))$ and $M(Q) \approx V(J(Q))/Zz$. In both cases, the spanning orthogonal idempotents in $M(Q)$ are given as above by $e_p = \sum \mu(r, p) r \;(r \leq p)$ and $p = \sum e_q$ $(q \leq p)$ for each p in Q.

Proposition 13. *An order morphism $\beta\colon P \to Q$ of finite posets induces ring homomorphisms $\beta'\colon V(J(Q)) \to V(J(P))$ and $\beta'\colon M(Q) \to M(P)$ given by $\beta'(e_q) = \sum e_p$ $(\beta(p) = q)$, where the sum is taken to be 0 if the index set is empty.*

Proof. We are identifying Q, P with the nonzero join irreducible elements of $J(Q)$, $J(P)$. Define a map $\lambda\colon V(J(Q)) \to V(J(P))$ by $\lambda(e_q) = \sum e_p \;(\beta(p) = q)$. Then for each $A \in J(Q)$, $A = \sum e_q \;(q \in A)$ and so $\lambda(A) = \sum e_p(\beta(p) \in A) = \beta'(A)$. Thus λ is the map induced on $V(J(Q))$ by the (u, z)-lattice homomorphism $\beta'\colon J(Q) \to J(P)$. Factoring out $z = \emptyset$ merely deletes $e_z = z$ from the expressions for e_p, e_q and β', so that in the quotient $\beta'\colon M(Q) \to M(P)$ is given by the same formula.

The conclusion of Prop. 13 still holds if we suppose β is only defined on an order ideal in P, or equivalently if β maps P into Q with a unit u' adjoined. If moreover P and Q have zeros and if β preserves zero, then $\beta'\colon \hat{J}(Q) \to \hat{J}(P)$ is a lattice homomorphism and by Prop. 9 every lattice homomorphism $\hat{J}(Q) \to \hat{J}(P)$ comes from exactly one such $\beta\colon P \to Q \cup \{u'\}$. From Prop. 13 it follows that $\beta'\colon V(J(Q)) \to V(J(P))$ is surjective iff β is injective and $\beta(P) \subseteq Q$ (nothing maps onto u') while β' is injective iff β maps onto Q (some elements could still go onto u'). Finally, if β maps onto Q, let $\alpha\colon Q \to P$ be any function such that $\beta\alpha = \mathrm{id}$ (i.e. $\alpha(q) \in \{p \mid \beta(p) = q\}$), and let T be the subgroup of $V(\hat{J}(P))$ spanned by $\{e_p \mid p \notin \alpha(Q)\}$. The subring $\beta'(V(\hat{J}(Q)))$ is spanned by $\{\beta'(e_q)\}$ and clearly $V(\hat{J}(P)) = \beta'(V(\hat{J}(Q))) \oplus T$. Furthermore, it is easy to see that when any y in $V(\hat{J}(P))$ is expressed as a linear combination of the $\{\beta'(e_q)\}$ and the basis described for T, then every coefficient is 0 or ± 1. Hence if $y \notin \beta'(V(\hat{J}(Q)))$, we may replace some e_p in the basis for T by y to get a basis for another direct complement of $\beta'(V(\hat{J}(Q)))$. These results combined with the duality in Prop. 9 and our subsequent discussion of mapping properties yields the following.

Proposition 14. *Let $\lambda\colon K \to L$ be a homomorphism of finite distributive lattices and $\lambda^e\colon V(K) \to V(L)$ the induced ring homomorphism. Then λ is an epimorphism iff λ^e is surjective, and λ is an injection iff λ^e is an injection. In any case, $\lambda^e(V(K))$ is a direct summand of $V(L)$. Moreover for any $y \in L$ with $y \notin \lambda^e(V(K))$ there is a direct complement of $\lambda^e(V(K))$ with a basis which contains y.*

22*

Corollary. *Let K be a sublattice of a finite distributive lattice L. Then every valuation v of K into an abelian group A can be extended to a valuation of L into A. Moreover, for any element $y \in L$ such that $y \notin V(K)$, there is an extension of v to L for which $v(y)$ is any prescribed value in A.*

We will see later precisely what the condition that $y \in L$ and $y \notin V(K)$ means in terms of the lattices K and L. The following interesting result will be used to extend part of Proposition 14 and its Corollary to infinite distributive lattices.

Proposition 15. *Let $B_1, \ldots, B_r$ be elements of a distributive lattice L. Every linear relation among the B_k which holds in $V(L)$ also holds in $V(M)$ where M is the finite sublattice generated by the B_k.*

Proof. Suppose that in $V(L)$, $\sum a_k B_k = 0$ where the a_k are integers. From our construction of $V(L)$ this means that $\sum a_k B_k$, when considered back in $Z^{(L)}$, is a linear combination of a finite number of elements of the form

$$C_i \vee D_i + C_i \wedge D_i - C_i - D_i.$$

Let N be the finite sublattice generated by all the B_k, C_i, D_i. Then in $V(N)$ the relation $\sum a_k B_k = 0$ holds. But N is finite and $M \subseteqq N$ so by Proposition 14 the induced map $j\colon V(M) \to V(N)$ is an injection. Thus any relation among elements of M which holds in $V(N)$ also holds in $V(M)$.

Remark. This immediately yields a proof of Proposition 4 which does not depend on the existence of enough prime ideals, that is, does not make use of Zorn's Lemma.

Proposition 16. *For every distributive lattice L and sublattice M, the natural map $j\colon V(M) \to V(L)$ induced by inclusion is an injection.*

Proof. If $B_1, \ldots, B_r$ are elements of M, any linear relation among them which holds in $V(L)$ also holds in the valuation ring of the sublattice of L generated by the B_k by Proposition 15. But since this sublattice is contained in M the relation also holds in $V(M)$.

Corollary. *Any valuation of M into any rational vector space A (or divisible abelian group) can be extended to a valuation of L into A.*

Stronger versions of this were proved by Horn and Tarski [6] and Pettis [9] for bounded real-valued modular functions.

4. Combinatorial Applications.

The Characteristic Valuation. For a finite distributive lattice L it is easily shown that the rank function r is given by $r(y) = |\{p \in P(L) \mid z < p \leqq y\}|$. Using the representation $y = \sum e_p$ $(p \leqq y)$ in $V(L)$ we see that r is the unique valuation on L for which $r(p) - r(p^0) = r(e_p) = 1$ for all $p \in \hat{P}(L)$ and $r(z) = 0$. The valuation χ on L for which $\chi(p) = 1$ for all $p \in \hat{P}(L)$ and $\chi(z) = 0$ is called by Rota the *characteristic valuation* of L.

Proposition 17. *For every* $y \in L$, $\chi(y) = -\sum \mu(z, q)$ ($q \in P(L)$ *and* $z < q \leqq y$). *For the element* p^0 *covered (in L) by some* $p \in \hat{P}(L)$, $\chi(p^0) = 1 + \mu(z, p)$.

Proof. Just apply the homomorphism χ on $V(L)$ to the expression $y = \sum \mu(p, q) p$ ($q \leqq y$) given in Theorem 2 to get $\chi(y) = \sum \mu(p, q)$ ($z < p \leqq q \leqq y$). Then $\chi(y) = -\sum \mu(z, q)$ ($z < q \leqq y$) and when $y = p^0$,

$$\chi(p^0) = -\sum \mu(z, q) \; (z < q < p) = \mu(z, z) + \mu(z, p).$$

Now suppose $P(L)$ is a lattice and y is the join in L of $p_1, \ldots, p_r \in \hat{P}(L)$, then

$$(\chi - 1)(y) = (\chi - 1)(\vee p_i) = \sum (\chi - 1)(p_i) - \sum (\chi - 1)(p_i \wedge p_j) + \cdots.$$

Since $\chi - 1$ is the valuation which takes the value 0 on $\hat{P}(L)$ and -1 on z, we get $\chi(y) - 1 = c_2 - c_3 + c_4 - c_5 + \cdots$ where c_k is the number of k element subsets of $p_1, \ldots, p_r$ whose meet is z. When $y = q^0$ for some $q \in \hat{P}(L)$ this yields a result due to Rota [18].

Corollary. *Suppose* P *is a finite lattice and* $\{q, p_1, \ldots, p_r\} \subseteq \hat{P}$ *is such that all* $p_i \leqq q$ *and every element covered by* q *is among the* p_i. *Then* $\mu(z, q) = c_2 - c_3 + \cdots$ *where* c_k *is the number of* k*-subsets of* $\{p_i\}$ *whose meet is* z.

Proof. In $\hat{J}(P)$, q^0 is the join of the p_i, hence by Proposition 17, $\chi(q^0) - 1 = \mu(z, q)$.

Following Rota, we indicate how the characteristic valuation is related to the Euler characteristic of combinatorial topology [19]. For any finite totally unordered set S, the nonzero elements of $J(S) = 2^S$ are called simplexes and the nonzero elements of $J^2(S)$ are the finite simplicial complexes with vertices in S. In this setting we usually treat 2^S simply as a poset and ignore its lattice structure, but joins and intersections of subcomplexes are important. For any finite distributive lattice $L = J(P)$ we would like to consider the elements of P as simplexes and the elements of $J(P)$ as simplicial complexes. We can approximate to this by constructing an analogue of the barycentric subdivision operator. The (first) barycentric subdivision of a simplicial complex $K \in \hat{J}(2^S)$ may be described as the simplicial complex $B(K)$ whose k-simplexes are all chains $A_0 \subset A_1 \subset \cdots \subset A_k$ of $k + 1$ (non-empty) simplexes A_i of K. By analogy, for any finite poset M let $B(M)$ be the set of all (non-empty) finite chains of elements of M ordered by inclusion. $B(M)$ is an ordered simplicial complex [10] and $J(B(M))$ is the lattice of all finite subcomplexes of $B(M)$. To each $A \in J(M)$ we associate the subcomplex $B(A)$ whose simplexes are those chains in $B(M)$ all of whose elements are in A, that is, the full subcomplex with vertex set A. Then B is a lattice monomorphism of $J(M)$ into $J(B(M))$. Note that for $m \in M$, $B((m])$ is usually not a simplex or even a subdivided simplex but it is topologically as simple since it is the cone with vertex m and base $B(\{m' \in M \mid m' < m\})$ and is thus contractible. When M consists of all elements of a finite lattice L except for the unit and zero, then the homology of the complex $B(M)$ is called the *order homology* of L [5, 18, 19]. Now if $\lambda\colon L_1 \to L_2$ is a (u, z)-homomorphism of finite distributive lattices, λ^* carries each chain in $\hat{P}(L_2)$ into a possibly smaller chain in $\hat{P}(L_1)$ so that λ^* may also be considered as an order homomorphism of $B\hat{P}(L_2)$ into $B\hat{P}(L_1)$. Taking preimages we get a (u, z)-homomorphism $B(\lambda)$ of $JB\hat{P}(L_1)$ into $JB\hat{P}(L_2)$.

The association of $JB\hat{P}(L)$ to L and of $B(\lambda)$ to λ is a functor from the lattice category into itself, and the lattice monomorphism $B\colon L \to JB\hat{P}(L)$ given by

$$B(y) = B(\{p \in \hat{P}(L) \mid p \leqq y\})$$

provides a natural transformation from the identity to this functor. It is well-known [10, 21] that the Euler characteristic E in combinatorial topology is a modular function from the lattice of subcomplexes of a simplicial complex into the integers, which is 0 on the empty subcomplex and takes the value 1 on any contractible subcomplex. Thus for any finite distributive lattice L, the composite

$$EB\colon L \to JB\hat{P}(L) \to Z$$

is a valuation on L which takes the value 0 on z and the value 1 on $\hat{P}(L)$, that is, $EB = \chi$ the characteristic valuation on L. From the classical formula for the Euler characteristic, for $y \in L$, $\chi(y) = EB(y) = a_0 - a_1 + a_2 - \cdots$ where a_k is the number of chains of $k+1$ elements in $\hat{P}(L)$ which are less than or equal to y.

Klee's Identity and Extensions of Valuations. In a paper on the Euler characteristic [11], Klee proves the following identity.

Proposition 18. *Suppose S is a $\wedge$-semilattice, (a_i) and (b_j) are finite indexed families of elements in S, and $u \in S$ such that $u \geqq a_i, b_j$ for all i, j. Then in the semigroup ring $Z[S, \wedge]$,*

$$\prod (u - a_i) + \prod (u - b_j) - \prod (u - a_i \wedge b_j) = \prod (u - a_i) \prod (u - b_j).$$

Proof. We may assume S is finite. From the Corollary to Theorem 1 we see that the identification of an element of S with the ideal it generates yields an isomorphism of $Z[S, \wedge]$ with $V(\hat{J}(S))$. But in $V(\hat{J}(S))$, the identity above reduces by Proposition 7 to the simple statement

$$(u - \vee a_i) + (u - \vee b_j) - (u - \vee \{a_i \wedge b_j\}) = u - (\vee a_i) \vee (\vee b_j)$$

where $\vee$ means join in $\hat{J}(S)$. And this holds because $\vee \{a_i \wedge b_j\} = (\vee a_i) \wedge (\vee b_j)$ in $\hat{J}(S)$.

Klee derived the relation above using the counting result below. For positive integers c, m, n let $P_c(m, n)$ denote the number of relations of cardinality c in $\{1, 2, \ldots, m\} \times \{1, 2, \ldots, n\}$ with domain $\{1, 2, \ldots, m\}$ and range $\{1, 2, \ldots, n\}$ (subsets of size c projecting onto all of $\{1, \ldots, m\}$ and $\{1, \ldots, n\}$ respectively).

Corollary. $-\sum\limits_c (-1)^c P_c(m, n) = (-1)^{m+n}$.

Proof. Choose a semilattice S and elements a_i, b_j such that all of the meets $a_{i_1} \wedge \cdots \wedge a_{i_p} \wedge b_{j_1} \wedge \cdots \wedge b_{j_r}$ are distinct. Then in the identity in Proposition 18, the expressions above are the coefficients on each side of the identity for any term which is a product of m of the a_i and n of the b_j.

The following is a generalization of a theorem by Klee in the same paper [11]. Let L be a lattice and K a $\wedge$-subsemilattice such that every element of L is a finite join of elements of K. For any function f from K into an abelian group, and for any finite family $(a_1, \ldots, a_n)$ of elements of K, let

$$f(a_1, \ldots, a_n) = \sum f(a_i) - \sum f(a_i \wedge a_j) + \sum f(a_i \wedge a_j \wedge a_k) - \cdots (i < j < k \cdots).$$

Proposition 19. *The function f on K can be extended to a distributive valuation on L iff for all families (a_i), (b_j) in K, if $\vee a_i = \vee b_j$ then $f(a_1, \ldots, a_n) = f(b_1, \ldots, b_r)$.*

Proof. By our earlier remarks this condition holds if f extends to a distributive valuation on L since then $f(\vee a_i) = f(a_1, \ldots, a_n)$. So assume f satisfies the condition. Then for any $a \in L$, define $f(a)$ to be $f(a_1, \ldots, a_n)$ where the $a_i \in K$ are chosen so that $\vee a_i = a$. Extend this function on L linearly to $Z[L, \wedge]$, then

$$f(a) = f(a_1, \ldots, a_n) = f\left(\sum a_i - \sum a_i \wedge a_j \ldots\right).$$

For (a_i) and (b_j) in K by Klee's identity,

$$f(\vee a_i) + f(\vee b_j) - f(\vee \{a_i \wedge b_j\}) = f((\vee a_i) \vee (\vee b_j)).$$

Finally, given a, b in L we can find (a_i), (b_j) in K such that $\vee a_i = a$, $\vee b_j = b$ and $a \wedge b = \vee \{a_i \wedge b_j\}$, hence f is a valuation on L and it is obviously distributive.

Corollary. *If L is distributive, f can be extended to a valuation on L iff for all (a_i) in K, if $\vee a_i = a$ is again in K then $f(a) = f(a_1, \ldots, a_n)$.*

Proof. One can easily show for any (a_i) in K, if $a_{n+1} \leqq \vee a_i$ then

$$f(a_1, \ldots, a_n) = f(a_1, \ldots, a_{n+1})$$

and from this that if $\vee a_i = \vee b_j$ then

$$f(a_1, \ldots, a_n) = f(b_1, \ldots, b_r).$$

See Klee [11] for this and further results.

Factorization in Möbius Algebras. Suppose P is a finite poset with zero and unit and K is the set of all elements covered by u. Let $t \in P$, $t \neq u$ and $C = \{k \in K : k \geqq t\}$ and identify as usual P with the join-irreducible elements of $\mathcal{J}(P) = L$. In $V(L)$ or $M(P)$, $e_u = u - u^0 = u - \vee K = \prod_K (u - k)$ and $e_u = \sum \mu(r, u)\, r$. For all c in C, $u - c$ is idempotent and $(u - c) \wedge r = 0$ for all $r \leqq c$. Thus

$$(u - c) \wedge e_u = (u - c) \wedge \Bigl(\sum_{r \nleq c} \mu(r, u)\, r\Bigr)$$

and continuing for all c in C,

$$e_u = \prod_C (u - c) \wedge e_u = \prod_C (u - c) \wedge \Bigl(\sum \mu(r, u)\, r\Bigr)$$

where the sum is over only those $r \in P$ for which $r \nleqq \vee C$. But $r \nleqq \vee C$ in $\mathcal{J}(P)$ iff in P, $\sup\{r, t\} = u$. Now in the Möbius algebra of the interval $[t, u]$ as poset, we have $\prod_C (u - c) = \sum_{r \geqq t} \mu(r, u)\, r$, but this may not hold in $M(P)$. In $M(P)$ we have

$$\prod_C (u - c) = u - \vee C = \sum e_p (p \nleqq \vee C)$$

and
$$\sum_{r \geqq t} \mu(r, u)\, r = \sum_{r \geqq t} \mu(r, u) \Big(\sum_{p \leqq r} e_p \Big).$$

The coefficient of e_p in this expression is 1 when $p \nleqq \vee C$, 0 when t and p are comparable and $p \neq u$, and $\sum \mu(r, u)$ ($t \leqq r$ and $p \leqq r$) when p and t are incomparable and $p \leqq \vee C$. It follows that, if for every $p \in P$, $\sup\{p, t\}$ exists in P, then in $M(P)$ we have $\sum_{r \geqq t} \mu(r, u)\, r = \prod_C (u - c)$. This yields the factorization theorem due to Greene [8].

Proposition 20. *Suppose P is a finite poset with unit and zero and t in P has the property that for every p in P, the join $p \vee t$ exists in P. Then in the Möbius algebra of P,*

$$\sum_{r \in P} \mu(r, u)\, r = \Big(\sum_{r \geqq t} \mu(r, u)\, r \Big) \wedge \Big(\sum_{r \vee t = u} \mu(r, u)\, r \Big).$$

Note that the condition on t is equivalent to requiring the injection $[t, u] \to P$ to be residuated, that is, the preimage of every principal filter in P is a principal filter in $[t, u]$. When P is a lattice, if Q is the $\wedge$-subsemilattice generated by C and u, then the inclusion $Q \to P$ induces a homomorphism $M(Q) \to M(P)$. Thus $\prod_C (u - c)$ can be computed in $M(P)$ as $\sum \mu_P(r, u)\, r$ ($r \in P$, $r \geqq t$) and in the sub-algebra $M(Q)$ as $\sum \mu_Q(r, u)\, r$ ($r \in Q$). Greene and Stanley [8] use the latter expression in applications of the factorization theorem to geometric lattices.

Residuated Maps. For a finite poset P, even if P has a zero, adjoin a new element $\bar{z}$ to get a poset $\bar{P} = P \cup \{\bar{z}\}$ with $\bar{z}$ as zero. Enlarge the Möbius algebra $M(P)$ to $M(P) \oplus Z\bar{z}$ and extend multiplication by defining $(\bar{z})^2 = \bar{z}$ and $M(P) \wedge \bar{z} = 0$. In this algebra it is easily checked that the elements $\{p + \bar{z} \mid p \in P\} \cup \{\bar{z}\}$ multiply precisely as do the elements of $\bar{P}$ in $M(\bar{P})$ so that we may identify $M(\bar{P})$ and $M(P) \oplus Z\bar{z}$. Now if $\varphi\colon P \to Q$ is an order morphism of finite posets and each of P and Q is enlarged by adding a zero $\bar{z}$, then obviously $\varphi\colon M(P) \to M(Q)$ is a $\wedge$-homomorphism iff $\varphi\colon M(\bar{P}) \to M(\bar{Q})$ is a $\wedge$-homomorphism. If $\bar{\varphi}$ is a $\wedge$-homomorphism, since $x \vee y = x + y - x \wedge y$ for x, y in $\mathfrak{J}(\bar{P})$ or in $\mathfrak{J}(\bar{Q})$, then $\bar{\varphi}$ is also a lattice homomorphism of $\mathfrak{J}(\bar{P})$ into $\mathfrak{J}(\bar{Q})$ taking only $\bar{z}$ in $\bar{P}$ onto $\bar{z}$ in $\bar{Q}$. But following Prop. 10 we saw that $\varphi\colon P \to Q$ extends to a lattice z-homomorphism $\bar{\varphi}\colon \mathfrak{J}(\bar{P}) \to \mathfrak{J}(\bar{Q})$ iff φ has the property that the preimage of each principal filter in Q is a principal filter in P or is empty. This completes the proof of the following proposition.

Proposition 21. *A map $\varphi\colon P \to Q$ of finite posets extends to a homomorphism $\varphi\colon M(P) \to M(Q)$ of the Möbius algebras iff the preimage of every principal filter in Q is a principal filter in P or is empty.*

Note that this condition is satisfied if φ is inclusion and P is either an ideal in Q or P is a $\wedge$-subsemilattice if Q is a $\wedge$-semilattice. If φ satisfies the condition in Prop. 21, φ is half of a Galois connection and the other part is the map $\beta\colon Q \to P \cup \{\bar{u}\}$ given by $\beta(q) = \min\{p\colon \varphi(p) \geqq q\}$. From Prop. 13 it follows that the map $\varphi = \beta'$: $M(P \cup \{\bar{u}\}) \to M(Q)$ is given by $\varphi(e_p) = \sum e_q (\beta(q) = p)$. But $\varphi(e_p) = \sum \mu_P(r, p)\, \varphi(r)$ and $\sum_q e_q (\beta(q) = p) = \sum_{t,q} \mu_Q(t, q)\, t\, (\beta(q) = p)$. Comparing coefficients of any t in Q yields Rota's principal theorem on Galois connections [8, 18].

Corollary. *Suppose $\varphi: P \to Q$ and $\beta: Q \to P$ are order morphisms of finite posets such that $\beta(q) = \min\{p: \varphi(p) \geqq q\}$ for each q in Q. Then for each t in Q and p in P,*

$$\sum_{\beta(q)=p} \mu_Q(t, q) = \sum_{\varphi(r)=t} \mu_P(r, p)$$

where a sum is taken to be 0 if the index set is empty.

References

[1] R. Balbes, Projective and injective distributive lattices. Pacific J. Math. **21**, 405–420 (1967).
[2] G. Birkhoff, Lattice Theory. Third ed., Providence 1967.
[3] H. Crapo and G.-C. Rota Combinatorial Geometries. M.I.T. Press 1970.
[4] R. L. Davis, Order Algebras. Bull. Amer. Math. Soc. **76**, 83–87 (1970).
[5] J. Folkman, The homology groups of a lattice. J. Math. and Mech. **15**, 631–636 (1966).
[6] L. Geissinger and W. Graves, The category of complete algebraic lattices. J. Combinatorial Theory (A) **13**, 332–338 (1972).
[7] G. Gratzer, Lattice Theory. San Francisco 1971.
[8] C. Greene, On the Möbius Algebra of a Partially Ordered Set. Proc. Conf. on Möbius Algebras, University of Waterloo 1971.
[9] R. R. Halmos, Measure Theory. Princeton 1950.
[10] P. Hilton and S. Wylie, Homology Theory. Cambridge 1960.
[11] V. Klee, The Euler characteristic in combinatorial geometry. Amer. Math. Monthly **70**, 119–127 (1963).
[12] A. Horn and A. Tarski, Measures in Boolean algebras. Trans. Amer. Math. Soc. **64**, 467–497 (1948).
[13] R. Larson and M. Sweedler, An associative orthogonal bilinear form for Hopf algebras. Amer. J. Math. **91**, 75–94 (1969).
[14] H. M. MacNeille, Partially ordered sets. Trans. Amer. Math. Soc. **42**, 416–460 (1937).
[15] B. Pettis, Remarks on the extension of lattice functionals. Bull. Amer. Math. Soc. **54**, 471 (1948).
[16] B. Pettis, On the extension of measures. Ann. of Math. **54**, 186–197 (1951).
[17] H. Rasiowa and R. Sikorski, The Mathematics of Metamathematics. Warsaw 1963.
[18] G.-C. Rota, On the foundations of combinatorial theory. I. Z. Wahrscheinlichkeitstheorie Verw. Gebiete **2**, 340–368 (1964).
[19] G.-C. Rota, On the combinatorics of the Euler characteristic. In: Studies in Pure Mathematics, pp. 221–223. London 1971.
[20] L. Solomon, The Burnside algebra of a finite group. J. Combinatorial Theory **2**, 603–615 (1967).
[21] E. Spanier, Algebraic Topology. New York 1966.

Eingegangen am 2. 10. 1972

Anschrift des Autors:
Ladnor Geissinger
Mathematics Department
University of North Carolina
Chapel Hill, North Carolina 27514, USA

Sonderabdruck aus
ARCHIV DER MATHEMATIK
Vol. XXIV, 1973 BIRKHÄUSER VERLAG, BASEL UND STUTTGART Fasc. 5

Valuations on Distributive Lattices III

By

Ladnor Geissinger

Introduction. In Parts I and II [Arch. Math. 24, 230–239, 337–345 (1973)] we were principally interested in combinatorial applications of the valuation ring of a distributive lattice. We now show how this ring provides a natural setting for some elementary results in measure theory as well as some classical results on representations of distributive lattices. Specifically, in the valuation ring $V(L)$ of a distributive lattice L it is easy to identify various extensions of L as well as prime ideals of L and so arrive at some theorems of Pettis, Birkhoff, and Stone. For any faithful representation of L as a lattice of sets the extension of $V(L)$ by real (or complex) scalars is naturally isomorphic to the algebra of simple functions, and the sup norm on the functions comes from an intrinsic norm on $V_R(L)$. The Stone space of L corresponds to the spectrum of $V_R(L)$ with the Zariski topology.

5. Extensions of a Distributive Lattice. It is well-known [9] that the ring (family closed under union and difference) generated by a lattice L of sets consists of all the finite disjoint unions of differences $E - F$ of elements of L. The existence and categorical properties of this and other minimal extensions of a distributive lattice L can be easily deduced using the valuation ring $V(L)$. We noted before that if $w \leqq x \leqq y$ in L then the element $w + y - x$ in $V(L)$ acts like the relative complement of x in the interval $[w, y]$ even if such a relative complement does not exist in L. If w, x_i, y_i are elements of L and $w \leqq x_i \leqq y_i$ then $(y_1 - x_1) \wedge (y_2 - x_2) = y_1 \wedge y_2 - (x_1 \wedge y_2) \vee (y_1 \wedge x_2)$ and $w \wedge (y_i - x_i) = 0$ in $V(L)$ so that both $y_i - x_i$ and $w + y_i - x_i$ are idempotent elements, $y_1 - x_1$ is orthogonal to $y_2 - x_2$ (the analogue of disjoint sets) iff $y_1 \wedge y_2 \leqq x_1 \vee x_2$, and

$$(w + y_1 - x_1) \wedge (w + y_2 - x_2) = w + y_1 \wedge y_2 - (x_1 \wedge y_2) \vee (y_1 \wedge x_2).$$

Let $R(L)$ be the set of all elements of $V(L)$ of the form $s = w + \sum (y_i - x_i)$ where $w \leqq x_i \leqq y_i$ in L for $1 \leqq i \leqq n$ and the $y_i - x_i$ are mutually orthogonal. If $v \leqq w$ then also $s = v + (w - v) + \sum (y_i - x_i)$ and $w - v$ is orthogonal to all the $y_i - x_i$. For another element $r = w + \sum (q_j - p_j)$ in $R(L)$ expressed as above, $r \wedge s = w + \sum (y_i - x_i) \wedge (q_j - p_j)$ is again in $R(L)$ since the $(y_i - x_i) \wedge (q_j - p_j) = y_i \wedge q_j - (y_i \wedge p_j) \vee (x_i \wedge q_j)$ are orthogonal and all elements are above w in L. Thus $R(L)$ is closed under the idempotent operation $\wedge$. To show that $R(L)$ is closed under $\vee$, first note that since the $y_i - x_i$ are orthogonal $w + \sum (y_i - x_i) = \bigvee (w + y_i - x_i)$

and dually if $t \geqq y_i \geqq x_i$ for all i then $t - \sum (y_i - x_i) = \bigwedge (t - y_i + x_i)$ and this element is also in $R(L)$. Now if $q \geqq p \geqq w$, and s is as above,

$$\begin{aligned}(w + q - p) \vee (w + \textstyle\sum (y_i - x_i)) &= \\ = w + \textstyle\sum (y_i - x_i) + (q - p) - (q - p) \wedge \textstyle\sum (y_i - x_i) &= \\ = \textstyle\sum (y_i - x_i) + w + (q - p) \wedge (t - \textstyle\sum (y_i - x_i))&\end{aligned}$$

for any t in L such that $t \geqq q$ and $t \geqq \bigvee y_i$, and since

$$w + (q - p) \wedge (t - \textstyle\sum (y_i - x_i)) = (w + q - p) \wedge (t - \textstyle\sum (y_i - x_i))$$

is in $R(L)$ and $(q - p) \wedge (t - \sum (y_i - x_i))$ is orthogonal to all the $y_i - x_i$ then $(w + q - p) \vee s$ is in $R(L)$. It then follows that for any r, s in $R(L)$ above w in L, $r \vee s$ and $r \vee s - s + w = v$ are both again in $R(L)$ and above w. Thus $R(L)$ is a distributive lattice. In fact, $R(L)$ is relatively complemented because if $t \leqq s \leqq r$ in $R(L)$ there is a w in L such that $w \leqq t$ and $v = r - s + w$ is in $R(L)$ so $t \vee v = r - s + t$ is in $R(L)$ and is the complement of s in $[t, r]$.

Proposition 22. *For every distributive lattice L, $R(L)$ is the unique minimal relatively complemented distributive extension of L. Every lattice homomorphism of L into a relatively complemented distributive lattice L' extends uniquely to a lattice homomorphism of $R(L)$ into L'.*

Proof. If $\varphi\colon L \to L'$ is a lattice homomorphism it extends uniquely to a ring homomorphism $\varphi\colon V(L) \to V(L')$. So φ extends to a lattice homomorphism of $R(L)$ into $R(L') = L'$ and the extension is unique since $R(L)$ is generated by relative complements $w + y - x$ with $w \leqq x \leqq y$ in L and $\varphi(x)$ has a unique relative complement in the interval $[\varphi(w), \varphi(y)]$ in L'.

Note that for the augmentation homomorphism $\varepsilon\colon V(L) \to Z$, $\varepsilon(R(L)) = 1$. Also, if L has a zero and unit (or just a zero) then $R(L)$ is a Boolean (generalized Boolean) algebra with the same zero and unit. If L does not have a zero or unit and we adjoin them to L to get L' then $R(L')$ will be the minimal (generalized) Boolean algebra generated by L, and $V(L')$ will have rank one or two more than $V(L)$. By our previous discussion of the universal properties of the map $L \to V(L)$ or by propositions 16 and 22 it follows that the embedding $L \to R(L)$ induces an isomorphism $V(L) \approx V(R(L))$. This gives part of a result due to Pettis [16] (for real valuations see also Smiley, Trans. Amer. Math. Soc. 48 (1944)).

Corollary. *If λ is a valuation on L into an abelian group A it has a unique extension to a valuation on $R(L)$ into A. If A is a partially ordered abelian group and λ is monotone on L then its extension to $R(L)$ is also monotone.*

Proof. $\lambda\colon L \to A$ extends uniquely to a group homomorphism $\lambda\colon V(L) \to A$ which restricts to a valuation on $R(L)$. For $s \leqq t$ in $R(L)$ there is a w in L such that $w \leqq s$ and so $r = t - s + w = w + \sum (y_i - x_i)$ with the $y_i - x_i$ orthogonal and $y_i \geqq x_i \geqq w$ in L. Thus $\lambda(t) - \lambda(s) = \sum (\lambda(y_i) - \lambda(x_i))$ so if λ is monotone on L then it is also on $R(L)$.

6. Representations and Prime Ideals. A form of the following statement seems to be a folk theorem of measure theory.

Theorem 3. *If L is a lattice of nonempty subsets of a set X then $V(L)$ is naturally isomorphic to the ring of simple functions $S(L)$ generated over Z by the characteristic functions of the sets in L.*

Proof. For each $A \in L$ let $C_A\colon X \to Z$ be the characteristic function of the subset A of X. Then $A \to C_A$ is a modular function from lattice L into the ring $S(L)$ which is multiplicative and hence extends uniquely to a ring homomorphism of $V(L)$ onto $S(L)$. To show it is a monomorphism it is necessary to show that if a relation $\sum d_i C_{A_i} = 0$ holds in $S(L)$ then also $\sum d_i A_i = 0$ holds in $V(L)$. If M is a finite sublattice of L containing $A_1, \ldots, A_n$, it will be enough by proposition 15 to show that $V(M)$ is isomorphic to $S(M)$. If B is a join-irreducible element of M and B^0 is the maximal element of M properly contained in B, then an element $x \in B \setminus B^0$ is not in any $A \in M$ for which $A \ngeq B$. So C_B is independent of C_A for all $A \in M$, $A \ngeq B$. It follows as in the proof of Theorem 1 that the C_B for all $B \in P(M)$ are independent and so $V(M)$ and $S(M)$ have the same rank. Since $V(M)$ maps onto $S(M)$ it follows that the map must be an isomorphism.

Another version of this result states that a valuation on such a lattice of sets L extends uniquely to a group homomorphism on $S(L)$.

The algebra of simple functions $S_R(L)$ generated over the real numbers R by $S(L)$ is a subalgebra of the Banach algebra $B(X)$ of all bounded real-valued functions on X with the sup-norm. The norm on $S_R(L)$ then yields a norm on $V_R(L) = V(L) \otimes R$ which we shall show is intrinsic, that is, can be defined using only the embedding of L into $V_R(L)$. We defer the definition until after a discussion of prime ideals.

If T is a proper prime ideal in the ring $V_R(L)$ then $T \nsupseteq L$ since L generates $V_R(L)$ as an R-algebra and if T and L were disjoint then for all $A \leqq B$ in L, $A \wedge (B - A) = 0$ so $B - A$ is in T which means that $T = I$ the augmentation ideal. Thus if T is not the augmentation ideal then $T \cap L$ is a proper (i.e. not L and nonempty) prime ideal of L. Also for all $A \leqq B$ in L if A is not in T then $B - A$ is in T so that in $V_R(L)/T$ all elements of the prime filter $L \setminus (T \cap L)$ are identified to the unit of the quotient algebra, which is then isomorphic to R. Thus all prime ideals of $V_R(L)$ are maximal and are the kernels of multiplicative linear functionals $V_R(L) \to R$. If P is a prime ideal of L, the valuation v_p which takes the value 0 on P and 1 on $L \setminus P$ extends to a multiplicative linear functional on $V_R(L)$. Thus the prime ideals of $V_R(L)$ other than I correspond bijectively to the proper prime ideals (or filters) of L, and hence also of $R(L)$. The next proposition implies the existence of enough prime ideals to separate elements of L or $R(L)$ and even stronger separation properties.

Proposition 23. *If M is a sublattice of L then every prime ideal of $V_R(M)$ (or L) lifts to a prime ideal of $V_R(L)$ (or L).*

Proof. Let T be a prime ideal of $V_R(M)$ not the augmentation ideal and $A \in M \setminus (T \cap M)$ and U the ideal in $V_R(L)$ generated by $T \cap M$. If A were in U then A would be a linear combination of elements $B_i \wedge E_i$, $i = 1, \ldots, n$ where

$B_i \in T \cap M$ and $E_i \in L$. But $\bigvee (B_i \wedge E_i) \leqq \bigvee B_i$ and $\bigvee_i B_i \ngeqq A$, so from our earlier results about independence, A must be independent of the $B_i \wedge E_i$. Thus no element of $M \backslash (T \cap M)$ is contained in the ideal U. Also $M \backslash (T \cap M)$ is closed under $\wedge$ since it is a filter in M. It is well-known, and easily proven, that an ideal in a commutative ring which is disjoint from a multiplicatively closed system is contained in a prime ideal with the same property. Thus there is a prime ideal W of $V_R(L)$ which contains U and is disjoint from $M \backslash (T \cap M)$. Since $W \cap V_R(M)$ is a prime ideal in $V_R(M)$ which contains $T \cap M$ and is disjoint from $M \backslash (T \cap M)$, it is clear that $W \cap V_R(M) = T$. Of course if T is the augmentation ideal of $V_R(M)$ it is contained in the augmentation ideal of $V_R(L)$.

Corollary (Birkhoff-Stone [2,22]). *If A is an ideal and B a filter in a distributive lattice L, and if A and B are disjoint, there is a prime ideal T such that $T \supseteqq A$ and $(L \backslash T) \supseteqq B$.*

Proof. Apply the theorem with $M = A \cup B$ and T the prime ideal of $V_R(M)$ corresponding to the prime ideal A in the lattice M, that is, T is generated by A and all $b' - b$ with $b, b' \in B$.

From this corollary with A, B single elements, the non-topological part of Stone's representation theorem [17, 22] follows immediately. That is, if we associate with each element b of L the set of proper prime ideals of L not containing b then this is a faithful representation of L as a lattice of subsets of the set of all prime ideals.

We shall now compare the topology introduced by Stone [22, 17] on the set of prime ideals of L with the usual Zariski topology on the prime spectrum of the ring $V_R(L)$. For any subset $A \subseteqq V_R(L)$ let $\mathcal{O}(A)$ be the set of all prime ideals of $V_R(L)$ which do not contain A. Then $\mathcal{O}(A) = \mathcal{O}(B)$ if either B is the ideal in $V_R(L)$ generated by A, or, in case $A \subseteqq L$, B is the ideal in L generated by A. The $\mathcal{O}(A)$ are precisely the open sets in the Zariski topology on the set X of prime ideals of $V_R(L)$. A base for this topology consists of the sets $\mathcal{O}(\alpha)$ for all $\alpha \in V_R(L)$. Suppose M is a finite sublattice of L and let e_p for all $p \in P(M)$ denote the orthogonal idempotents z_M and $p - p^0$ introduced earlier. It follows from Theorem 2 and Prop. 22 that $R(M)$ consists of all elements $e_z + \sum e_p (p \in A)$ for all subsets $A \subseteqq \hat{P}(M)$. For $\alpha = \sum d_p e_p$ in $V_R(M)$ let $e_\alpha = e_z + \sum e_p (p \neq z,\ d_p \neq 0)$, then $e_\alpha \in R(M) \subseteqq R(L)$. Moreover, if $\alpha \notin I$ then α and e_α generate the same ideal in $V_R(M)$ and hence also in $V_R(L)$, whereas if $\alpha \in I$ the same is true of α and $e_\alpha - e_z$. Thus for $\alpha \in V_R(M)$, if $\alpha \notin I$ then $\mathcal{O}(\alpha) = \mathcal{O}(e_\alpha)$, and if $\alpha \in I$ then $\mathcal{O}(\alpha) = \{T \in X \colon z_M \in T \text{ and } e_\alpha \notin T\}$. In the latter case if L has a zero we may assume z_M is the zero of L, then

$$\mathcal{O}(\alpha) = \mathcal{O}(e_\alpha) \backslash \{I\}.$$

Theorem 4. *Let L be a distributive lattice with zero and let Y be the set of all prime ideals of $V_R(L)$ except for the augmentation ideal, that is, Y is the prime spectrum of the ring $V_R(L)/(z)$. For each nonempty ideal A of the lattice $R(L)$ let*

$$U(A) = \{T \in Y \colon T \nsupseteq A\}.$$

Then the map $A \to U(A)$ is an isomorphism of the lattice of ideals of $R(L)$ onto the *lattice of all open sets in the Zariski topology on Y. The sets $U(a)$ for all $a \in R(L)$ are precisely the open compact subsets of Y, and the topology of Y is Hausdorff, locally compact, and totally disconnected. Y is compact iff L has a unit.*

Proof. Our computation above shows that for any $\alpha \in V_R(L)$ there is an $e_\alpha \in R(L)$ such that $U(\alpha) = U(e_\alpha)$. Now for any ideal A in $R(L)$, A is the union of principal ideals generated by the elements $b \in A$ and $U(A) = \bigcup U(b)$. It is easy to see that every open set is of the form $U(A)$ for some ideal A in $R(L)$ since for a finite collection $b_1, \ldots, b_n$ of elements of $R(L)$, $U(\bigvee b_i) = \bigcup U(b_i)$. The corollary to proposition 23 implies that the correspondence is one-to-one. The fact that finitely generated ideals are principal translates into the statement that the only open compact sets are the $U(b)$ for b in $R(L)$. The topology is Hausdorff because $R(L)$ is a generalized Boolean algebra and the remaining assertions are easily checked.

This representation of L or $R(L)$ by the subsets $U(a)$ of Y differs slightly from the situation described in Theorem 3 since $U(z)$ is the empty set. Here $V_R(L)/(z)$ is isomorphic to the ring $S_R(L)$ of simple functions on Y generated over R by the characteristic functions of the $U(a)$ for all $a \in L$. For $\alpha \in V_R(L)$ or $V_R(L)/(z)$ let C_α denote the corresponding function in $S_R(L)$. Then for any prime ideal $y \in Y$, $C_\alpha(y) = \lambda_y(\alpha)$ where λ_y: $V_R(L)/(z) \to R$ is the algebra homomorphism with kernel y. Thus the sup-norm on $S_R(L)$ yields a norm on $V_R(L)/(z)$ given by

$$\|\alpha\| = \max\{|\lambda_y(\alpha)|, y \in Y\}.$$

Clearly the functions in $S_R(L)$ are continuous on Y, separate points, have bounded support, and for every point of Y there is a function which does not vanish there. Thus by the Stone-Weierstrass theorem, the completion of $S_R(L)$ is the space $C_0(Y)$ of all continuous functions which vanish at ∞. Note that if the characteristic function of a set $A \subseteqq Y$ is in $C_0(Y)$ then A must be open and compact and so $A = U(a)$ for some $a \in R(L)$, and the function is already in $S_R(L)$. This shows that $R(L)$, but usually not L, can be recovered from $V_R(L)$, namely as those elements $\alpha \in V_R(L)$ for which $\lambda_y(\alpha) = 0$ or 1 for all $y \in Y$ and $\varepsilon(\alpha) = 1$. Now it is well-known that the continuous linear functionals on $C_0(Y)$ correspond to bounded regular Borel measures on Y, but we can describe them more simply as follows. Any element $\alpha \in V_R(L)$ can be expressed as $\alpha = \sum_1^n d_i b_i + dz$ where $b_i \in R(L)$, $b_i \neq z$, and $b_i \wedge b_k = z$ for $i \neq k$. Then for any prime ideal $y \in Y$, at most one of the b_i is not in y, in which case $\lambda_y(\alpha) = d_i$, and for each b_i there is a prime ideal which does not contain it. Thus $\|\alpha\| = \max\{|d_i|\}$ and so the unit ball in $V_R(L)/(z)$ consists of elements $\alpha = \sum_1^n d_i b_i$ with the b_i as above and $|d_i| \leqq 1$.

Proposition 24. *The linear extension of a bounded R-valued modular function v on $R(L)$ with $v(z) = 0$ is continuous on $V_R(L)/(z)$, hence extends uniquely to a continuous linear functional on the completion of $V_R(L)/(z)$.*

Proof. Now v is the difference of two bounded nonnegative finitely additive measures, thus we may assume v is nonnegative, finitely additive, and $v(z) = 0$.

If $\|\alpha\| \leqq 1$ with $\alpha = \sum d_i b_i$ as above, then

$$|v(\alpha)| = |\sum d_i v(b_i)| \leqq \sum |d_i| v(b_i) \leqq \sum v(b_i) = v(\vee b_i).$$

So a bound for v on $R(L)$ is also a bound for the linear extension of v on $V_R(L)/(z)$.

Theorem 5 (Tarski-Pettis [12,15]). *Let M be a relatively complemented sublattice of a distributive lattice L and suppose both contain z. Then any bounded finitely additive function $v: M \to R$ with $v(z) = 0$ can be extended to a function on L with the same properties. Moreover, for any $b \in L \backslash M$, the value $v(b)$ may be prescribed arbitrarily.*

Poof. The injection $V(M) \to V(L)$ induces an isometry $V_R(L)/(z) \to V_R(M)/(z)$ by Prop. 23. Apply the Hahn-Banach theorem to extend the bounded (by Prop. 24) functional v on $V_R(M)/(z)$ to a bounded functional on the completion of $V_R(L)/(z)$. Finally, since M is a generalized Boolean algebra, $M = R(M)$ and as we saw above, no other element of $R(L)$ except those already in $R(M)$ can be in the completion (closure) of $V_R(M)/(z)$.

Corollary. *Even if M is not relatively complemented the conclusion holds provided v is nonnegative monotone and bounded.*

Proof. From the Corollary to Prop. 22 the unique extension of v to $R(L)$ is non-negative, and it is easily seen that it is bounded.

Finally, if we return to the situation in Theorem 3 where L is (or is represented by) a lattice of nonempty subsets of a set, then $V_R(L) \approx S_R(L)$. For any elements $A_1, \ldots, A_n$ of L, any finite sublattice M of L containing all A_i, and any $x \in B \backslash B^0$ where $B \in P(M)$ and B^0 is maximal in M less that B (or any $x \in B$ if $B = z_M$)

$$\sum d_i C_{A_i}(x) = \sum d_i (x \in A_i) = \sum d_i (B \leqq A_i) = v_B(\sum d_i A_i),$$

where, as before, $v_B(A)$ is 1 if $A \geqq B$ and 0 otherwise, that is, the v_B are precisely all multiplicative linear functionals on $V_R(M)$. Furthermore, for any x which is in some A_i, the minimal element of M which contains x is such a join-irreducible element B. Thus the values of the function $\sum d_i C_{A_i}$ are the numbers $v_B(\sum d_i A_i)$ for all $B \in P(M)$. Since by Prop. 23 each of these extends to all of $V_R(L)$, then the sup-norm of $S_R(L)$ can be defined intrinsically in $V_R(L)$ by

$$\|\sum d_i A_i\| = \max\{|v_B(\sum d_i A_i)| : B \in P(M)\}$$

for any finite sublattice M of L containing the A_i. If $\varphi: L \to L'$ is any lattice homomorphism of distributive lattices then it is easy to see that the induced map $\varphi: V_R(L) \to V_R(L')$ is norm-decreasing and by Prop. 23 it is an isometry iff φ is an injection. Completing $V_R(L)$ for each L yields a functor from the category of homomorphisms of distributive lattices to the category of (norm-decreasing) homomorphisms of commutative Banach algebras.

References

[1] R. Balbes, Projective and injective distributive lattices. Pacific J. Math. **21**, 405–420 (1967).
[2] G. Birkhoff, Lattice Theory. Providence 1967.
[3] H. Crapo and G.-C. Rota, Combinatorial Geometries. Cambridge 1970.
[4] R. L. Davis, Order Algebras. Bull. Amer. Math. Soc. **76**, 83–87 (1970).
[5] J. Folkman, The homology groups of a lattice. J. Math. and Mech. **15**, 631–636 (1966).
[6] L. Geissinger and W. Graves, The category of complete algebraic lattices. J. Combinatorial Th. (A) **13**, 332–338 (1972).
[7] G. Gratzer, Lattice Theory. San Francisco 1971.
[8] C. Greene, On the Möbius Algebra of a Partially Ordered Set. Proc. Conf. on Möbius Algebras, University of Waterloo 1971.
[9] P. R. Halmos, Measure Theory. Princeton 1950.
[10] P. Hilton and S. Wylie, Homology Theory. Cambridge 1960.
[11] V. Klee, The Euler characteristic in combinatorial geometry. Amer. Math. Monthly **70**, 119–127 (1963).
[12] A. Horn and A. Tarski, Measures in Boolean algebras. Trans. Amer. Math. Soc. **64**, 467 –497 (1948).
[13] R. Larson and M. Sweedler, An associative orthogonal bilinear form for Hopf algebras. Amer. J. Math. **91**, 75–94 (1969).
[14] H. M. MacNeille, Partially ordered sets. Trans. Amer. Math. Soc. **42**, 416–460 (1937).
[15] B. Pettis, Remarks on the extension of lattice functionals. Bull. Amer. Math. Soc. **54**, 471 (1948).
[16] B. Pettis, On the extension of measures. Ann. of Math. **54**, 186–197 (1951).
[17] H. Rasiowa and R. Sikorski, The Mathematics of Metamathematics. Warsaw 1963.
[18] G.-C. Rota, On the foundations of combinatorial theory. I. Z. Wahrscheinlichkeitstheorie Verw. Gebiete **2**, 340–368 (1964).
[19] G.-C. Rota, On the combinatorics of the Euler characteristic. In: Studies in Pure Math., pp. 221–233. London 1971.
[20] L. Solomon, The Burnside algebra of a finite group. J. Combinatorial Th. **2**, 603–615 (1967).
[21] E. Spanier, Algebraic Topology. New York 1966.
[22] M. H. Stone, Topological representations of distributive lattices and Brouwerian logics. Casopis Math. Fys. **67**, 1–25 (1937).

Eingegangen am 29. 1. 1973

Anschrift des Autors:
Ladnor Geissinger
Mathematics Department
University of North Carolina
Chapel Hill, North Carolina 27514, USA

Permissions

Birkhäuser Boston thanks the original publishers and authors of the following papers for granting permission to reprint specific papers in this collection.

[1930] Reprinted from *Proc. London Math. Soc.* (2) **30,** © 1930 by the London Mathematical Society.
[1932] Reprinted from *Trans. Amer. Math. Soc.* **34,** © 1932 by American Mathematical Society.
[1935a] Reprinted from *Compositio Math.* **2,** © 1935 by Martinus Nijhoff Publishers BV.
[1935b] Reprinted from *J. London Math. Soc.* **10,** © 1935 by the London Mathematical Society.
[1935c] Reprinted from *Amer. J. Math.* **57,** © 1935 by the Johns Hopkins University Press.
[1940] Reprinted from *Duke Math. J.* **7,** © 1940 Duke University Press.
[1941] Reprinted from *Proc. Cambridge Phil. Soc.* **37,** © 1941 by Cambridge University Press.
[1943] Reprinted from *Bull. Amer. Math. Soc.* **49,** © 1943 by American Mathematical Society.
[1947] Reprinted from *Proc. Cambridge Phil. Soc.* **43,** © 1947 by Cambridge University Press.
[1950a] Reprinted from *Ann. of Math.* (2) **51,** © 1950 by Princeton University Press.
[1950b] Reprinted from *Amer. J. Math.* **72,** © 1950 by the Johns Hopkins University Press.
[1951] Reprinted from *Simon Stevin* **28,** © 1951 by Simon Stevin.
[1952] Reprinted from *Canad. J. Math.* **4,** © 1952 by Canadian Journal of Mathematics.
[1956a] Reprinted from *Bull. Amer. Math. Soc.* **62,** © 1956 by the American Mathematical Society.
[1956b] Reprinted from *Canad. J. Math.* **8,** ©1956 by Canadian Journal of Mathematics.
[1956c] Reprinted from *Amer. Math. Monthly* **63,** © 1956 by The Mathematical Association of America.
[1957a] Reprinted from *Pacific J. Math.* **7,** © 1957 by Pacific Journal of Mathematics.
[1957b] Reprinted from *Canad. J. Math.* **9,** © 1957 by Canadian Journal of Mathematics.
[1959] Reprinted from *Canad. J. Math.* **11,** © 1959 by Canadian Journal of Mathematics.
[1961a] Reprinted from *Physica* **27,** © 1961 by North-Holland Physics Publishing.
[1961b] Reprinted from *Canad. J. Math.* **13,** © 1961 by Canadian Journal of Mathematics.
[1962] Reprinted from *Proc. Amer. Math. Soc.* **13,** © 1962 by American Mathematical Society.
[1963a] Reprinted from *Trans Amer. Math. Soc.* **106,** © 1963 by American Mathematical Society.
[1963b] Reprinted from *Proc. Cambridge Phil. Soc.*, **59,** © 1963 by Cambridge University Press.
[1964] Reprinted from Z. Wahrscheinlichkeitstheorie und Verw. Gebiete **2,** © 1964 by Springer-Verlag Heidelberg.
[1965] Reprinted from *Canad. J. Math.* **17,** © 1965 by Canadian Journal of Mathematics.
[1966a] Reprinted from *Proc. of the Colloquium held at Tihany, Hungary, Sept. 1966,* © 1968 by Akadémiai Kiado, Publishing House of the Hungarian Academy of Sciences.
[1966b] Reprinted from *J. Combinatorial Theory* **1,** © 1966 by Academic Press.
[1968] Reprinted from *Arch. Math.* **19,** © 1968 by Birkhäuser Verlag.
[1969] Reprinted from *J. Combinatorial Theory* **7,** © 1969 by Academic Press.
[1970a] Reprinted from *J. Math. Phys.* **11,** © 1970 by American Institute of Physics.
[1970b] Reprinted from *Advances in Math.* **5,** © 1970 by Academic Press.
[1972a] Reprinted from *Advances in Math.* **8,** © 1972 by Academic Press.
[1972b] Reprinted from *J. Combinatorial Theory Ser. B* **13,** © 1972 by Academic Press.
[1972c] Reprinted from *J. Combinatorial Theory Ser. B* **13,** ©1972 by Academic Press.
[1973a] Reprinted from *Discrete Math.* **5,** © 1973 by North-Holland Elsevier Science Publishers B.V.
[1973b] Reprinted from *Arch. Math.* **24,** © 1973 by Birkhäuser Verlag.
[1973c] Reprinted from *Arch. Math.* **24,** © 1973 by Birkhäuser Verlag.
[1973d] Reprinted from *Arch. Math.* **24,** © 1973 by Birkhäuser Verlag.